Polymers, Liquid Crystals, and Low-Dimensional Solids

PHYSICS OF SOLIDS AND LIQUIDS

SUPERIONIC CONDUCTORS
Edited by Gerald D. Mahan and Walter L. Roth

HIGHLY CONDUCTING ONE-DIMENSIONAL SOLIDS
Edited by Jozef T. Devreese, Roger P. Evrard, and Victor E. van Doren

ELECTRON SPECTROSCOPY OF CRYSTALS
V. V. Nemoshkalenko and V. G. Aleshin

MANY-PARTICLE PHYSICS
Gerald D. Mahan

THE PHYSICS OF ACTINIDE COMPOUNDS
Paul Erdös and John M. Robinson

THEORY OF THE INHOMOGENEOUS ELECTRON GAS
Edited by S. Lundqvist and N. H. March

POLYMERS, LIQUID CRYSTALS, AND LOW-DIMENSIONAL SOLIDS
Edited by Norman March and Mario Tosi

A Continuation Order Plan is available for this series. A continuation order will bring delivery of each new volume immediately upon publication. Volumes are billed only upon actual shipment. For further information please contact the publisher.

Polymers, Liquid Crystals, and Low-Dimensional Solids

Edited by

Norman March
University of Oxford
Oxford, England

and

Mario Tosi
International Center for Theoretical Physics
Trieste, Italy

Plenum Press • New York and London

Library of Congress Cataloging in Publication Data

Main entry under title:

Polymers, liquid crystals, and low-dimensional solids.

(Physics of solids and liquids)
Bibliography: p.
Includes index.
1. Polymers and polymerization. 2. Liquid crystals. 3. Solids. I. March, Norman H.
(Norman Henry), 1927–   . II. Tosi, Mario, 1932–   . III. Title: Low-dimen-
sional solids. IV. Series.
QD381.P6125   1984   530.4′1   84-13445
ISBN-13: 978-1-4612-9448-1   e-ISBN-13: 978-1-4613-2367-9
DOI: 10.1007/978-1-4613-2367-9

©1984 Plenum Press, New York
Softcover reprint of the hardcover 1st edition 1985

A Division of Plenum Publishing Corporation
233 Spring Street, New York, N.Y. 10013

Contributors

S. *Chandrasekhar*, Raman Research Institute, Bangalore, India

G. *Durand*, Laboratoire de Physique des Solides, Université de Paris-Sud, Orsay, France

A. *Keller*, Physics Laboratory, University of Bristol, England

J. *Ladik*, Chemistry Department (and Laboratory of the National Foundation for Cancer Research), University of Erlangen-Nürnberg, Federal Republic of Germany

S. *Lundqvist*, Theoretical Physics Department, Chalmers University of Technology, Göteborg, Sweden

N. H. *March*, Theoretical Chemistry Department, University of Oxford

G. *Marrucci*, Istituto di Prinicipi di Ingegneria Chimica, Universita di Napoli, Italy

M. *Seel*, Chemistry Department (and Laboratory of the National Foundation for Cancer Research), University of Erlangen-Nürnberg, Federal Republic of Germany

F. *Stern*, IBM Thomas J. Watson Research Center, Yorktown Heights, New York, U.S.A.

R. B. *Stinchcombe*, Institut Laue-Langevin, Grenoble, France; and Theoretical Physics Department, University of Oxford, England

S. *Strässler*, Cerberus AG, Männedorf, Switzerland

S. *Suhai*, Chemistry Department (and Laboratory of the National Foundation for Cancer Research), University of Erlangen-Nürnberg, Federal Republic of Germany

J. *Vannimenus*, Groupe de Physique des Solides, Ecole Normale Supérieure, Paris, France

P. *Wyder*, Research Institute for Materials, University of Nijmegen, The Netherlands

Preface

This book deals with three related areas having both fundamental and technological interest. In the first part, the objective is to provide a bird's-eye view on structure in polymeric solids. This is then complemented by a chapter, directly technological in its emphasis, dealing with the influence of processing on polymeric materials. In spite of the technological interest, this leads to some of the current fundamental theory.

Part II, concerned with liquid crystals, starts with a discussion of the physics of the various types of material, and concludes with a treatment of optical applications. Again, aspects of the theory are stressed though this part is basically phenomenological in character.

In Part III, an account is given first of the use of chemical-bonding arguments in understanding the electronic structure of low-dimensional solids, followed by a comprehensive treatment of the influence of dimensionality on phase transitions. A brief summary of dielectric screening in low-dimensional solids follows. Space-charge layers are then treated, including semiconductor inversion layers. Effects of limited dimensionality on superconductivity are also emphasized.

Part IV concludes the volume with two specialized topics: electronic structure of biopolymers, and topological defects and disordered systems.

The Editors wish to acknowledge that this book had its origins in the material presented at a course organized by the International Centre for Theoretical Physics, Trieste.

N. H. March
M. P. Tosi

Contents

2. Crystallinity and Kinetics of Crystallization 33

A. Keller

3. The Basic Crystal Unit 71

A. Keller

III. LOW-DIMENSIONAL SOLIDS

12. Chemical Bonding 291

N. H. March

16. Superconductivity via Electron-Phonon and Electron-Exciton Interactions 475

S. Strässler and P. Wyder

References for Part III 509

IV. SPECIAL TOPICS

17. Biopolymer Electronic Phenomena 523

J. Ladik, S. Suhai, and M. Seel

I

POLYMERS

Introduction to Polymeric Structure and Properties

A. Keller

1.1. Classification

Polymers are large molecules of a long sequence of units. The basic units are the "monomers," which are joined together by chemical bonds in the course of the chemical reaction constituting the synthesis (polymerization). The monomer itself can be anything from a simple molecule, consisting of a few atoms, to a large and complex molecule. Its nature will define the chemical identity of the polymer. In the simplest case it is one single kind of unit repeating itself in the final chain — it can also be a multiplicity of units forming more complicated repeat sequences or no repeating sequence at all (see later). The essential feature of the monomer is that it must be multifunctional, i.e., it must contain more than one potentially reactive chemical group, potential "hooks" so to speak, by which the monomer units can be joined up.

1.1.1. Linear Chains and Networks

1.1.1.1. Linear Chains

If the monomer is bifunctional, the polymerization will lead to a linear chain molecule. This is represented schematically in Figure 1a. Here ✕——✕ represents the monomer, where ✕ stands for the functional

A. Keller · Department of Physics, University of Bristol, England.

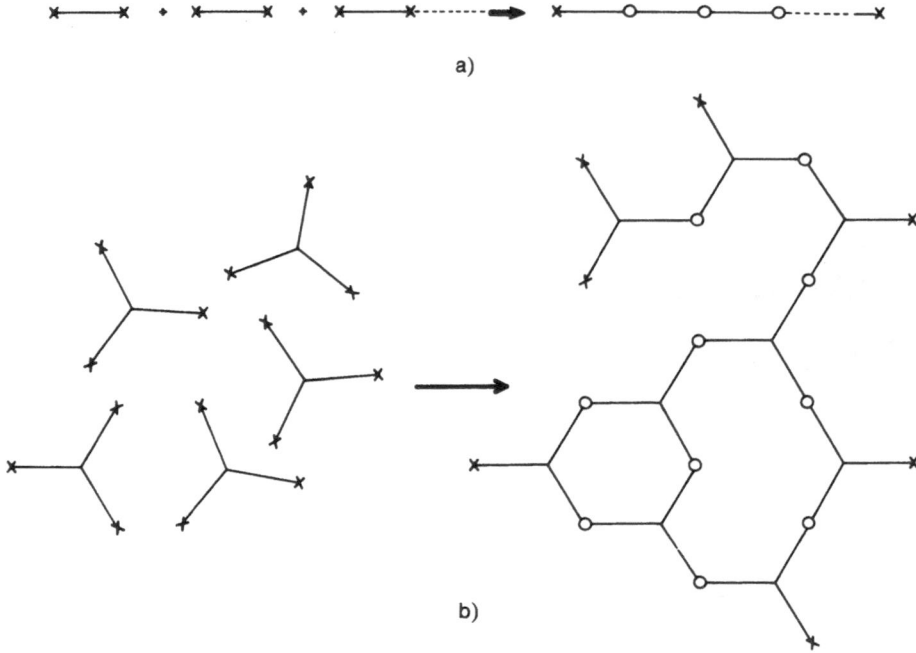

Figure 1. Polymerization scheme of (a) a bifunctional monomer leading to chains and (b) a trifunctional monomer leading to branched chains and eventually networks.

group drawn at each end, and the rest of the monomer is represented by a straight line. In the course of the polymerization reaction the monomers join up at their functional groups, the sites of combination, i.e., the newly formed chemical bonds, indicated by the open circles representing the junction sites.

1.1.1.2. Networks

The polymerization of polyfunctional monomers (i.e., with more than two reactive groups) leads to networks, such as that represented schematically by Figure 1.1(b) for a trifunctional monomer.

1.1.1.3. Intermediate Cases

a. Open Trees — Branched Molecules. Here, the branches have not yet become joined up sufficiently to form a continuous network. The stage at which the network becomes continuous is very distinct experimentally (the so-called "gel point" in the stage of polymerization) and represents mathematically a percolation problem handled by the appropriate mathematical techniques.

b. Linear Chains with Latent Functionalities. In this case, two functional groups within a monomer are more reactive than the rest. These can be made to react first, leading to linear chains. Reacting some of the remaining functional groups will subsequently join these chains into a network.

Linear chain and network polymers display extremely different properties. The former are soluble and fusible, the latter insoluble and infusible (the latter in the sense that the polymer does not flow and is thus unmoldable). Linear chain materials can thus be obtained in desired shapes by molding and/or casting and can be reformed in other shapes subsequently if so desired. In the technological sphere these form the class of thermoplastic materials. Materials of network chains set in a permanent, unalterable shape in the course of the polymerization reaction; hence once polymerized, they cannot be processed further. In the technological sphere these form the class of thermosetting materials. The intermediate class in subsection b above clearly combines the characteristics of both, a salient technological example being the vulcanization of rubber, where the vulcanization corresponds to the formation of the network by linking the linear chains already present.

The present discussion, including the continuing classification below, will be confined to the linear polymers mentioned above, as these lend themselves more to physical structure studies and application of physics in general.

1.1.2. Periodic and Aperiodic Polymers

1.1.2.1. Periodic Polymers

This category comprises polymers that consist of a single identical monomer unit repeating along the chain, or of a repeating pattern of monomer units. This class embraces the most important synthetic polymers and such biopolymers as possess structural function in the organism (e.g., cellulose, chitin, and other polysaccharides). For most people the repeating feature in question is the essence of a polymer, in fact synonymous with the concept of a polymer itself as reflected by conventional textbooks. It should be recognized nevertheless, that this is merely a limiting case of a spectrum, ranging from a strict repetition (implied by the present heading) to such an imperfect repeat pattern that for most purposes the molecule is virtually aperiodic.

1.1.2.2. Aperiodic Polymers

In this case the monomer units along the chain are not identical chemical entities. Such material comprises the functional biopolymers, globular proteins in particular. Here, although the monomers all have the same backbone (α-amino acids — see later), they have disparate appendages (side groups) so that the resulting chain is better described as an aperiodic rather than a periodic system. In this case the determination of the precise sequence of the monomers is a primary research task, either by methods of organic chemistry or crystallography. Here the exact sequence and biological function are closely linked. Important as they are, such materials will only concern us in passing.

1.1.2.3. Intermediate Cases between Periodic and Aperiodic Polymers

a. One case is a repeating *pattern* along the chain backbone but not exact identity, such as collagen where there is a repeating triplet of amino acids: glycin, prolin, and X, where X can stand for a variety of amino acids.

b. Another case is regular repeat along the chain backbone but no periodicity in the *limited variety* of side groups attached to the backbone. There are many examples of this in synthetic polymers, but perhaps the most important one is DNA in biology, where the aperiodic sequence of four bases attached to the sugar-phosphate backbone contains the genetic code.

c. A third case is exact repetition only along portions of the chain with interruptions by different monomer species. This is the situation with copolymers (see the next section), which can be both synthetic and natural.

The real significance of classification under Section 1.1.2 for our purposes lies in its consequences for crystallizability and for the nature of the resulting crystal structure. A fully periodic chain, when stretched out, represents a one-dimensional periodicity, which in the crystalline state will be one, and for that matter the most important crystal axis in the three-dimensional lattice. If the periodicity is imperfect for any of the above reasons, this will be reflected in the final crystal. As long as there is a strict periodicity along the chain direction (such as case b) there can be a correspondingly well-developed periodicity along one crystal direction even if along the other directions the exact order may be deficient. If there is an interruption of the periodicity (such as in case c) this may become incorporated in the lattice as a defect, provided this interruption is short. If it is long (block copolymers), it may remain an uncrystallizable region or form a crystal region of its own. It the chain is aperiodic (see Section 1.1.2.2) the monomer sequence along the chain cannot be the underlying periodicity in the sense implied above. If, however, the chains within an assembly, although aperiodic, are exactly equal, then they may still form a crystal but in a totally different manner, to be described later.

1.1.3. Homopolymers and Copolymers

1.1.3.1. Homopolymers

These arise from the polymerization of a single monomer species. Thus

$$A + A + A + A + A \cdots \rightarrow AAAAA\text{---}$$

where A stands for a particular species of monomer and AAA--- for the resulting chain.

1.1.3.2. Copolymers

These arise from polymerization of a multiplicity of monomer species that can link chemically with each other. There are three broad classes:

1. Random copolymers. Here, say for simplicity, two monomer species A and B join up to form a chain in a random manner leading to a random distribution of A and B along the chain, such as

   ---AABABAABBBBAABAB---

2. Block copolymers. Consider again the two species A and B, and let them form a short repeating sequence AAAAA and BBBBB separately at first, and then join up the two, by subsequent polymerization, into a sequence such as

   ---AAAAABBBBBAAAAA---

   One can have two-block, triblock, or any desired multiple-block chains.

3. Graft copolymers. Here we start with a homopolymer of, say, species A and graft onto it polymers from a second species B. This can be done, for instance, by inducing polymerization of species B along the parent chain, usually at random positions, leading to a branched chain such as

```
   ---AAAAAAAAAAA---
           B           B
           B           B
           B           B
           B           B
           ⋮           ⋮
```

The general feature of copolymers is that they combine the properties of those possessed by the homopolymers corresponding to the individual monomer species. Exactly in what way will depend on which type (1), (2), or (3) of copolymer is formed, and within each type on the gross ratio of the species, and in cases (2) and (3) on the distribution of the blocks and grafts. (The relevance of the block copolymeric nature on the crystallizability has been mentioned in the previous section.) Copolymers therefore offer a virtually limitless possibility of tailoring polymers to particular needs and are very widespread among industrial products. They are also frequent in nature (e.g., structural polysaccharides).

1.1.4. Single-Component Polymers — Polymer Blends

1.1.4.1. General

Copolymerization (see Section 1.1.3) is the chemist's method of tailoring polymers. There is now a rapidly developing new trend of physical tailoring, which consists of blending different polymer species that are chemically unconnected. There is a broad analogy in this respect with alloying in the metallurgical field. There is, however, a very basic difference between alloying metals (or other simple materials) and the blending of polymers, in that polymers are largely incompatible even in the liquid (molten) state. Crystallization may accompany or follow the primary liquid–liquid phase separation in cases where the components are crystallizable under the given circumstances.

The origin of the intrinsic incompatibility of polymers rests in the very small entropy of mixing, which in turn has its source in the large size of the molecules. Hence compatibility will only be achieved when the affinity between the different species is sufficiently large to provide the driving force for mutual dissolution. (Beyond stating this generality the present review will not develop the argument quantitatively.) Thus, in general, with very few exceptions, chemically different polymers will not mix. The few exceptions known to date have important applications.

Present trends in the polymer blend field can be summarized as follows:

1. Search for compatible systems: the practical activity of finding compatible systems more or less empirically; and the theoretical attempts to understand the conditions of compatibility.
2. Understanding and optimization of the phase segregation within incompatible systems: the achievement of desirable states of dispersion of phases in a controlled manner; and the study of the condition of segregation and of the particulars of the phase-segregated structures.

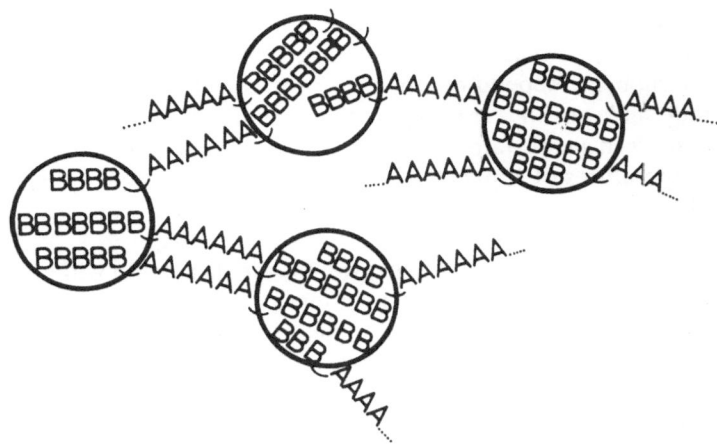

Figure 1.2. Schematic illustration of microphase segregation in the case of a triblock copolymer. Here the AAA--- blocks constitute the matrix and the BBB--- blocks the (in this case) spherical dispersed phase.

1.1.4.2. Block Copolymers as Blends — A Special Case

The components of block copolymers (see Section 1.1.3.2 above) are usually intrinsically incompatible even in the liquid state. Thus the individual blocks will have the tendency to phase-segregate. However, since the different phases in a given chain are molecularly limited, such segregation can only occur on a microscale where the phases will remain molecularly connected. Let us consider the triblock system AAAA---ABBBBB---BAAAA---A. The resulting microphase is represented schematically in Figure 1.2. Here A is the matrix and B the (in this case) spherical, dispersed phase, where the B units in the spheres come from the B block components of a number of different chains. Dependent on the ratio of A and B in the system as a whole, the microphases will be spheres (for the most disparate ratios), cylinders (less disparate ratios), or lamellae (comparable amounts of A and B). In the case of spheres and cylinders the components in larger and smaller amounts will form the matrix and the dispersed phase, respectively. Typical dimensions of microphases are in the range of a few hundred angstroms.

1.1.4.3. Phase-Segregated Polymers as Composites

All phase-segregated systems, whether blends or block copolymers, are essentially composites. Their special properties arise from the nature of the individual components and from the way they are interconnected. As polymers can be, mechanically speaking, solid (crystalline, or glassy) or liquid (or rubbery, depending on the temperature — see later) the most interesting and useful cases are those where one of the composite components is in a stiff state, the other in a compliant state (such as rubber) at the temperatures of application. Some block copolymers are specifically designed to serve this purpose (thermoelastomers: e.g., the triblock polystyrene-polybutadiene).

Having stated the broader background, the discussion will be confined to the simplest class of all, namely the linear homopolymer: Its purpose will be to provide an appreciation of the structural features within a polymeric material when confined to the above class as a model for the more complicated and specialized systems. "Structure" is to be understood in the widest generality, any discussion of which must start from the nature of the individual molecule. The chain molecules are chemical compounds so a certain amount of chemical background will therefore be invoked in what follows.

1.2. Main Types of Polymerization Reaction

While chemistry does not lie within the scope of this review, the appreciation of an absolute minimum of the chemical reactions involved in the synthesis is deemed necessary.

There are two main routes to the synthesis of a polymer chain according to the nature of the monomer.

1.2.1. Condensation Polymerization

Here the functional groups are COOH and OH leading to polyesters, and COOH and NH_2 leading to polyamides with the splitting out of water.

As an example, let us consider the case of a technologically important polyamide, the most widely used nylon (nylon 66):

$$HOOH-CH_2-CH_2-CH_2-CH_2-COOH$$
dicarboxylic acid

$$+ H_2N-CH_2-CH_2-CH_2-CH_2-CH_2-CH_2-NH_2 \rightarrow$$
diamine

$$HOOC-CH_2-CH_2-CH_2-CH_2-\overset{\overset{\displaystyle O}{\|}}{C}O-\overset{\overset{\displaystyle H}{}}{N}-CH_2-CH_2-CH_2-CH_2-CH_2-CH_2-NH_2 + H_2O$$

Here, the functional group of the new entity remains, as before, capable of reacting with corresponding groups in other entities leading to the polymer

$$\begin{matrix} O & & O & & & \\ \parallel & & \parallel & H & & H \\ \text{---C---(CH}_2)_4\text{---C---N---(CH}_2)_6\text{---N---} \end{matrix}_n$$

where n is the number of repeats (monomer units in the molecule).

1.2.2. Addition Polymerization

In this case the monomer units link up without any product splitting off. This occurs by an initially unreactive electron pair becoming available for chemical bonding. For this to happen initial activation is necessary, a step termed "initiation." We illustrate this initiation and the polymerization to follow by the example of a free-radical-initiated unsaturated monomer, propylene, leading to the technologically important polymer, polypropylene: Propylene monomer,

$$\begin{matrix} CH_2{=}CH \\ | \\ CH_3 \end{matrix}$$

Free radical, R·, where R is a general symbol and the dot stands for an unpaired electron.

1.2.2.1. Initiation Step

$$R^{\cdot} + \underset{\underset{CH_3}{|}}{CH_2{=}CH} \rightarrow R{-}CH_2{-}\underset{\underset{CH_3}{|}}{C^{\cdot}{\overset{H}{\overset{|}{}}}}$$

Here the radical becomes attached to the monomer by leaving one unpaired electron resulting from the opening of the double bond at the terminal carbon atom, which thereafter will be an active site. This can react with a still uncreated monomer and thus propagate the polymerization reaction.

1.2.2.2. Propagation Step

$$R{-}CH_2{-}\underset{\underset{CH_3}{|}}{\overset{\overset{H}{|}}{C^{\cdot}}} + \underset{\underset{CH_3}{|}}{CH_2{=}CH} \rightarrow R{-}CH_2{-}\underset{\underset{CH_3}{|}}{\overset{\overset{H}{|}}{C}}{-}CH_2{-}\underset{\underset{CH_3}{|}}{\overset{\overset{H}{|}}{C^{\cdot}}}$$

This reaction can repeat itself many times and thus lead to the polymer. The chain, however, will not grow indefinitely even in the presence of unreacted monomer, but will become terminated in a variety of ways.

1.2.2.3. Termination Step

Termination can occur by the reactive chain end combining with a free radical, with an active monomer, with an active polymer, or with a reactive impurity. In all these cases polymerization ceases and a reactive center is removed from the system. In another method of termination the reactivity of a center is not removed altogether but is transferred elsewhere. This method of termination is named "chain transfer." Here, e.g., a hydrogen atom is extracted from another polymer chain, which will thus inactivate the reactive carbon atom along the growing chain. At the same time this process will activate one of the carbon atoms along another chain (or at a remote part of the same chain — termed "backbiting"), which will lead to branched chain growth. An appropriate monovalent atom of the solvent may also be extracted, thus leading to a reactive solvent atom that can then act as a free radical and initiate polymerization in itself.

For the present purposes two consequences of the nature of the polymerization reaction are noteworthy: first, it determines the nature of the end group of the final polymer chain, and second, it determines the length of the chain or, more precisely, the distribution of chain lengths in the final assembly. Both of these features will be referred to again below.

1.3. Imperfection Types in a Linear Homopolymer Chain

As stated earlier we shall confine ourselves to linear homopolymers for the rest of the review. It is important to realize, however, that a material consisting even of this simplest type of polymer is still not uniform. A given chain will contain various sources of imperfection and nonuniformity; also, all the chains within an assembly will not be equal with respect to these nonuniformities and imperfections. It is important to bear this in mind, and to be aware of the imperfection types when performing experiments and interpreting the results, in particular when working toward some theoretical objective. I shall use this theme as a vehicle for the introduction of some of the most important issues in polymer science.

1.3.1. End Groups

The ends of a chain are necessarily different from the rest, hence they represent a source of nonuniformity in themselves. The preceding section on polymerization reactions has more or less set the scene. Accordingly, chain ends can be functionally active species as in condensation polymers (e.g., COOH, OH, or NH_2 groups), which given the chance could react further; or the ends can be foreign groups, such as arise in addition polymers. In the latter case, as we have seen, they can be the constituents of the initiator radicals (or ions in ionically initiated polymerization) themselves, certainly at one end of the chain, and may be the same at both ends depending on the method of termination. Different methods of termination will introduce different terminal entities, which could thus be impurity atoms, or in the case of chain transfer some constituent atom of the solvent, or simply an additional hydrogen. Alternatively it could also be a double-bonded group.

In a long chain the end group will only represent a very small fraction of the total. For many purposes therefore its influence may be totally negligible (e.g., for mechanical behavior). For some purposes, however, it may be all-important, e.g., polar end groups in the electrical properties of an otherwise insulating polymer, or if the ends initiate further chemical reactions (polymerization or degradation), say in the course of heat treatment.

1.3.2. Molecular Length (Weight) Distribution

A polymerization reaction as conducted in a laboratory flask or within an industrial reactor always leads to nonuniformity in chain length. (This is in contrast to nature, which can produce biopolymers where each chain is strictly of identical length.)

Finite and, as a rule, nonidentical chain lengths arise for the following reasons:

1. A thermodynamic limit of the growing chain set by the law of mass action, a limit which however is seldom, if ever, reached.
2. Kinetic limitations. Even if the end groups remain functional (condensation polymers) the chain growth will slow down as the chains get longer, and thus the reactive end groups fewer. The chain length distribution at any stage is then determined by statistics and may be predictable mathematically.
3. By the competing effects of initiation, propagation, and termination rates in cases where such pertain. Again, knowledge of the individual rate constants may enable the final length distribution to be predictable.

Ideally one would like to know the full molecular weight distribution. Indeed there are methods for achieving this. Molecular weight distribution can be characterized by averages, which are often invoked both for theoretical and practical considerations. Often such averages can be determined directly by experiment without knowledge of the actual distribution. Two averages are particularly important:

Number average:
$$M_n = \frac{\sum n_x M_x}{\sum n_x}$$

Weight average:
$$M_w = \frac{\sum n_x M_x^2}{\sum n_x M_x}$$

where M_x is the molecular weight of a molecule corresponding to a degree of polymerization x (i.e., consisting of x monomer units of molecular weight M_0, thus $M_x = x M_0$) and n_x is the number of such molecules.

For many purposes the width of the distribution is characterized by M_w/M_n. For a homopolymer this ratio is of course equal to unity. For a random distribution resulting, e.g., from the statistically determined combination of diacid and diamine, as in the above-quoted example of nylon 66, this ratio is ideally two. Many industrial processes (e.g., for polyethylene) result in distributions that are much broader and can have M_w/M_n values of approximately 15. In synthetic polymers M_w/M_n values of about 1.1 are usually considered as corresponding to very sharp distributions and are taken as a good approximation for a uniform (homodisperse) polymer and, as a rule, are quite difficult to obtain.

1.3.3. Isomerism

A chain, which is totally uniform as regards chemical constitution and has a well-defined molecular weight (or forms an assembly of uniform chains all of the same molecular weight), is still not fully characterized in view of the possibility of isomeric differences and isomeric imperfections. Exactly which kind of isomerism needs to be taken into account will depend on the polymer in question. In what follows some common and important types of isomerism will be listed briefly.

1.3.3.1. Branches

The possibility of branching has been raised previously as one consequence of chain-transfer reactions. Thus, a chain of given molecular weight can be straight or branched with a variety of branch distributions, e.g.,

| linear | one long branch | two short branches |

This form of isomerism is particularly important in the case of polyethylene, where the different technological products differ with respect to branching.

1.3.3.2. Isomerism of Unsaturated Polymers (Polyenes)

This important class of polymers comprises the conventional elastomers (rubbers). Here, two kinds of isomerism are important and widespread.

1. The monomer contains more than one double bond, where only one is used for polymerization to form a linear molecule. Isomerism arises according to which is used. An example is butadiene:

$$\text{Butadiene monomer} \qquad \underset{(1)}{CH_2}\!\!=\!\!\underset{(2)}{CH}\!-\!\underset{(3)}{CH}\!\!=\!\!\underset{(4)}{CH_2}$$

where (1), (2), etc., number the C atoms.

Polymerization may occur either via (1), (4) or (1), (2) C atom:

$$\text{1:4 polymer} \qquad \{CH_2-CH\!\!=\!\!CH-CH_2\}_n$$

$$\text{1:2 polymer} \qquad
\begin{array}{c}
CH_2 \\
\| \\
CH \\
| \\
\{CH_2-CH\}_n
\end{array}$$

2. *Cis–trans* isomerism. This kind of isomerism based on the absence of free rotation around double bonds, well known in basic organic chemistry, has drastic consequences for poly(isoprene), the basic constituent of two important natural compounds:

$$\text{Isoprene monomer:} \qquad
\begin{array}{c}
CH_3 \\
| \\
CH_2\!\!=\!\!CH-C\!\!=\!\!CH_2
\end{array}$$

$$\text{cis polymer:} \qquad
\begin{array}{c}
H \quad CH_3 \\
| \quad\; | \\
\{CH_2-C\!\!=\!\!C-CH_2\}_n
\end{array}$$

$$\text{trans polymer:} \qquad
\begin{array}{c}
CH_3 \\
| \\
\{CH_2-C\!\!=\!\!C-CH_2\}_n \\
| \\
H
\end{array}$$

The *cis* polymer is natural rubber, the *trans* polymer is guttapercha. Guttapercha is a hard solid at room temperature as opposed to the familiar behavior of rubber. The difference rests in the better crystallizing ability (higher melting point) of guttapercha, which (if the above formulas are drawn out sterically) has the straighter chain conformation.

1.3.3.3. Isomerism in Polyolefins

Polyolefins have the general formula

$$\begin{array}{c} R \\ | \\ -\!\!+\!\!CH_2\!-\!CH\!-\!\!]_n \end{array}$$

where R can be a variety of chemical groups. (In the case of polypropylene quoted earlier R is CH_3.) Two kinds of isomerism will be mentioned, the second of which is of utmost importance.

   a. Head-to-Tail and Head-to-Head Isomerism. Let us consider the monomer

$$\begin{array}{c} R \\ | \\ CH_2\!\!=\!\!CH \\ (1)\quad (2) \end{array}$$

where (1) and (2) again label the C atoms. In the head-to-tail polymer carbon (1) links up with carbon (2) in the course of polymerization, leading to a chain such as

$$\begin{array}{ccc} R & R & R \\ | & | & | \\ \cdots\!-\!CH_2\!-\!CH\!-\!CH_2\!-\!CH_2\!-\!CH_2\!-\!CH\!-\!\cdots \end{array}$$

In the head-to-head polymer the carbons link as (1) to (1) and (2) to (2), leading to a chain such as

$$\begin{array}{cccc} R & R & & R & R \\ | & | & & | & | \\ \cdots\!-\!CH_2\!-\!CH\!-\!CH\!-\!CH_2\!-\!CH_2\!-\!CH\!-\!CH\!-\!\cdots \end{array}$$

   The head-to-tail sequence is the one which occurs far the most frequently and is thus considered "normal" for polyolefins. Nevertheless, "faulty" higher-energy sequences can occur along a given, otherwise head-to-tail chain with much lower frequency, more exactly dependent on the type of substituent R and on the conditions of polymerization.

  b. Stereoisomerism. This very important type of isomerism relies on the presence of an asymmetric carbon atom (to be denoted C*). It is known from basic chemistry that if all the four carbon valences link up with different groups (sterically the carbon is at the center of a tetrahedron, where the four substituents are at the vertices of the tetrahedron), two stereochemically different conformations arise from the otherwise identical constituents, which are sterically mirror images of each other. In simple organic compounds the two correspond to the *d*- and *l*-configurations rotating the plane of plane-polarized light in the right and left directions, respectively.

  In a polyolefin the tertiary carbon atom is asymmetric

$$\begin{array}{c} R \\ | \\ +CH_2-C^*+_n \\ | \\ H \end{array}$$

because it links to R, H, and to the chain, portions on its right and left (as drawn in the formula above), which in general are nonequivalent, hence the tetrahedra of which C* is the center can be of two different kinds that are mirror images of each other. The same applies to each consecutive C* atom along the chain. Which of the two configurations pertains is determined during the polymerization reactions, when the new monomer attaches itself to the growing chain. In general there is no control over which of the two forms occurs, and hence the sequence of *d*- and *l*-configurations will be mixed, determined by random chance.

  However, special catalyst systems (Ziegler–Natta catalysts) enable the controlled formation of steric isomers. In this way it is possible to obtain chains where each consecutive C* atom gives rise to identical (i.e., *d*- or *l*-) conformation. The consequences of this can be most readily visualized as follows. Consider the C–C chain stretched out completely (disregarding whether the substituents R sterically permit this or not). Then all the C–C atoms will lie in a plane forming a zigzag corresponding to the tetrahedral angle. The substituents R and H will then lie above or below this plane as determined by the tetrahedral geometry (Figure 1.3). The two conformations *d* and *l* will then differ in as far as R lies above or below this plane. If all C* atoms are stereochemically equivalent the Rs will all be on the same side. Such a polymer has been termed "isotactic" by Natta (see Figure 1.3a). In the random sequence the Rs will be above and below the C–C plane in a haphazard fashion. Such a polymer is termed "atactic" (see Figure 1.3c).

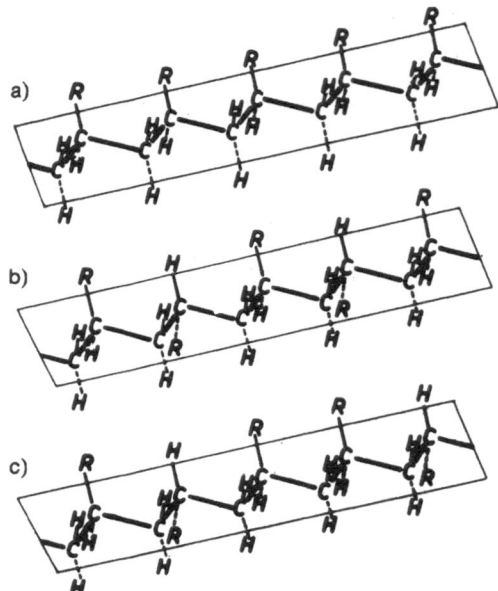

Figure 1.3. Representation of (a) an isotactic, (b) a syndiotactic and (c) an atactic polyolefin (after Natta[1]).

Of course, other types of regularities can also be envisaged; the next simplest one is the *dldl---* alternating sequence. In the above representation the Rs will be alternately above and below the C–C plane (Figure 1.3b). Such a polymer is termed "syndiotactic."

The kind and degree of stereochemical regularity is termed "tacticity" and is an important parameter in the characterization of the chain with far-reaching technological consequences. The latter arises from the fact that regular sequences (isotactic, syndiotactic) can crystallize while the atactic cannot. (Polypropylene owes its commercial usefulness to the fact that it can be obtained in the isotactic form.)

Control of tacticity (by means of appropriate catalysts in the course of polymerization) and the determination of the kind and degree of tacticity that pertains to a particular product (carried out by NMR techniques) is a most important activity both academically and technologically.

1.4. Formulas of Some Important Polymers

A minimum appreciation of what the real polymers actually are is clearly required for any discussion of polymeric structure. The chemical formulas of some important synthetic materials will be listed below.

1.4.1. Condensation Polymers

1.4.1.1. Polyesters

a. Polyethylene Terephthalate (Trade names: Terylene, Dacron (fibers), Melinex, Mylar (films))

$$\begin{matrix} O && C \\ \parallel && \parallel \\ +C-\langle O \rangle - C-O-CH_2-CH_2-O\}_n \end{matrix}$$

arising from

$$HO-\overset{O}{\overset{\parallel}{C}}-\langle O \rangle - \overset{O}{\overset{\parallel}{C}}-OH + OH - CH_2-CH_2-OH$$

terephthalic acid   ethylene glycol

b. Polycarbonate (Trade name: e.g., Lexan)

$$\begin{matrix} O && CH_3 \\ \parallel && | \\ +C-O-\langle O \rangle - C-\langle O \rangle -O\}_n \\ && | \\ && CH_3 \end{matrix}$$

arising from

$$HO-\langle O \rangle - \overset{CH_3}{\overset{|}{\underset{|}{C}}}-\langle O \rangle - OH + \overset{OH}{\overset{|}{\underset{|}{C}}}{=}O$$
$$CH_3 \qquad OH$$

1.4.1.2. Polyamides (Trade name: nylons)

a. From Diacid and Diamine

$$+NH-(CH_2)_p-NH-\overset{O}{\overset{\parallel}{C}}-(CH_2)_r-\overset{O}{\overset{\parallel}{C}}+$$

arising from

$$H_2N-(CH_2)_p-NH_2 + HO-\overset{O}{\overset{\parallel}{C}}-(CH_2)_r-\overset{O}{\overset{\parallel}{C}}-OH$$

diamine   diacid

where p and $r+2$ signify the total number of C atoms in the amine and acid components, respectively. The most common nylons (trade name) of this type are nylon 66 and 610.

    b. From ω-Amino Acids

$$\underset{}{\begin{array}{c} O \\ \parallel \end{array}}$$
$$+\!\!\!\!-C\!\!-\!\!(CH_2)_s\!\!-\!\!NH\!\!-\!\!\!\!\mid_n$$

arising from

$$\underset{}{\begin{array}{c} O \\ \parallel \end{array}} \qquad\qquad \underset{}{\begin{array}{c} O \\ \parallel \end{array}}$$
$$HO\!\!-\!\!C\!\!-\!\!(CH_2)_s\!\!-\!\!NH_2 + HO\!\!-\!\!C\!\!-\!\!(CH_2)_s\!\!-\!\!NH_2 + \cdots$$

    c. From α-Amino Acids. These are the polypeptides and proteins, the basic constituents of living matter

$$\begin{array}{c} R \\ \mid \end{array}$$
$$+\!\!\!\!-C\!\!-\!\!C\!\!-\!\!NH\!\!-\!\!\!\!\mid_n$$
$$\begin{array}{c} \parallel \\ O \end{array}$$

$$\qquad\qquad\qquad \begin{array}{c} R \\ \mid \end{array}$$
(α-amino acid is $HO\!\!-\!\!C\!\!-\!\!C\!\!-\!\!NH_2$).
$$\qquad\qquad\qquad\qquad \begin{array}{c} \parallel \\ O \end{array}$$

    Here R is one of the 21 different organic groups defining the naturally occurring α amino acids. In the so-called sequential polypeptides the Rs are all alike. With a few exceptions these are synthetic laboratory products. In the naturally occurring proteins the Rs along a given molecule are different, and in general do not form a repeating sequence. The enormous potential number of combinations of Rs is the source of the very large variety of proteins and of their specificity.

1.4.2. Addition Polymers

1.4.2.1. Polyethylene (Trade name: Polythene)

$$+\!\!\!\!-CH_2\!\!-\!\!\!\!\mid_n$$

arising from

$$CH_2\!\!=\!\!CH_2$$
ethylene

1.4.2.2. Polyolefins

General formula:

$$R \atop {+CH_2-CH}{\large\}}_n$$

Important representatives:
  Polypropylene (PP)

$$CH_3 \atop {+CH_2-CH}{\large\}}_n$$

The technological product is in the isotactic form.
  Polyvinylchloride (PVC)

$$Cl \atop {+CH_2-CH}{\large\}}_n$$

The technological product has some preponderance of syndiotacticity but is otherwise atactic.
  Polystyrene (PS)

$$+CH_2-CH{\large\}}_n$$

The technological product is atactic.
  Polyacrylonitrile (PAN; trade names: Orlon, Acrylan)

$$N \atop C \atop {+CH_2-CH}{\large\}}_n$$

The technological product is largely atactic.
  Polymethylmethacrylate (PMMA; trade name: Perspex)

$$CH_3 \atop {+CH_2-C}{\large\}}_n \atop COOCH_3$$

The technological product is atactic.

1.4.2.3. Polyethers

Polyoxymethylene (Trade name: Delrin)

$$\text{+CH}_2\text{—O}\text{+}_n$$

Polyoxyethylene (Trade name: Polyox)

$$\text{+CH}_2\text{—CH}_2\text{—O}\text{+}_n$$

Polyphenylene Oxide

This is compatible with polystyrene and is being used as a blend component.

1.4.2.4. Fluoropolymers

Tetrafluoroethylene (Trade names: Teflon, Fluon)

$$\text{+CF}_2\text{+}_n$$

Trichlorofluoroethylene

1.4.2.5. Inorganic Polymers

The silicon-based ones are the most important: General formula:

$$
\begin{array}{c}
R_1 \\
| \\
\text{+Si—O+}_n \\
| \\
R_2
\end{array}
$$

Here R_1 R_2 can be a variety of organic groups.

1.5. Melting Range of Polymers; Specialty Materials

For structural purposes a polymeric material needs to retain its shape and support load. This it can only do in the solid state, the upper temperature limit of which is determined by the melting point. The melting range has been one of the principal considerations in the design of polymeric materials, its rise to increasingly higher temperatures being one of the main objectives. When listing polymers in order of melting points T_m we shall not only provide additional information, but shall also add to our list new polymeric materials, and through them introduce new polymeric properties associated with these materials, which are in the forefront of current research interests.

First, it must be stated that as polymers do not melt sharply (more about this later) but over a temperature range, it is more appropriate to refer to a melting range rather than a melting point, at least for practical characterization. The T_m values to be quoted are representative temperature figures within this melting range.

1.5.1. Conventional Polymers

1.5.1.1. Polyethylene

There are two classes: low density and high density:

1. Low-density polyethylene (LDP). $T_m \sim 100\ °C$. An important technical drawback is that it becomes unusable at the temperature of boiling water. It is very tough and pliable and has the familiar waxy feel.
2. High-density polyethylene (HDP). $T_m \sim 135\ °C$. It clearly overcomes the handicap of LDP. Also it is much more rigid.

A few words on the origin of the differences between LDP and HDP are in order. LDP, historically the first, is obtained by polymerization at high temperature and pressure where frequent internal chain transfer takes place ("backbiting") within the same chain. The result is branching within the chain. The branches reduce the crystallizability of the molecule leading to smaller and less perfect crystals, which are responsible for the properties listed under (1). In contrast, HDP consists of completely linear chains that can crystallize with greater facility, and hence gives rise to larger and more perfect crystals, and more of them, accounting for the increase in T_m and rigidity. The absence of branches arises from the fact that, thanks to the use of Ziegler–Natta catalysts (mentioned earlier), polymerization can be conducted at low (room) temperature and pressure when side reactions during chain growth are minimized. It is noteworthy that here, there being no asymmetric carbon atom, the question of tacticity does not arise; nevertheless the Ziegler–Natta catalyst has a profound effect for the reasons just stated.

1.5.1.2. Others

Polypropylene (isotactic): $T_m \sim 170\ °C$.
Technological polyesters and polyamides: $T_m \sim 220–280\ °C$.
Polytetrafluoroethylene: $T_m \sim 320\ °C$.

The last entry represents the upper limit for common polymers. Clearly many users require the withstanding of higher temperatures. This prompted the current research toward high-temperature polymers.

1.5.2. Specialty Materials

1.5.2.1. High-Temperature Polymers

The common underlying principle is that the chains have to be rigid, which is best obtained through the linking-up of aromatic monomers.
A chemically simple example is the unfusible polyphenylene

which degrades above 530 °C without melting. A compound of the kind

with $T_m > 900\ °C$ is obtained by heating polyacrylonitrite which, drawn out with appropriate valence angles, is

On further graphitization this gives rise to the much-publicized carbon fiber, which is essentially a "polycrystalline" form of one-dimensional graphite, an important specialty product of exceptional properties.

The aromatic analogue of nylons of types 66 and 610 is the recent Kevlar (trade name)

$$\left[\!-NH-\!\left\langle\bigcirc\right\rangle\!-NH-\overset{\displaystyle O}{\overset{\|}{C}}-\left\langle\bigcirc\right\rangle\!-\overset{\displaystyle O}{\overset{\|}{C}}\!-\right]_n$$

(note the aromatic rings instead of $(CH_2)_p$ sequences), which is an exceptionally strong and stiff fiber. In common with many other polyaromatic chains of its class it is totally unfusible and can only be dissolved (which is necessary for processing) in very special and corrosive solvents. The solutions thus formed can be liquid crystals.

1.5.2.2. Liquid-Crystal Polymers

Kevlar is one example of a new class of polymeric materials that can be obtained as liquid crystals. The underlying science is still in its infancy. Here only a few comments will be made as regards classification.

The liquid crystals are usually the results of rigid chains or chain portions, a property assured by the multiple aromatic grouping. If the full main chain is liquid-crystal forming, the chain is usually so stiff that it will not fuse in any sense and, as Kevlar, becomes a liquid crystal through appropriate solvents (lyotropic). If the chain is a copolymer with both flexible and rigid groupings, the corresponding material may fuse and pass into the liquid-crystal state in that way (thermotropic). In another class of compounds the main chain may be totally flexible so that it can exist normally in the isotropic molten phase, but it is the appropriate side groups that form liquid crystals.

1.5.2.3. Conducting Polymers

Although conducting polymers deserve a chapter of their own, they will only be mentioned here because polymers in this class usually have very stiff chains and thus also possess high-temperature stability; in fact, they do not melt. It is owing to this last fact that they are mostly unprocessable, a deficiency that obstructs practical utilization. For instance, the totally untractable polyparaphenylene becomes conducting on appropriate doping, say with AsF_5.

Currently, the fully conjugated polyacetylene

$$\left[\!-CH\!=\!CH\!-\right]_n$$

has attracted much attention, becoming again conducting on suitable doping with donor molecules. Again unfusible and insoluble, it is only available as formed during the polymerization itself, which is a rather ill-defined fibrous web. While an obvious candidate for technological exploitation (to replace metals!) and for fundamental studies of one-dimensional conductivity, this unfavorable texture and its untractability presently bar progress.

As a quite recent announcement the compound

$$\left[S - \bigcirc \right]_n$$

may overcome some of the difficulties. It is processable and is said to become conducting after oxidizing in the presence of As_2F_5.

The inorganic polymer referred to as $(SN)_x$, probably with the formula

$$\left[S = N \right]_n$$

is highly conducting along the chain direction. (It clearly does not satisfy the normal rules of valency and is thus supposed to have delocalized electrons along the chain.) Quite exceptionally for polymers, it can be obtained as macroscopic single crystals in the form of needles in the course of polymerization itself, and is thus eminently suitable for structure and conductivity studies. It is a truly one-dimensional pseudometal. Again, quite exceptionally and extraordinarily, it is superconducting below 0.3 K.

The discussion of the conductivity itself is beyond the scope of this review.

1.6. The Physical State

Beyond the classification according to chemical constitution adopted so far, polymers can be classified and/or characterized according to their physical state. The two broadest categories are: "amorphous" and "crystalline." We shall see that this classification is not hard and fast, as most "crystalline" polymers also contain an amorphous component, and also there are gradations of order within the crystalline component. Nevertheless, as a broadly adopted criterion a polymer is termed "crystalline" when recognizable Bragg reflection can be identified in its X-ray diffraction patterns, otherwise it is classified as amorphous. At this stage it is important to recognize that a given polymeric sample may be amorphous either because although intrinsically crystallizable, it was not given the chance to crystallize, or because it is intrinsically uncrystallizable *per se*. The former will, of course, be the case with all polymers above their crystalline melting point, and in such crystallizable polymers that have been quenched from the melt into the glassy state where crystallization is unmeasurably slow (see later). Amorphous polymers of this class may be converted into crystalline polymers by appropriate heat treatment. Amorphous polymers, termed above "intrinsically uncrystallizable," can exist only in the amorphous state, fluid or rigid (glassy, see below), by virtue of their chemically (including stereochemically) irregular constitution, which prevents them from forming a crystal lattice.

Obviously, structural features are expected mainly in crystalline polymers. For this reason the latter will be treated comprehensively in the following chapters devoted essentially to structural aspects. However, as most crystalline polymers also possess a certain amount of amorphous material, a very brief survey of the amorphous state is required for an appreciation of what is to follow. This will be accommodated under the present heading.

1.6.1. Resumé of the Amorphous State

1.6.1.1. The Random Coil — A Reminder

A brief mention will be made of the random coil concept, which is well known to be the abstraction of a flexible molecule adopted as a model in macromolecular physics. A given flexible macromolecular chain owes its flexibility to the fact that there is rotation around the main chain bonds. If this rotation is sufficiently free and the chain is sufficiently long, the path of an isolated chain can be described by a three-dimensional random walk. A random walk is known to be characterized by the mean-square end-to-end distance $\overline{r^2}$ given by

$$\overline{r^2} = nl^2 \tag{1.1}$$

where n denotes the number of links of length l, and the links are freely jointed.

The "links" are neither the main chain bonds nor the monomer units, but the minimum length of chain in terms of which a chain can be described by equation (1.1) on the basis that they are totally freely jointed. They are termed "statistical segments," and the corresponding n value is the number of such segments. Of course the chain cannot cross its own path, but this "excluded volume" problem can be ignored for the present. It should suffice to say that an isolated flexible chain can be regarded as a random coil and can be readily defined mathematically as such. This random coil is the most probable state to which the system will tend to return if perturbed externally (entropy elasticity).

1.6.1.2. Structure of Amorphous Polymeric Matter

By definition, there should be no structure in the amorphous state other than the statistical short-range ordering required by local packing considerations. This situation is familiar from simple liquids and glasses. The analogous situation in polymers corresponds to amorphous material consisting of chains in their random configuration, as defined above for an isolated chain, with the addition that in the condensed state the chains are freely interpenetrating. The consequence of the latter will be that the expansion due to self-exclusion in the case of the isolated chain is compensated by the effect of mutual exclusion between different chains, the net result being for each chain an ideal random configuration as if excluded volume effects were absent. This is the so-called Θ-condition, commonly realizable in dilute solutions and postulated by Flory to pertain also to the amorphous state. Short-range order in the sense pertaining to a liquid should still exist, and is in principle describable and experimentally accessible by radial distribution functions derived from diffraction patterns.

However, as in simple liquids and glasses, the possibility of larger structural units (supermolecular structures, crystal precursors, etc.) has arisen also in polymers and the existence or otherwise of such ordering has been subject to much debate. Experimental evidence quoted in favor of the latter includes nodular structures seen in the electron microscope, X-ray scattering effects at small angles and, most undeniably, changes in the properties of amorphous materials (glasses) on heat treatment. (If the structure is ideally random, clearly there should be nothing to change.)

The decisive evidence, however, in favor of Flory's totally random model is claimed to have emerged from recent neutron scattering experiments. The essentials of the method, to be referred to again in connection with crystalline material, are as follows. The aim is to characterize the configuration of a chain in an environment of chains of its own kind. This is achieved by neutron scattering with the aid of labeled probe molecules, which are distinct as regards scattering of neutrons. Such suitable probe molecules are polymers containing deuterium as opposed to hydrogen. If these are mixed in small quantities with the main mass of amorphous material, the latter will be like a dilute solution of the deuterium containing polymer as solute within a proton containing polymer as solvent. Here the neutrons will "see," so to speak, the solute molecules. Then the size (radius of gyration) and finer details, depending on the angular range of the scattering explored, can be obtained from low-angle neutron scattering patterns. The general result that has emerged from studies on PMMA and PS is that the molecule in the molten and glassy state (see below) is in the random configuration corresponding to the Θ-state. This fully agrees with Flory's predictions and seems to disclaim the existence of larger-scale organizations. For most people this is where the subject stands.

Nevertheless, *all* the claims for larger-scale structures cannot be ignored even if some are undoubtedly influenced by artefacts. We have to distinguish between the ideal amorphous state undoubtedly realized by suitably chosen materials and samples, and materials and samples that are conventionally regarded as amorphous, but where the amorphous state is not fully achieved. The latter may arise in materials that are intrinsically crystallizable, in contrast to materials that are not (atactic PS and PMMA are of the latter kind). In the former, experiment may detect residual or incipient crystallinity, which can be important for the sample in question but is irrelevant or even misleading when relating to the true nature of the amorphous state. Some of the arguments in the literature may arise from a failure to recognize this distinction.

1.6.1.3. The Five States of Amorphous Matter

The consistency of amorphous polymers is greatly affected by temperature. This is most readily expressible in terms of a mechanical property. Following Tobolsky we shall take the modulus of atactic polystyrene (a typical amorphous polymer) as measured after 10 s loading (as polymers are viscoelastic materials, the modulus is time-dependent), and examine its behavior as a function of temperature (Figure 1.4).

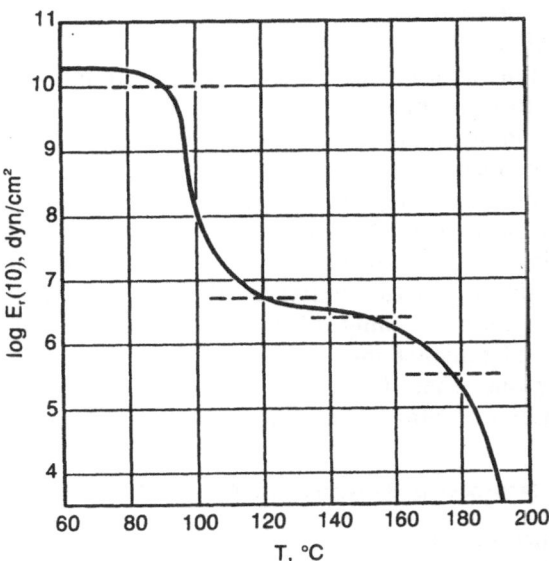

Figure 1.4. Five regions of viscoelastic behavior of an amorphous polymer after 10 s loading showing modulus ($E_r(10)$) vs temperature in the case of polystyrene (after Tobolsky[2]).

We identify the following five regimes with increasing temperature:

I. Modulus $\sim 10^{10}$ dyn cm^{-2}. The material here is a glass, rigid, and stiff.

II. The modulus drops sharply from 10^{10} dyn cm^{-2} to about 10^6 dyn cm^{-2}. This is in the range of the "glass transition" usually characterized by a single temperature, the glass transition temperature T_g.

III. Modulus $\sim 10^6$ dyn cm^{-2} varying only little with temperature. This low-modulus region is associated with very high and (on an appropriate time scale) recoverable extensibility. This regime corresponds to the "rubbery" state.

IV. The modulus drops again significantly with temperature from 10^{-6} dyn cm^{-2} to about 10^5 dyn cm^2. Here we have a combination of rubbery behavior with viscous flow.

V. Modulus less than 10^5 dyn cm^{-2}, becoming increasingly undefinable on the time scale of the measurement. Here we proceed to a viscous liquid, the truly molten polymer.

Regimes I–V are readily interpreted in molecular terms. Little need be said about the glass. Here the main chains are rigid, their motion frozen in; the modulus measures the extension and deformation of covalent bonds in common with glasses from simpler substances. Regime III is unique to polymers. In this regime rotational motion around the main chain bonds is taking place and the chain is thus capable of changing its conformation by thermal motion without, however, any noticeable displacement (on an appropriate time scale) of its center of gravity (micro-Brownian motion). On external loading this micro-Brownian motion is biased and the chain will stretch out without net displacement of the center of gravity. On removal of the load the chain will retract to its random configuration, the restoring force thus being entropic (the basis of rubber elasticity), hence the high elastic extensibility coupled with low modulus, which is the characteristic of rubbers. In Regime V the center of gravity of the whole molecule is shifting in the course of thermal motion (macro-Brownian motion). On external loading this motion is biased resulting in true flow, the property of polymer melts. Regimes II and IV are the appropriate transition regions, where Regime II (T_g) is the most dramatic, corresponding to the onset of rotation around the valence bonds of the main chain.

1.6.2. The Usefulness of Polymers in Terms of Their Physical State

For structural purposes a polymer needs to retain its shape and support the load required. In the light of the foregoing we can now broadly identify the circumstances when this pertains and also classify the polymers in daily use (and thus most readily available also for scientific investigation, and conversely, most in demand for such investigations) accordingly.

It follows that a noncrystalline polymer satisfies the above criterion only up to and including Regime III and, if rigidity is required, only in Regime I, i.e., in the form of a glass. This means that if glassy properties are required T_g must be above the temperature of intended application. Where T_g is much above, say, room temperature, intrinsically uncrystallizable atactic polymers will be glassy over a temperature range relevant to everyday use and can thus be useful commercial products, important examples being PMMA and PS. Being glasses they are also transparent, but at the same time, while rigid, they are brittle to varying extents.

If the temperature of application is above T_g or, conversely, T_g is below, say, room temperature, amorphous polymers can be useful as elastomers, like the familiar rubbers (Regime III). Unsaturated polymers possessing the highest chain flexibility are the most suitable in this mode of application (*cis* polyisoprene, polybutadiene, see Section 1.3.3.2). Nevertheless, even such materials are not suitable as technological rubbers because on protracted loading bodily displacement of chains will occur and the extension will not be fully recoverable (they will "creep"). This is prevented by permanently linking the individual chains with chemical bonds, thus forming a loose network, where the individual network elements still display the entropy elastic properties characteristic of rubbers. The above unsaturated polymers provide opportunities for the introduction of such chemical cross-links by activating very few of the double bonds along a given chain (vulcanization).

If the chain is not to be cross-linked, then for use above T_g the polymer must be crystalline to serve as a viable structural material. This applies to all those polymers, excluding those used as rubbers, where T_g is below the ambient temperature.

It is a fact that a common polymer is only semicrystalline, i.e., it will contain both amorphous and crystalline regions (more of this later). As such it combines the characteristics of crystalline and amorphous matter. As amorphous polymeric matter can exist in the form of either a glass or a rubber, this distinction will also apply to cases where it is in combination with crystals. Above T_g the amorphous component is, of course, in a rubbery consistency and the corresponding semicrystalline solid will display the properties of a stiff and strong crystalline material coupled with the elasticity and toughness of rubbers, more precisely dependent on the ratio and on the nature of connectedness of the two components. This is the consistency in which most crystalline polymers find practical application. Such a material can be considered as a rubber reinforced with crystals (in fact, the crystals also act as the (physical) cross-links preventing creep), or as a crystalline solid plasticized with a rubbery component. Polyethylene is a widespread example of this class as a whole, LPE being more like the former and HDP like the latter. Nevertheless, even when the amorphous component is in the glassy state crystallinity still adds to stiffness and strength, but now the material, while stiff and strong, will be more brittle, i.e., it will support little strain.

In the case of crystalline materials the ultimate limit of applicability is determined by the melting point of the crystals T_m. The melting ranges of crystalline polymers have already been listed and discussed earlier. More on the physics of melting will be presented later.

It will be apparent that T_m and T_g are the most important characteristics of polymers determining their use. It is noteworthy that when a polymer can exist either in the crystalline state, or in the amorphous state, or in both states, T_m is always significantly higher than T_g.

Crystallinity and Kinetics of Crystallization

A. Keller

2.1. Basic Classifications

2.1.1. Generalities

Crystallinity implies three-dimensional periodicity. In this respect we have to distinguish between crystallinity of periodic and aperiodic polymeric molecules. This distinction has already been implied in the introductory sections above; it is fundamental and self-evident yet, surprisingly, not usually pointed out even in basic texts.

2.1.1.1. Periodic Molecules

In the case of periodic molecules the inherent periodicity of the molecule itself represents one direction in the crystal lattice. As stated earlier, this direction is a particularly important one because it corresponds to the direction of the covalently bonded chain. It is implicit that in order for this intrinsic periodicity to be utilized for the building up of a lattice, the chain direction must be uniquely defined (whether the chain is straight or helical is then a further question; see later). The parallel association of an assembly of chains with well-defined registry along their length then defines the full three-dimensional lattice (Figure 2.1a). The chains them-

---

A. Keller · Department of Physics, University of Bristol, England.

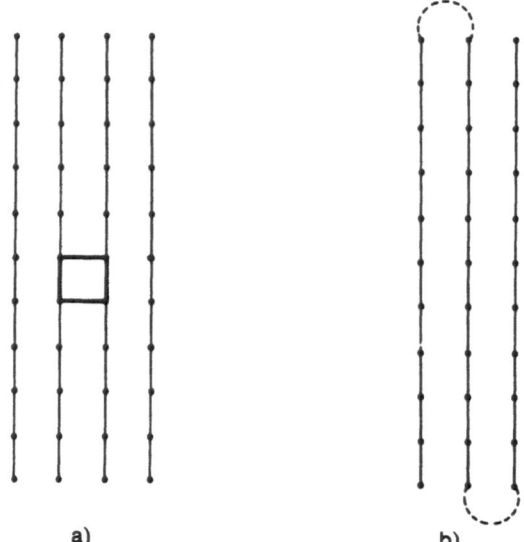

Figure 2.1. Concept of a lattice from parallel periodic chains, dots representing geometric repeating units. (a) Schematic illustration of the meaning of the unit cell (one drawn boldly). (b) Chain folding to form the lattice. In both cases the representation is two-dimensional.

selves need not be of uniform length provided they contain a sufficient number of repeats.

2.1.1.2. Aperiodic Molecules

For aperiodic molecules to form a crystal lattice the molecules must be strictly equal, both as regards sequence of monomeric constituents and chain length. This is satisfied by the all-important globular proteins. These molecules assume a complex convoluted configuration so as to satisfy the requirements of delicate interactions between different monomeric units along the *same* chain. (Much of the biological function is related to these interactions.) The overall configuration will therefore be a globular coil of specific internal structure, all exactly the same for each molecule. It is these coils themselves which pack in a regularly repeating fashion to form the crystal lattice (Figure 2.2). It is noteworthy that this lattice bears no relation to the chain direction. The momentous achievements of crystal structure analysis of globular proteins (hemoglobin, enzymes, etc.) consist of the determination of the internal structure of the globule, not of the determination of the arrangement of the globules, i.e., the lattice, which in fact is usually trivial.

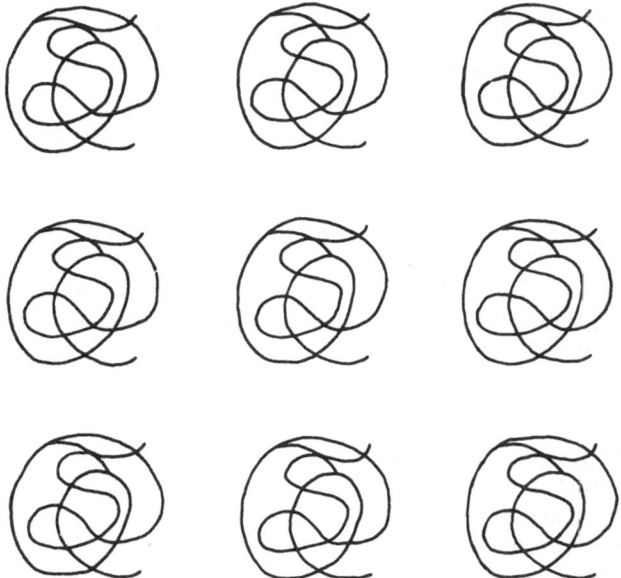

Figure 2.2. Schematic representation of a lattice from globular (protein) molecules. Here each molecule, although aperiodic, is strictly identical. The convoluted chain path is the "protein folding."

2.1.1.3. Chain Folding: An Aside

The folding of chains features frequently in the subject of crystallinity of chain molecules. It is important to realize (it is in fact self-evident, although nowhere explicitly stated) that it has a different meaning in the fields of periodic and aperiodic molecules.

In the case of periodic molecules folding is one way in which the one-dimensional periodicity of the chains achieves a periodicity in a second dimension, thus leading to a sheet. Consequently, as sketched in Figure 2.1b, the folding itself contributes to the building up of a lattice.

In the case of aperiodic molecules, globular proteins in particular, the term "folding" refers to the establishment of the convoluted globular structure of the coil. Here the fold is at some specified localities of the chain, so as to satisfy some specific intramolecular interaction. This is the "protein folding," which accordingly does not contribute to the building up of the lattice *per se*.

In what follows we shall confine ourselves to the crystalline state associated with periodic molecules.

2.1.2. Sources of Lattice Imperfections in Polymers

Crystal imperfections represent a major facet of the physics of the solid state. Lattices are not, as a rule, ideally perfect but contain imperfections such as point defects and dislocations. This is also true for polymer crystal lattices. However, the lattices of polymer crystals contain defects over and above the conventional defects of the solid state in simpler substances and these are usually more significant. These additional defects arise from the chemical (including stereochemical) imperfections within the chain itself. In what follows we shall briefly review all those chemical imperfection types discussed in a previous section that could be the source of potential lattice imperfections:

a. End groups. As noted already all chains have ends. The chain lengths are nonuniform, so these ends cannot be in register throughout the lattice and hence cannot be part of the lattice periodicity.
b. Chemical irregularities along the main chain. These irregularities are either accidental, due to a "wrong" step in the polymerization reaction (e.g., a double bond in polyethylene), or deliberate, as in the case of a random copolymer.
c. Stereochemical irregularity, i.e., nonuniform tacticity.
d. Irregularly placed side groups — branches. As in category (b), these can be either accidental (e.g., branches in low-density polyethylene), or deliberate, as achieved by copolymerization (such as in the case of currently frequently used ethylene-propylene, or ethylene-butylene copolymers).

The above chain imperfections will impair crystal perfection in two ways: (1) they may become incorporated into the lattice, formed by the dominant periodic component of the chain, giving rise to lattice defects of various kinds, and (2) they may become ejected from the crystals into the amorphous regions, which will limit the size of the crystals that can develop. Whether (1) or (2) occurs is the subject of long-standing debates. There is not likely to be a unique answer as this will depend on the system in question and on the mode of crystallization. In the latter respect, in particular, it will depend on whether the crystallization is slow enough for the imperfect chain portion to become ejected from the growing crystal, if indeed this is thermodynamically favored.

In "real life" there are always some unavoidable chain imperfections of the kind listed above under (a)–(d). Such imperfections are often deliberately created in attempts to chemically tailor molecules, a trend of ever-increasing topicality.

If there is an excessive number of irregularities along the chain, the existence of crystallinity proper (as evidenced by well-recognizable X-ray reflections) becomes questionable and we enter the gray area of when a polymer is to be regarded as still crystalline or as being amorphous. There are two categories of such a high level of crystal imperfections, or incipient states of ordering (depending on the end of the ordering spectrum from which the issue is approached):

i. The lattice is in principle more or less perfect, but the crystals are so small that they do not give rise to sharp diffraction effects. PVC of normal commercial use, while largely atactic yet possessing a certain amount of crystal-forming syndiotacticity, may well be an example.

ii. The lattice itself is not a completely regularly repeating assembly of the basic motive but incorporates systematic departures. This includes the much quoted and argued concept of paracrystallinity [R. Hosemann and S. N. Bagchi, *Direct Analysis of Diffraction by Matter*, North-Holland Publ. Co., Amsterdam (1962)], where the mean position of one motive with respect to the next does not correspond to that of an exact lattice repeat. A possible example is commercial PAN.

Finally, if chain irregularities exceed a certain proportion, the molecule becomes uncrystallizable and the corresponding material will be amorphous on any standard.

Having introduced all the above imperfection types, we shall now confine ourselves to the ideal lattice of ideally periodic molecules.

2.1.3. Modes of Crystallization

There are essentially three basic modes by which crystallinity can develop in a polymer.

1. *Concurrent with polymerization.* Here the crystallinity develops in the course of the polymerization reaction itself. As the chains grow in the course of the reaction, under appropriate circumstances the system becomes supercooled and the growing chains precipitate in the form of crystals (taking the frequent example that polymerization starts in solution) while the chains continue growing. The resulting product is the so-called "nascent polymer." Many widespread technological polymers are obtained in this form (polyethylene, polypropylene). In very special instances a macroscopic monomer crystal can be converted into a macroscopic polymer crystal by polymerization in the solid state (see later).

2. *Orientation-induced crystallization.* In this case, if long chains are aligned first while in the amorphous state they can be induced to

crystallize under conditions (temperature) where normally they would not. This includes the historically most important case of stretching-induced crystallization of rubber (detected by X-ray diffraction in 1925).

3. *Crystallization from the completely random state.* Here a random system, stationary melt or solution, is being supercooled sufficiently for it to pass into the crystalline state.

It is to be noted that (1) and (2) are typically "polymeric" phenomena without counterparts in the crystallization of simple substances. Case (3) is the equivalent of the normal supercooling-induced crystallization in simple atomic or low-molecular-weight materials. However, even here the long-chain nature of the molecule gives rise to many intriguing and novel effects. While some of them are necessarily much more complicated than in simpler substances, this need not always be the case: namely the one-dimensional periodicity intrinsic to the basic building units can represent a simplification when compared to the truly three-dimensional processes in simpler materials. Some references to this point will be made later.

In what follows we shall be largely concerned with case (3). Case (2) will have a smaller, but nevertheless decisive part to play, while only brief reference will be made to case (1) in the rest of this review.

2.2. Crystal Structure

Traditionally, the principal purpose of crystallographic research is the determination of the crystal structure. Knowledge of the crystal structure requires identification of the repeating entity, the unit cell, which is traditionally followed by determination of the cell content in atomic detail. Even if in later trends of macromolecular crystallography this sequence is often reversed (see further below), we shall proceed with the unit cell first.

2.2.1. The Unit Cell

The essentials have already been implied by the foregoing and are illustrated in Figure 2.1. Accordingly, the repeating unit does not involve the molecule as a whole but merely its repeating constituents. Thus one of the unit cell dimensions (usually denoted by c) corresponds to the repeat unit *along* the chain, while the other two cell edges (a and b) correspond to lateral repeat distances between different chains in equivalent positions (see Figure 2.3 for PE). Thus along one direction the lattice periodicity is defined by valence bonds in contrast to that along the other directions, where it correponds to distances between otherwise unconnected atoms.

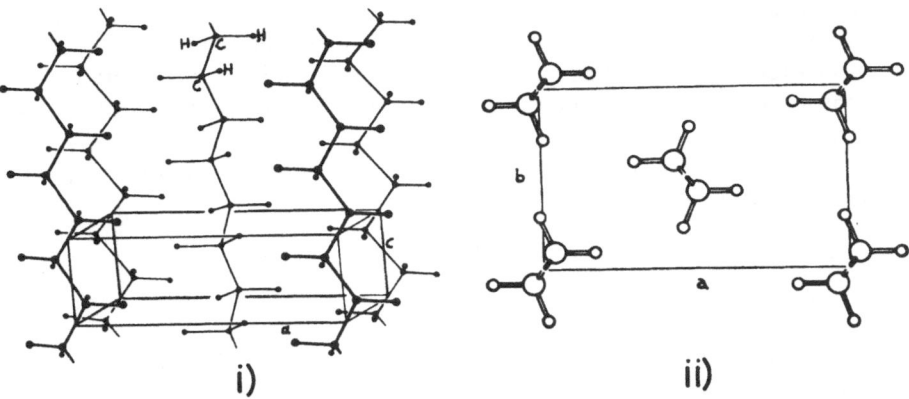

Figure 2.3. The unit cell of polyethylene. (i) Perspective view. (ii) View along the chain direction (c axis) (after Bunn[3]).

In view of this last emphasized feature the issue of crystal-structure determination is subdivided into two facets:

1. *Chain conformation*, corresponding to the shape and path taken up by the individual chain directly responsible for the geometric periodicity.
2. *Chain packing*, corresponding to the mode of packing of the different, but parallel, chains which is related to their lateral separation.

Clearly chain conformation is the more important of the two as it is an intrinsic property of the chain molecule. The above subdivision makes the crystal-structure issue of a long-chain molecule distinct from that of simple substances, say metals or small molecules, where the different directions are not as fundamentally different physically because they all correspond to distances between unconnected entities.

In what follows we shall treat (1) and (2) in turn.

2.2.2. Chain Conformations

The determination of chain conformations is one of the principal activities in structural polymer science. It can be done in two ways: (1) by analysis of diffraction patterns (X-ray or electron), and (2) by *a priori* predictions from the chemical formula, or most productively by a combination of (1) and (2). Method (2) in particular can again be pursued in two ways: (a) by building space-filling models with which the sterically impossible or unfavorable clashes can usually be spotted, and (b) by calculation, the so-called conformational analysis, by which the energetically most favorable conformation can be determined, provided sufficiently realistic parameters (bond lengths, bond angles, atomic radii) and potential functions are used. In what follows some basic results will be quoted using the example of the simplest of all carbon chains.

Let us consider the simplest hydrocarbon chain, i.e., that corresponding to polyethylene:

$$\begin{array}{cccc} \text{H} & \text{H} & \text{H} & \text{H} \\ \text{---C---C---C---C---} \\ \text{H} & \text{H} & \text{H} & \text{H} \end{array}$$

where each C atom is at the center of a tetrahedron with its four valences leading to the four vertexes. Suppose full rotation around each C—C bond is possible and we wish to find the energetically most favorable configuration arising from these rotations.

Let us select any one C—C bond and regard it as vertical to the plane of the paper, so that we are looking along the bond. For any one of the two overlapping carbon atoms we shall then see the three remaining valences (lines connecting the center of the tetrahedron to the three apexes of one of its faces) at 120° to each other. Figure 2.4 shows an arrangement for the ethane molecule, where the three valences that belong to the upper carbon are denoted by solid lines, those that belong to the lower carbon are denoted by dashed lines, and the bold central circle corresponds to the two overlapping carbon atoms. All rotation angles are possible, so the three solid and dashed lines in Figure 2.4 could assume any rotational relation with respect to each other. It turns out that the particular arrangement in Figure 2.4 where the solid and dashed lines are at maximum angular separation in the presented projection is energetically the most favorable. This result is the principle of "staggered bonds," which expresses the fact that substituents, of adjacent carbon atoms are trying to avoid each other as much as possible.

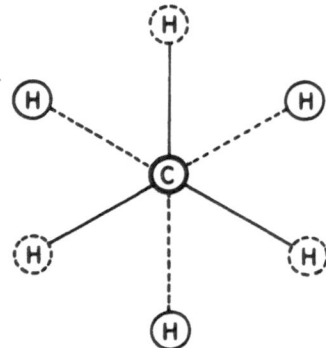

Figure 2.4. Illustration of the staggered bond principle in the case of an ethane molecule. View along the C—C bond.

In alkanes there are three possible such staggered positions, as is apparent from Figure 2.5. One of them is termed trans (T) and, as will be apparent, ensures that the main-chain carbon atoms all lie in one plane within which the C—C bonds form a zigzag line. The other two conformations in Figure 2.5 are termed gauche (G), distinguished by G^+ and G^-. Here the C—C bonds will not lie in a single plane but will describe a convoluted path. Calculation reveals that for a simple CH_2 chain (hydrocarbons, polyethylene) the T conformation is energetically the most favorable. There are energy minima also for the other two staggered positions G^+ and G^-, as shown by the function of potential energy V vs angle of rotation ϕ, but these are not as deep as for the *trans* state (Figure 2.6).

When replacing the H atoms by other constituents, which will necessarily all be bulkier than H, these may clash to varying extents. In order to minimize this clash there will be rotation around the C—C bonds away from the trans position which, if always in the same sense in a given chain, will transform the planar zigzag into a helical path. This can occur to different degrees according to the substituents.

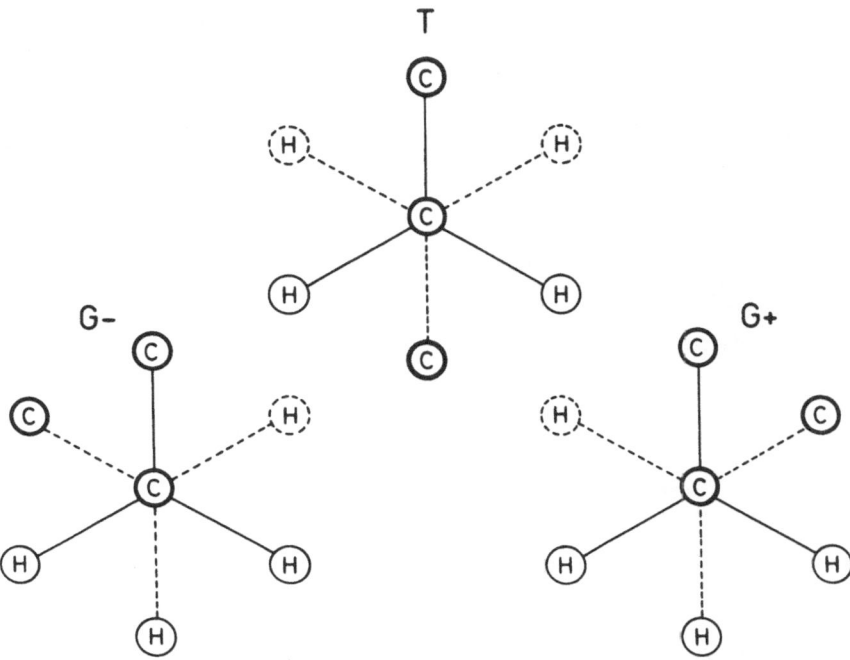

Figure 2.5. Three staggered bond configurations in the case of an alkane chain: T is the all-planar trans configuration, G^+ and G^- are the two gauche configurations.

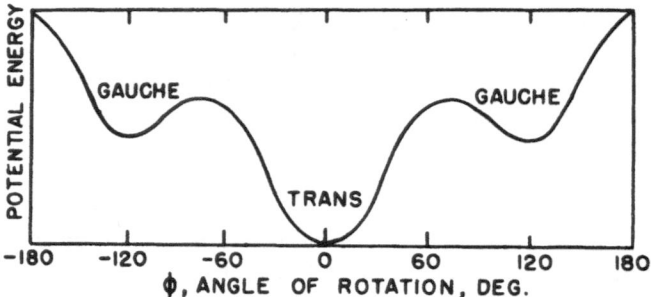

Figure 2.6. Potential energy as a function of rotation around a C—C bond. The deepest central minimum corresponds to the T state and the two smaller minima at $-120°$ and $+120°$ correspond to the two G states in Figure 2.5.

In PTFE each H atom is replaced by an F (fluorine) atom causing a rotation around the C—C bonds. This rotation is only slight, yet essentially within the trans trough of the potential surface, even if displaced from $\phi = 0°$ (or 180°). The result will be a slowly winding helix, which at room temperature is a 13_1 helix (13 repeat units in 1 turn of the helix; see Figure 2.7).

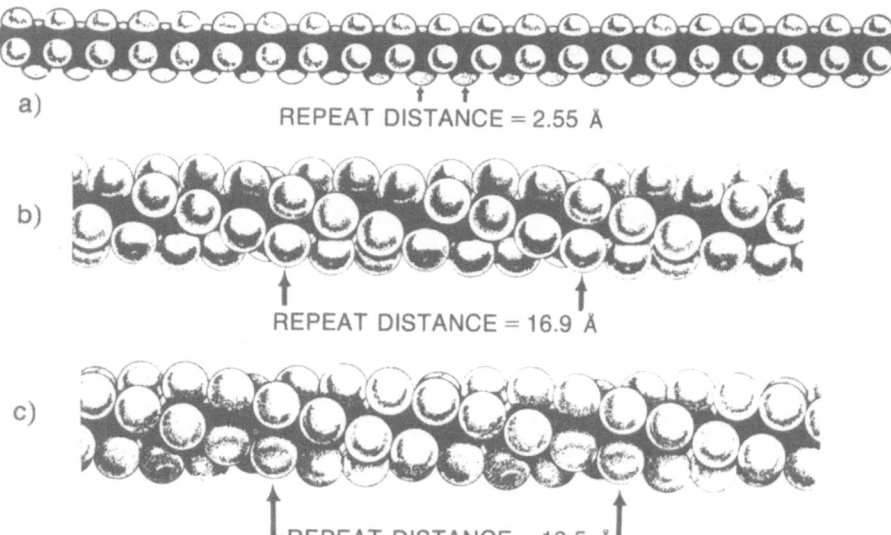

Figure 2.7. Space-filling models showing (a) the planar zigzag of polyethylene, and (b), (c) two slowly winding helical conformations of polytetrafluoroethylene. Dark balls represent carbon atoms and white balls the hydrogens in (a) and the fluorines in (b) and (c) (after Geil[4]).

In an isotactic polyolefin every second C atom will possess a more or less bulky constituent other than H, and when considering the chain stretched out in a planar zigzag (T configuration) they will all lie on the same side with respect to the plane of the zigzag, as shown in Figure 1.3. If these substituents are CH_3 (polypropylene) or larger, they will clash in this all-T configuration, consecutive bonds requiring a rotation around every second bond by 120°, i.e., into one of the G troughs. The most favorable conformation along a given chain will therefore be TG^+TG^+--- or TG^-TG^-. This will define a 3_1 helix (3 monomer repeats in 1 turn of the helix). This 3_1 helix with near variants (such as the 7_2 helix) is the standard chain formation for polyolefin (Figure 2.8), first predicted from models by Bunn [C. W. Bunn, *Proc. R. Soc. London, Ser. A* **180**, 67 (1942)] before isotactic polymers were available, later abundantly verified on the actual systems by Natta and Corradini [G. Natta and P. Corradini, *Nuovo Cimento* **15**, Suppl. No. 1, 68 (1960)]. (It should be stated that while the above comments appear in all textbooks and are essentially correct, latest experimental findings and subsequent calculations reveal also conformations other than the 3_1 helix as being of low energy, which under certain circumstances can determine the behavior of the system. The causes and consequences of these alternatives have not yet been fully evaluated. Even so they point to the extreme caution required as regards the *uniqueness* of the predicting power of conformational analysis.)

2.2.3. Chain Packing

2.2.3.1. Geometric Considerations

Provided there are no specific interactions between chains but only van der Waals forces, the packing of long-chain molecules is most readily envisaged with reference to the packing of cylinders. Circular cylinders, of course, pack in regular hexagonal arrays. Chains with helical conformations are in fact close approximations to circularly cylindrical rods and many of them indeed pack hexagonally (e.g., isotactic polystyrene). Packing other than regular hexagonal can still be profitably considered in the same terms because even here the packing can be usually recognized as pseudohexagonal.

As an example let us consider polyethylene. The unit cell projection along the chain directly reveals the pseudohexagonal nature of the chain packing (Figure 2.3). The cross-sectional view of the all-T planar zigzag is approximately an ellipse, and the particular packing in Figures 2.3 and 2.9 corresponds to the optimum packing of ellipses [A. I. Kitaigorodsky, *Organic Chemical Crystallography*, Consultants Bureau, New York (1961)].

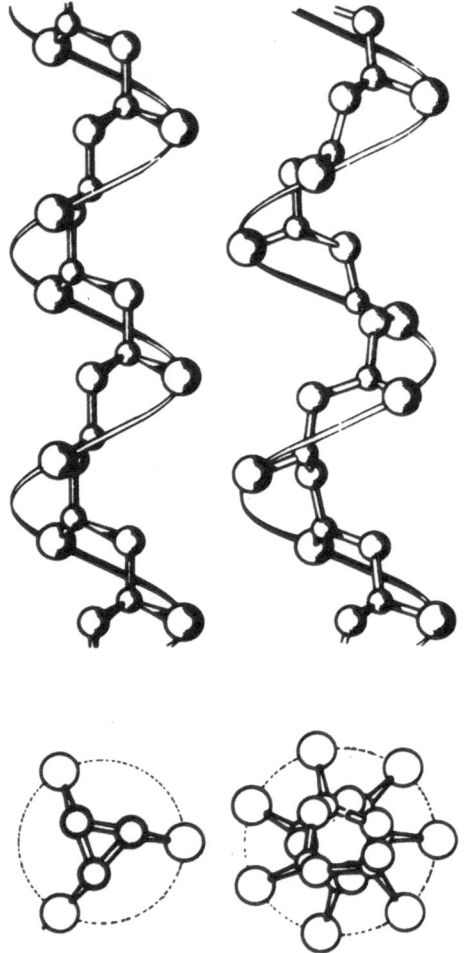

Figure 2.8. Two polyolefin chain conformations based on TG sequences. Left: 3_1 helix (an example is polypropylene); right: 7_2 helix (an example is poly-4-methyl-pentane). Top and bottom diagrams represent views perpendicular and parallel to the helical axes, respectively (after Natta[5]).

Looking at the packing problem in the above way is not only simple, but can be helpful when considering changes in packing caused by physical and/or chemical circumstances.

Let us examine thermal expansion. It is general knowledge with all paraffinoid substances (paraffins, fatty acids, polyethylene) that on increasing the temperature the orthorhombic lattice of Figure 2.3 will change toward becoming truly hexagonal, a state that, depending on the materials or other circumstances, may only be approximately or actually attained (rotary phase in paraffins). The reasons for this will be evident if we remember that the chain, or segments of it, will perform increasing thermal vibrations around the chain axis and will lead to an overall circular envelope, the packing of which will approach that of hexagonal rods (Figure 2.9).

The same happens under the influence of chemical imperfections such as side branches (copolymers, low-density polyethylene) and occasional cross-links (e.g., those that can be introduced by irradiation). Whatever the description in detailed molecular and atomic terms, such imperfections reduce the overall ellipticity of the chain cross-section in the sense of Figure 2.9 with the result that the system will approach hexagonal packing.

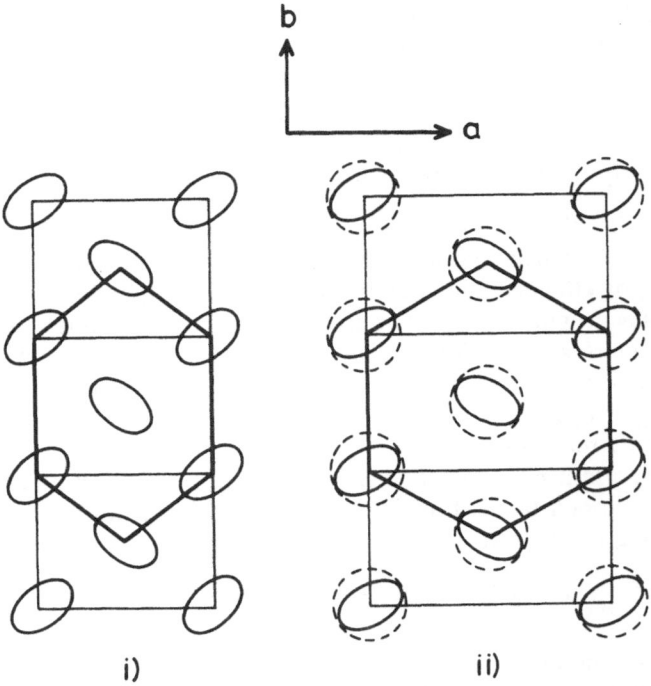

Figure 2.9. Schematic illustration of pseudohexagonal packing of chain molecules with elliptical cross section (i), which changes toward regular hexagonal packing (ii), on increasing thermal vibration so causing expansion along the *a* direction. The respective hexagons are drawn by solid lines. (The lattice with the given dimensions does not correspond to any particular polymer.)

2.2.3.2. A Note on Crystal Polymorphism in Polymers

Polymorphism is a widespread occurrence among crystalline polymers. With very few exceptions the polymorphic structures differ as regards mode of chain packing but not chain conformation. This follows from the fact that chain packing is determined by interchain forces, which in general are not only much weaker than the intrachain forces but also much less specific. As a result there will be a number of ways in which chains can pack with only comparatively small differences in stability, i.e., corresponding to only comparatively shallow potential wells with only slight differences between them, leading to the possibility of several crystal structures. In simple substances differences in crystal structure usually have a profound effect on the material. This is generally not so with polymers. Although possibly interesting for the crystallographer, the unspecific nature of chain packing and the small energetic differences associated with the different polymorphs, as well as differences in crystal structure are usually not of primary significance for polymeric behavior. (This assertion may need to be qualified when specific forces between chains are operating (see below), especially when the effect of polarity on electric properties is involved as, e.g., in the case of the piezoelectric polyvynylidene fluoride.)

2.2.3.3. Specific Interchain Interactions

In the above discussion on chain packing only geometric packing factors were considered without the specific forces between the chains, for which polyethylene and the most common polyolefins are appropriate examples. In polymers containing polar groups the interchain interactions can be more specific. Such groups could be ions (in polyelectrolytes they form ion bridges between chains), dipoles, and whatever are capable of forming hydrogen bonds. Below we shall consider examples from the last-mentioned category.

Important examples of hydrogen bonding are found in polyamides. Here hydrogen bonds form between the carbonyl and the amine groups belonging to adjacent chains

$$\overset{\displaystyle \diagdown}{\underset{\displaystyle \diagup}{C}}=O \text{---} HN\overset{\displaystyle \diagup}{\underset{\displaystyle \diagdown}{}}$$

where --- denotes the hydrogen bond. In such polymers the adjacent chains will occupy positions so as to satisfy these interactions. In favorable instances hydrogen bonding between all groups capable of such bonding can be achieved, in which case the mutual arrangement of the chains can be obtained virtually by inspection. This will be illustrated by two examples: nylon 66 and nylon 6.

The arrangement in nylon 66 is shown by Figure 2.10, where all hydrogen bonds are seen to be formed. As the ↑ and ↓ directions are equivalent, any parallel chain can hydrogen-bond with its neighbors. This is not so with nylon 6 where the chain direction has polarity, hence directions ↑ ≠ ↓. Nevertheless, it is seen from Figure 2.10(b) that hydrogen bonding can be fully satisfied if and only if alternate chains point up and down, i.e., the arrangement is ↑ ↓ ↑ ↓ ---.

As in both cases the chains are planar zigzags, the hydrogen-bonded structures are sheets. The rest of the chain-packing problem will then be confined to the mode of packing of these sheets. As there are no specific forces between the hydrogen-bonded sheets the optimum sheet packing is not very distinct from other packing possibilities. It follows that the different, energetically very similar modes of sheet packing form a source of polymorphism in polyamides. This again exemplifies the theme already stated in connection with the packing of individual chains and polymorphism, namely the less specific the interactions, the more difficult it is to predict or to determine the corresponding packing geometry. On the other hand, the distinction between the various possibilities (polymorphism) is correspondingly less significant for the behavior of the system as a whole.

We wish to mention that the presence of strongly interacting groups, if closely spaced along the chain, can affect not only the chain packing but even the configuration of a given chain itself. This is the situation with polypeptides, where hydrogen bonds can form internally within a chain leading to the famous Pauling α-helix, one of the starting points of modern structural molecular biology.

2.2.4. Crystal Structure Determination

For the kind of chain molecules under discussion, crystal structures are usually determined from X-ray fiber diffraction patterns. These correspond to single-crystal rotation patterns with the chain (fiber) axis as axis of rotation. These diffraction patterns display layer lines perpendicular to the fiber axis. The layer-line periodicity, directly read off the patterns, corresponds to the geometric periodicity along the chain. In a straight zigzag chain this will be the monomer periodicity (or a higher order of it in the case of systematic absences). If the chain is helical or has some convoluted conformation the geometric repeat will be correspondingly different, but as long as the staggered-bond principle holds (i.e., only T and G configurations apply) the conformational possibilities will be finite. Hence the actual chain conformation, the most important structure feature, is usually directly identified (by model building, simple geometric calculations, or even by inspection) from the layer-line periodicity.

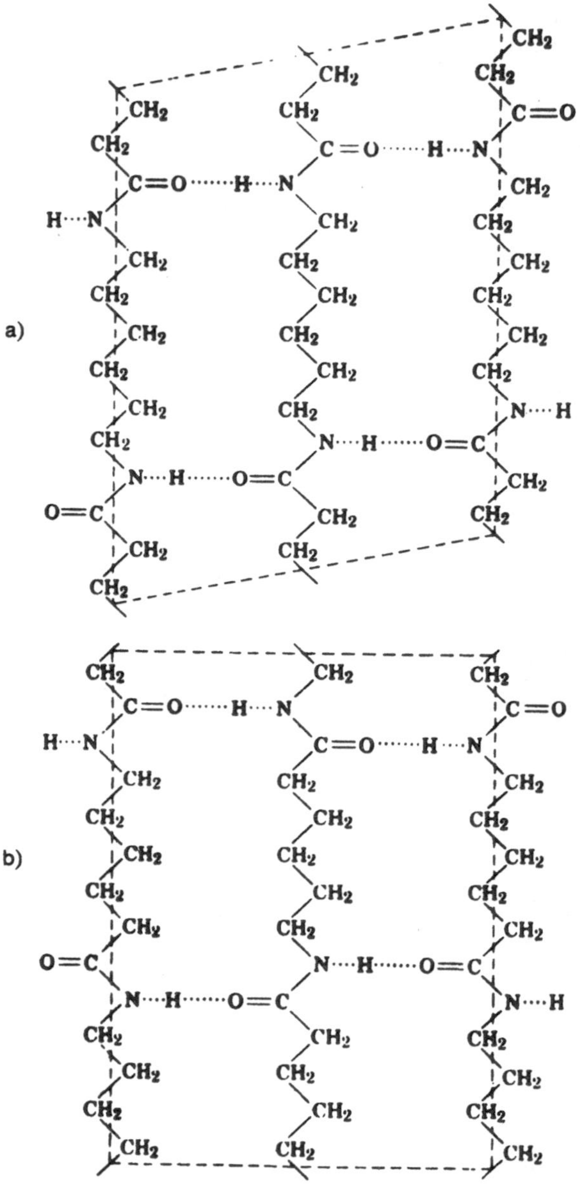

Figure 2.10. Examples of sheet formation in nylons via establishment of hydrogen bonds between chains: (a) nylon 66 and (b) nylon 6.

For more complicated chains, as in the case of many biopolymers, the Fourier transform method is particularly useful, namely the chain itself often has a readily recognizable, highly structured transform (the diffraction, which would be produced by a single chain, or a parallel array of uncorrelated chains) and the reflections actually correspond to the sampling of this transform by the interference function due to the full three-dimensional crystal lattice. The Fourier transform of some conjectured chain conformation can be readily calculated and then, after cylindrical averaging in the case of fibers, compared with the diffraction pattern; or conversely, certain typical chain conformations have readily recognizable fingerprints (arrangements of strong reflections or absences) in the diffraction patterns.

The trend toward hexagonal packing can be a useful guide for the determination of chain packing. Regular hexagonal packing is readily recognized where the strongest reflections on the equator (0 layer line) correspond to that of the hexagonally close-packed planes. If the structure is pseudohexagonal there will be two or three such strong reflections instead of one (e.g., 110 and 200 in polyethylene). This is the orgin of the characteristic prominent single ring (hexagonal) or doublet ring (orthorhombic, etc.) in the unoriented (i.e., powder-type) diffraction patterns of most simple synthetic polymers.

2.3. Degree of Crystallinity

As stated earlier, materials consisting of long-chain molecules are in general not fully crystalline and contain a not-inappreciable amount of amorphous constituent. A most self-evident manifestation of this is the fact that, say, a piece of plastic such as polyethylene, while displaying clearly defined X-ray reflections, does not have properties like that of a conventional crystalline solid but behaves to some extent like an amorphous rubber or glass. If crystallinity is only partial, then clearly knowledge of the "degree of crystallinity" is required.

2.3.1. The Principle of the Determination

For a quantitative determination of the degree of crystallinity, often referred to as the "amorphous–crystalline ratio," the following general method is usually adopted. A property that is crystallinity-sensitive is chosen. If its dependence on crystallinity is known for a polymer, it can then serve as an indicator of the degree of crystallinity for any given sample of that polymer. Such properties fall broadly within three classes: thermodynamic, diffraction-based, and spectroscopic.

2.3.2. Methods of Determination

2.3.2.1. Thermodynamic

a. *Specific Volume v or Density* ρ. If the degree of crystallinity is denoted by χ_c, then

$$\chi_c = \frac{v_a - v}{v_a - v_c} \equiv \frac{\rho - \rho_a}{\rho_c - \rho_a} \tag{2.1}$$

where v and ρ refer to the relevant property of the sample under examination, v_c and ρ_c to that of the purely crystalline state, and v_a and ρ_a to that of the purely amorphous state. If the crystal structure (i.e., at least the unit cell dimensions and the number of chains within this cell) is known, v_c or ρ_c are also known quantities. The determination of v_a or ρ_a can present a problem if the polymer is unobtainable in a fully amorphous state at the temperature at which χ_c is required (usually ambient temperature), which is frequently the case. In this situation v_a must be extrapolated from its value in the melt assuming knowledge of its thermal expansion coefficient in the molten state. This is a long extrapolation with all the related uncertainties, not to speak of the inherent assumption that $v_a(T)$ is a linear function over the temperature range concerned. Hence for this method, v_a is the limiting factor to the accuracy that can be attained, in addition to such apparently trivial, but by no means easily avoidable problems as the influence of voids (which can be submicroscopic) on the measurements.

b. *Heat of Fusion* (ΔH). Here $\chi_c = \Delta H / \Delta H_c$, where ΔH is the heat of fusion of the sample under consideration and ΔH_c that of the fully crystalline material; ΔH_c is the basic limitation in view of the fact that a fully crystalline material is not generally available. Therefore in this case ΔH_c has to be obtained theoretically and/or by extrapolation from the behavior of shorter chain oligomers (e.g., paraffins as oligomers for polyethylene).

2.3.2.2. Diffraction Methods

These essentially rely on the assessment of the ratio of the amorphous to the crystalline scattering intensities in the X-ray diffraction patterns of random specimens. The technical problems encountered mainly pertain to the separation of crystalline reflections from amorphous halos in the usual case where they overlap to some extent. Problems of principle comprise the issues of how far out one needs to go in reciprocal space (scattering angle) for the information to be meaningful and of how small a crystal is still a crystal (i.e., how broad a reflection still counts as a Bragg reflection).

2.3.2.3. Spectroscopic Methods

a. Infrared Spectroscopy. Certain features in the infrared spectrum are sensitive to the configuration and/or packing of the chains in the crystal and can thus serve for the assessment of crystallinity. These features fall into two categories.

The first kind relies on the observations that certain adsorption bands appear in the crystalline state and disappear on melting, and are accordingly then classified as "crystalline" bands. Similarly, "amorphous" bands can also be identified by their disappearance, or at least reduction, on crystallization. The intensity of such bands can then be used as a measure of the degree of crystallinity. In the first place this may only be a purely empirical exercise, which at a later stage may or may not acquire theoretical foundation.

The second class of spectral features relies on the fact that in a crystal structure containing symmetrywise nonequivalent neighbor chains, such chains can vibrate inphase or antiphase while in the crystalline state corresponding to two adsorption bands at slightly different frequencies. The result will be doublets in the spectrum. This distinction disappears in the amorphous phase where only a singlet will be seen. The evaluation of such doublets, where they exist, provides a theoretically well-founded method for the measurement of amorphous–crystalline ratios.

b. NMR Spectroscopy (Broad Line). This method does not measure crystallinity but mobility, and can be used for crystallinity determinations only as far as mobility can be equated with the amorphous phase and immobility with the crystalline phase. It relies on the fact that in the immobile (crystal) phase the external magnetic field will be locally modified by the magnetic moments of the neighboring nuclei (protons), and differently so by the different neighbors, with the result that the resonance envelope will be broad. In the mobile (amorphous) regions there will be an averaging out of the effect of neighboring nuclei and the resonance peak will be sharp. The method then consists of measuring the ratios of the sharp and broad resonance peaks. It should be noted that this method is sensitive to small amounts of mobile material.

2.3.3. An Appreciation of the Different Methods

There exists much literature concerned with the correct absolute value of crystallinity and with the comparison of results obtained by different methods, including claims for agreement and disagreement. The first point to be appreciated is that the different methods are measuring different properties, and therefore even if correctly measured and interpreted (some of the potential uncertainties have already been noted), they need not necessarily give exactly identical numerical results. It therefore follows that the measurement and definition of any absolute value of the crystallinity is problematic. Even if it were possible to obtain a reliable unique absolute value this would still be of questionable significance *per se* because, as will be apparent below, materials possessing the same degree of crystallinity as determined by one particular method can still have grossly different properties depending on the route along which the particular degree of crystallinity was attained. This is because this route affects the crystalline fine structure, i.e., the morphology, of which much more will be said below.

More meaningful than the absolute values themselves are the comparative differences between different samples as regards degree of crystallinity and, in particular, changes within the same sample produced by, say, certain treatments. In this respect the different methods should at least indicate the same trends, which they generally do.

One of the most straightforward applications of comparative crystallinity measurements on a given sample is that which records the development of crystallinity as a function of time. This serves to determine the rates of crystallization, and in general represents the basis for the study of crystallization kinetics.

2.4. Kinetics of Crystallization

2.4.1. Rates of Crystallization

The kinetics of crystallization is important in technological applications as it determines, e.g., the rate of solidification of a molded object during cooling or that of a thread emerging from a spinarette. It is also important, of course, for the fundamental understanding of crystallization.

The basic investigations consist of measuring χ_c as a function of time at preselected constant temperatures. In what follows we shall take the specific volume as an indicator of crystallinity. This can be most readily measured with a dilatometer. We express the results in terms of $v_a - v \equiv \Delta v$ as a function of time at a given temperature T. These lead to sigmoidal curves, such as that drawn schematically in Figure 2.11. The rate R can then be defined either as the steepest slope at the inflection point, or as the half-time of the full volume change Δv_{max}.

In general, the curve of R vs T will be bell-shaped, such as that in Figure 2.12, because the rate will be zero at the melting point T_m (in fact, it will not assume a measurable value until well below T_m, i.e., polymer crystallization only starts at considerable supercooling) and will become vanishingly slow again at the glass transition T_g. In between T_m and T_g there will necessarily be a maximum at some temperature T_{max}. This

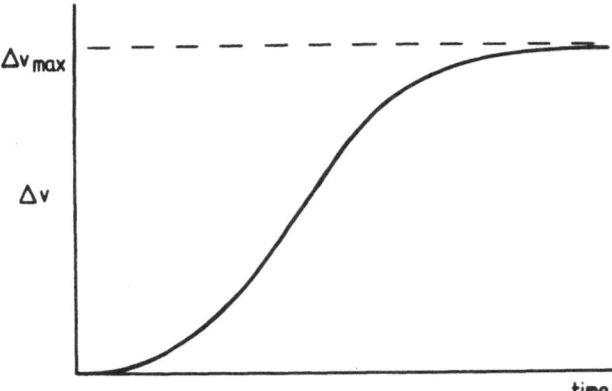

Figure 2.11. A schematic crystallization isotherm obtained in the study of crystallization kinetics. In this instance volume change Δv is plotted against time.

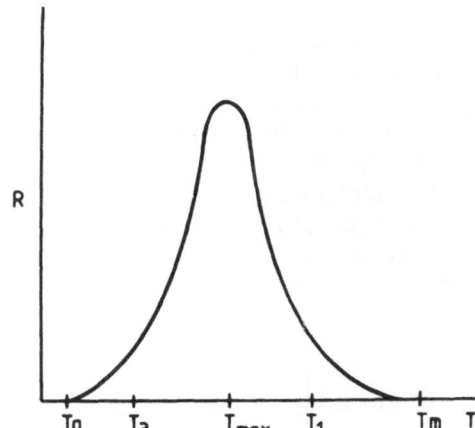

Figure 2.12. A typical curve representing rate of crystallization R as a function of crystallization temperature T.

maximum is due to two opposing effects: the increasing thermodynamic driving force with increasing supercooling ΔT ($\Delta T \equiv T_m - T$) and the decreasing mobility at the correspondingly lower temperature.

The different crystallizable polymers differ as regards the magnitude of the maximum in the R vs T curve. In practice one may place them in two groups: those where R_{max} is sufficiently low so that a normal sample *can* be cooled through T_{max} without crystallization setting in, and those which *cannot.* The former group can be obtained in an amorphous state, hence in the glassy state, and if so required can be crystallized at temperatures below T_{max}. PET, polycarbonates, and isotactic polystyrene are examples. In fact crystallization rates in the latter two cases are so slow that prolonged holding at T_{max} is required to achieve any crystallization at all. This is why polycarbonates, though crystallizable, are or can be used in the glassy amorphous state and are mostly used as glasses by virtue of being transparent in this state, while owing to their particular chemical constitution they possess a high T_g (180 °C).

2.4.2. The χ_c vs t Curves

It is most important to realize that Δv_{max} does not correspond to full crystallinity, i.e., to $\chi_c = 1$. It corresponds to the maximum χ_c achievable at that particular temperature T, which is < 1. Thus the kinetically fully crystallized material is not fully crystalline (which reflects the fact that complete crystallinity is not normally achievable).

We shall therefore define the kinetic crystallinity χ_c' in the form

$$\chi_c' = \frac{v_0 - v_t}{v_0 - v_\infty}$$

where the subscripts of v refer to the time elapsed after the sample has attained the intended crystallization temperature. It follows that χ_c' at $t = \infty$, denoted $\chi_{c\infty}'$, is always less than unity.

Next we consider the Δv vs t curves as a function of T. When these curves can be obtained over the full range of crystallization temperatures, the resulting situation corresponds to that sketched diagrammatically in Figure 2.13. Three T values are represented with $T_1 > T_2 > T_3$, where T_2 corresponds to T_{max}. It is noteworthy that $\chi_{c\infty}'$ depends on T for values of T below T_{max}. In particular $\chi_{c\infty}'$ (the maximum value for Δv) becomes

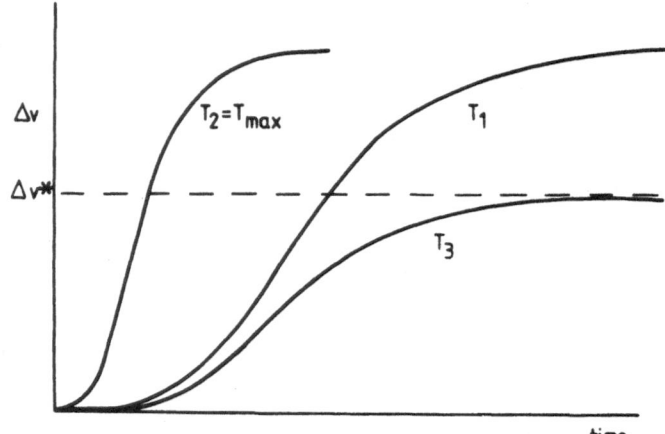

Figure 2.13. Crystallization isotherms corresponding to crystallization temperatures at the high-temperature side T_1, low-temperature side T_3, and maximum region in Figure 2.12.

increasingly smaller with decreasing temperature. This means that the maximum achievable crystallinity at these temperatures will be lower even if, kinetically defined, the sample is fully crystallized.

2.4.3. Morphological vs True Crystallinity

When a polymer melt crystallizes, spherical crystal aggregates nucleate and grow (so-called spherulites; more about this later!). At each stage of growth the sample volume will be partitioned between the spherulites and the unaltered melt, the fraction of spherulite-filled volume becoming increasingly larger with time. When all the spherulite-free amorphous volume is consumed, the crystallization (primary, see later) comes to an end. At this stage Δv_{max} is attained, hence $\chi'_c \equiv \chi'_{c\infty} = 1$. Here the sample is fully crystallized and in morphological terms fully spherulitic. We can say that the sample is *morphologically* (but not truly) fully crystalline. Thus the Δv vs t curves record the development of morphological crystallinity.

It follows from the fact that $\chi'_{c\infty}$ can be different for different temperatures of crystallization (e.g., for T_2 and T_3 in Figure 2.13) that in absolute terms the degree of crystallinity of the spherulites themselves must be different for the two different crystallization temperatures. (Thus spherulites formed at the lower temperature T_3 are *less* crystalline.)

Accordingly, the properties of polymers can show a twofold dependence on crystallinity: a dependence on the degree of morphological crystallinity and on the degree of true crystallinity. Conversely, it follows that a single (true) crystallinity value does not fully characterize the crystalline state of a sample.

The last statement will be illustrated schematically as follows. Let us consider again Figure 2.13 and take the development of crystallinity corresponding to Δv^*. We see that along curve T_3 this corresponds to Δv_{max}, hence to $\chi'_{c\infty}$, i.e., to the morphologically fully crystallized sample, while along curve T_1 the crystallization is far from complete. If we compare sections of the two samples under the microscope we see images such as in Figure 2.14a and b corresponding to Δv^* along curves T_1 and T_3, respectively. Figure 2.14a is meant to represent isolated spherulites within an amorphous matrix, while in Figure 2.14b the spherulites are all impinging and no untransformed material is left. Clearly, the two morphologies are very different, a difference reflected by the physical properties. Yet the absolute crystallinity as defined by equation (2.1) will be identical. Conversely, identical morphological crystallinities (i.e., fractional spherulite occupancy) attained along the different isotherms will correspond to different true crystallinites; the properties here again will be largely different. (The above significant exposition does not feature in the conventional literature; it is based on unpublished researches by the author.)

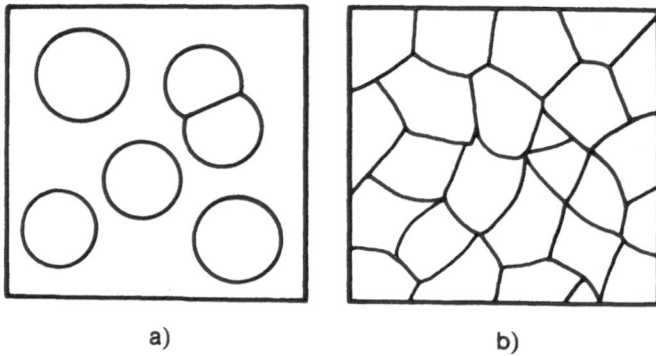

a)   b)

Figure 2.14. Schematic representation of two stages in the course of crystallization. (a) Spherulites are largely isolated within an untransformed amorphous matrix. (b) Sample is fully pervaded by spherulites. Such a sample is fully crystallized and corresponds to the stage Δv_{max} in Figure 2.11, or to Δv^* along curve T_3 in Figure 2.13.

2.4.4. Primary and Secondary Crystallization

The Δv vs t curves do not often level off smoothly but can take one of the two courses shown in Figure 2.15. This behavior is the consequence of secondary crystallization following the initial primary crystallization discussed earlier. In Figure 2.15a the first plateau signifies the termination of the primary crystallization, and the second plateau that of the secondary crystallization. In this case the rates of the two processes are sufficiently disparate for them to appear separately in the Δv vs t curve. If both processes occur at comparable rates, i.e., the primary process is still in progress when the effect of the secondary process becomes significant, their

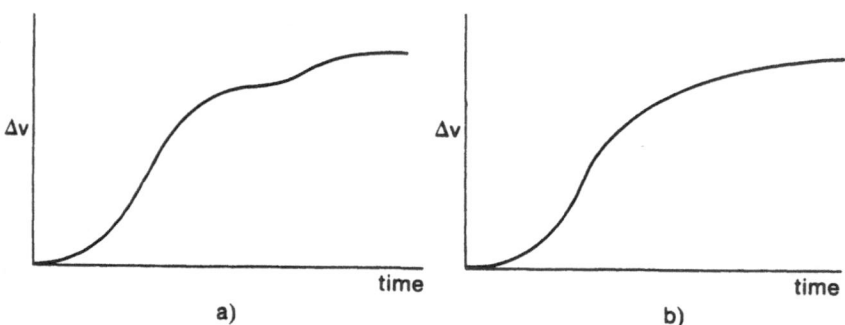

Figure 2.15. Crystallization isotherms with two different manifestations of secondary crystallization.

resultant effect will merge and curves such as those in Figure 2.15b will result.

In structural terms, secondary crystallization represents the increase in the true degree of crystallinity of the spherulites (which themselves are the products of the primary crystallization). Thus the spherulites improve their crystallinity with time! If this improvement is slow compared to the growth of the spherulites themselves, we have the behavior demonstrated in Figure 2.15a, otherwise that in Figure 2.15b. This is an important illustration of the many-stranded relation between crystallinity and crystalline texture. Of the fine structural features underlying this effect more will be said later.

2.4.5. Textures

The overall process of crystallization consists of two separate processes: (1) nucleation of new crystals and (2) growth of crystals already present. The overall crystallization rates, as in Figure 2.12, result from the compounding of these two separate processes. The rate of each of the separate processes (1) and (2) has a maximum with temperature (the maxima for the two do not coincide, that for (1) being at lower temperatures). For a given crystallization temperature, the comparative rates of the two processes and the number of nuclei present to begin with determine the overall scale of the resulting crystal entities within the final polycrystalline material. This feature is termed *crystal texture* as opposed to crystal structure on the molecular level.

Nucleation itself can be of two types: (1) predetermined and (2) spontaneously forming.

1. The predetermined nuclei are heterogeneities, the effect often being referred to as heterogeneous nucleation. In polymers there can be a further subdivision: extraneous heterogeneities (catalyst residues, impurities, etc.) and residual crystal fragments of the polymer itself giving rise to so-called "self-seeding." This latter mode of nucleation is unique to polymers and arises from the fact that, for a variety of reasons to be dealt with later, polymers have a broad melting range. The last crystal fragments to melt may be present in such a small proportion that they are undetectable in what otherwise appears to be a homogeneous melt, and may even be in a superheated form above the conventionally identified melting range. These will serve as nuclei for crystallization on cooling.

2. Spontaneously forming nuclei in principle result from spontaneous fluctuations that exceed the critical nucleus size. It is often termed *sporadic nucleation* or *homogeneous nucleation*.

In general, large numbers of nuclei lead to a fine-grained texture and few nuclei to a coarse-grained texture. This is true for all polycrystalline materials. In addition, with polymers we have a two-tiered texture. The texture as determined by primary nucleation in the above sense relates not to single-crystal grains (as in metals or rocks, etc.) but to the spherulites. As we shall see, these spherulites are not single crystals but are themselves crystal aggregates, hence possess a texture. The fact that the spherulites themselves can possess different degrees of crystallinity relates to differences in their texture; in particular, the increase in crystallinity on secondary crystallization relates to the coarsening of this spherulite texture, more of which later.

It is a general feature of all polycrystalline materials that the macroscopic (e.g., mechanical) properties are highly texture-dependent. Coarsely textured materials are generally stiffer (high modulus) but more brittle (support lower strain before fracture), and conversely for the finer-scale texture. This is true also for polymers, with the important addition that it applies to textures on both levels, the spherulitic and intraspherulitic. In fact, we shall see that there can be more than two levels. The existence of such texture hierarchies is a unique feature of macromolecular substances and becomes extremely intricate with biological materials.

In general, the crystallization of polymer melts is practically always induced by predetermined nuclei. Extraneous heterogeneities (mostly catalyst residues) can virtually never be avoided. For this reason the temperature regime of homogeneous nucleation (except for some very specialized experiments) cannot be reached because crystal (spherulite) growth emanating from these predetermined centers pervades the whole sample, hence completes the primary crystallization, before temperatures low enough for homogeneous nucleation to occur at appreciable rates are attained in the course of cooling. The predetermined nature of the nucleation is usually apparent from the fact that the spherulites are all of equal size, signifying that growth started from given centers simultaneously. This is not to say that sporadic nucleation cannot be observed, i.e., spherulites continually appear in the still untransformed portion of the material, giving rise to a final texture of widely differing spherulite sizes. However, there is evidence that even in this latter case nucleation starts from predetermined centers where the nucleating efficiency of these centers is different, resulting in a sporadic emergence of new spherulites. The latter is to be envisaged as a reservoir of predetermined nuclei, the nuclei having a range of incubation periods, the predetermined nature of the centers only becoming apparent when the exhaustion of the reservoir can be observed before the primary crystallization is complete. A more rigorous understanding of all these nucleation phenomena has yet to be achieved.

2.4.6. Analytical Treatment of Crystallization Kinetics

2.4.6.1. Analysis in Terms of Expanding Spheres (the Avrami Equation)

The recognition that crystallization of polymers is represented on a macroscopic scale by nucleation and growth of spherulites leads to a simple mathematical description of isotherms, such as that in Figure 2.13. This relies on the calculation of the volume fraction as yet uncovered (i.e., the amorphous fraction) by the newly forming and growing spherulites at any particular time of crystallization. The information required initially is the rate of volumetric spherulite growth (alternatively, this can be obtained from the experimental isotherms using the theory to be outlined below).

Let us first consider the growth rates. An important basic fact is that, at a given temperature, spherulites are observed to grow at a constant linear velocity, and are thus characterizable by a single radial growth rate $\dot{r}$ in the mathematical treatment.

As regards the formation (nucleation) rate, the usual treatments embody the two alternatives discussed in the preceding section: (1) the spherulite centers are predetermined (heterogeneous nucleation) and (2) new spherulites form at a constant rate within the still untransformed material (sporadic nucleation).

In treating dimensionality and shape, the spherulites in three-dimensional samples are regarded as spheres. Two-dimensional samples (thin films) can also be important for basic laboratory studies under the microscope, in which case the spherulites are considered as disks. Other shapes and their combination with different dimensionalities also feature in the literature. It will be shown that except for special, restricted conditions of application, the formulas derived for these are erroneous.

The treatment of expanding spheres and circles to be outlined below is usually referred to as the *Avrami treatment*, although it has been applied in several instances prior to Avrami, notably as far back as Poisson.

Consider spherical bodies $1, 2, \ldots, n$ of volume $V_1, V_2, \ldots, V_n$ in some state of expansion before they impinge, with total volume V. The probability p_i that a randomly chosen point P lies outside V_i is given by

$$p_i = 1 - V_i/V$$

The probability p that the point lies simultaneously outside bodies $1, 2, \ldots, n$ is expressed by

$$p = p_1 p_2 \cdots p_n = \prod_{i=1}^{n} (1 - V_i/V) \qquad \text{or} \qquad \ln p = \sum_{i=1}^{n} \ln(1 - V_i/V)$$

Since $V_i \ll V$, powers of V_i/V higher than the first can be neglected in the expansion of $\ln p$, hence

$$\ln p = -\sum_{i=1}^{n} (V_i/V) = \delta \bar{V} \qquad \text{or} \qquad p = e^{-\delta \bar{V}}$$

were $\delta = n/V$, the number of entities per unit volume or the nucleus density, and $\bar{V} = \Sigma V_i/n$, the mean volume of entities present.

An expression for p is also given by the volume fraction of material not engulfed by the "bodies," namely the volume fraction of the remaining nonspherulitic (hence totally amorphous) material. Thus

$$p = 1 - \chi_c' = e^{-\delta \bar{V}} \tag{2.2}$$

The main problem beyond this simple geometric derivation is how to treat the situation when different bodies abut. Avrami does it in a very elaborate way. However, this situation can be treated very simply (essentially going back to Poisson, but revived by U.R. Evans [*Trans. Faraday Soc.* **41**, 365 (1945)] for the treatment of the spread of rust on a metal surface) by considering two potentially overlapping spherical bodies, as presented in Figure 2.16, where 1 and 2 refer to the unobstructed spheres. If a point P is situated within either of the unobstructed spheres, then it will lie within the overlapping composite body as well. If it is outside the unobstructed spheres (point P') it will also be outside them when they abut (see Figure 2.16) as long as neither body outgrows the boundaries of the other in the hypothetical absence of overlap. Thus we can assume that P' lies in the nonspherulitic material irrespective of overlap. Accordingly, equation (2.2) will still hold.

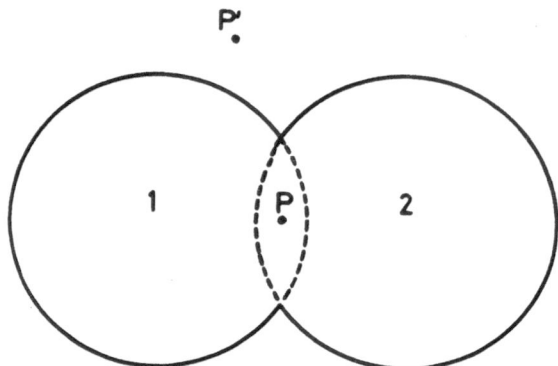

Figure 2.16. Effect of spherulite overlap as referred to in the derivation of the Avrami equation.

The exponent in equation (2.2) can be expressed as a function of time t (see standard textbooks on polymers embracing crystallization) in the form

$$1 - \chi_c' = e^{-Kt^n} \qquad (2.3)$$

where n and K are constants, n depending on the dimensionality (two- or three-dimensional growth) and on the type of nucleation and K including the parameters relevant to the rate of growth, rate of nucleation, and number of nuclei. This is shown more explicitly in Table 2.1.

Equation (2.3) usually provides a good description of the observed Δv, or $1 - \chi_c'$, the kinetic amorphous fraction into which Δv is readily converted, as t varies (Figures 2.11 and 2.13) with appropriate constants. The constants can then be determined by plotting $\ln \ln(1 - \chi_c')$ against $\ln t$, which usually yields a straight line of slope n and intercep K. Hence the nature of the nucleation (predetermined or sporadic) and the growth and nucleation rates, respectively, can be assessed. These quantities can be observed directly under the optical microscope in suitably prepared specimens, and agreement with those derived from the Δv vs t curves via equation (2.3) can be checked.

In a number of cases correspondence between the kinetic isotherms and the measurements on the microscopic image was sufficiently encouraging to provide confidence when employing the above method of kinetic analysis, which is used extensively as a consequence. Nevertheless a few reservations, not to be found in textbooks, may be appropriate.

2.4.6.2. Limitations of the Avrami Analysis

Measurements giving χ_c' as a function of t (particularly via v or ρ) can be very simple and the result is always informative *per se* as regards the rate of crystallization. Possession of these isotherms then naturally invites evaluation via equation (2.3), hence the widespread popularity and frequent misuse of the Avrami analysis.

Table 2.1. Parameters in the Avrami Equation (2.3)[a]

Dimensionality	Constants	Predetermined nuclei	Sporadic nucleation
Two dimensional	n	2	3
(circles)	K	$\pi \dot{r}^2 g$	$\frac{1}{3}\pi k \dot{r}^2$
Three dimensional	n	3	4
(spheres)	K	$\frac{4}{3}\pi \dot{r}^3 g$	$\frac{1}{3}\pi k \dot{r}^3$

[a] The number of nuclei per unit volume or area is denoted by g, the rate of radial growth by $\dot{r}$, and the rate of nucleation by k.

Table 2.1 is confined to circles (disks in the case of two-dimensional, i.e., thin, samples) and spheres. Textbooks often give more extensive tables for other geometries too. These can be misleading or outright erroneous. The assumption underlying the derivation was that point P' remains outside the body irrespective of whether the growth of body 1 (see Figure 2.16) is being obstructed or not. For spherical objects growing at a constant rate, this is obviously true (sphere 1 in Figure 2.16 could never outgrow sphere 2). It will not, however, be generally true for disks in any arbitrary orientation within a three-dimensional specimen or, as often quoted, for arbitrarily oriented rods growing longitudinally. For both these cases this will only hold if they are all parallel (which is satisfied for the two-dimensional case in Table 2.1) but not otherwise. For other, more complex shapes (such as sheaves) sometimes quoted, the above condition relating to P' will never hold, hence the Avrami treatment for these situations will be invalid.

Even for spheres, the above treatment is confined to primary crystallization only, i.e., when the spherulites do not subsequently change their crystallinity. Although a treatment for the inclusion of secondary crystallization exists, this is not normally noted in most published works and would in any case be very difficult to incorporate in actual practice, thus removing much of the simplicity, and hence the attraction of the method. By the simple considerations featuring in Table 2.1 integral n values are expected. Early works indeed have claimed such values, usually the value 3. However, subsequent accumulated experience has shown that nonintegral ns can be just as frequent. Lack of attention to secondary crystallization may well be responsible for these departures.

Finally, whatever information may be correctly obtained by the Avrami-type analysis it will leave unanswered the basic questions regarding the intimate fine structure of the crystallization. For this a direct attack on the structure is required.

2.5. Spherulites

Spherulites represent the highest level in the morphological hierarchy of a crystalline polymer. In the previous sections we treated crystallization in terms of formation and growth of spherulites readily identifiable in the optical microscope, and defined the resulting texture in terms of these spherulites. In this respect, spherulites are analogous to the grains in, say, a polycrystalline metal with the fundamental difference, however, that they themselves are not single crystals but a particular type of aggregate of crystals. A spherulite is usually defined as a birefringent object with spherical (circular in thin layers of material) symmetry (Figure 2.17). The

Figure 2.17. Photomicrograph of a particular spherulitic sample displaying spherulites in the process of impingement during growth. The spherulites are banded, corresponding to rotation of a biaxial crystal (Figure 2.20). The polymer is poly (β-hydroxybutirate). Crossed polaroids (Barham *et al*[22]).

developed objects themselves are spheres (or circles in thin layers) as long as they are isolated. When they abut, they acquire polygonal outlines (see the examples in Figure 2.17), the internal spherical symmetry nevertheless being maintained.

In what follows we shall discuss the optical properties, and following this, the morphology.

2.5.1. Optical Properties of Spherulites

The optical characteristics are the foremost features of spherulites via which they are normally identified. Two principal features are usually observed when viewed between crossed polaroids in parallel light, and are readily visible in Figure 2.17: a dark Maltese cross and concentric dark banding. The former is practically always present and in fact forms part of the definition of spherulites, while the latter is usually, but not necessarily, associated with it.

2.5.1.1. The Maltese Cross

The Maltese cross is parallel to the polarizer and analyzer directions, as in Figure 2.18. It arises from the spherically (or circularly) symmetrical arrangement of the index ellipsoid, which in turn corresponds to the anisotropic entity (crystal) constituting the spherulite as shown by Figure 2.18. The origin of the cross should be apparent: it is due to extinction arising wherever a transmission direction coincides with the direction of the polarizer or analyzer. The transmission directions themselves corres-

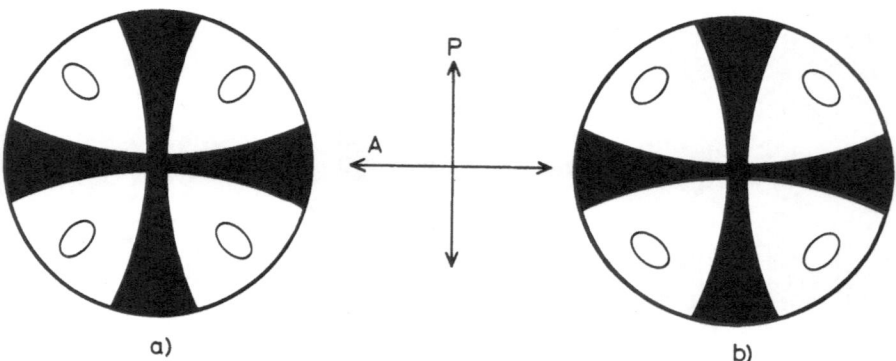

a)   b)

Figure 2.18. Illustration of the origin of the Maltese extinction cross in spherulites. Small ellipses represent appropriate sections of the index ellipsoids of the optically anisotropic constitutents. (a) Direction of slow vibration is radial. (b) Direction of fast vibration is radial. P and A denote polarizer and analyzer directions.

pond to the long or short axes of the elliptical projections (in a plane perpendicular to the direction of viewing) of the index ellipsoids (Figure 2.18).

2.5.1.2. Concentric Banding

This banding represents a periodicity in the orientation of the index ellipsoid (hence that of the underlying crystal) along any given spherulite radius, the periodically varying orientations along different radii being in phase at any given radial distance throughout the whole spherulite. The black bands arise through an optic axis becoming periodically parallel to the direction of viewing, when the birefringence is zero, hence complete extinction will arise at the corresponding locality. This implies that the birefringent units are rotated around a radial direction (while being simultaneously displaced radially) so that an optic axis is perpendicular to this radial direction. (This is usually the case but not always. An optic axis can also lie at an angle to the axis of rotation, in which case more complex periodic extinction effects result in the microscopic image.) If the crystal is uniaxial the dark bands will be equidistant (Figure 2.19). If the crystal is biaxial and the rotation axis (the spherulite radius) is normal to the optic axial plane, the banding will be constituted by two sets of nonequidistant

Figure 2.19. Sketch of banded spherulite with banding due to helicoidally varying orientation of a uniaxial crystal with optic axial direction perpendicular to the spherulite radius, which is the rotation axis.

Figure 2.20. Sketch of banded spherulite with banding due to helicoidally varying orientation of a biaxial crystal with optic axial plane normal to the spherulite radius, which is the rotation axis. Alternating band spacings reflect the inequality of optic axial angles, narrow and wide spacings corresponding to acute and obtuse angles, respectively.

rings, where the relative ring separations correspond to the ratio of the optic axial angles (see textbooks on crystal optics; Figure 2.17). The micrograph in Figure 2.17 corresponds to such a situation.

The ultimate reason for the helicoidal arrangement of the index ellipsoids and of the underlying crystals, giving rise to the above concentric bandings, is not known and represents one of the major outstanding problems in polymer crystallization and beyond that in crystal growth. In the latter context it is most important to realize that the phenomenon is not confined to polymers. Virtually any crystallizable substance can display the effect. In fact, it was first identified in silicate minerals late in the 19th century followed by extensive studies of the phenomena in a descriptive manner, but virtually forgotten thereafter. While in small molecular substances the effect is regarded rather as a curiosity, in high polymers it is typical. Impurities are believed to be a requirement for both spherulites and the associated helicoidal crystal arrangements within them. As polymers are never absolutely pure (e.g., distribution of molecular lengths, lack of perfect tacticity, etc.) they are clearly candidates for this kind of crystal growth on that score. This is clearly a fruitful area for further research, but no clear direction of research is evident.

2.5.2. Morphology of Spherulites

The crystal texture and the underlying growth of a spherulite are always fibrous. Morphologically therefore they need to be visualized as a radiating array of fibrous crystals, which through appropriate branching will fill out spherical space. It follows that there must be a central discontinuity. The center could be a heterogeneity that nucleates fibrous crystals, which will thus grow out from it radially. Such uncoordinated radial growth does occur under certain circumstances and the resulting entity will display the Maltese cross, even if not the concentric bands. Such objects (Figure 2.21) nevertheless are not typical spherulites.

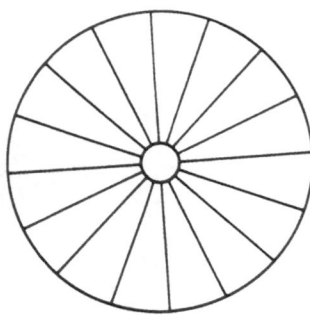

Figure 2.21. Sketch of a spherulite nucleated by heterogeneity. The radially growing fibrils are uncoordinated.

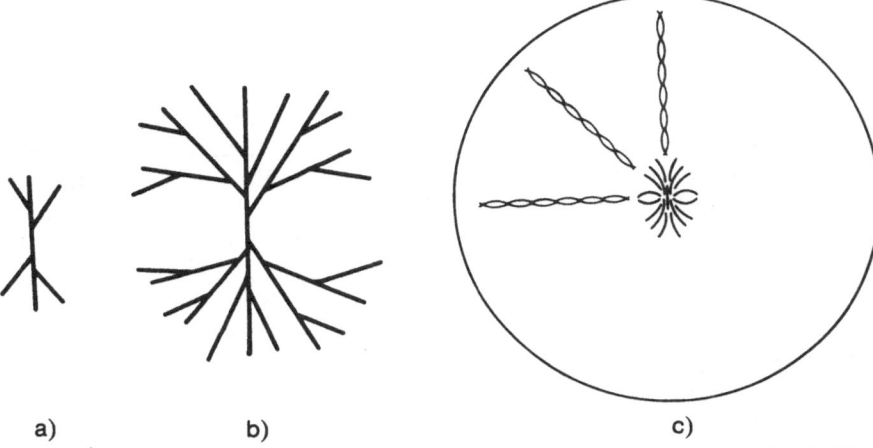

a)   b)   c)

Figure 2.22. Development of a typical spherulite from a single-crystal nucleus: (a) branching fibrillar parent crystal; (b) developing sheaf; (c) final spherulite with sheaf-like origin recognizable in the center.

Typical spherulites develop through sheaving crystal growth of a single fibrous parent crystal, represented in two dimensions by Figure 2.22 (without any attempt to incorporate helicoidal winding in stages (a) and (b) of the sketch). When developed sufficiently, the sheaf will take on approximately spherical (circular in two dimensions) outlines. The first essential point to note is that here each point within the spherulite is generically related to the initial fiber situated in the center, hence also to all other fibers in the spherulite. Such a spherulite could therefore be regarded generically (even if not geometrically) as a single crystal, as opposed to the type of growth in Figure 2.21 where the different fibers are uncoordinated. The second point to note is the characteristic shape of the central discontinuity, namely double-leaf-shaped (or a corresponding three-dimensional form), by which this type of spherulitic growth is readily identified. If this double leaf is small compared to the final spherulite, it will hardly impair the spherical symmetry of the latter. If it is large it may even make unrecognizable the spherical symmetry, as manifest by the Maltese extinction cross. Whether the former or the latter applies, depends on the scale of the fibrosity relative to the final spherulite size, where the latter in turn depends on the nucleation density.

The above distinction leads to two types of definition of spherulitic structure:

1. Polarizing optical definition: This is the historical definition quoted at the commencement of this section and relies on an identifiable Maltese cross.

2. Morphological definition: This is based on the sheaf development, as in Figure 2.22.

It is noteworthy that a sample can be "spherulitic" by (2) without displaying the optical effects in (1) (e.g., the sheaves have not been given the chance to develop into spheres). This is often a source of confusion in the literature as to whether a given sample crystallizes in a spherulitic or some other mode. Definition (2) is more general and appropriate, even if historically the subject owes its existence to definition (1) by which the spherulitic crystallization was first identified.

If we superpose helicoidal windings onto the fibrils, as in Figure 2.22c, three-dimensional space filling creates additional topological problems so far unsolved. (One is tempted to compare this situation with droplets of cholesteric liquid crystals formed by appropriate simple substances, which display both a Maltese cross and periodic banding, the origin of the latter being fully understood. However, liquid crystal drops are low-energy equilibrium shapes (without any internal fibrous texture) while crystalline spherulites result from a particular mode of fibrous crystal growth, the resulting entities being far from equilibrium. Hence the topological arguments for the liquid-crystal case cannot be simply transferred.)

As regards the origin of spherulitic growth a particular quantifiable scheme is finding widespread acceptance, according to which fibrous growth develops due to impurities causing constitutional supercooling, analogous to cellular growth in metals [H. D. Keith and F. J. Padden, *J. Appl. Phys.* **34**, 2406 (1963)]. To this should be added a noncrystallographic branching scheme (for which there is no *a priori* reason, even less so for the helicoidal winding when the latter is present). One advantage of the scheme, however, is that it defines the scale of the fibrosity in terms of assessable parameters, namely the diffusion coefficient of the crystallizing molecule D and the linear growth rate G. Accordingly, the scale of the fibrosity should be defined by D/G. There are differing views and experimental results regarding the appropriateness of the ratio D/G for characterizing the intraspherulitic texture. The most recent fine structure results [D. C. Bassett and A. M. Hodge, *Proc. R. Soc. London, Ser. A* **377**, 25, 39, and 61 (1981)] in fact explicitly refute the expected correspondence between D/G and the dimension of the relevant structure element, with corresponding implications for the underlying theoretical scheme.

2.5.3. The Fine Structure of Spherulites

Spherulites are highly organized crystal aggregates and are so compact that they do not yield readily to a direct attack on their finer architecture. In particular, questions as to both the basic crystal units constituting them and their mode of space filling are not readily answered by their continued direct examination. However, in polymers, we know through an indirect approach (see later) that the basic crystalline entities constituting the spherulite are ribbonlike lamellae (see, e.g., Figure 3.19 in the next chapter). It is very plausible therefore to visualize a radial entity within a spherulite as a continuous helicoidally twisting ribbon sprouting branches so as to occupy spherical space, as implied by Figure 2.22. However, such a continuous helicoidal path along a given ribbon has not yet been followed through morphologically (examinations are necessarily confined to fracture surfaces, usually etched so as to provide a relief); this deficiency is to be contrasted with the unassailable polarizing optical evidence of a continuous and periodic orientation variation of the index ellipsoid. There are running disputes in the literature on the explicit morphological realization of these orientation variations owing to the above-mentioned missing morphological information.

Whatever the explanation for spherulites and the banding within them, it must have general validity for crystallization and cannot be restricted to polymers alone. For polymers certainly, some specific features need considering, such as the lamellar ribbon-shaped basic crystal entity, and the fact that (as observed by micro-X-ray work not to be elaborated here) the overall direction of the chain molecules is perpendicular to the spherulite radius.

Even though so many features about spherulites are unknown, their study has served many useful purposes toward understanding and describing polymer crystallization. Thus their number in a given sample is an indicator of nucleation in polymer melts and their rate of radial growth (G, which as noted earlier is a constant at a given temperature) provides a measure of crystal growth, both of these quantities having served a useful purpose in the analyses presented in Section 2.4.6. In addition, G can be measured as a function of supercooling ΔT: it has been found that

$$G \propto e^{-Q/\Delta T} \tag{2.4}$$

where Q is a constant to be specified later. This relation suggests that the growth-rate-determining factor is secondary (i.e., surface) nucleation along the growing crystal front, a long-standing observation the significance of which will become apparent later.

The Basic Crystal Unit

A. Keller

3.1. Single-Crystal Lamella

Historically, the basic crystal entity was observed on crystallization from solution, primarily because in the dilute system the basic entities could be obtained in isolation, in contrast to crystallization from the melt, where the individual crystal entities are intimately interlocked and thus escape identification by straightforward inspection. It has taken many years of subsequent development until the same entities, which were already familiar from solution crystallization, could be identified also in the bulk melt-crystallized product where they are usually components of more complex aggregates (such as sheaves or spherulites).

3.1.1. Discovery and Description

The polymer, polyethylene in the first instance, was dissolved in hot solvent and then precipitated by cooling. The crystalline precipitate formed a turbid suspension where the suspended particles could be examined by optical (phase or interference contrast) and electron microscopy after sedimention on suitable support. The particles proved to be lamellae with well-defined crystallographic facets usually forming lozenges or various truncated versions of them (Figure 3.1), more exactly depending on the crystallization temperature. The crystal increases its overall thickness via the familiar spiral terrace screw dislocation growth mechanism. The thickness of the individual layers, as readily assessed in the micrographs

A. Keller · Department of Physics, University of Bristol, England.

Figure 3.1. Electron micrograph of a typical solution-grown single crystal of polyethylene (after Keller[6]).

themselves by the shadow length of suitably metal-shadowed specimens, was in the 100 Å range.

Electron diffraction confirmed the single-crystal nature of the object (Figure 3.2) and identified the lattice orientation with respect to external morphology (Figure 3.2a, c). It is seen from Figure 3.2c that the chains (the c direction) are perpendicular to the lamellar plane: in some instances they can also lie at a specific large angle (around 60°) to the lamellar surface. The prism faces of the lamellae can also be readily assigned; for the specific case of polyethylene in Figure 3.2 this is {110}. While the present illustration is for polyethylene, the situation is analogous for all other flexible polymers, namely the crystals are lamellae, which for different

a)

b)

Figure 3.2. (a) Electron micrograph of a monolayer crystal of polyethylene (after Keller and Organ). (b) Corresponding electron diffraction pattern (after Keller and Organ). (c) Lattice orientation with unit cell in c projection corresponding to (a) and (b). It follows that the chain direction is perpendicular to the lamellar surface.

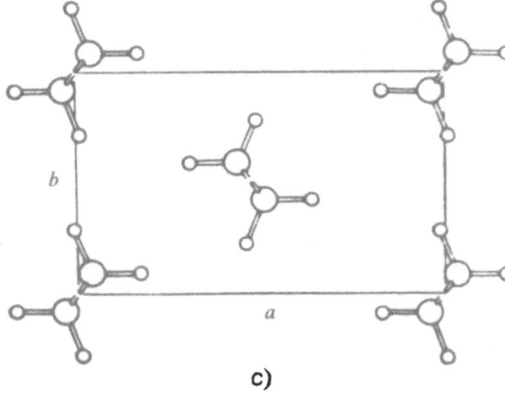

c)

polymers will follow different external habit patterns with the chain direction always perpendicular (or at a specified large angle) to the lamellar surface.

3.1.2. The Chain-Folding Model

At this point it should be noted that the chains are long, say $10^3 - 10^6$ Å, and are of nonuniform length in a given material. The following problem arises: how can such chains be perpendicular to layers the thickness of which ($\sim 10^2$ Å) is much less than the length of the chains, and be of uniform thickness? The answer is that the chains must fold back on themselves at regular intervals (as illustrated in Figure 3.3), the lattice register within the crystal, i.e., between the fold stems, being retained (as mentioned earlier in connection with Figure 2.1b). This is the argument on which the existence of chain folding is based. It is a nonmathematical commonsense argument, altogether convincing and irrefutable in its essentials. Any arguments about it (and, as we shall see, there are many) relate to the nature of the folding (see later) and not to its existence! (There are unfortunately several misleading popularizations to do with these controversies of which the reader is duly forewarned.) In particular, the neatness of the folding and the strictly adjacent deposition of the stems as expressed in Figure 3.3 are subject to debate. Nevertheless the representation in Figure 3.3 will be used in order to develop more readily some of the arguments to follow.

It is worth recalling the total unexpectedness of the developments just outlined. The findings are unexpected because (1) the basic crystal units

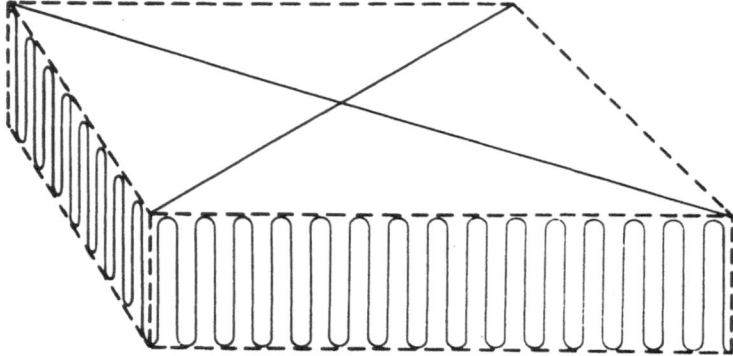

Figure 3.3. Schematic illustration of a chain-folded crystal structure to account for evidence in Figure 3.2.

turned out to be lamellae instead of fibrils, which might have been anticipated from long chains, and (2) the chains are along the shortest (not the longest!) crystal direction, a fact that leads directly to the model of chain folding.[†]

Let us consider the unprecedented situation these crystals create in the study of crystals in general. In such a situation we have an external habit feature, the lamellar thickness l, which corresponds to a molecular structure feature, the fold length. Hence the descriptive study of external shape and size (morphology) and that of internal fine structure on a molecular level (traditional "crystal structure analysis") merge into a more generalized investigation of the total crystal structure.

The fold length l, as we have seen, has both morphological and molecular significance so we shall concentrate on its measurement, behavior, and interpretation in what follows.

3.1.3. Fold Length l

Various questions arise: Is the fold length invariant or variable? In either case what are the factors determining it? How is chain folding to be accounted for in terms of molecular behavior? First, however, we must clarify the methods of measuring l.

3.1.3.1. Methods of Measurement

There are essentially three methods:

1. Electron Microscopy (EM). This technique relies on the measurement of the shadow length of appropriately metal-shadowed samples on the electron micrographs themselves.

2. Small-Angle X-Ray Diffraction (SAX). The crystalls that precipitate as a suspension can be sedimented into coherent macroscopic sheets, so-called mats, which contain the lamellae in an overall parallel orientation. Such a mat of layers acts as a one-dimensional grating to X-rays falling on it edgewise, giving rise to X-ray diffraction maxima in accordance with Bragg's law ($l = \lambda/2 \sin \theta$, where λ is wavelength and 2θ scattering angle). As the layer periodicity is large (100 Å and more) the diffraction angle is small, hence the diffraction maxima are observed at small angles requiring appropriate collimation techniques. This is the most widely used method, the lamellar thickness thus obtained often referred to as "long spacing."

[†] It should be noted that chain folding is superposed onto the primary chain conformation. For instance, if the primary conformation is helical, such as in polyolefins (Figure 2.8), then the whole helical chain itself folds.

3. Low-Frequency Raman Spectroscopy (referred to as LAM, longitudinal accordion (or acoustic) mode). This method is comparatively recent and has its origin in the longitudinal accordion-type vibration of the straight stems with nodes in the center and antinodes at the lamellae surfaces (Figure 3.4). If the vibrating stems are approximated by straight rigid rods, the fundamental Raman frequency ν is given by

$$\nu = \frac{1}{2l}\sqrt{\frac{E}{\rho}} \qquad \text{hence} \qquad l = \frac{1}{2\nu}\sqrt{\frac{E}{\rho}} \qquad (3.1)$$

where E is the modulus (force constant in molecular terms) and ρ the density. While more elaborate expressions than the simple approximation of equation (3.1) have been derived, this equation describes adequately what is required for most purposes. It follows that for large l, ν will assume very small wave numbers, hence very high degree of monochromatization is required in the experimentation so as to detect peaks close to the exciting frequency.

3.1.3.2. Comparison of Methods

1. EM. Advantages: qualitatively, it shows in real space what the entities actually are, all-important information not directly obtainable by

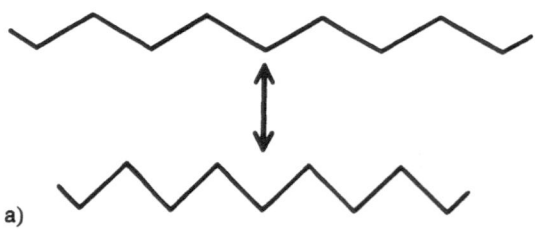

a)

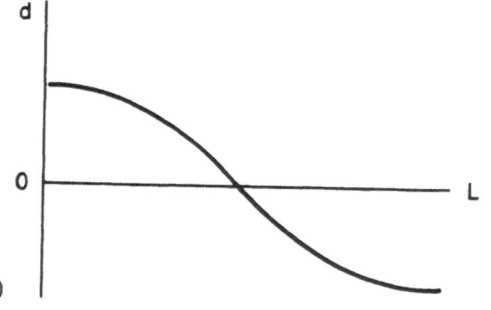

b)

Figure 3.4. Origin of the low-frequency longitudinal accordion-type Raman band (LAM). (a) Accordion-type vibration of the chain. (b) Corresponding longitudinal total displacement (d) along the vibrating chain with origin at $L = 0$.

any other method (e.g., without EM we would never know that we have lamellae!), and quantitatively, it measures the layer thickness directly and at the localities of potential interest. One disadvantage is that information derived pertains to a very small fraction of the material and thus may not be representative; also accuracy is limited.

2. SAX and 3. LAM. An advantage is that the information is representative of the whole or of large portions of the sample, and can thus be used for systematic quantitative work and hence to lay the foundation for theories. A disadvantage is that these methods are not self-contained but require EM [method (1) above] for knowledge of what the units are which actually produce the SAX and LAM effects. Hence the message is clear: the combination of *all* techniques is essential!

There are various distinctions between SAX and LAM. SAX relies on interference effects due to a periodic structure, hence on the existence of a periodicity (in addition to the existence of the layers as such). Such a periodicity is produced by the regular stacking of lamellae, which accordingly is a prerequisite for the application of SAX, at least in the present simplest form utilizing Bragg's law. In contrast, LAM relies only on the existence of the individual layers as such. This can be important, e.g, in cases where we have a mixture of lamellar thicknesses. Here complicated, smoothed-out interference effects result by SAX, while the LAM effect may remain clear, giving rise to separate Raman peaks corresponding to each lamellar thickness in the mixture.

SAX gives the total interlayer periodicity (stem plus fold; plus interlayer gap, if any) while LAM is expected to depend primarily on the stem length only. Hence in LAM we have a potential tool to distinguish between crystal core and fold surface — a topical issue (see later). However, efforts in this direction have encountered considerable difficulties, insofar as LAM, by equation (3.1), usually gives the same value for l as does SAX by Bragg's law. At present this is an open problem, which invites further in-depth theoretical considerations.

SAX (and recently LAM) effects in the form of discrete peaks have been observed also in melt-crystallized bulk polymers even when lamellae could not be directly, or at least readily, observed by EM. By analogy with solution-grown crystals, such effects are nevertheless being interpreted in terms of lamellar thickness and fold length. Indeed, such "jumps" in the argumentation proved to be justified in most cases when bulk samples became accessible to EM experimentation through later improvement in techniques. Even so a certain caution is advised in the indiscriminate identification of SAX maxima and LAM peaks with lamellar thickness in the absence of explicit EM evidence. Examples prompting this statement exist, but will not be elaborated here.

3.1.3.3. Factors Determining Fold Length

While applying methods (1)–(3) above, it was observed that the fold length l is not invariant but is affected by the temperature of crystallization T_{cr}, and by the temperature of heat treatment subsequent to crystallization, i.e., heat annealing T_A, where $T_A > T_{cr}$.

a. Temperature of Crystallization. At the first level of experimentation l vs T_{cr} curves, such as those in Figure 3.5, were obtained with polyethylene. Accordingly higher T_{cr} values yield higher values of l, increasingly so toward higher T_{cr}. Figure 3.5 shows two curves, one from a poorer solvent (octane) the other from a better solvent. The identical l values for different T_{cr} values along the two curves, i.e., the horizontal shift of the two curves by an approximately constant factor, means that it is not the absolute T_{cr} but the supercooling $\Delta T = T_m^0 - T_{cr}$ that is the factor determining l (where T_m^0 is the dissolution point of the infinitely extended chain crystal or the melting point for melt crystallization). For a poorer solvent T_m^0 is higher, hence the same T_{cr} corresponds to smaller ΔT.

The following important additional observation should be added to the findings expressed by Figure 3.5. If T_{cr} (or ΔT) is being changed during crystal growth, the crystal continues to grow with an altered l appropriate to the new T_{cr}, or ΔT, as expressed by Figure 3.5. Morphologically, this will be apparent by a step in the crystal lamella concentric with the crystal boundaries, where the crystal has become thinner or thicker according to whether ΔT has been decreased or increased, respectively. This means that the fold length is only affected by ΔT but remains unaffected by the thickness of the substrate onto which the molecule deposits.

Curves such as those in Figure 3.5 have formed the basis of all theories until recently. In subsequent developments, suitable choice of polymers (other than polyethylene, e.g., isotactic polystyrene) have enabled higher

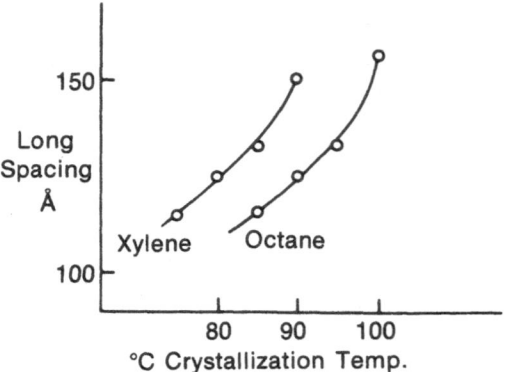

Figure 3.5. Crystal thickness ("long spacing" as measured by small-angle X-ray diffraction) as a function of crystallization temperature in polyethylene single crystals grown from two different solvents (xylene, octane) (after Kawai and Keller[7]).

values of ΔT to be attained for such l vs T_{cr} studies. The results now extending over a much wider range of T_{cr} (hence ΔT) values are expressed schematically by Figure 3.6, which can be taken as the general l vs T_{cr} relationship as known today. It is characterized by the same behavior as in Figure 3.5 for comparatively small supercoolings (high T_{cr}) followed by a horizontal plateau at high ΔT (low T_{cr}), i.e., at high values of ΔT l becomes insensitive to the temperature of crystallization.

The last feature was not included in the original formulation of the crystallization theories. Subsequent adaptation of theories to incorporate the horizontal portion of the l vs T_{cr} curves is a subject of topical interest (see later).

b. The Annealing Phenomenon (T_A). First we make the following observation: When a crystal is heated beyond T_{cr}, the temperature of its formation, it thickens irreversibly (l increases according to all methods of assessment) without any change in the overall lattice, in particular the chain orientation.

Subsequent interpretation leads to the inescapable fact that the chains have refolded in order to give rise to a new increased fold length, represented schematically (with the provisos expressed in connection with Figure 3.3) by Figure 3.7.

This extraordinary chain-refolding effect underlies all annealing behavior of crystalline polymers, including heat treatment in technological practice. It raises many questions and touches on many issues (not to be treated further in the present review), including the mechanism of refolding; the kinetics of refolding; the issue of chain mobility, particularly in the

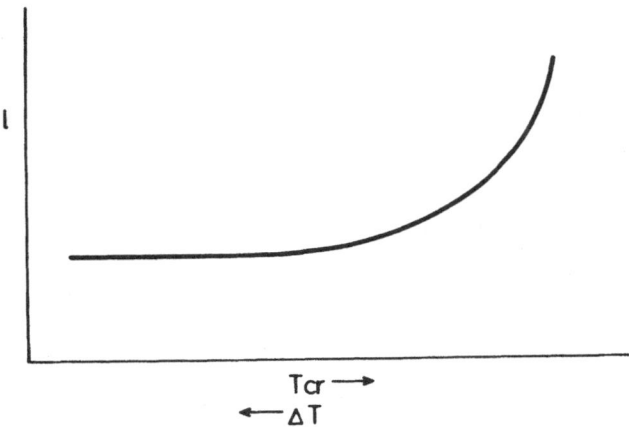

Figure 3.6. Fold length vs crystallization temperature over a very large range of supercoolings (ΔT) in case of crystallizations from solution (a combination of behavior observed for polyethylene and isotactic polystyrene).

solid state; the issue of full or partial melting followed by recrystallization, as opposed to pure solid-state transformation with the chain threading through the crystal.

The range of l values achievable by refolding can be very large: from the initial ~ 100 Å it can attain several thousand angstroms, particularly when hydrostatic pressure is applied. For polyethylene, the normally realized upper limit by standard heat treatments on a laboratory time scale is approximately 600 Å.

c. Isothermal Thickening. It was recognized early that under certain circumstances a crystal can thicken even while it is growing isothermally, i.e., crystals form with a particular l, and then this l increases while the edges of the crystal are continuing to grow and incorporate more, still uncrystallized material. Thus older crystals, or crystal portions, will be thicker than the younger ones in such isothermally crystallized material. So far this effect has only been observed in melt crystallization, and I personally believe this is a consequence of the fact that here T_{cr} (absolute) is higher than in the case of crystallization from solution. This phenomenon has a very important role in current developments, both as regards knowledge of what is the primary l value of the crystal as formed, and as regards the detailed fold structure of the crystal still unaffected by subsequent refolding. Further, isothermal thickening is *the* principal source of "secondary crystallization," i.e., the perfectioning of the crystals during crystallization, a concept that has featured in the earlier chapter dealing with crystallization kinetics.

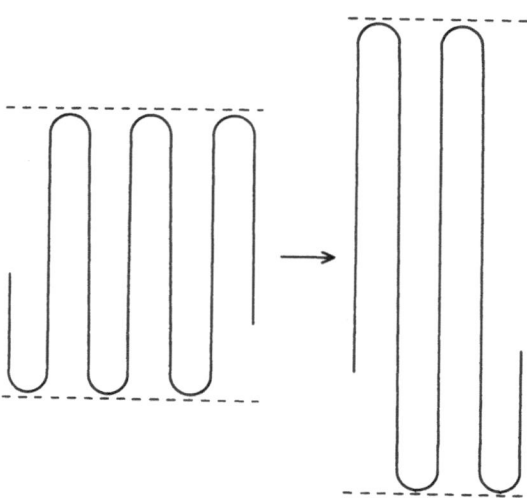

Figure 3.7. Schematic illustration of chain refolding to yield greater fold length as induced by heat annealing of single crystals.

3.2. Theories of Chain Folding

A theory with any claim of validity would have to provide answers to the following questions: (1) Why chain folding? (2) Why uniform fold length? (3) Why the observed dependence of l on ΔT as expressed by Figure 3.5 (when the theories were first postulated the full l vs T_{cr} dependence in Figure 3.6 was not yet known) together with the observation that l remains uninfluenced by a preexisting substrate? (4) Why the observed irreversible fold-length increase on annealing?

Initially theories were pursued along two lines:

I. *Equilibrium Approach.* In this approach, the crystallization occurs by chain folding because the folded chain corresponds to the state of lowest free energy and hence is thermodynamically in the stablest form. Hence reversibility of l with temperature would be expected, but was never observed. Even if excuses for the latter were found, this class of theories, while not actually disproved, never made their mark on the subject and are at present only of historical interest.

II. *Kinetic Approach.* Here it is the fully extended chain crystal that has the lowest free energy. Nevertheless the chains crystallize by folding because in this way crystallization is most rapid. Thus chain folding is due to kinetic reasons, and the resulting crystals are not in their stablest state but will tend toward it whenever they have a chance, e.g., on subsequent heating. The latter immediately explains the trend toward high l values on heat annealing. From now on we shall be concerned with this class of theories only.

3.2.1. Framework of the Kinetic Theories

3.2.1.1. The Basic Scheme

First we invoke the basic fact that l is not influenced by a preexisting substrate. This means that as long as we are only connerned with accounting for l, we can ignore primary nucleation and consider crystal growth only, i.e., we consider the deposition of chains along a preexisting substrate of any thickness. For simplicity one may take this thickness as infinite.

We shall therefore consider the deposition of a chain along an infinitely thick crystal face of the polymer. In the first stage the chain is regarded as infinitely long (so that chain ends will not matter) and the system as dilute, so that a given chain can deposit unimpeded by others. Further, and this is most important, it should be remembered that the system is supercooled ($T_{cr} < T_m^0$). Pictorially the deposition is represented by Figure 3.8, and the reaction path by Figure 3.9 where ϕ is the free energy and n the number of depositing chain atoms.

Now let us consider deposition of the first straight stem $P–Q$. This will be associated with an increase in ϕ due to the creation of new surfaces. (For simplicity the pathway $P–Q$, etc., is drawn as a straight line in Figure 3.9; the finer-scale representation, embodying discontinuities corresponding to the deposition of successive individual chain members, is ignored.) Thus the continuing deposition of this straight segment will lead to an increasingly less stable state. If at some point Q the chain happens to fold back on itself (such as $Q–R$), for which there is a finite statistical

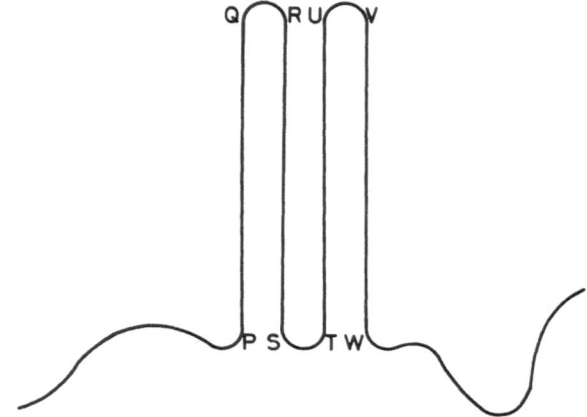

Figure 3.8. Folding scheme of a chain depositing along a crystal face of infinite thickness (not shown).

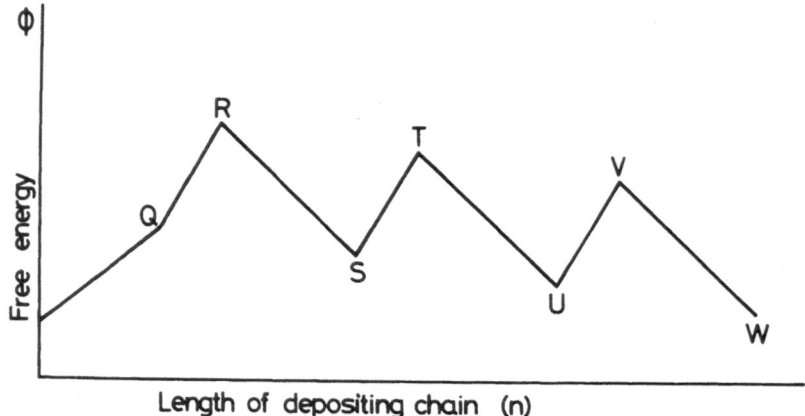

Figure 3.9. Free-energy pathway corresponding to chain deposition in Figure 3.8.

probability, ϕ increases sharply (Q–R in Figure 3.9). However, after completion of the fold, continuing deposition R–S will cover up newly created surface while creating more crystal. Consequently ϕ decreases. Further deposition would create new surface again, hence raise ϕ, but if the chain subsequently folds (S–T) then after a sharp initial rise (S–T in Figure 3.9) ϕ will drop again during continuing deposition (T–U), and so on. As long as ϕ decreases successively for stages Q, S, U, W, etc., continuing folding will increase the stability of the new chain-folded patch along the preexisting crystal face until it becomes a stable critical nucleus with ϕ lower than in the starting state, and the crystal will be capable of continuing growth. (It is noteworthy that the critical nucleus just mentioned corresponds to a secondary growth nucleus along a completely smooth crystal face in growth theories of conventional crystals.)

It will be seen that a plausible model is provided to explain why folding occurs in the first place. The next question is why this should lead to a predominant fold length. In broad outlines this can be seen as follows:

If P–Q (hence the first segment) is long, namely l is large, there is a high increase in ϕ, hence the attachment probability of such a segment will decrease with its length. However, once the chain folds over there will be a larger decrease in ϕ on steps R–S, T–U, etc., for a long initial segment P–Q, hence the detachment probability of such a longer folded-over segment will be lower. For a shorter initial length l (short P–Q) the probability of the first-segment attachment will be higher, but so will be the probability of detachment after the chain has folded over (R–S, T–U, etc.). With two opposing trends there will exist a most probable fold length (l^*) optimizing the resultant of the corresponding attachment and detachment rates. The purpose of the detailed theories is to calculate this value of l^*.

Mathematically, the first steps of this process are as follows: Consider each stem as a prism-shaped box of length l, width a, and thickness b (Figure 3.10). Let the side surface energy be σ, the end surface energy (that of the fold surface) be σ_e, where $\sigma \ll \sigma_e$, and suppose Δf is the change of free energy involved in creating unit volume of crystal. The net free-energy change $\Delta\phi$ on depositing the first segment l is then given by

$$\Delta\phi = 2bl\sigma + 2ab\sigma_e - abl\Delta f \tag{3.2}$$

where the first two terms on the right-hand side represent the work involved in creating the new side and end surfaces, respectively, while the third term is the free-energy change in creating the corresponding volume of crystal. Clearly, the formation of surfaces increases the free energy of the system while the formation of a volume of crystal (in the supercooled state) decreases it. For a ribbon to become stable the volume term (namely $abl\Delta f$) must exceed the sum of the surface terms.

Now side surface is produced by the first segment only; subsequent segments cover up as much surface as they have produced. Thus continuing segment deposition will create new end surface only, plus new crystal. The limiting condition for segment deposition to proceed is that the newly formed lattice must at least compensate for the creation of the new end surface, i.e.,

$$abl_{min}\Delta f = 2ab\sigma_e$$

hence

$$l_{min} = 2\sigma_e/\Delta f \tag{3.3}$$

where l_{min} is the lowest stable stem length.

Conditions for stability are represented schematically in Figure 3.11, where the spikes in Figure 3.9 have been smoothed out and the lines connect points Q, S, U, V, etc. For $l = l_{min}$ there is no gain in stability as n increases, i.e., as the crystal grows, consequently there will be no driving force for growth. In the case $l < l_{min}$ continuous growth will increase the free energy of the system, hence growth will not occur *a fortiori*. It follows that, for finite growth rate, l must be larger than the critical length (l_{min}).

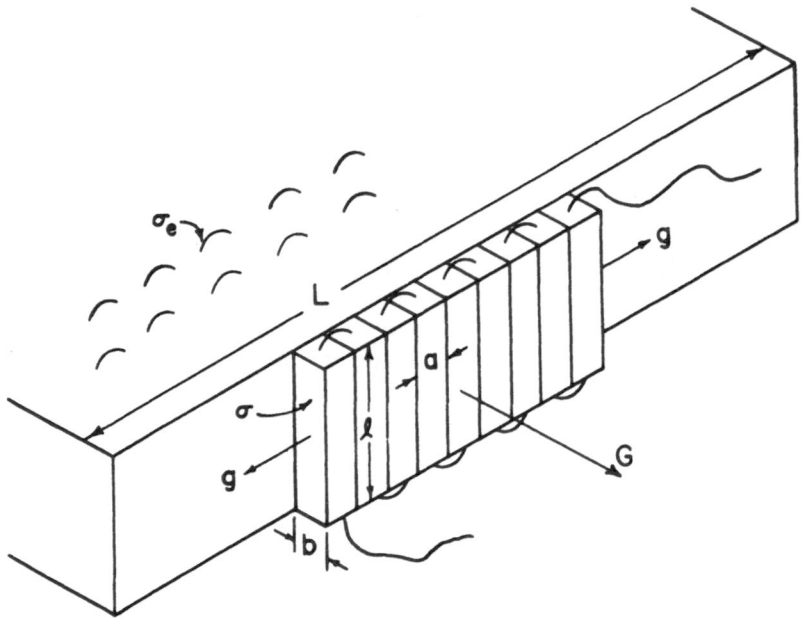

Figure 3.10. Crystal strip depositing along the prism face of a crystal with spreading rate g resulting in an advance rate G of the crystal face (after Hoffman *et al.*[8]).

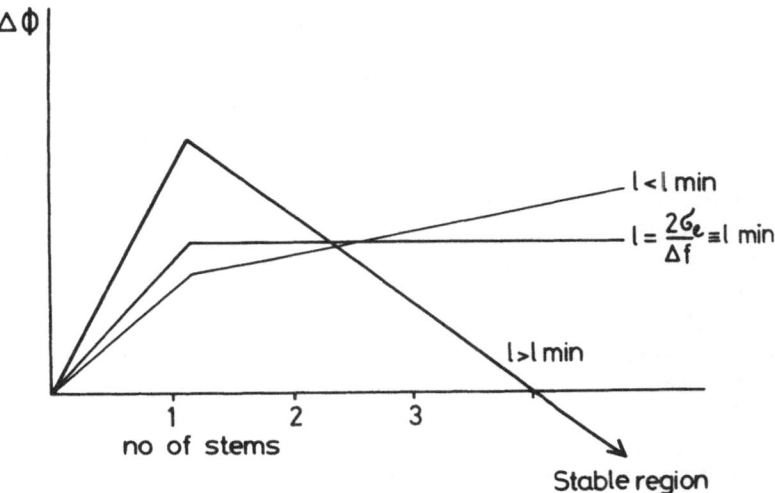

Figure 3.11. Simplified free energy pathway for chain deposition showing existence of a limiting lowest fold length (l_{min}).

Next we derive an expression for Δf. By definition $\Delta f = \Delta H - T\Delta S$, where ΔH is heat of fusion and ΔS entropy of fusion. In case of equilibrium (i.e., at the melting or dissolution point T_m^0) $\Delta f = 0$, hence $\Delta S = \Delta H / T_m^0$. If ΔS is assumed to remain unchanged for a particular supercooling corresponding to $T < T_m^0$ we have

$$\Delta f = \frac{\Delta H \cdot \Delta T}{T_m^0} \tag{3.4}$$

where $\Delta T = T_m^0 - T$.

If expression (3.4) is inserted into (3.3) for l_{min}, we derive

$$l_{min} = \frac{2\sigma_e T_m^0}{\Delta H \cdot \Delta T} \tag{3.5}$$

3.2.1.2. Assumptions

The theories assume that secondary nucleation along an otherwise flat lateral crystal surface is the rate-determining step in the growth of the crystal and proceed to show that this growth is maximum for one particular fold length l^*. In the first stages of the theories the nucleation of a surface strip itself (at a rate i) determines the crystal growth rate G: once a strip is nucleated it will spread along the full face (or a specific portion of it) at a rate g where $i \ll g$, so that $G \propto i$ (see Figure 3.10). (In later stages of the theory this mode of growth was denoted as Regime I, pertaining to low supercoolings — see below.) Here the pertinent value of l is that for which G is fastest.

In the very first versions the spread of l values in the distribution calculated was regarded as corresponding to separate crystals with different ls, in each of which there was a maximum population with l^*. Subsequently it was assumed that there exists a distribution of ls within a given crystal, the value l^* pertaining to the maximum number. This implies that l can vary (within certain limits) as a strip deposits. Accordingly, such theories incorporate fluctuations in l during deposition. It was by no means obvious that fluctuations, once permitted, would converge, which is required if the theory is to explain the facts. Such a convergence could in fact be demonstrated within the framework of the theories [F. C. Frank and M. Tosi, *Proc. R. Soc. London, Ser. A* **263**, 323 (1961); also the review by J. D. Hoffman, G. T. Davis and J. I. Lauritzen, in: *Treatise on Solid State Chemistry* (N. B. Hannay, ed.), Vol. 3, Chapter 6, Plenum Press, New York (1976)]. As a consequence it does not matter whether one starts from a thicker or thinner substrate layer: l will always converge to the same l^*, as required by the experimental findings.

3.2.1.3. Most Probable Fold Length

The main part of the theories is to find a steady-state expression for the flux S over the barrier of nucleation in terms of A_0 (the rate constant for deposition of the first stem), B_1 (that of the corresponding backward reaction), A (rate of deposition of all subsequent stems), and B (that of the corresponding backward steps). In the notation of Hoffman and Lauritzen (see review by Hoffman, Davis, and Lauritzen quoted above) we have (corresponding to the pathway in Figure 3.12; see also later):

$$A_0 = \beta \exp\left(-\frac{2b\sigma}{kT} + \frac{4abl\Delta f}{kT}\right) \tag{3.6}$$

$$B_1 = \beta \exp\left[-\frac{(1-\psi)abl\Delta f}{kT}\right] \tag{3.7}$$

$$A = \beta \exp\left(-\frac{2ab\sigma_e}{kT} + \frac{\psi abl\Delta f}{kT}\right) \tag{3.8}$$

$$B = \beta \exp\left[-\frac{(1-\psi)abl\Delta f}{kT}\right] \tag{3.9}$$

Here β is essentially a retardation factor that contains the transport terms, incorporating effects like those resulting from viscosity and surface transport, geometric factors, etc.; ψ is an apportioning factor, which apportions the change in volume free energy Δf between the forward reactions (ψ)

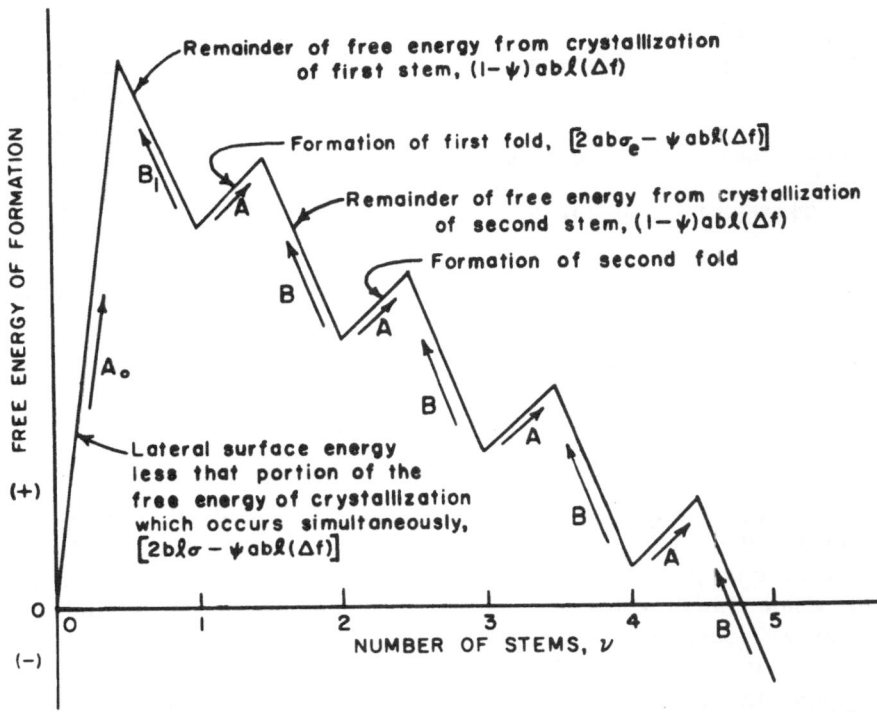

Figure 3.12. Free-energy pathway for chain deposition incorporating a factor $\psi < 1$ (after Hoffman et al.[8]).

and backward reactions $(1 - \psi)$ (see Figure 3.12). (Thus the monotonic increase in ϕ on depositing the first segment in Figure 3.9, i.e., along P–Q, corresponds to $\psi = 1$; see later.)

Theory shows that for steady-state flux

$$S(l) = N_0 A_0 (A - B)/(A - B + B_1)$$

The total flux S_T is obtained by summing the fluxes over all l.

Of special interest is the mean lamellar thickness, which should be the observed l^*. In this case one forms the expression $\int l S(l)\,dl / \int S(l)\,dl$. The result of the calculation is

$$l^* = \frac{2\sigma_e}{\Delta f} + \delta l \tag{3.10}$$

which, on substitution for Δf [by equation (3.4)], becomes

$$l^* = \frac{2\sigma_e T_m^0}{\Delta H \cdot \Delta T} + \delta l \tag{3.11}$$

Here the first term on the right-hand side is l_{min} [see equation (3.5)], the smallest l that is still stable, and δl is the additional term in l that enables the crystal to grow. At moderate supercoolings δl is a small fraction (10–15 %) of l^*, hence the detailed form of δl is of no tangible consequence. The different theories give different analytical expressions for δl.

The first Lauritzen–Hoffman theory gives

$$\delta l = \frac{kT}{2b\sigma} \tag{3.12}$$

According to a later version

$$\delta l = \frac{kT(4\sigma/a) - \Delta f}{2b\sigma(2\sigma a) - \Delta f} \tag{3.13}$$

The dominant term in equation (3.11) for l^*, namely $2\delta_e T_m^0/\Delta H \cdot \Delta T$, provides the inverse ΔT relationship, which had been the principal observation expressed by Figure 3.5 that the theories had originally set out to explain. The fit with experiment in this regime (where δl is small in any case) was found to be very satisfactory with reasonable values for the parameters σ_e, ΔH, and T_m^0. The values of ΔH and T_m^0 are known by extrapolation from other experiments (T_m^0 corresponds to the melting or dissolution of crystals from infinitely long and infinitely extended chains, not the melting of the actual crystal). Quantity σ_e contains the work to form a fold and can be estimated in a variety of ways: it is approximately 100 ergs/cm², a value which agrees well with that emerging from the theories by applying the experimentally observed values of l.

The above agreement can be regarded as a major success of the kinetic theories and of their underlying models. As the model is currently under considerable criticism from certain circles [see below and also *Faraday Discussion* No. 68 (1979)] it is worth remembering that no alternative theory is being proposed by its critics. Without this model there would be no explanation why polymers crystallize as uniformly thick lamellae with lamellar thicknesses that correspond to basic observation and fact.

3.2.2. Further Developments and Problems

3.2.2.1. Avoidance of the "δl Catastrophe"

The above theories predict that at high supercoolings δl in equation (3.13) will not remain small; actually, in a rather narrow range of ΔT, δl will increase abruptly and tend to infinity causing $l^* \to \infty$ accordingly. This predicted upswing (referred to as the "δl catastrophe") at low T_{cr} (i.e., high ΔT) has never been observed. Initially, the correspondingly high range of ΔT was experimentally inaccessible with the much-studied polyethylene system and thus an explicit refutation of the "δl catastrophe" remained outside the scope of experimentation. Later, however, poorly crystallizable polymers enabled researchers to reach the ΔT range where the "δl catastrophe" was expected. No upswing of l was found; on the contrary, l leveled off to a constant value as in Figure 3.6. Accounting for the absence of the "δl catastrophe" and of an l value independent of crystallization temperature at high ΔT forms the next stage in the theoretical development currently in progress.

Let us begin with a few words on the origin of the upswing of l. In nonmathematical terms this upswing is due to the removal of the activation barrier for the deposition of the first stem. The path $P-Q$ in Figure 3.9 would lead to decreasing (not increasing) ϕ, i.e., the line $P-Q$ will be inclined in the opposite sense. In other words, at very high supercoolings the depositing single-chain portion itself is capable of leading to a stable regime. Hence there would be no need to first form a secondary nucleus.

There are two approaches in any modification of the theory to make it fit experiment at high ΔT.

1. Hoffman and Lauritzen (see review by Hoffman, Davis, and Lauritzen quoted above) make ψ in equations (3.6)–(3.9) sufficiently small (less than $\frac{1}{2}$, possibly close to 0). This will not remove the δl upswing altogether (except for $\psi = 0$) but can keep it down to unrealistically high ΔT values. Formally, this arises because making ψ small promotes the activation barrier to the forward step in the deposition of the first stem; in fact such a barrier can be created where, at the high supercoolings in question, it would otherwise not arise. This is readily seen from the representation in Figure 3.12. (Actually, the choice of reaction path in the form shown in Figure 3.12 was introduced in preference to that in Figure 3.9 so that this possibility became apparent!)

The above procedure acquired physical meaning by considering the possibility that a long-chain molecule would adsorb first prior to commencing its deposition according to the scheme in Figure 3.10, there being good reason to believe that such adsorption could take place. Here the activation barrier will correspond to this first step of the adsorption (pinning down of the chain with associated loss of entropy without gain in crystallization-free energy).

2. Irrespective of the possible effect of the adsorption, Point [J. J. Point, *Macromolecules* **12**, 770 (1979)] has quite recently developed a deposition pathway that avoids the "δl catastrophe" and leads to the horizontal plateau (such as that in Figure 3.6) at high ΔT, while correctly reproducing the l vs T_{cr} curves at low values of ΔT (Figure 3.5), similar to

the other kinetic theories with which the new theory merges at low values of ΔT. The details are complicated. The essential point can perhaps be conveyed by stating that the pathway subdivides the stem deposition (represented as a single event, such as $P-Q$, by Figures 3.8 and 3.9) into a number of discrete subevents, the stem depositing in small increments each having its own activation barrier to surmount. The chain is then given the chance to fold back on itself at each substage. At very high ΔT, where the overall path ($P-Q$ in Figure 3.9) slopes downward, the chain available for deposition will gradually be "consumed," so to speak, by such a potential back-folding possibility at each step, thus limiting the final chain extension that can be attained. In this way the "blow-up" in l is avoided. In addition, and chiefly, it is shown that the mean length converges to a fixed value irrespective of further increase in ΔT, thus accounting for the horizontal plateau in Figure 3.6.

3. A third, somewhat negative view may be added to the above two. In this approach, at such high supercoolings the quasi-equilibrium processes underlying all the kinetic theories are no longer pertinent and a totally new theoretical framework may be needed. No work in this direction has yet been carried out.

3.2.2.2. Existence of Different Growth Regimes

In all the above considerations the nucleation of a new strip (with rate i along a preexisting substrate) was taken as the growth-rate-determining step, this being much slower than the spreading rate (g) of the strip along the substrate (see Figure 3.10). Subsequently this was termed Regime I pertaining to low ΔT.

With increasing ΔT (lowering of crystallization temperature) the nucleation rate becomes increasingly faster compared to the spreading rate. When i becomes comparable to g, there will be multiple nucleation along a given growth face, i.e., i and g will compete. The ΔT range when this occurs has been termed Regime II.

The criterion for Regimes I and II is that $Z < 0.01$ and $Z > 0.01$, respectively, where

$$Z = iL^2/4g \qquad (3.14)$$

L being a spreading length along the substrate.

The expression for l^* in equation (3.10) remains unaffected by a change in regime, but the growth rate will be sharply affected, as revealed by kinetic studies on the melt (see also the next section). In a most recent theoretical treatise [J. D. Hoffman, *Polymer* **24**, 3 (1983)] at still larger undercooling, a further growth regime (Regime III) has been postulated, where nucleation along the growth face becomes so abundant that the stem deposition, constituting the formation of the closely spaced nuclei, itself represents most of the lateral growth of the crystals. This change from Regime II to Regime III should again lead to a sharp change in the temperature coefficient of the lateral growth rate (see below). Change in the growth regime is expected to affect the lateral habit type, Regime I giving rise to smooth prism faces, Regime II to less regular, possibly microfaceted prism faces, with further *kinetic* roughening of the growth face for Regime III.

3.2.3. Growth Rates

It has been observed directly that the linear growth rate of single crystals (G in Figure 3.10), like the radial growth rate of spherulites (see end of Chapter 2), is a constant at a given temperature [namely $G(T) = K$]. By theory, if surface (i.e., secondary) nucleation is the rate-controlling step

$$G = G_0 e^{-\Delta F/kT} \cdot e^{-\Delta \phi/kT} \tag{3.15}$$

where G_0 is a constant for the system, ΔF is the activation to interfacial transport (related to β by the flux equations (3.6)–(3.9), while $\Delta \phi$ is the work of forming a critical nucleus. In solutions, the effect of the first exponential should be small and hardly affected by temperature; $\Delta \phi$ will depend inversely on ΔT yielding an overall function G given by $G \propto e^{-1/\Delta T}$. Explicitly, by inserting $\Delta \phi^*$ for the most probable l value one gets

$$G = G_0 \exp \left(\frac{-\Delta E}{kT} \right) \cdot \exp \left(\frac{-m\sigma_e \sigma T_m^0}{kT\Delta H \cdot \Delta T} \right) \tag{3.16}$$

where m is equal to 4 for Regime I and to 2 for Regime II (the effect of regime type on growth rate referred to above!). For the newly postulated Regime III m again becomes 4.

The $G \propto e^{-1/\Delta T}$ relation has been verified both for solution-grown single crystals and for melt-grown spherulites by direct morphological measurement of their diameters as a function of time at different temperatures of growth. (See also equation (2.4) in the section on spherulites.) In the case of spherulites, a sudden transition was observed from Regime I to Regime II in accordance with m changing from 4 to 2 in equation (3.16). In fact, latest results in our own laboratory [P. J. Barham, J. Martinez-Salazar, and A. Keller, *J. Polym. Sci. Phys. Ed.*, in press] have quite unexpectedly identified a reversal from $m = 2$ to $m = 4$ with further increase in ΔT in accord with the change from Regime II to the recently postulated Regime III (see above). Values for $\sigma\sigma_e$ could also be obtained from plots of $\ln G$ vs $\ln[1/(T\Delta T)]$. Not only were these values consistent with values of σ and σ_e obtained by alternative ways, but in the few cases where the growth rates were measured on both single crystals and spherulites for the same material $\sigma\sigma_e$ assumed very close values for both (i.e., for solution and melt crystallization).

3.2.4. Melting Behavior as a Function of l

In very small crystals the melting temperature (like that of other phase transitions) will be lower than the value pertaining to crystals of macroscopic (in practice, infinitely large) size; more exactly, it depends on size and shape. For the crystal platelets under discussion, the corresponding melting-point depression is readily derived. It should be recalled that at the melting point of the actual crystal T_m the difference in free energy $\Delta\phi$ between crystal and melt is zero. For a crystal of lateral dimensions p and q and thickness l, we have

$$\Delta\phi = 2pl\sigma + 2ql\sigma + 2pq\sigma_e - pql\Delta f \tag{3.17}$$

where again σ and σ_e are side- and end-surface (basal plane) energies, respectively, and Δf is the change in bulk free energy. The first three terms on the right-hand side correspond to the work required in the creation of the surfaces (the first two that of the sides, the third that of the basal surface), and the fourth corresponds to the free-energy change associated with the formation of the crystal lattice. As side surfaces are small compared to basal surfaces, terms involving σ can be neglected. On equating the rest of the right-hand side to zero and substituting for Δf from equation (3.4), we derive

$$T_m = T_m^0 \left(1 - \frac{2\sigma_e}{l\Delta H}\right) \tag{3.18}$$

for the melting point of the crystal of thickness l. This is the reduced melting point compared to that of the infinite crystal T_m^0.

The value of l depends on the crystallization temperature in accordance with equation (3.11), so the above result means that the melting point will also depend on the crystallization temperature. Thus crystals formed at a higher temperature will also melt at a higher temperature, and vice versa.

At this point reference will be made to a long-standing observation of general validity according to which the melting point of a crystalline polymer is a function of its crystallization temperature in the above sense.

It is a rather satisfactory development that in equation (3.18) a long-known observation has found a natural explanation in terms of morphology in general and chain folding in particular.

The following additional consequences of equation (3.18) are noteworthy. Commercial polymeric objects do not usually melt sharply but have a melting range. Also, such samples are not crystallized isothermally but on cooling. The latter means that they crystallize over a range of temperatures and accordingly will have crystals with a range of l values. It follows that, by equation (3.18), melting will occur over a broad temperature range in agreement with experience.

In the absence of direct structural information on l, equation (3.18) can be used to determine this important parameter, a frequently adopted practice. Nevertheless, two important qualifications are involved in such a use of melting points:

1. The chains may refold during heating, hence affecting the melting-point measurement itself. The final melting point will then correspond to that of the refolded crystal, not to that of the original crystal. The occurrence of such a situation is usually apparent from the fact that the ultimate melting point is dependent on the rate of heating: the faster the heating the lower the melting point (there is less chance for chain refolding!).
2. The use of equation (3.18) for determining l assumes implicitly that the internal perfection of the crystals under comparison is identical; in fact, if the ideal ΔH is used, the crystal is assumed to be defect-free, which may not always be the case.

Finally, T_m vs l^{-1} curves can serve both for extrapolating to T_m^0 and for determining σ_e. It is a remarkable and reassuring fact that the values of σ_e obtained in this way usually agree well with the values of σ_e obtained by relating observed l values to the crystallization temperature according to the kinetic theories of crystallization via equation (3.11), and further, that this applies to crystallization both from solution and from the melt.

3.2.5. Comparison of Crystallization from Solution and Melt

When the derivation of the kinetic theories was first outlined above, it was stated that deposition of individual chains is assumed and that this process occurs singly, chain by chain, unimpeded by others. It follows that dilute systems were implied. Nevertheless, in the course of what followed crystallization both from solution and from the homophase melt was invoked. This inconsistency reflects fairly the present practice in the literature. In view of the numerous controversies (see below) it is important to understand this in the right light, hence the following comments.

Initially, the kinetic theories were formulated to account for folding and for the observed l values in the case of crystallization from solutions, where the assumption of isolated chain deposition is largely valid. In the meantime the lamellar nature of the melt-crystallized material has gained increasing support (see, e.g., Section 3.3.2.2 and Figure 3.19 below) hence the results of the theory, initially formulated for crystallization from solution, became gradually transferred also to the melt. The important point to appreciate is that in many respects this transfer was found to work and bear fruit. Some consequences have been mentioned above: for instance, equation (3.16) for growth rates and equation (3.18) for melting behavior were found to be equally valid for solution and melt crystallization; in fact, the distinction between Regimes I, II, and III crystallization was explicitly established in connection with the melt (there has been no corresponding attempt for solution crystallization). Further, the values for $\sigma\sigma_e$ and σ_e derived from equations (3.16) and (3.18), respectively, proved to be very similar for crystallizations from solutions and melts.

However, identification of the molecular mechanisms in the dilute and condensed systems is subject to much debate. In the melt the chains may be expected to impede each other on the assumption that they are entangled, and also because a given growing crystal along the crystal–melt interface is clearly in permanent contact with segments belonging to many different chains. In addition, an all-important difference in the available experimental material needs to be pointed out. In contrast to solution crystallization, material on melt crystallization is confined essentially to information concerning growth rates and not directly to the fold length. There are good reasons for this, because isothermal thickening (Section 3.1.3.3c) makes a direct assessment via equation (3.11) difficult if not impossible, and thus the l values recorded, by whatever method, correspond to already refolded crystals and not to the primary crystals in the case of melt crystallization. Even so equation (3.11) has provided the most direct test for the theories in the case of solutions. In the case of melt crystallization, however, in spite of all the satisfactory equating of the two systems regarding growth rates and melting behavior, no such direct test has been provided.

It will only be stated that at the time of writing, this last-mentioned handicap is in the process of being overcome. In a series of works the effect of isothermal thickening in melt crystallization is being gradually reduced and the initial fold length due to primary crystallization progressively approached [P. J. Barham *et al.*, *J. Polym. Sci. Lett. Ed.* **19**, 539 (1981); *J. Polym. Sci. Phys. Ed.* **20**, 1717 and 1733 (1982)]. In fact, in the latest work it appears to have been actually attained (P. J. Barham, R. A. Chivers, J. Martinez-Salazar, S. J. Organ, and A. Keller, unpublished). From the last results the satisfying conclusion is in the process of emerging that, for a given ΔT, l^* is always very similar, irrespective as to whether crystallized from solution or melt. The full consequences of this result for all previous works and ideas on melt crystallization, which in the past have relied on growth rates alone, are waiting to be assessed. Whatever the outcome of such an assessment, a welcome unification of such widely disparate material comprising both melt and solution crystallization is hopefully anticipated. Even as matters stand at the stage of writing it appears as a firm trend that l^*, the primary fold length, is determined by the supercooling ΔT and subsequent fold-length increase (whether on isothermal thickening or heat annealing) by the absolute temperature T.

3.2.6. Some New Perspectives in Crystallization Theories

It may be salutary to recall that crystal growth in general (not merely confined to polymers) can proceed by three mechanisms as described by the classical theories, in the first stages associated with the names of Kossel, Volmer, Stranski, and Kaishew and at a later stage with Frank, Burton, and Cabrera [for explicit references see general texts on crystal growth, such as J. J. Gilman (ed.), *The Art and Science of Growing Crystals*, John Wiley, New York–London (1963)]. These mechanisms are: (1) by secondary nucleation, (2) by dislocation, and (3) without either (1) or (2) but through the agency of equilibrium surface roughness of the growth face. Mechanism (2) through the familiar screw-dislocation mechanism is operational in the growth of the large majority of crystals and is manifest through the resulting spiral-terrace topography of the growth face. While such spiral terraces abound in polymer crystals (see Figures 3.1 and 3.17), the advance of the corresponding basal faces (with the Burgers vector of the corresponding screw dislocation being equal to the lamellar thickness) is not the rate-determining factor for the lateral growth of the crystal, nor is it responsible for chain folding. Both these latter factors are governed by chain deposition *along the lateral surfaces* (Figure 3.10). The kinetic theories outlined in the preceding sections rely on mechanism (1), i.e., on secondary nucleation as the ultimate cause, and thus rate-determining factor, of the lateral growth of the layer, hence in the ultimate analysis, of polymer crystal growth and associated chain folding. These theories have taken no serious account of the exact lateral habit type involved, e.g., whether it gives a pure lozenge (Figure 3.2) or a truncated lozenge (Figure 3.14), neither did they need to for the purposes in question.

Nevertheless, at increasing crystallization temperatures (absolute) qualitative changes in habit occur: the straight prism faces become gradually rounded; in fact, solution-grown crystals obtained at the highest temperatures become "leaf shaped." Further, such leaf-shaped crystals seem to be the rule for crystals grown from the melt, wherever lateral habits can be identified (just about distinguishable in Figure 3.19). Bearing in mind that crystallization from the melt normally occurs at higher absolute temperatures than for solutions (even for comparable supercoolings), it is tempting to generalize that curved lateral habits are characteristic of crystals grown at high absolute temperatures. It was this trend that has directed attention to crystal-growth mechanism (3) [D. M. Sadler, *Polymer* **24**, 1401 (1983)]. To recall, it is familiar among simple nonpolymeric crystals that mechanism (3) can take over at sufficiently high temperatures where the *equilibrium* roughness of the crystal surface becomes appreciable and enables an advance of the growing surface as a result of a net gain in deposition, as opposed to detachment, of the atomic or molecular entities constituting the crystal. It is well known that, in contrast to mechanisms (1) and (2), the consequence of growth by mechanism (3) is the nonfaceted nature of the resulting crystal [H. J. Leamy, G. H. Gilmer, and K. A. Jackson, in: *Surface Physics of Materials* (J. M. Bakeley, ed.), Vol. I, Academic Press, New York (1975)] that manifests itself in curving crystal faces (possibly the clearest example being ^4H_e). As just cited, curving faces are being observed in polymer crystals grown at the highest temperatures where growth is still possible, which suggests the invoking of mechanism (3).

However, in view of the large bonding enthalpies involved when depositing a full fold stem to a growing face (Figure 3.10), the applicability of mechanism (3) to polymer crystal growth is by no means obvious *a priori*, and even less so that this will lead to lamellae with chain folding where, in addition, the fold length and the lateral growth rates of the crystals obey the experimentally observed supercooling dependence [equations (3.11) and (3.16)]. At the time of writing it has just been demonstrated that by considering the deposition of fractional fold stem lengths, one at a time, the application of mechanism (3) to polymer crystal growth could be justified. Moreover, it could be shown that the formation of lamellae of finite thickness (hence chain folding) and the appropriate functional relations involving lamellar thickness, as well as growth rates on the one hand and growth temperature on the other [as in equation (3.11) and (3.16)], can be upheld on the basis of mechanism (3) by treating it as an Ising model (D. M. Sadler and G. H. Gilman, in preparation). The impact of this new approach is too early to assess, in particular, whether it represents an extension to the usual secondary nucleation-based theories into growth temperature regimes where the former could not be applied previously, or whether the new approach will take over as the more appropriate one in at least some temperature ranges where the nucleation-based theories have been applied so far. (The problems arising as a result of applying the theories, such as that underlying Figure 3.10, to crystals with rounded surfaces will be immediately apparent.) Whatever the future reveals, the approach based on mechanism (3) has the merit of introducing lateral habit considerations for the first time in theories of chain-folded crystal growth, and, more generally, it has expanded the conceptual framework of the whole subject, which was about to settle down to a more sedate course. However, viewed from the other side, the notable predictive power of the conventional theories along approach (1) must be kept in sight, in particular their verified prediction of the existence of growth regimes I, II, and III, and to a certain extent the temperatures of transition between the regimes. Finally, all theories will require an underlying molecular mechanism, which should be feasible, to say the least. The secondary nucleation-based theories have gone a long way to provide this, even if some aspects are being argued, and some of the most recently observed, astonishingly fast growth rates (approximately 1 m s^{-1} for polyethylene) are presenting them with new challenges. Mechanisms along route (3), being much more recent, have not yet been given the same chance.

3.3. Morphology of Chain-Folded Crystals

The basic unit is the single layer, although in "real life" crystals are mostly multilayer structures. The basic observations, however, have been made on individual monolayers, which will be reviewed first.

3.3.1. Monolayer Crystals

3.3.1.1. Sectorization — The Basic Idea

According to the foregoing sections chain-folded crystals grow by the deposition of folded ribbons along the prism faces of the lamellae (Figures 3.3 and 3.10). If this is so, it should give rise to a unique structural consequence, namely the fold-plane direction should be preserved by the structure after the growth front has passed, which means that the crystal should consist of structurally distinct sectors defined by the fold-plane direction, as indicated by Figure 3.3. It will be apparent that within a single crystal there should be as many distinct sectors as there are prism faces. In purely lozenge-shaped crystals (as in Figure 3.3) all prism faces are {110}, all four of which are structurally equivalent but differ in terms of the orientation of the plane of folding. A frequent habit in polyethylene is the truncated lozenge with six prism faces, which are {110} and {100} (Figure 3.13). It is apparent that such a crystal should consist of six sectors where {110} and {100} are not equivalent. In a first approximation the unit cell is identical within all sectors (some subtle distinctions have been observed but these will be disregarded for the present) and the sectors are distinct merely by virtue of the direction of folding.

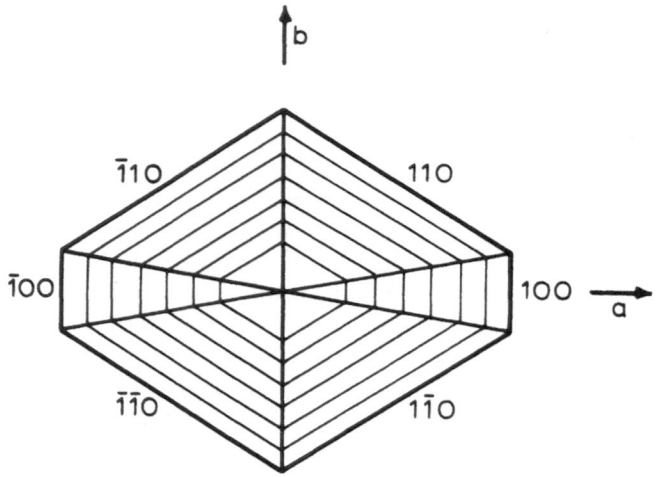

Figure 3.13. Principle of sectorization for a polyethylene crystal displaying {110} and {100} prism faces (truncated lozenge).

All the above was originally a surmise on the assumption that the chain-folding picture is correct. It was therefore most gratifying when these structurally unique predictions were verified experimentally.

3.3.1.2. Verification

The existence of sectorization was verified most dramatically and conclusively by a number of different approaches of which three will be mentioned here, as they can be simply described in brief.

a. Morphological Evidence of Sectors. The different sectors within a given layer can be distinct morphologically. This is most conspicuous in crystals that are hollow pyramids, rather like tents, where the different sectors form the different panels of the tents. During an ordinary examination in the dried state the nonplanarity is apparent through specific pleats in the collapsed structure and certain diffraction effects, but the fully three-dimensional tentlike object becomes evident when viewed (by dark-field optical microscopy) while suspended in the liquid (Figure 3.14). The straight-stem direction is always identical throughout all sectors and is parallel to the pyramid axis. The origin of this morphology lies in the fact that the folds do not all pack in a level manner but prefer to be staggered, which in turn must be related to the space requirement, hence possibly to the detailed shape of the fold itself.

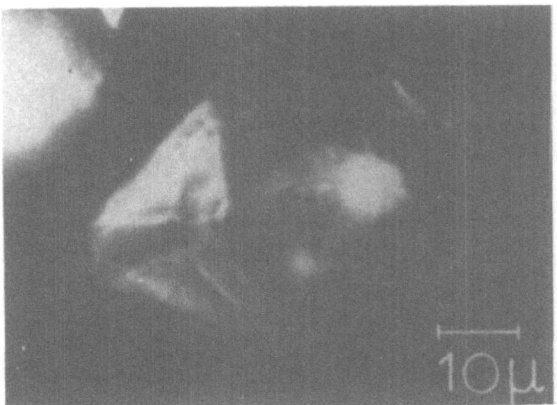

Figure 3.14. Dark-field optical micrograph of morphological manifestation of sectorization. The crystal displays a hollow-pyramid habit while floating in its mother liquor (after Bassett et al.[9]).

   b. Thermodynamic Evidence of Sectors. Sectors that are structurally different, such as {110} and {100} (see Figure 3.13), melt as different temperatures. This can be actually observed (see Figure 3.15). Such a striking visual observation is also supported by calorimetric evidence: crystals like that in Figure 3.15 display two endothermic melting peaks, one for each sector.
   c. Mechanical Evidence of Distinct Sectors. The behavior of single-crystal fracture is clearly influenced by the fold-plane direction, e.g., there is ready cleavage parallel to a fold plane. Thus there will be a clean crack along a ⟨110⟩ direction within a {110} sector. However, when such a crack reaches a sector boundary and passes, say, from a 110 to a 1$\bar{1}$0 sector, in the latter sector, the same ⟨110⟩ direction that was previously parallel will now lie at a large angle to the fold plane. A crack, if it continues into such a sector, will cut across fold planes and hence become bridged by threads. All this has, in fact, been observed (Figure 3.16).

3.3.2. Multilayer Crystals

3.3.2.1. Disposition of Consecutive Layers

   This subject is essentially descriptive and the variety of pertinent observations will not all be enumerated now. The point to be made here will relate to some generalities concerning the relative disposition of consecutive layers within a multilayer structure, and mainly to the fact that

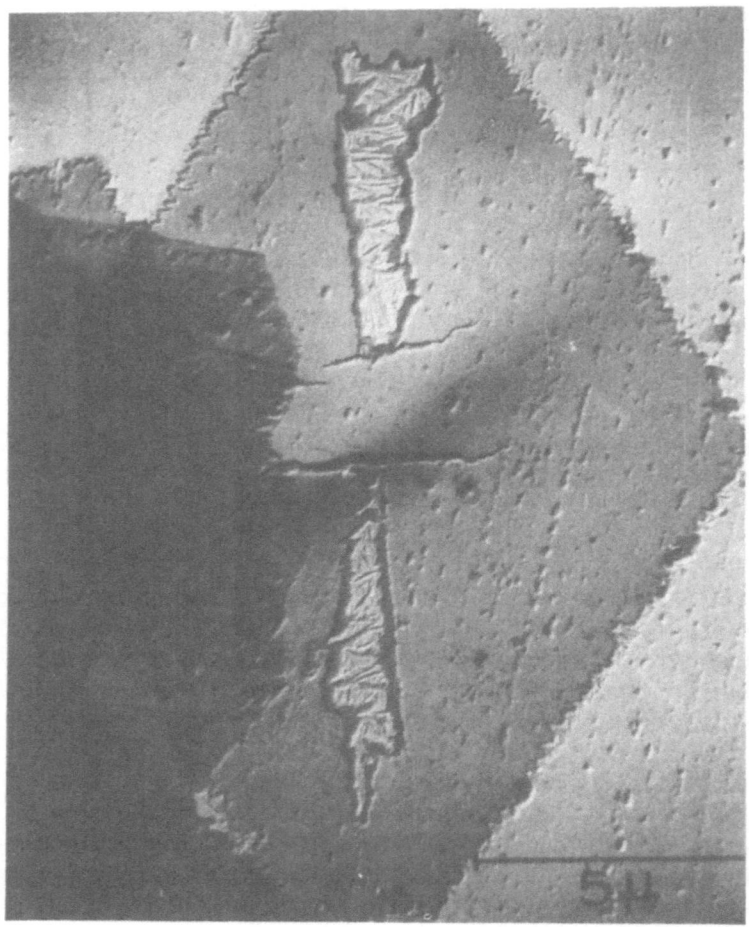

Figure 3.15. Thermal manifestation of sectorization. The {100} sectors melt at a lower temperature than the rest of the crystal. Electron micrograph.

consecutive layers in general are not strictly in crystallographic register. There are two classes of departure from perfect-layer register.

1. Rotational displacement of consecutive layers. Consecutive layers within a given crystal, when seen flat-on, are as a rule not in parallel orientation but rotated with respect to each other, usually by small amounts (1–3°) in an irregular manner. Under very exceptional cir-cumstances such a rotation can be regular and always in the same sense, in

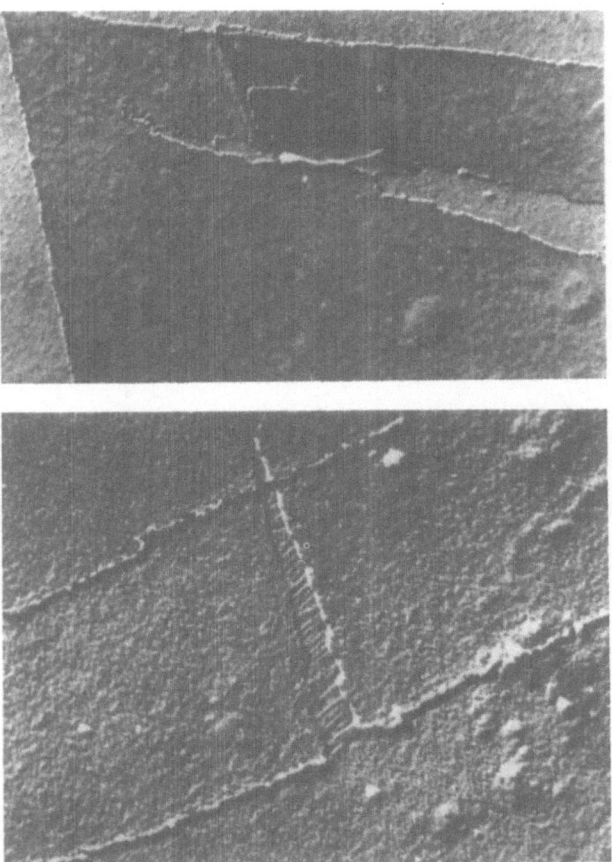

Figure 3.16. Fractographic manifestation of sectorization. The crystal cleaves cleanly parallel to the prism face within the corresponding sector where it runs along fold planes (top), but the crystallographically identical cleavage direction (along {110}) pulls threads in a sector where it cuts across fold planes (bottom) (after Lindenmeyer[11]).

which case some striking appearances result. Such an example is illustrated in Figure 3.17 mainly for its aesthetic appeal. If rotationally displaced crystal layers are in close contact (often they are not — see below) interfacial dislocation networks may arise at the layer boundaries (twist boundaries).

2. Splay of consecutive layers. When multilayer crystals are viewed edgewise (only possible by optical microscopy while suspended in the

Figure 3.17. Electron micrograph of an exceptional example of a multilayer crystal of polyethylene displaying regular rotation of successive growth terraces (after Keller[12]).

liquid) the layers are often seen to splay apart. Figure 3.18 is such an edge-on view of a crystal shown in Figure 3.1.

Perhaps the most important physical consequence of (1) and (2) above is that a usual multilayer crystal, while a "single crystal" generically, is not a single crystal geometrically speaking. In other words, we cannot define a lattice vector that repeats throughout the whole object in view of the large discontinuities across the layer boundaries. This sets a natural limit to the growing of *large* chain-folded single crystals in the sense of conventional solid-state physics.

3.3.2.2. Approaching Spherulites

The splay in class (2) above can become increasingly irregular as the multilayer character of the crystal increases. This in turn is promoted by increasing the concentration of the solution in which the crystallization is conducted. Such crystals become increasingly "bushy," the layers curve, and in an edge-on view give the impression of sheaving fibrils. This leads naturally to the sheaf forms described previously as the precursors of spherulites (see Figure 2.22). Hence this splaying, sheaving multilayer development described so far in solution promises to be a natural bridge between solution-grown lamellar crystals and the melt-crystallized spherulites discussed in Chapter 2.

We previously alluded to the fact that lamellae have also been identified as the basic structure element in melt-crystallized spherulitic polymers. Now we shall illustrate this with an example (Figure 3.19). Here,

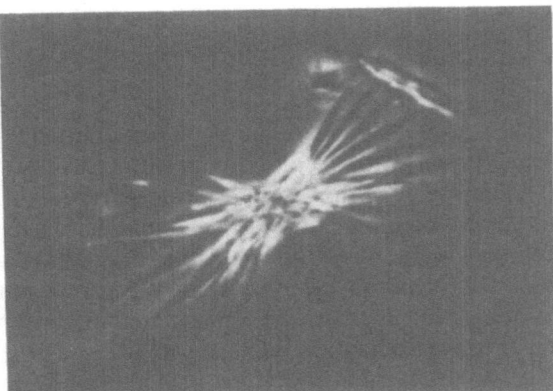

Figure 3.18. Optical micrograph of a multilayer crystal seen edgewise in its mother liquor revealing splaying of layers (after Mitsuhashi and Keller[13]).

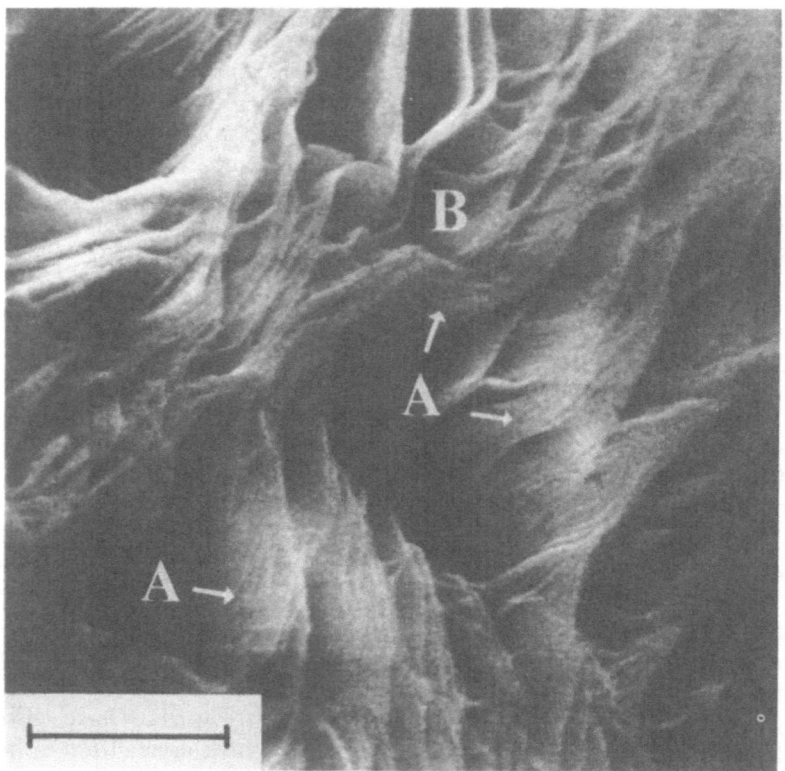

Figure 3.19. Scanning electron micrograph illustrating the lamellar structure of a melt-crystallized spherulitic polyethylene. The cavernous texture enabling the lamellae to be seen in depth arises as follows: the lower-molecular-weight material segregates on crystallization in separate lamellar packets, which can be removed subsequently by selective dissolution treatment (after Winram *et al.*[14]).

as in many other examples, the lamellae are not obtainable in isolation and could be identified only as some etching or disintegration products, or in some cross-sectional view within microtome sections. Much less is therefore known about the detailed nature of such lamellae, even less about the way they fit together to build up the spherulite. Nevertheless, the common denominator of the basic lamellar unit invites some generalizations as regards the chain-folded structure (even with modifications — see later) that also pertain to crystallization from the melt. Further, the multilayer development of single crystals seems to provide continuity between the simplest crystal unit, the monolayer on the one hand, and the spherulite, the characteristic crystal element from the melt on the other.

3.4. Fold Structure — Nature of Amorphous Material

3.4.1. The Issues

The subject of the present section is highly controversial and originates in two different, nevertheless interconnected, enquiries: What is the structure of the fold? How is the amorphous content of a single crystal to be visualized?

3.4.1.1. Structure of the Fold

The fold is such an important part of the chain-folded crystal that its structure on the atomic level is of some consequence. Nevertheless, the problem remains outside the scope of traditional X-ray crystal-structure analysis because we cannot obtain a macroscopic chain-folded crystal. The nearest achievement is the structure determination of a cyclic paraffin. This is a closed ring-shaped paraffin of 34 C atoms. It crystallizes with the ring collapsed, so that most of the ring is in a close-packed form, such as that realized by linear paraffin chains (a slightly modified form of the structure is shown in Figure 2.3). Here adjacent straight portions are necessarily bridged by the rest of the ring, the bridge involving four C—C bonds. Such a bridge is clearly a close analogue of the fold. It does show that an adjacently reentrant sharp fold (a much contested issue nowadays) is clearly possible sterically when required by the constraint of the closed-ring shape of the molecule. Whether a long linear polyethylene chain will do the same on its own accord, however, is a more open question. On the assumption that it does, a number of conformational analyses were carried out to determine the exact fold conformation and have led to a variety of not grossly dissimilar fold conformations, involving three, four, or five C—C bonds. However, the whole problem has been, at least temporarily, superseded by an overriding issue: do chains fold in a specifiable regular manner? The latter question has its origin in the second enquiry posed above, to which we now turn.

3.4.1.2. Amorphous Content of a Chain-Folded Single Crystal

It was noted earlier that a synthetic polymer is usually only partially crystalline, hence the existence of the whole subject of "degree of crystallinity" treated phenomenologically in Section 2.3 of Chapter 2. With the recognition of morphologically identifiable single crystals, the question has arisen whether such an entity possesses any "amorphous content" in the traditional sense. By applying the usual methods of crystallinity determination to such single crystals a not inappreciable crystallinity deficiency was found; in the case of polyethylene this amounted to about 20%, which was then attributed to the existence of amorphous material. For a variety of reasons, which will not be enumerated here, this supposedly amorphous material was envisaged to be located along the lamellar surface and thus attributed to an irregularly, loosely looped structure of the folds (Figure 3.20). This was given more explicit form by Flory [P. J. Flory, *J. Am. Chem. Soc.* **84**, 2857 (1962)], who envisaged the fold surface as a telephone switchboard (switchboard model). There is evidence for and against such a view. Accounting for the amorphous content is certainly a point in its favor. Nevertheless, the remarkably regular features observed in such crystals (sectorization and its consequences, for example) and, chiefly, the issue of how the overall fold length should be defined unambiguously by the crystallization temperature in the case of so much randomness at the fold surface (not to speak of how such a structure should form; there is no alternative class of theory other than the one outlined earlier, requiring a sequential deposition of fold stems) raises more questions (in the author's opinion) than the switchboard model can answer. In addition, there is an overriding geometrical problem, namely the prohibitive overcrowding that the switchboard model would create at the fold surface, where the density of the amorphous region would have to exceed that of the crystal (recognizable even in the sketch of Figure 3.20). The overcrowding can be relieved by making some of the chains fold back sharply in order to provide more room for other folds to form loose loops of an overall amorphous character. The necessity for such partially sharp

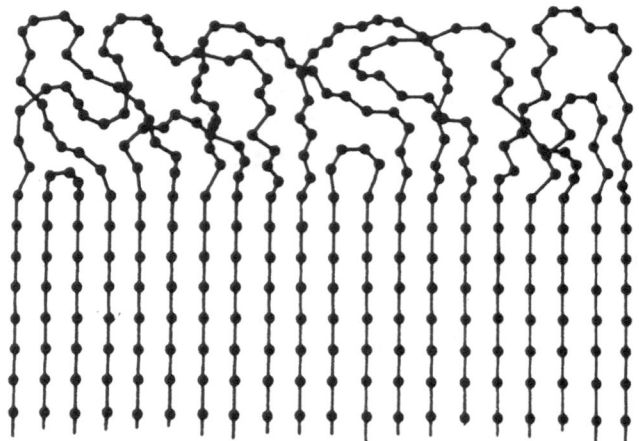

Figure 3.20. Sketch of a disordered fold surface: the switchboard model in a two-dimensional representation (after Fischer[15]).

refolding is nowadays more or less agreed on by all concerned. The issue about which the controversies are currently polarized turns around the question of how much such sharp refolding is required. On this, in turn, would depend whether the overall character of the fold surface is to be regarded as an essentially random or an essentially regular structure. In the present author's view, there is no *a priori* reason why the fold surface should always be the same in all crystals under all circumstances, hence that there is necessarily a unique answer to a question posed in the above way. Accordingly, it may make more sense to pose the question in the form as to whether or not a chain *can* fold in a regular manner, given the appropriate conditions. If the answer is "yes," there will be many reasons why departures from maximum regularity may occur in a given sample.

3.4.2. Experimentation in Aid of the Fold-Surface Problem

The fold-surface problem can be subdivided into two distinct structural issues, which are unfortunately often confused or inextricably telescoped together in much of the controversial literature:

1. Adjacent vs nonadjacent stem reentry (irrespective of the nature of the fold).
2. Regular vs irregular folds.

Only six of the experimental techniques employed in the arguments will be enumerated. A further one, the recent neutron scattering, will be dealt with in a separate section below.

1. Interchain interactions of isotopically labeled species in an isotopic mixture can be assessed by infrared spectroscopy. Thus one can mix a small amount of deuterated species of polyethylene into normal hydrogenated polyethylene and explore the nearest-neighbor environment of a given deuterated stem as reflected by the splitting of appropriate infrared bands. In this way adjacency and nonadjacency should in principle be distinguishable.

2. A chain-folded layer can be subdivided into fold surface and crystal core by small-angle X-ray scattering (SAX). This is done by detailed analysis of the intensity in the SAX pattern (in contrast to merely using Bragg's law to determine *l*, the overall periodicity, referred to so far).

3. Crystal-core thickness can be determined directly from line-profile analysis of suitable wide-angle X-ray reflections corresponding to a periodicity along the lamellar normal.

4. Raman LAM analysis could be used. In principle it should define the straight-stem length, hence the core thickness. (In fact, rather disappointingly it does not seem to do so but gives the same *l* value as SAX, at least by the simple analysis — see Section 3.1.3 on the LAM technique.)

5. Chemical methods can be used that rely on selective etching involving removal of the fold surface coupled with physical (SAX) and chemical (molecular weight) analysis of the residue.

6. Use can be made of detailed electron microscopy of effects relating to the fold surface.

Intentionally, no conclusions are drawn with respect to any of the above techniques since some of the interpretations are still being debated and no justice can be conveyed by a single statement pertaining to each. Some broad, overall statements will be made in what follows.

3.4.3. Outcome of the Enquiries

Certain evidence favors the switchboard model while other evidence requires the existence of regularity; the individual points will not be pursued here. Though there are obviously areas in which final conclusions cannot be drawn at the time of writing, nevertheless the material already presented, together with that of the following section on neutron scattering, occupies a considerable portion of the relevant literature. The present coverage should, at the very least, prepare the reader for approaching the original literature in the areas covered. However, in spite of uncertainties two specific outcomes will be mentioned in general terms.

The first pertains to the nature of the amorphous component. Instead of merely referring to the "amorphous fraction" as in the earlier chapter on crystallinity, arguments on the fold-surface structure have focused attention on the multitude of variants that would normally be covered collectively by the term "amorphous." This arises because disordered material, such as that in a chain-folded crystal, can now be classified in terms of its relation to the lamellar crystal core. Thus it can consist of loops that, even if not all regular, can correspond to different degrees of looseness. Alternatively, there can be loose hairs emanating from the layers that have failed to become incorporated at their other end, or they become incorporated into a different lamella, in which case they will not be loops but interlamellar ties. All this is not as purely qualitative as it may appear, because the different ways of incorporating loose "amorphous" portions into the crystal can be quantified in terms of loss of configurational entropy due to the loose chain or portion being confined at one or both ends with a specific distance between the ends.

The second outcome is that these modes of constraints can significantly affect the behavior of such chains, hence the macroscopic properties. In particular, tie molecules will be load-bearing (all important for mechanical behavior) while loops, at least on their own, will not; this again is relevant to mechanical properties (relaxation behavior). Further, when the effect of the constraints, and the associated entropy changes, are considered in combination with the crystals, they will potentially influence the melting behavior of the crystals themselves (premelting, superheating — see later).

An awareness of all the above factors provides a more explicit molecular basis on which to define and treat the so-called amorphous component of a semicrystalline polymer.

It also follows that the generalization of an "amorphous content" or "amorphous–crystalline ratio" is an oversimplification. The experimental methods listed above are leading increasingly to the overall conclusion that what at first sight had been termed "amorphous" component really corresponds to a state of intermediate order, which in the broadest generality expresses the fact that a spectrum of material with different constraints is involved.

3.5. Neutron Scattering Experiments: The Chain Trajectory

In recent years the subject has received a major boost by the application of neutron scattering, made possible by the availability of new high-flux neutron sources. The results have extended the scope of the enquiries, but have also further aggravated existing controversies. For an impression of the subject and an in-depth study involving all aspects of the controversies, the most up-to-date and comprehensive reference is *Faraday Discussion* No. 68 (1979).

3.5.1. Technique and Potential

The technique in question involves coherent elastic scattering of neutrons at small angles. The idea is to use scattering data to obtain information on the overall dimensions of, and the path described (trajectory) by, a chain molecule in an environment of its own kind. In such an approach a minority of the molecular population needs to be distinguishable from the rest as regards scattering, i.e., there must be contrast, which is achieved by isotopic doping. Hence a small fraction of the molecules must be isotopically different in a way that makes them distinct as regards scattering of neutrons, but not otherwise. For the present issue this is achieved by mixing deuterated guests to the normal proton-containing host, the scattering power of the two isotopes of hydrogen being significantly different. Thus individual molecules of the guest species become amenable for studies in the condensed phase, as in the dilute phase (solutions), by light scattering.

The obtainable information depends on the angular range of the scattering: the smallest angles provide information on the large-scale features, the global dimensions of the molecule as a whole, while increasingly larger angles yield information on increasingly smaller features of the molecular trajectory. While there is no intrinsic discontinuity in terms of angles, nevertheless some distinctions arise in actual working practice due to the technicalities of the experiment and the various convenient approximations used in the interpretation.

3.5.2. Angular Ranges

1. Smallest angles ("Guinier" range). For the present type of polymer problem this range is given by

$$10^{-3} < q < 5 \cdot 10^{-2} \text{ Å}^{-1}$$

where $q = (4\pi \sin \theta)/\lambda$; here θ is the scattering angle and λ the wavelength.

This range provides a measure of the radius of gyration R_g of the molecule

$$I/I_0 = 1 + q^2 R_g^2/3 \tag{3.19}$$

or its approximation by the Guinier law

$$I = I_0 \exp(- R_g^2 q^2/3) \tag{3.20}$$

where I is the intensity scattered at a particular q and I_0 at $q = 0$, respectively.

2. Intermediate (but still small) angles. The range here is

$$5 \cdot 10^{-2} < q < 5 \cdot 10^{-1} \text{ Å}^{-1}$$

Scattering in this range gives a certain amount of information on the details of the chain trajectory, particularly pertinent to the chain-folding problem.

Results are most informatively expressed in terms of Iq^2 vs q^2 plots (so-called Kratky plots).

3. Larger angles. The range is given by

$$5 \cdot 10^{-1} < q \to 1 \text{ Å}^{-1} \quad \text{and beyond.}$$

Scattering in this range gradually approaches the usual crystallographic information on chain conformation and (in the case of crystalline polymers) interplanar dimensions.

3.5.3. Some Results

The first studies were carried out on amorphous polymers, melts, and glasses. Perhaps the most significant result was obtained in the region of smallest angles. Here R_g was found to correspond to the value for an unperturbed random coil (as in Θ solvent) with $R_g \propto M^{1/2}$ as expected from a random coil, a result quoted earlier in connection with the amorphous state (Section 1.6.1).

The next stage was to follow what happens to the molecular dimensions and trajectory on crystallization. Here a significant complication was encountered in the form of isotopic segregation, at least in the best-examined, and in other respects best-explored system of polyethylene. This means that the isotopic guest molecules do not quite behave as their hosts with respect to crystallization, as initially expected, but form isotopically enriched regions or clusters (there are different views as to which of the two are formed). Hence the neutron-scattering technique will not see a given guest molecule in isolation, and so can no longer provide the information required, at least at sufficiently low angles that enable R_g to be identified. (This became apparent from excess intensity at the extrapolated zero angle, excess in regard to the expectation from a single molecule of known M.) To avoid this segregation, crystallization must be carried out rapidly, i.e., at high supercooling. This is a very serious limitation as it restricts the technique to samples where the chains have not been given a chance to sort themselves out. Apparently this disturbing segregation effect is absent with isotactic polystryrene and polypropylene, where measurements can and have been carried out at (from the crystallization standpoint, most fundamental) low supercoolings. Nevertheless, these latter polymers are less crystalline and their morphologies are less understood. Consequently the information from neutron scattering does not yet carry the same weight as regards the abstract basic question of the chain trajectory in crystalline polymers, the issue over which existing controversies have now centered.

The actual results will only be very briefly mentioned (see *Faraday Discussion* No. 68 for details). In melt-crystallized polyethylene, at any rate, R_g did not change on crystallization (at the high supercoolings to which experiments were restricted), hence the much publicized notion that globally nothing happens to the molecule: it merely "freezes in" with some local order representing crystallization. This, however, disregards all the morphology, i.e., the question as to why lamellae form with well-defined thickness, not to speak of why the chains fold with the uniformity and supercooling dependence observed. However, in solution-grown crystals R_g is certainly observed to change on crystallization; in fact it gets smaller with some interesting additional effects not to be detailed here.

In the angular range given in (2) Section 3.5.2, the results seem to be most clear-cut with solution-grown crystals: they are consistent with the scattering object having the form of a sheet with dimensions corresponding to a chain-folded ribbon deposited along a crystal face (see, e.g., Figure 3.10). This is in agreement with morphological studies; nevertheless there is much argument as to what extent the stems belonging to a given molecule are arranged adjacently. It appears at present as if there were numerous gaps in the ribbon formed by a given chain, and the point being argued is whether the basic picture of adjacent folding (with some defects) can still be retained, or whether the description of a more random folding pattern is appropriate. In melt-crystallized (rapidly cooled by necessity) samples, the stems scatter neutrons more like individual isolated rods (by some claims), or as very short rows of at least partially adjacent stems (by other, most recent claims!).

I have no intention of exploring the arguments further; I shall merely indicate their scope. Clearly this all leads back to the fold-surface problem of the previous chapter. A few general questions will be posed for the sake of perspective when trying to evaluate individual claims:

1. In any particular sample examined, have the chains been given adequate opportunity to realize a sufficiently representative mode of deposition to serve as a model for crystallization behavior?
2. Is the interpretation of a particular scattering pattern sufficiently unique for a far-reaching generalization?
3. In a given model are elementary space-occupation problems being adequately dealt with? (See the issue of overcrowding at a lamellar surface in the preceding chapter.)

3.6. Alternative Morphologies

The morphologies to be listed here are still part of the general class C (Section 2.1.3), i.e., they correspond to the crystallization of random chain molecules in the quiescent state.

3.6.1. Extended-Chain-Type Crystals

In the cases of several polymers, crystallization from the melt under elevated pressures ($3 \cdot 10^3$–$9 \cdot 10^3$ atm) leads to brittle solids (as opposed to the usual pliable solids, such as a normal polyethylene type) with near-100% crystallinity [Wunderlich, Bassett. For specific references see textbooks by above authors: N. Wunderlich, *Macromolecular Physics*, Vol. 1, Academic Press, New York (1973); D. C. Bassett, *Principles of Polymer Morphology*, Cambridge University Press, Cambridge (1981)]. Fracture surfaces reveal that they consist of very thick lamellae with the chains normal to the lamellar planes (Figure 3.21). The lamellar thickness can extend up to several μm and become comparable to the chain length. For molecular weights M of about 15,000 (for polyethylene), the lamellar thickness corresponds to the chain length and the spread of lamellar thicknesses over a given sample to the actual molecular weight distribution of the material, indicating among other points surprisingly accurate fractionation during crystallization by molecular weight. (This gives rise to the unanswered question as to how accurately the chains group themselves into fractions and how they can achieve this.)

It follows that the chains in the above crystal type are fully extended within the crystal, hence the term "extended-chain" crystals. (It is noteworthy that *these* extended-chain crystals formed from unoriented melts are lamellae, not fibers; see later.) With higher molecular weights ($M = 10^4$–10^5 for polyethylene) there is no longer an exact correspondence between M and lamellar thickness: the lamellae can be thinner by factors of 2–3 X. Hence in such cases there must be *some* folding. Nevertheless, such samples remain distinctly different from those crystallized at atmospheric pressure in the normal manner. Hence the term "extended-chain-type" crystals.

The full story of the reasons for such crystallization is only partially understood. It has been observed that such samples, at least with

Figure 3.21. Electron micrograph of extended-chain-type crystal of polyethylene crystallized under 4800 atm pressure. (By courtesy of Wunderlich based on Wunderlich *et al.*[16].)

polyethylene, crystallize first in a hexagonal phase while under high pressure (as opposed to the usual orthorhombi͡c phase to which it transforms on cooling and pressure removal). It is claimed that this is a general requirement for extended-chain-type crystallization [D. C. Bassett, in: *Developments in Crystalline Polymers* (D. C. Bassett, ed.), Vol. 1, p. 115, Applied Science Publ., London (1982)]. Further, it was observed that the lamellae forming under pressure are thin to begin with, but subsequently thicken following initial growth. Accordingly, this is a case of fold-length increase corresponding to an extreme case of isothermal thickening (see Section 3.1.3.3c), which here proceeds to *full* or *nearly full* chain extension. Accordingly, it is this extreme chain refolding that seems to be promoted by the high hydrostatic pressure. Among possible reasons are the conformational changes in the melt induced by the pressure and/or the changed properties of the crystals actually formed. In any event we know that the crystal, when formed, is not the usual one, namely that the chains within it pack in a hexagonal lattice. It is known that hexagonal phases are more mobile than the usual orthorhombic ones (see Section 2.2.3), hence refolding will be facilitated. Also, the possibility that a liquid-crystal-type phase intervenes has been suggested.

Whatever the reasons for its formation, the properties of the resulting product are quite exceptional for a polymer. As stated earlier, they are virtually fully crystalline and thus contain no amorphous component. Their melting point is very high, and can attain values beyond the theoretical value for fully extended chains because of superheating. For all these reasons samples with this extended-chain-type texture can serve as models of the corresponding infinite polymer crystal as regards several properties, often quoted in theoretical works.

3.6.2. Micellar Crystals (Crystal Gels)

The fringed micelle (Figure 3.22) was the traditional model for crystallization before being superseded by the chain-folded lamellae. It is now having a comeback under the very special circumstance of crystallization at very high supercooling in systems capable of such supercooling. This class of effects is currently becoming apparent through the phenomenon of thermoreversible gelation from solutions.

As an example, we consider a very highly supercoolable polymer, such as isotactic polystyrene. When an appropriate solution of it is being cooled, the normal crystal suspension of chain-folded platelets forms first. At very high supercooling, however, such solutions display a very conspicuous effect: they may set as a gel. X-ray diffraction indicates that such gels consist of crystals. On heating the gel dissolves, hence it is thermoreversible.

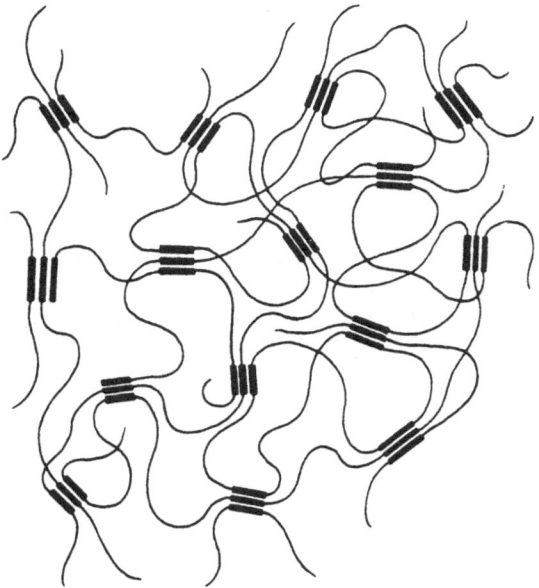

Figure 3.22. Model of fringed, micellar gel-forming crystallization (after Keller[17,23]).

Gelation is due to connectedness. Structurally, a gel is a swollen network. In the network under discussion the junctions are the crystals. For them to be junctions, several chains have to come together, hence they must be largely of micellar character as in Figure 3.22.

The above type of micellar crystals are very small, possibly approximately 100 Å linear dimensions, and melt at much lower temperatures than lamellae. The smallness of the micelles is a structural necessity for this type of crystal form. Greater lateral extension creates the overcrowding problem at the interface where the chains emerge from the lattice (mentioned in Section 3.4.1.2). This problem can only be relieved if some of the chains fold back, which in turn would bring us back to chain-folded lamellae. The above argument suggests a connection between micelles, hence gel formation, and the high supercooling: at very high supercooling the critical crystal nucleus is only required to be very small — thus it would be sufficient for only a very few chains to come together, which then "stick," — so to speak — forming a stable nucleus with lateral dimensions still too small to create cumulative strain at the interface where the chains emerge.

There is evidence that in an appropriate T_{cr} range, micelles and lamellae can compete and form simultaneously. To mention one characteristic, such samples display the properties of a composite morphology and accordingly possess two distinct melting points.

Gelation is a direct result of connectedness. Accordingly, such connectedness can only be diagnosed by this straightforward test in crystallization from solution. If the same network generating crystallization were to occur also from the melt, it would remain undetectable by such a simple test for connectivity (because here gelation would not apply). There is evidence that crystals formed from the melt at very high supercooling (e.g., close to T_g (see Figure 2.12) as opposed to those at low supercooling close to T_m^0) are very different (e.g., they possess a very much lower melting point), which suggests that they may not merely be smaller but may have a qualitatively different character compared to those formed at low supercoolings, i.e., the lamellae, the main subjects of study in crystallization behavior so far. There is therefore a likelihood that they have a micellar character without substantial amounts of chain folding.

We shall now revert to the other classes of crystallization listed in Section 2.1.3, where we stated that they have no counterparts, not even conceptually, among nonpolymeric substances.

Other Classes of Crystallization

A. Keller

4.1. Crystallization Concurrent with Polymerization (Nascent Polymers)

Most polymers, including technologically important ones (such as polyethylene, polypropylene, and polyoxymethylene), are crystalline when polymerized. So are most natural polymers (cellulose, wool, etc. — biosynthesis!). They crystallize concurrently with the polymerization reaction itself. The significance of this lies in the fact that here the polymerization reaction and crystallization may interact with each other: the polymerization reaction influences the crystal morphology, while crystallization affects the reaction, molecular weight, and its distribution in particular. For instance, chains grow up to a certain length and then crystallize (precipitate from solution), or the monomers add to the polymer, which is already in the crystalline state. Little is known about this area. The "nascent" crystal morphology itself is in general little explored (e.g., that arising in a reaction vessel), but it is known that such morphology can be very specific with special properties, which the polymer will lose on customary reprocessing.

In some instances, when the polymer cannot be processed at all (infusible, insoluble), this nascent morphology is the final one. To make such a polymer useful, the actual nascent morphology has to be controlled. (Normally the morphology is determined by the subsequent processing operation.)

A recent important example of the above situation is polyacetylene (see Section 1.5.2.3). As stated there, after suitable doping this polymer

A. Keller · Department of Physics, University of Bristol, England.

conducts electricity, raising the possibility of becoming a commercially viable conductor, optimistically replacing copper. However, currently its usefulness is limited by the uncontrolled polycrystalline, randomly oriented nascent morphology, which cannot be altered subsequently, at least at present.

In general, nascent polymers are polycrystalline. This usually applies also to the case of *solid-state polymerization.* Here one usually starts with a macroscopic monomer single crystal and induces polymerization of the monomer while within the crystal (e.g., by heat or radiation). The monomers then join up into chains, when usually the crystal falls apart into submicroscopic polymer crystals. In a few exceptional cases, however, the monomer crystal converts into a macroscopic polymer single crystal with the polymer chains running through the whole crystal, apparently without interruption. Very special geometric conditions must be satisfied for this to happen. The first such instance was the formation of polyoxymethylene from tetroxane, the cyclic tetramer of formaldehyde.

Of greater significance are the polydiacetylenes [G. Wegner, see, e.g., Faraday Discussion, No. 68, *Organization of Macromolecules in the Condensed Phase*, p. 494 (1979)]. These are polymers of the monomer of type

$$\begin{array}{cc} R & R \\ | & | \\ C\equiv C - C\equiv C \end{array}$$

where R stands for a variety of chemical groups. The polymer is

$$\begin{array}{cc} R & R \\ | & | \\ \left. \vphantom{X}\right\vert C - C\equiv C - C \left.\vphantom{X}\right\vert_n \end{array}$$

and provides practically perfect macroscopic crystals. Such crystals have special macroscopic properties (Figure 4.1): first, because they have a one-dimensional character with valence bonds all along a given direction in a continuous sequence; second, because of the alternating triple–double bonds along the chain. The latter gives rise to many unusual optical and electrical properties, which form the subject of many physical studies.

Another example is $(SN)_x$ forming from S_2N_2. (The monomer is a $\begin{smallmatrix} S-N \\ N-S \end{smallmatrix}$ ring that opens up to form chains.)

The S_2N_2 crystal forms first by sublimation, but polymerization

50 hr at 60 °C

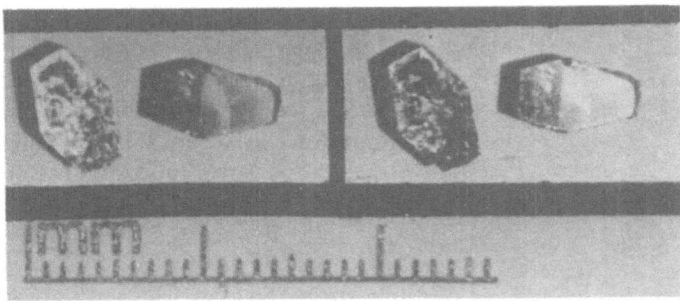

Figure 4.1. Macroscopic single crystals of a polydiacetylene obtained by solid-state polymerization (displaying anisotropic optical properties in different orientation with relation to plane-polarized light) (by courtesy of D. Bloor, unpublished).

immediately starts to produce the —S≡N—S≡N— chain. Macroscopic crystals of a variety of shapes (prisms, needles) up to 1 cm in size can be obtained. As already stated in Section 1.5.2.3, they are quasi-one-dimensional metals with strong metallic conductivity along the chain direction and are superconducting below 0.3 K. It is an important substance for the study of one-dimensional systems. However, it is likely to be a unique "one-off" compound and not a member of a family as originally hoped.

All macroscopic single-crystal polymers are very special cases. Nevertheless, they form the closest link between traditional solid-state physics and polymer science and are potential models for theoreticians. They have found no practical application so far, nonetheless they have the potential for specialty devices due to their special electronic behavior (optical and electric).

4.2. Orientation-Induced Crystallization

4.2.1. General

This is the second, specifically polymeric mode of crystallization (Section 2.1.3.2). Historically this was one of the first observations of polymer crystallization, when a stretched rubber was examined for X-rays in 1925 [M. Katz, *Naturwissenschaften* **13**, 410 (1925)].

4.2.1.1. Thermodynamic Considerations

Orientation reduces the entropy of the random chain, hence on crystallization ΔS is also reduced. Accordingly the melting point is raised, because $T_m = \Delta H/\Delta S$. This means that for given T the supercooling is increased, hence so is the driving force for crystallization. It is even possible to produce crystallization in this way when T is above the melting point in the unstretched state; here the crystals melt on stress removal.

4.2.1.2. Kinetic Considerations

A chain in its fully, or partially stretched out, oriented state is closer to the configuration it will adopt in the crystalline state. Entropy considerations apart, this means that less conformational rearrangement is needed for a chain to fit into a crystal lattice, hence it can proceed faster, less hindered by retarding factors such as viscosity.

4.2.2. Morphological Background

We have seen above that a random chain will form chain-folded platelets if allowed to crystallize. If the chains are being stretched while in the amorphous state, they will crystallize as fibers (Figure 4.2). It will be reinvoked that before the recognition of lamellar crystallization, fibers were the products expected from crystallization of long chains *a priori*. It is for this reason that lamellar crystals were so unexpected. We now see that

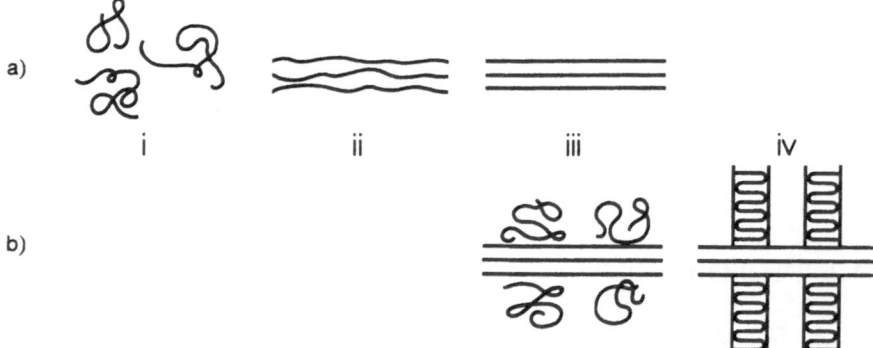

Figure 4.2. Scheme of orientation-induced crystallization. (a) Formation of smooth, extended chain fibers: (i) random coil, (ii) chains become oriented by external influence, (iii) preoriented chains have crystallized. (b) Formation of fiber–platelet composites ("shish-kebabs"). Here not all the chains have been aligned and at stage (iii) only the oriented chains crystallize. The remaining random chains use the already-formed fibers as nuclei for the formation of chain-folded platelets (iv) (after Keller[17]).

fibers are indeed obtained during oriented crystallization, nevertheless it should be noted that to obtain fibrous crystals the chains need to be oriented first, otherwise chain-folded platelets will result.

However, this is not the full story. Usually it is not possible to orient *all* the chains within and assembly. Some chains will become stretched out while others will be left more or less random. The stretched-out chains will form the fibrous crystals as above. The unoriented chains will use these fibers as nuclei and deposit onto them epitaxially as chain-folded platelets, as shown by the sketch of Figure 4.2 and observed experimentally (Figure 4.3). This composite fiber–platelet structure is the usual product of orientation-induced crystallization and is termed "shish-kebab" for obvious reasons.

The above fiber-platelet morphology illustrates the principle expressed in Section 3.1.3.3a of Chapter 3 in a rather extreme way: namely even when presented with an infinite substrate the newly depositing, unaligned chains will crystallize onto them by folding, with a fold length that corresponds to the prevailing supercooling. We recall that this behavior has featured prominently in the formulation of the kinetic theories.

The subject will now be subdivided into the following four aspects: (1) mode of chain extension, (2) structure of shish-kebabs, (3) properties of shish-kebabs, and (4) practical consequences.

4.2.3. Mode of Chain Extension

4.2.3.1. Static Chain Extension: Networks

Amorphous melts can be stretched, and held stretched, until crystallization sets in, provided the chains remain stretched and do not meanwhile relax. This is only possible on a sufficiently extended time scale if the polymer is cross-linked so as to form a network. The classical example is vulcanized rubber. In fact it is known that the crystallization of rubber on stretching (contrary to historical models) occurs along the shish-kebab route. In the case of such networks, chains can also be stretched out while in the form of a solution (which will then correspond to a swollen gel).

4.2.3.2. Dynamic Chain Extension: Flow

The chains in a flowing solution or melt may become stretched out, which brings us into the realm of hydrodynamics. To achieve such a situation the flow must be of a special kind where the extensional component dominates over the rotational component. Such flow is termed "extensional." In the usual kind of flow, such as capillary flow, which is

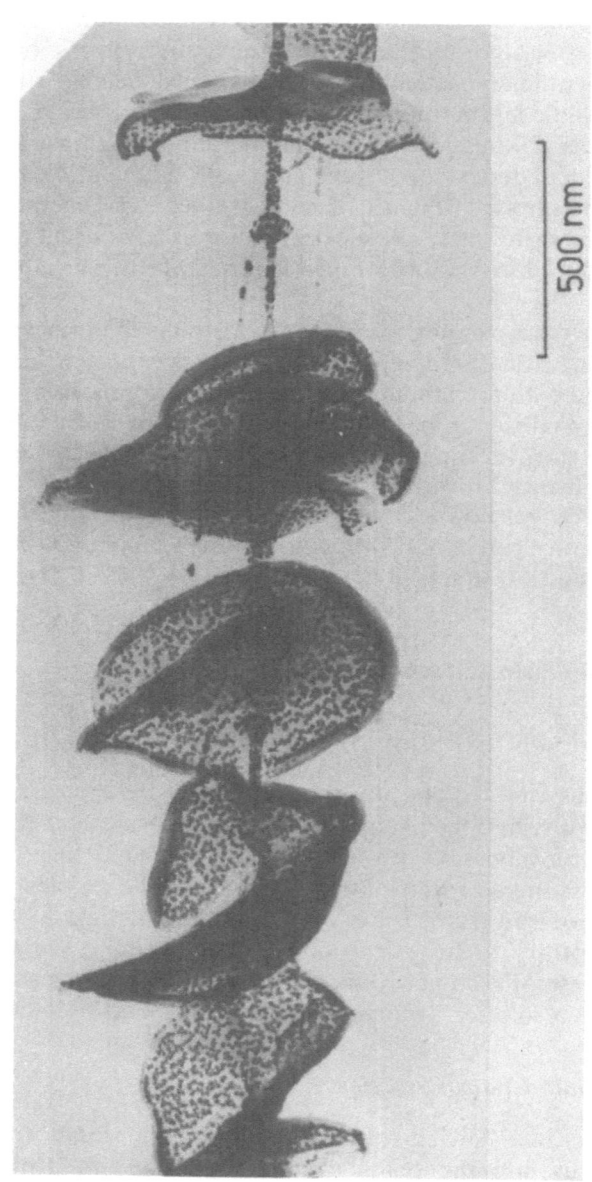

Figure 4.3. Electron micrograph of a shish-kebab-type crystal of polyethylene grown by oriented crystallization from solution (after Hill *et al.*[18]).

simple shear flow, the rotation and extension rates are of equal magnitude. In such a case the fluid element, and the chain contained by it, cannot attain a high degree of extension. The extensional component will dominate, hence flow becomes extensional if, e.g., the flow accelerates or decelerates. Mathematically

$$|S|^2 - |\omega|^2 \equiv |\sigma|^2 > 0$$

where S is the extensional and ω the rotational strain rates. Only if $|\sigma|^2 > 0$ will there be persistent extension, where σ is the persistent extension rate. The simplest extensional flow (to quote an example) is pure uniaxial stretching flow defined by the velocity gradient tensor

$$\dot{\boldsymbol{\nu}} = \begin{vmatrix} 1 & 0 & 0 \\ 0 & -\frac{1}{2} & 0 \\ 0 & 0 & -\frac{1}{2} \end{vmatrix}$$

Chain extension in elongational flow is an increasing function of molecular weight; for a given strain rate, the longest chains stretch out most. In fact, for a given strain rate beyond a certain critical chain length the chains will all be practically fully extended, and below this strain rate virtually unextended. Thus in the case of a distribution of molecular weights we have fully extended and unextended chains in a system undergoing elongational flow. This in itself should suffice to account for the situation outlined in Figures 4.2 and 4.3, and hence for the composite fiber–platelet structure of shish-kebabs arising when such a system crystallizes.

The above molecular-weight dependence implies that each chain is isolated, as it would be in a sufficiently dilute solution. If chains become entangled or associated in any other way (e.g., by localized micellar or fibrous crystallization) then it is the relaxation time of the whole aggregate entity, which in turn depends on its molecular weight, that will determine the chain extension. If the association/entanglement is beyond a critical value, and hence extends over the whole system, this will then become a network (a gel in solution) and the effect of the orientational influence will essentially amount to the stretching of a network, i.e., correspond to Section 4.2.3.1 above. According to current developments, many structures that were believed in the literature to arise through the stretching of *individual* chains have their origin in the aforementioned gel stretching. It follows that the boundary between the stretching modes dealt with in Sections 4.2.3.1 and 4.2.3.2 has become blurred. This whole subject is therefore in a very fluid state at the time of writing.

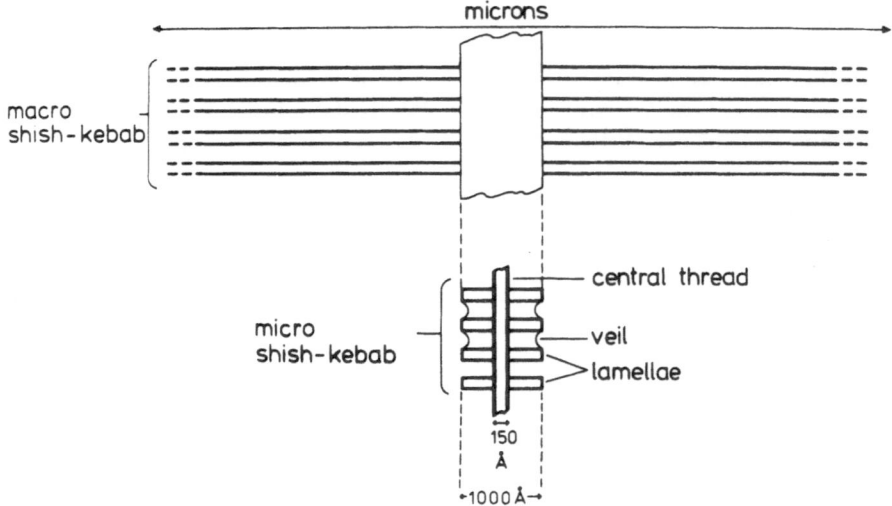

Figure 4.4. Schematic representation of the relation between macro- and micro-shish-kebabs (after Keller and Barham[19]).

4.2.4. Structure of Shish-Kebabs

According to the foregoing a shish-kebab is a two-component structure consisting of a central fiber and a platelet overgrowth.

4.2.4.1. The Platelet Overgrowth

The platelets can be of two types.

1. Large-scale removable platelets. These platelets correspond to epitaxial overgrowth of separate molecules, usually formed when a partially crystallized melt or solution, already containing the fibrous backbones, is being cooled to room temperature. They can be removed by selective dissolution achieved by reheating the suspension, or prevented from forming altogether by exchanging the hot solvent at the original temperature of fiber formation. It is helpful to term this whole shish-kebab entity a macro-shish-kebab (Figure 4.4).

2. Small, molecularly attached platelets. The fiber core of the above macro-shish-kebab is, however, not a smooth fibrous crystal but, contrary to all expectations, was found to display a shish-kebab character itself on a much smaller scale (sketched in Figure 4.4 and termed micro-shish-kebabs). Such micro-shish-kebabs cannot be denuded of their platelet population, which accordingly must be molecularly connected to the central fiber core. An "artists impression" of this connectedness, due to Pennings (who developed so much of our current knowledge on shish-kebabs), is presented in Figure 4.5. According to recent works the origin of the attached platelets is as follows: The fibers as formed are "hairy" to begin with, the "hairs" loose, but attached molecules dangling freely in solution. It is these hairs that crystallize by forming the attached chain-folded platelets on further storage or on subsequent cooling. Proof for this model is provided by the fact that the scale and separation of platelets, in fact the entire external appearance of micro-shish-kebabs, can be reversibly altered by heating and cooling the whole assembly while in the solvent. Here the platelets dissolve, reverting to loose but attached hairs, which then renucleate and reform according to the condition prevailing under the conditions of this subsequent storage or cooling. Thus the appearance of a micro-shish-kebab can be affected and controlled by what we term a "hairdressing" procedure.

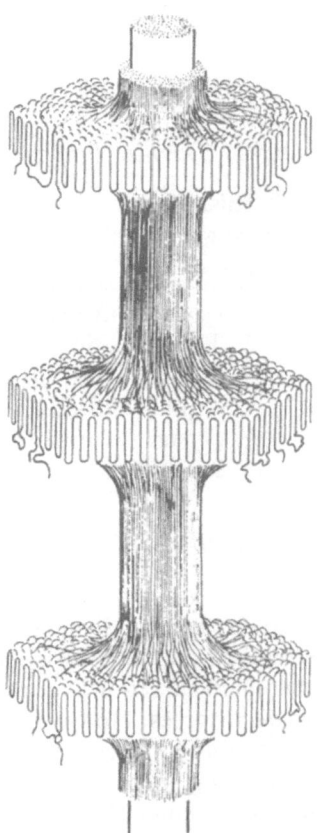

Figure 4.5. Diagram of a micro-shish-kebab showing molecular connections between chain-folded platelets and central core (after Pennings[20]).

In the last paragraph we consistently referred to solution crystallization only, because only in this case can micro-shish-kebabs be obtained in isolation. In practice, melt crystallization always yields macro-shish-kebabs as in Figure 4.4, where the large platelets cannot be removed by ready methods.

4.2.4.2. The Backbone Fiber

These are essentially of the chain-extended type as implied by Figure 4.2. Nevertheless the latest electron-microscope (and some termal-shrinkage) evidence implies that the crystals in such backbones are not continuous throughout. In fact the backbones are rather segmented into amorphous and crystalline, or rather, perfectly and less perfectly crystalline regions connected in series. Current theoretical attempts regard these structures as resulting from multiple crystal nucleation occurring in series along several localities of the chains that are already aligned by the flow.

4.2.5. Properties of Shish-Kebabs

4.2.5.1. Thermal Properties

Shish-kebabs are a composite texture and therefore display a correspondingly complex melting behavior. The main features are as follows:

1. Shish-kebabs possess a multiple melting behavior, the platelet component melting at a lower temperature than the core fiber.
2. The core fibers not only melt at higher temperatures than the platelets, but in fact may do so at temperatures beyond the equilibrium melting temperature of the infinitely extended chain crystal (as denoted earlier by T_m^0), i.e., such fibers are prone to superheat.

4.2.5.2. Mechanical Properties

For an appreciation of the issues the following preamble, significant in its own right, is deemed necessary. Ideally fibrous material consisting of stretched-out and aligned long chains should be very stiff and strong along the chain direction, because here the external force acts against primary valence forces. However, most disappointingly in practice, the maximum achievable stiffness and strength with the usual technological fibers are not even being approached (for figures see below). The reason is the presence of chain folding. In the usual technological fibers the chain may be highly oriented by appropriate tests (X-ray diffraction, birefringence), which does not mean, however, that it is also fully stretched out (see also later, Figure 5.2), because a parallel stack of chain-folded platelets would produce the symptoms of a high degree of chain orientation, as in fact it does in the technological fibers, without realizing the full potential of the fully extended chain.

Achievement of full chain extension is more than a trivial matter. One approach is through oriented crystallization that leads to shish-kebabs. Here the platelets detract from the final properties. Hence the objective is to produce backbones that are as platelet-free as possible, a target of many recent endeavors. To this condition must be added the further, most recently realized objective, that the structure of the backbone be as defect-free as possible, i.e., the segmented nature of the crystal sequences (referred to above) must be reduced. Recently special preparation conditions have enabled the theoretical values for modulus and strength to be approached. For example, in the case of polyethylene the theoretical stiffness is 250–400 GPa (according to the mode of estimate). The conventional technologically drawn fiber or film has a modulus of approximately 5–10 GPa. Good solution-grown shish-kebab fibers have moduli of approximately 100 GPa. In one special case 280 GPa has already been achieved. To appreciate this achievement it suffices to say that the typical modulus of a steel wire is of the order of 200 GPa.

4.2.6. Some Practical Consequences

The modulus and strength issue has already been mentioned. At present, industrially oriented research is endeavoring to create elongational flow fields or to stretch networks in order to achieve ultrastiff, ultrastrong fibers. This is purposeful utilization of fundamental knowledge.

The consequences of what has been described above are manifest, even if unintentionally so, in practically all processed materials where melts or solutions are at some stage in a state of flow. In the preparation of injection-molded articles the melt passes through constrictions and orifices and meets obstacles where the flow, even if only locally, will accelerate or decelerate, and hence will assume an elongational character and correspondingly undergo chain stretching. This in turn will give rise to shish-kebabs. Such shish-kebabs, occurring only in the appropriate localities, will represent heterogeneities within the final molded object. This will influence the properties of the technological products, usually in a deleterious manner. It will also affect test samples such as those prepared for scientific purposes. An awareness of such possibilities is thus clearly required also in academic studies.

The traditional spinning processes always involve a certain amount of elongational flow (melts or solutions passing through spinarets!), hence a certain amount of, often uncontrolled, fibrous crystallization will occur and affect the final structure, and correspondingly the properties.

An important manufactured product is the melt-extruded film (often obtained by blowing large cylindrical bubbles from the melt in a continuous manner). The usual polyethylene or polypropylene wrapping sheet is a familiar example. Here crystallization occurs while the expanding melt solidifies. This creates shish-kebab structures of the macro kind — see Figure 4.3 — with the following, further important variant.

In such films the stresses during solidification are low. This means that the elongational strain rate characterizing the flow, the degree of network stretching, and hence all extensional influences will be correspondingly moderate. As a consequence, the resulting concentration of fibrous crystals will be low and the fibers will lie far apart, but still parallel to the orienting influence. Under the low stress the overgrowth platelets will tend to crystallize, as they do in the unoriented melt, hence they will tend to form spherulites, where the ribbon-type overgrowth crystals will twist (see Figure 2.22c). With nucleation centers close together along the core fiber the spherulite growth will be confined to essentially two dimensions, the resulting spherulitic "disks" being stacked parallel while strung along the central core (Figure 4.6), the whole assembly forming a columnar structure.

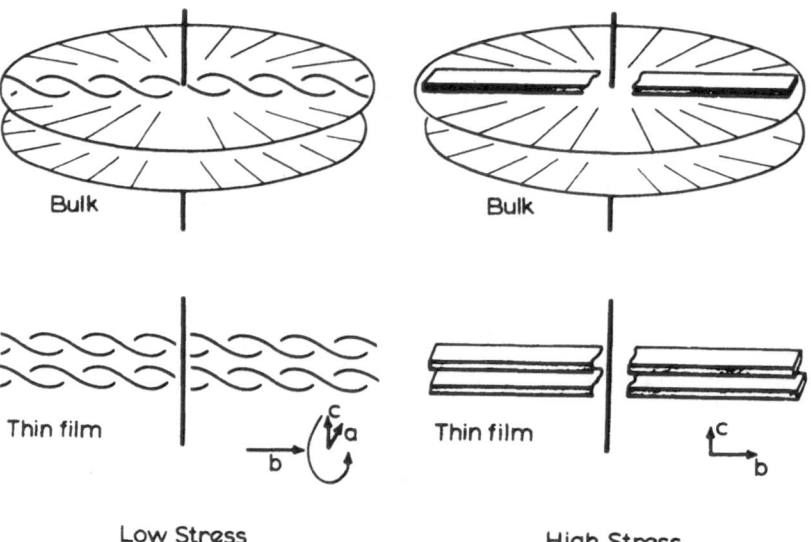

Low Stress   High Stress

Figure 4.6. Diagram of columnar structures arising during crystallization of flowing melts. Left: under low stress, with columns far apart, when the overgrowth platelets twist as they do in spherulites. Right: under high stress when the overgrowth platelets are all aligned parallel with chains along the stress direction (vertical) (after Keller and Machin[21]).

The usual melt-extruded film indeed consists of an assembly of such parallel columns with some explicit consequences as regards structure and properties. This example is not only a very widespread and important one in practice, but is also an instructive illustration of the hierarchical nature of the morphology of a crystalline polymer. The complex and intricate interrelation of the different morphological entities (fibers, platelets — the latter with twists, columns) should be noted.

In the next chapter the material already presented will be summarized, however, not in the manner of a conventional summary but by utilizing the previous material in connection with some enquiries pertinent to the field of polymers. The enquiries in question are important in their own right, but at the same time offer an opportunity to reiterate the multitude of structural information invoked in the preceding chapters within several different contexts.

Hierarchical Nature of Macromolecular Structure

A. Keller

5.1. Introduction

One of the most important themes throughout this subject has been the hierarchical nature of macromolecular structure. Explanation or interpretation of any property must take congnizance of the existence of this hierarchy, which forms the main topic of the present summarizing survey. First, the members of the hierarchies will be briefly recapitulated.

The basic building unit is, of course, the chain molecule. In amorphous material this will assume a more or less random configuration.

5.1.1. Crystalline Constituents

In the crystal the molecule will be the constituent of a *lattice*. The crystal lattice will then be the building block of the higher-level *morphological hierarchies*.

The basic morphological elements are *lamellae*, the as yet only diffusely identified *micelles*, and the *fibers*.

The lamellae build up a variety of lamellar aggregates culminating ultimately in the spherulites.

The fibers form the central core of a range of platelet–fiber composite structures which range from the fibrous shish-kebabs, as obtained in solutions, to the columnar entities in melt-crystallized material.

A. Keller · Department of Physics, University of Bristol, England.

5.1.2. Amorphous Constituents

As mentioned earlier, an awareness of the hierarchical nature of the crystalline architecture of a crystallizable polymer also enriches our conception of the amorphous component within a semicrystalline polymer. In other words, the relation of amorphous chains with respect to the crystal entities creates a finer distinction between what otherwise would be termed collectively amorphous material.

Let us recapitulate. First we have the pure amorphous component unconnected to any crystal entity, e.g., the matrix material not yet pervaded by spherulites.

All other forms of amorphous chains are constrained in some way by being molecularly connected to crystals. One example comprises amorphous chains associated with lamellae: in an isolated layer the noncrystalline chain portions are confined to a given lamella constituting the fold surface. The chains may be attached at both ends (loose loops) or at one end only with the other end free (cilium). In the case of amorphous material lying between two lamellae there will be chains with two ends confined to each. These are the interlamellar ties.

A second example consists of amorphous chains associated with fibers. They include two classes: first, those within the central core that interrupt crystal continuity (the source of the segmental nature of the core), and second, the chain portions tying the platelet overgrowth to the central core (originally loose hairs when the core had formed; see Section 4.2.4.1).

The absence or presence of constraints, or the detailed nature of the constraints arising from its relation to the crystal hierarchy, influences the mobility of such chains and can affect macroscopic behavior.

In what follows the consequences of the hierarchical nature of a crystalline polymer will be traced through three subject areas: (1) crystal defects, (2) thermal behavior, and (3) deformation behavior.

5.2. Crystal Defects

Defects have a prominent role to play in the physics of simple solids. This applies equally to polymers but, with the greater variety of structural elements, they span a much wider range of effects.

5.2.1. Defects Within the Crystal Lattice

Polymeric sources of defects arise specifically through the lack of perfect chemical, including stereochemical, regularity of the chain molecule itself. As discussed in Section 2.1.2 these include occasional branches, comonomers, cross-links, tactic inhomogeneities, etc. Only when these are taken into account do we come to the level of defects familiar from the physics of more conventional solids, such as point defects, dislocations and stacking faults. These latter can be of significance, nevertheless they are usually overshadowed by the much more prominent consequences of the imperfections in the molecular architecture referred to above, and by the much larger-scale disturbances on the higher hierarchical levels of the structure.

5.2.2. Defects Beyond the Level of the Lattice

5.2.2.1. Lamellar Structures

Let us first consider the single lamella. Here we have the fold surface in its various representations. From whichever way it is viewed, it represents an interruption of the lattice continuity and in addition, dependent on the kind of sample (and opinions held), it embodies various elements of disorder, or may even display features characteristic of amorphous material.

Next let us consider the multiple lamella. Here we have a natural discontinuity when passing from one lamella to the next. In addition to the possibility of interlamellar material there arises the question of imperfect layer register. These can be of two kinds: rotational mismatch, and splay of consecutive layers.

Finally we consider spherulites. These complex lamellar aggregates, in addition to all the sources of defects arising in multilayer crystals, embody new sources of structural imperfections of their own. Most prominent is the central discontinuity, next the unexplained twisting of the lamellae within spherulites, and all the discontinuities that must arise from the way the space is filled (not understood even in a descriptive sense). The overall, radially arranged fibrous texture is known to be a potential source of radial discontinuities and, of course, the junctions of different spherulites represent major fault lines in a melt-crystallized polymeric object.

5.2.2.2. Molecular Segregation during Crystallization

Molecular segregation during crystallization can be a major source of inhomogeneity, which can manifest itself in different forms.

1. Lower-molecular-weight but crystallizable species have lower melting points. They become ejected during the crystallization of the rest of the material and crystallize in isolated pockets, later on cooling with correspondingly thinner lamellae.

2. Noncrystallizable portions are ejected (such as atactic chains or chain portions) and form amorphous pockets of various size and disposition in the final material.

The principal accumulation sites of ejected material of either kind can be situated within spherulites along radial discontinuities and/or at the spherulite boundaries.

5.2.2.3. Defects in Fibrous Crystallization

These possibilities follow directly from what was said about fibrous structures. It was stated that the central fibrous core, while of extended-chain character, is not continuous crystallographically but contains interruptions.

There are a variety of defect structures relating to the overgrowth platelets of shish-kebabs, some of which are common to those associated with chain-folded crystals while others arise from the mode of their connection to the central core (e.g., "veils" between platelets and between platelet and fiber).

On a still larger scale are the discontinuities associated with the columnar boundaries in melt-crystallized material that have much in common with the spherulitic interfaces.

It is clear from this brief recapitulation that there are many sources of departure from ideal crystal properties and, chiefly, that these will not be easily attributable to a single cause but to the entire hierarchical nature of the morphology. Thus fracture, to take one example, may arise in structural terms from weakness at any level of the hierarchy, ranging from chemical weaknesses along the chain to those along the boundaries where spherulites meet.

5.3. Thermal Behavior

Let us consider what may happen when a semicrystalline polymer is being gradually heated up. Here again, the changes that take place will involve all elements of the structure hierarchy.

5.3.1. Amorphous Material

The amorphous components in all their variety will pass through the stages expressed by Figure 1.4. In particular, they will pass from glassy to rubbery behavior when going through T_g with correspondingly profound effects on the properties.

5.3.2. Crystal Lattice

As in any other crystalline material, polymorphic transitions can take place with change of temperature while still below the melting point. These will not be itemized here. We only make the general statement that in the case of chain molecules, increase in thermal vibrations leads to increased amplitudes in vibration corresponding to rotations around bonds, both of the main chain and the side group. The former in particular can lead to the so-called "rotary phase," which in appropriate polymers may correspond to a transition from an orthorhombic (or triclinic) to a hexagonal phase (see Section 2.2.3 in Chapter 2).

5.3.3. Melting Range

Crystalline polymers usually possess a broad melting range. The reason for this is best appreciated in terms of the molecular inhomogeneity and the morphological hierarchy usually present in polymeric materials, the two often being intricately interlinked.

5.3.3.1. Molecular Inhomogeneity

This factor arises through the existence of a distribution in molecular weights and perfections. The mere existence of such distributions, even in the case of complete, homogeneous mixing, would lead to a depression of the melting point. In addition, we have seen that the different species may segregate during crystallization leading to regions of different melting points. Even without the intervention of any other factor this will broaden the temperature interval within which the sample as a whole melts.

5.3.3.2. Morphological Factors

We can distinguish between the effect of premelting and partial melting.

Premelting. We stated above that the amorphous chains and/or chain portions can exist in different relations to the underlying crystals according to the morphology. This creates a variety of possible constraints on the amorphous chain with related effects on the configurational entropy, which in turn will affect the melting point of the whole crystal with which such constrained amorphous chain portions are associated. This can lead to premelting and under special circumstances, superheating phenomena [H. G. Zachman,, *Kolloid-Z. Z. Polym.* **231**, 504 (1969); E. W. Fischer, *Kolloid-Z. Z. Polym.* **231**, 458 (1969)].

Let us consider, e.g., the fold surface of a lamella. Loose loops, which may be present, are constrained at both ends, consequently their entropy will be reduced compared to the totally unrestricted chain. If now the crystal starts to melt from the fold surface downward, the loops will become longer (with end separation unaffected) and hence the effect of the constraint on the entropy (entropy tension) will be reduced. Without entering into details here, it will merely be stated that as a result of such considerations the overall entropic component (i.e., crystal plus amorphous) associated with such melting will be increased beyond that due to melting of the crystal lattice alone, giving credence to the uniquely polymeric process of interfacial premelting of lamellae.

Similar situations could be quoted in the case of fibrous crystals due to loosening up of the constrained tie chains during the gradual melting of crystals.

Partial Melting. The main point here is that small crystals melt at lower temperatures. Here we reinvoke what was stated in connection with equation (3.18), namely that crystals with small l have lower melting points. In the case of a range of l values, the existence of a melting range will follow. The origin of a range of crystal thicknesses may be manifold. The most self-evident case is where crystallization takes place during cooling, when different l values result at the different crystallization temperatures. Another source is the phenomenon of isothermal thickening during crystallization in the melt (see page 80), which will result in a range of crystal thicknesses in the final crystal product. And finally, as already stated in Section 3.2.4 of Chapter 3, as a crystalline sample is being heated in the course of melting-point determination, it may refold in the process to higher l values than present in the original crystal. This frequently takes the form of a multiple population of crystal thicknesses with corresponding multiplicity of melting points.

5.3.3.3. Superheating Phenomena

Superheating effects arise under a variety of circumstances, mostly in extended-chain-type morphologies. The essential reasons are due to two classes of effects. (1) The chains cannot transform into their random form in any other way than to peal off one by one from the outside, as in the case of morphologies shown in Figure 3.21. This takes time during which, under practicable heating rates, the crystals superheat. (2) The chain in the melt does not relax instantaneously, as in the case of the central cores of the shish-kebab crystals. As for the unrelaxed stretched chains ΔS (for melting) is smaller, thus T_m will be higher, with reference to the random melt. Hence the system (on a limited time scale) will superheat.

In summary, it is apparent that in a polymer there exists an intricate multistranded connection between the thermal behavior and the presence, detailed characteristics, and interconnectedness of the various structural entities that constitute it.

5.4. Deformation

This is the last example chosen to illustrate the role of hierarchies, even if only in the briefest terms. Deformation is particularly important for polymers, as most thermoplastics are deformed in the course of fabrication (drawing, rolling) in a plastic manner, while long-range elastic deformation is, of course, the characteristic feature of elastomers. Changes occur during deformation in all the regimes of the structural hierarchy, which in turn also influences the resulting properties (stiffness, fracture, etc.).

5.4.1. Polymers as Self-Structured Composites

5.4.1.1. Generalities

As a generalization, it may be useful to visualize a partially crystalline polymer as a kind of composite where the individual constituents are the amorphous and the crystalline components. Some overriding consequences follow from this viewpoint.

Above T_g the amorphous material is a rubber (particularly if and when the crystals act as cross-links) with the following characteristics: the modulus is low, but the deformation remains elastic up to very high strains (several 100%). In contrast, the crystal portions (like any crystal) have high moduli but remain elastic only up to very low strains (1% or less). In the composite treated here we have a combination of both. For instance, we can have a situation in which, for a given macroscopic strain, the crystals deform plastically while the deformation of the amorphous component, even if very substantial, is still within the elastic range. Suppose, e.g., the load is removed in such a case, then the amorphous portions will retract but not the crystals, which will remain in their plastically deformed state. In addition, the elastic retraction force due to the former can induce further plastic-deformation modes in the crystal.

The above serves to illustrate that a semicrystalline polymer is a composite with components of mechanically, grossly disparate properties where, in addition, the components may interact in a unique fashion. The exact mode of interaction will depend on the mode of coupling of the constituents, which in turn depends on the detailed morphology of the system under consideration. As a broad, phenomenological generalization one can have two extreme modes of coupling:

1. Series coupling where the stresses, but not the strains, are equal in both components (Figure 5.1a).
2. Parallel coupling where the strains, but not the stresses, are equal (Figure 5.1b).

Of course in "real systems," there will be a whole spectrum of cases intermediate between modes (1) and (2) as regards gradation of components (e.g., we have seen that there can be ranges of behavior even within what we grossly termed "amorphous"), and many intermediate states in the coupling pattern between modes (1) and (2). Further, if the different phases are not macroscopically separate blocks (as in Figure 5.1) but the components in question are mutually dispersed or interleaved in an intricate fashion, there will be mutual constraints influencing each other's behavior. (For example, a layer of rubber sandwiched between wide plates made from some stiff material to which it closely adheres — a close analogue to crystal lamellae in polymers — will not be able to extend according to the Young modulus it would possess were it on its own in response to tension being applied to the stiff plates, because of its inability to contract laterally owing to the constraints created by the plates.) In brief, we have a composite where the deformation behavior will be affected both by the nature of the components of the structure hierarchy at all levels, and by the way they are coupled mechanically.

We shall now briefly survey the hierarchy from the above point of view.

5.4.1.2. Amorphous Component and Crystal Lattice

Let us first consider the random amorphous chain. The essentials of its behavior, particularly above T_g, has been repeatedly defined: it is the seat of the long-range elastic or viscoelastic behavior of the system. If coupled in series with the crystal it will support nearly all of the strain (Figure 5.1a); if coupled parallel it will "dilute" the effect of the crystals as regards supporting stress (Figure 5.1b).

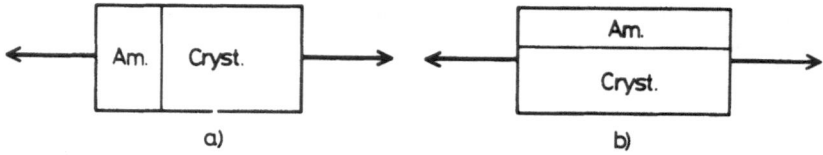

Figure 5.1. Simplest representation of (a) series and (b) parallel couplings of amorphous and crystalline components of a partially crystalline polymer for the interpretation of response to external loading (arrows).

As regards the crystal lattice, here the usual elements of crystal plasticity apply as in any other crystal and comprise slip, twinning, and phase transformation (martensitic type). There exist explicit works demonstrating the existence and consequences of each. Nevertheless, one may state in broad generality (and that is all I can do at this juncture) that these crystal-plasticity effects are not as important for polymer deformation as in the case of other crystalline substances. The reason for this has already been implied by the foregoing, namely that the main source of deformation in a polymeric solid lies elsewhere: it is the stretching-out of the amorphous random chain, or, as far as the crystal is concerned, the pulling-out of the fold (see below). Thus, just as lattice defects *per se* make only a small contribution to the total defect structure of a usual polymeric solid (see Section 5.2), so the generation and propagation of lattice defects play only a subordinate role in the deformation of the average semicrystalline polymer. In any event they are associated with small strains, and *vice versa*, while the most conspicuous characteristics of polymers are due to their capability of undergoing large extensions.

Nevertheless one feature of the polymer crystal lattice is of major importance, namely its extreme mechanical anisotropy, associated with the existence of valence bonds along the chain direction, as compared to the much weaker interchain forces perpendicular to it. This will favor chain slip over other slip modes and ultimately lead to the alignment of the chains (e.g., during uniaxial tensile deformation) even within the crystal. This is in addition to the stretching-out of chains from the random conformation within the amorphous portions. More will be said below about this chain alignment and chain stretching, which fall within the realm of large-scale deformation.

5.4.1.3. Larger-Scale Crystal Entities

The description of polymer deformation abounds with the mention of "anomalous orientations" by which is meant that the overall chain direction as assessed by, e.g., X-ray diffraction, is different from what one would expect *a priori*, namely chain alignment along the direction of the orienting influence. All such "anomalous" effects can be traced back to the influence of morphological factors. Accordingly, appropriate stress systems may not necessarily always tend to align the chain as such, but orient some larger morphological entity, the resulting chain orientation then conforming to this orientation of the morphological unit in question. While this principle is general, we shall invoke it specifically in connection with the lamellae.

a. The Lamellar Element. The lamellar entity has a role of its own to play in the deformation of a semicrystalline polymer. The lamellar surfaces themselves may act as slip planes, which in a suitable stress system can produce *interlamellar slip*. Thus it is possible to demonstrate that lamellar surfaces may align parallel and perpendicular to extensional or compressive stresses, respectively, even if this may lead to chain orientations that are grossly different from what one would expect if chain slip or chain extension were acting alone.

Rotation of otherwise unaffected crystal lamellae is being envisaged in the aforementioned interlamellar slip. Instead, or in addition, the lamellae themselves may undergo plastic deformation by chain slip within them. In the case of such *intralamellar slip* the chain direction (the direction of the straight chain traverse — fold stem) will be altered with respect to the lamellar surface.

Finally, the separation of the lamellae themselves can be affected. For example, a tensile component normal to the lamellar surface can pull the lamellae further apart (increasing the lamellar periodicity, by which such an effect is assessed) presumably by acting on the interlamellar tie molecules.

In a polycrystalline lamellar system all three of the above effects, namely interlamellar slip, intralamellar slip, and change in interlamellar separation, contribute to the total deformation up to small or moderate strains, where the latter may amount to 50%. Such deformation effects are fully or partially reversible.

b. The Spherulite. Everything mentioned earlier about lamellae applies to spherulites, composed of such lamellae. Here we take a micro- and a macrostructural approach.

First we examine the microstructural aspect. Let us consider a spherulite (as in Figure 2.22c) acted on by a tensile force that is, e.g., vertical. The first point to note is that the different spherulite radii will be situated at different angles to the direction of the deforming influence. As far as these radii have a structural existence of their own, the vertical radius will be stretched along the radial (ribbon) direction and the horizontal radius in a perpendicular direction, and correspondingly for the intermediate cases. It is readily seen that the resulting deformation along these different radii is expected to be very different.

The second point is the periodically varying orientation of the ribbon-shaped lamellae along each radius. This means that along, e.g., the horizontal radius in Figure 2.22c, consecutive portions of a twisting lamella will be stressed alternately parallel and perpendicular to the ribbon plane (which, in view of the fact that the chains are approximately perpendicular to the lamellar surface, means that the stress will be perpendicular and parallel to the chain direction, respectively). It follows that all three lamellar deformation modes invoked in the preceding section will become operative, but each to a different extent within the different spherulite localities, depending on the resolved shear stress in a given locality with respect to a particular deformation mode. Of course the spherulite itself is a contiguous body, so the locally different deformations will be correspondingly constrained by what happens in the other portion of the spherulite, an issue that has not yet been properly solved.

Now we consider the macrostructural aspect. Here it suffices to say that the deformation of the spherulite can and needs to be considered also at the level of what happens to it as a microscopic inclusion or grain. In the simplest case it deforms in affine relation to the macroscopic sample, i.e., a sphere converts into an appropriate ellipsoid. Frequently, however, this affine relation does not hold, and according to circumstances the spherulite may, e.g., on uniaxial tension, become contracted at its "waist" — as if it yields along radii perpendicular to the direction of the tension — or may stretch out more at its apex (radius parallel to the extension).

In conclusion, it should be apparent from the above examples that the deformation of each hierarchical entity must be treated at its own level first, while taking note of the fact that there will be interactions with the levels below and above in the morphological hierarchy. The orientation effects observed, say by diffraction, will represent the resultant, which however would be difficult to interpret, if at all overall possible, without an awareness of the underlying hierarchical structure.

5.4.1.4. Full Chain Extension

All the above cases referred to moderate extensions that may extend up to say 50%. The ultimate objective in deforming a macromolecular substance is to stretch the chains. This of course takes place at high strains, 100% or more (up to 3000%). At such high strains the random chain portions stretch out, and chain folds become aligned and pulled out to varying extents. In fact, the morphological entities discussed above become irreversibly disrupted in its course. The lamellae themselves break up into small chain-folded blocks and remain strung together to form beaded fibers. The whole oriented structure becomes an assembly of microfibrils [A. Peterlin, *J. Mater. Sci.* **6**, 490 (1971)].

At this point it is essential to distinguish between chain orientation and chain extension, as already alluded to in Section 4.2.5. A parallel array of chain-folded blocks, the remnants of the original chain-folded lamellae, such as those forming continuous microfibrils (Figure 5.2a), may be registered by diffraction methods as a highly oriented structure with the chains all parallel even if they are far from being fully extended (Figure 5.2a). The usual technological fiber drawing (cold-drawing) process only leads to the stage in Figure 5.2a, and very special, only comparatively recently adopted methods are required to approach the full extension in

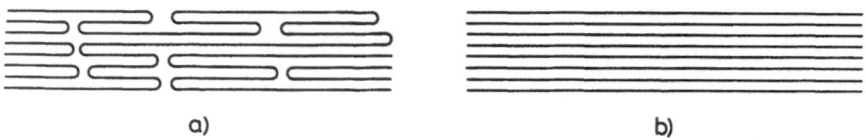

Figure 5.2. Oriented fibers (a) without full chain alignment and (b) with chains fully stretched. Usual tests for orientation cannot distinguish between cases (a) and (b), nevertheless the mechanical properties can be significantly different.

Figure 5.2b, currently referred to as *ultradrawing*. The implication of full chain extension (or the lack of it) for the modulus and strength was mentioned in Section 4.2.5.2, where the alternate method for achieving high chain extension by oriented crystallization (as opposed to deforming structures that are initially crystalline) was described.

Clearly, even within the restricted sphere of deformation behavior many of the morphological constituents of the structure hierarchy were left unmentioned. Nevertheless those that were mentioned should hopefully help one appreciate how intricately the response of polymeric materials to stress depends on its microstructure at different levels. And beyond deformation, in fact beyond the issues covered in the present summarizing survey, it should now be apparent that an understanding of polymeric matter requires an awareness of the whole fabric of structures constituting it, and an awareness of the multitudinous ways in which they are connected.

Before turning to the final chapter on polymers, concerned with the influence of processing on polymeric materials, it is worth reiterating that the objective of these five chapters has been to cover most, even if not all, of the elements that constitute a solid polymeric structure, from the molecules to the various multilevel organizations. It is hoped that the bewildering multitude of structural organizations will not discourage the theorist who may have expected more derivations from a few fundamentals. If the above chapters do not provide deepened understanding, in the sense meant by the theoretician, they at least indicate causal connections and interrelations that will doubtless serve to establish a theoretical framework in the future.

Influence of Processing on Polymeric Materials

G. Marrucci

In this chapter, the most common processes by which thermoplastic and fiber-forming polymers are formed into manufactured items are briefly surveyed. In most cases, solidification of the liquid polymer takes place when it is still flowing or when it has just stopped flowing. As a consequence of the flow, the flexible polymer molecules are stretched and oriented. The resulting items may then exhibit properties very different from those obtained when the polymer is solidified under quiescent conditions.

Existing theories of the mechanics of liquid polymers for large deformations and deformation rates are reviewed. They allow one to make predictions about the molecular orientations obtained under flow. Some consequences of the crystallization process are discussed.

Following this general outline, it is relevant in what follows to make some more specific introductory comments. Although the large variety of properties of polymeric materials, and indeed of all other materials, are ultimately rooted in the details of the chemical structure, there exists in many cases a distinct influence of their polymeric nature. This works in two ways. On the one hand, there are cases where the details of the chemical structure are largely irrelevant and the polymeric nature together with the topological organization of the material are determinant *per se*. A good example is rubber elasticity, which can be successfully studied both theoretically and experimentally in a unified way, i.e., independently of the

---

G. Marrucci · Faculty of Engineering, University of Naples, Italy.

specific chemistry. On the other hand, the polymeric nature of the material also works toward increasing the diversification of properties. For the same polymer, articles may be fabricated that have such different properties as to seem like different materials. One example among many is offered by isotactic polypropylene. Like similar crystallizable polymers, polypropylene solidifies from the melt into semicrystalline solids where the crystalline phase usually exhibits a spherulitic morphology. The resulting properties are more or less typical of a solid plastic with, e.g., a relatively narrow range of elastic deformations. However, under appropriate fabricating conditions, a film of the same polymer can be produced that is highly elastic like a rubbery material, though with a larger modulus than that of a typical rubber. There is definite evidence that the crystalline phase has a completely different morphology in such a case.

The example outlined above can be generalized to most thermoplastics and fiber-forming polymers. The properties in the solid state that determine their use as materials can be influenced to a lesser or greater extent by the conditions encountered in the processes by which the various articles are fabricated. These are physical processes that do not normally alter the chemical structure of the polymer, their main purpose being formation of the material into a manufactured article. In serving this purpose, the fabricating process also influences the way in which the polymer molecules organize themselves in the "structure" of the solid and this influence exists essentially because the substance is polymeric, i.e., highly anisotropic at the molecular level.

It is mainly this aspect that will be considered in the following. Being a feature common to many polymers, it is in part independent of the particular chemical structure and properly belongs to the realm of physics. We shall first briefly review the most common processes and then concentrate on some of their aspects, particularly on the influence of flow when the polymer is processed in the liquid state.

6.1. Polymeric Processing

Polymers that can exist in the liquid state, either as melts (thermoplastics) or in solution with a suitable solvent, are processed almost invariably according to the following simple scheme:

1. The raw polymer, plus other ingredients if any, is converted into a homogeneous liquid mass.
2. The desired shape is given to the polymeric liquid by a flow process.
3. The manufactured item is finally obtained by returning to the solid state.

If the polymer is a thermoplastic, the first and last stages are made by heating the material above the melting or softening point and then cooling it down again. Alternatively, if the polymer decomposes before melting or for some other special reason, the liquid state is obtained by use of a solvent that is then removed either by evaporation or by means of a coagulating liquid, usually water, in which the polymer is insoluble. The latter techniques are unsuitable for the fabrication of thick objects and are used for some kinds of fibers and membranes. In the rest of this section we shall refer mainly to the thermoplastic case, which admits a larger variety of processes.

A common way of obtaining the polymer in the molten state is by means of a *screw extruder*, depicted schematically in Figure 6.1. The polymer is fed to the extruder as a powder or in the form of pellets. The rotating screw conveys the material along the extruder, where it is progressively heated by means of electric heaters placed around the barrel and by the work of friction forces. In the second part of the extruder, after the polymer has been melted, the screw conveys a viscous liquid that is progressively pressurized up to the region of the die, where several hundred atmospheres (tens MPa) can be reached. The screw extruder not only melts the polymer but also thoroughly mixes it and provides the pressure required to drive the flow through the die or other channels. It is a versatile apparatus, which with minor variations is found in most operations of polymer processing.

The second step in the process, that of giving a shape to the liquid mass, is achieved in a variety of ways depending on the geometry of the articles to be produced and other requirements. By way of example, we list some of the techniques.

Extrusion is used for producing articles with a constant cross section, such as pipes, bars, and sheets. It does not require any special apparatus beyond the screw extruder equipped with an appropriate die. Of course, the geometry of the die determines that of the extruded article but in some cases the relationship between them is more complex than one might expect, due to the elasticity of the melt.

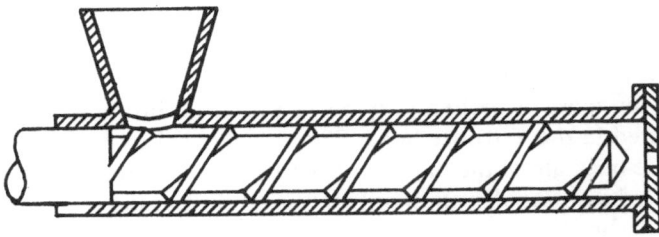

Figure 6.1. Schematic representation of a screw extruder.

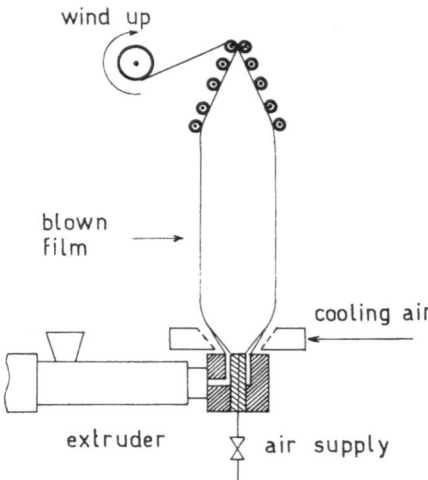

Figure 6.2. Blow extrusion process. Polymer melt extruded from an annular die is blown into a cylindrical film by the pressure of the air trapped in the bubble.

Blow extrusion is a variant of normal extrusion used for producing thin cylindrical films for bags and packaging in general. As shown in Figure 6.2, the molten tube emerging from the die is blown to a larger diameter by the pressure of the air entrapped in the "bubble" and, simultaneously, is drawn in the advancing direction by means of the collecting rolls. Due to the radial and longitudinal expansions, the thickness decreases to the desired value.

Blow molding is used for producing plastic bottles and other hollow noncylindrical articles. A cylindrical tube is extruded downward from a die. Intermittently, when the extruded tube (called a "parison") is of the appropriate length, it is clamped between the two halves of a mold that seals the end of the parison. Compressed air is then injected inside the parison via a needle or through the end still attached to the die, and the parison expands to assume the shape of the mold.

Injection molding is also an intermittent or cyclic operation, used for the production of a large variety of plastic articles, from small toys to whole bodies of refrigerators. A special extruder is commonly employed, equipped with a screw that can also move longitudinally along the barrel. A sturdy mold, which contains cavities corresponding to the articles being produced, is connected to the extruder head.

A given mass of molten polymer is prepared inside the extruder and accumulated in the extruder head. This is achieved by letting the rotating screw slide back along the barrel as it conveys the material forward. When the required amount of molten polymer is ready and the mold is empty, the screw is pushed forward by hydraulic pistons and, acting like a plunger, injects the polymer in the mold at an elevated pressure to fill the cavities.

While the liquid mass solidifies inside the mold, and the mold is opened to eject the articles, the extruder prepares the polymer for the next shot.

Calendering. Fabrication of sheets may be accomplished by extrusion (already mentioned) or by calendering, i.e., by using a system of rotating rolls which squeeze the plastic in their nip. The clearances between the rolls and their velocity determine the thickness of the sheet. Also, when extrusion through a thin flat die is used as the primary step in producing a sheet, this is often followed by a system of rolls and other means which stretch the sheet and decrease its thickness.

Sheet forming. In this operation, a preheated flat sheet is adapted to the rim of an open mold and forced to assume the shape of the mold by the action of, say, a vacuum (cf Figure 6.3). Of course, the sheet is stretched and thinned while deforming to match the surface of the mold. Trays, cups, and similar objects can be fabricated in this way.

Fiber spinning. Though this operation does not differ conceptually from simple extrusion, it has some distinguishing features of its own which justify a separate description.

One of these features is represented by the smallness of the transversal dimensions of the individual product. Each filament is obtained by extrusion through an orifice and, after solidification, is collected by a winding-up mechanism. Though the collecting velocity is usually the largest compatible with technological difficulties, mass production obviously requires that a very large number of filaments be extruded simultaneously due to the smallness of the fiber diameter. Thus the operation is performed typically in the following way. From a central screw extruder or similar apparatus, the liquid mass is conveyed to, and by means of, volumetric gear pumps equally subdivided into a set of parallel spinnerets. Each spinneret has a number of equal orifices from which the filaments emerge. Typically, all the fibers produced by a single spinneret are collected and twisted together to form a thread.

An important feature of this operation is that the cross section of the solid fiber at the collecting point is usually much smaller than that of the orifice from which the liquid filament emerges, i.e., a significant draw of the

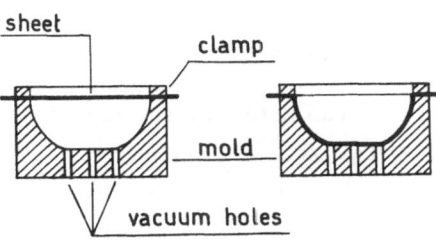

Figure 6.3. Schematic representation of a sheet-forming process.

liquid filament takes place outside the spinneret. In many cases, the solid fibers are further stretched in a subsequent drawing operation, where they assume the final dimensions and properties.

In all the above operations, the third stage of the process, namely reconversion to the solid state, requires little description. In all molding operations, it is the mold — cooled by running water — that takes up the heat. In extrusion processes, including fiber spinning, the liquid polymer emerging from the orifice is brought to the solid state by the action of a fluid medium, usually air or water. Finally, whenever rolls are involved in the process, they may be used to cool the polymer as well as to heat it, as the need arises.

It should be noted that, in all the aforementioned operations, solidification begins while the polymer is flowing or when it has just stopped flowing. This has important consequences, discussed below.

6.2. General Comments on Influence of Flow

Consider a viscous medium that undergoes a flow having a spatially uniform velocity gradient. Now assume that an anisotropic rigid object, such as a rod, is suspended in the medium and that only viscous forces are relevant. We can divide the motion of the rod into the motion of its center (a translation) plus that about the center (a rotation). It is readily shown that the rod translates "together with the fluid" in the sense that its center has the same velocity as that of the fluid particle which would occupy that position in the absence of the rod.

The motion about the center is more complicated and depends on the velocity gradient. We shall not go into details at this point but it can be accepted on an intuitive basis that the orientation of the rod will be influenced by the flow. We can make the general statement whereby a *flow has an orienting effect upon suspended anisotropic objects.*

Furthermore, if the object is not rigid but flexible, i.e., if it has internal degrees of freedom, *the flow also affects the conformation of the object.*

These statements, which are fairly obvious for macroscopic objects, can be extended with some caution also to molecules. The relevant differences are:

1. The molecules are subject to fast thermal motions.
2. At the scale in question, the medium surrounding each molecule can hardly be considered a continuum.

The second point does not appear to be important at all. Friction effects have been successfully described also at the molecular level in a continuum fashion, i.e., by using friction coefficients. A friction force may be viewed at the molecular level as a bias in the mean field of forces exerted upon a given molecule by the surrounding ones.

The first difference, i.e., the presence of thermal motions, is however very important. The orienting and deforming effects of the flow upon anisotropic flexible molecules are normally frustrated by the randomization effect of thermal agitation. The reason for this ineffectiveness of the flow is readily understood by comparing the relaxation time due to thermal motions (or the maximum relaxation time when there exists a spectrum of relaxations) with the reciprocal magnitude of the velocity gradient, which can be taken to represent the orienting effect of the flow.

For ordinary molecules, the former is extremely small, of order 10^{-8} s at most, whereas the latter is never smaller than 10^{-6} s, even for extremely fast flows, and is usually much larger. In other words, thermal agitation is normally fast enough to counteract completely the tendency of the flow to orient and deform the molecules.

However, the situation may be reversed in the case of macromolecules. The maximum relaxation time τ of a polymer molecule is several orders of magnitude larger than that of small molecules and grows rapidly with molecular weight. For a flexible linear polymer in a dilute solution τ is proportional to M^2 (in the free-draining or Rouse case) and also grows proportionally with the viscosity of the solvent. In the case of concentrated solutions and melts, more relevant to polymer processing, two kinds of behavior must be distinguished.

Below a critical molecular weight, or below a critical concentration, there is no special cooperative effect among the macromolecules. Each macromolecule behaves individually, as in a dilute solution, with the rest of the system acting as if it were a fairly viscous solvent. Above the critical values, however, the system becomes cooperative in some sense. The polymer chains entangle with one another and the relaxation time grows more rapidly with molecular weight and/or concentration. In the latter situation, often encountered in polymer processing, relaxation times of the order of seconds or even minutes may occur. This is particularly true when the temperature is just above the glass transition temperature T_g, where the free volume barely suffices to allow rearrangements of the chain conformation induced by thermal motions.

We may therefore expect that in a polymeric liquid undergoing a flow process, depending on the magnitude of the velocity gradient as compared with the relaxation rate, the molecules may be significantly oriented and deformed with respect to the isotropic equilibrium distribution prevailing in a stagnant liquid.

The elementary processes by which the polymeric liquid is brought to the solid state are also influenced by this situation. If the polymer is amorphous, since the glass transition is not a phase transition in a thermodynamic sense but only the result of a self-retarding volume-contraction process, nothing qualitatively different occurs at the transition beyond the fact that the anisotropy of the liquid that has survived the relaxation processes remains "frozen" in the glass.

Conversely, if the polymer is able to crystallize, the anisotropy of the liquid acts also at a thermodynamic level. In fact, the entropy of a polymeric liquid is largely made up of the conformational entropy of the chains. The negative entropy change, which occurs at the transition, corresponds to the loss of the conformational entropy. Now, if the macromolecules are already partially oriented in the liquid state, the entropy change at the transition is reduced, i.e., the formation of crystals becomes easier. The crystallization temperature correspondingly increases. Various other consequences of the flow-induced anisotropy may be mentioned. The kinetics of the crystallization process is also considerably accelerated and the morphology of the crystals produced is affected, more or less profoundly.

All the effects described so far are readily understood on a qualitative basis. More difficult is a quantitative description. This primarily requires a model of the liquid polymer able to predict molecular orientations resulting from fast flows and large deformations, i.e., under conditions far removed from equilibrium, which are characterized by a nonlinear response. The following sections will deal with this problem.

6.3. Dumbbell Model

We shall first consider the case where the molecules behave "individually," which means that they are free to move around without constraints throughout a "medium" treated as a viscous continuum. We know that this situation applies to solutions and melts only when the molecular weight and/or the concentration are below the critical values. Though this is not the most interesting case for applications, it is simpler and pedagogically useful and is therefore examined first.

The statistics of linear flexible macromolecules can be treated independently of their "chemistry" by considering an equivalent chain made up of equal rigid links freely jointed to one another. We call b the length of the link and n the number of links in the chain. The chemical structure of the chain, particularly its flexibility, determines b, which usually corresponds to a few monomer lengths. Of course n is proportional to the molecular weight. In the following we shall assume that n is a large number, which is the case for most polymers.

From the early results by Kuhn,[1] we know that the chain end-to-end vector r has an equilibrium distribution, which can be approximated by the

Gaussian

$$W(x, y, z) = \left(\frac{3}{2\pi nb^2}\right)^{3/2} \exp\left[-\frac{3}{2nb^2}(x^2 + y^2 + z^2)\right] \qquad (6.1)$$

where x, y, z are the Cartesian components of **r**.

From equation (6.1), it follows that the mean-square end-to-end distance is given by

$$\langle r^2 \rangle = nb^2 \qquad (6.2)$$

It also follows that if the chain is held fixed at the ends, an entropic force acts to bring the ends together. This force obeys the relation

$$\mathbf{F} = -\frac{3kT}{nb^2}\mathbf{r} \qquad (6.3)$$

Equation (6.3) is valid only as long as the chain is not too much extended, i.e., if $r \ll nb$. In practice, equation (6.3) represents a good approximation up to $r \simeq 0.3nb$, which is usually, though not always, a large enough range for practical applications.

When the liquid to which the chains belong is set in motion, the chains are acted upon by friction forces distributed all along the chain contour length. This is a difficult problem to analyze and is therefore simplified by assuming that the friction is concentrated at a discrete number of spots along the chain. The simplest yet effective model is obtained when the friction is concentrated at two points only, e.g., at the chain ends. This is the dumbbell model, depicted schematically in Figure 6.4.

The dumbbell is seen to be made up of two friction points, called beads, connected by a spring that is made to obey the elasticity law expressed by equation (6.3). The beads of the dumbbell are acted upon by three kinds of forces:

1. Friction forces proportional to the relative velocity between the beads and the surrounding medium.
2. Elastic forces due to the spring.
3. Random forces generating a Brownian motion.

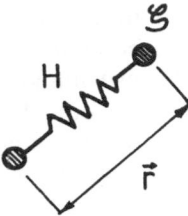

Figure 6.4 The dumbbell model. The value of the spring constant H is obtained from entropy considerations. The friction coefficient accounts globally for friction effects distributed along the chain.

We now wish to write the equation governing the change in the distribution function $W(x, y, z)$ due to the flow. In the space of vectors $\mathbf{r}$ we may define a flux $\mathbf{J}$, which results from the forces acting on the beads and is given by

$$\mathbf{J} = (\underset{\sim}{L} \cdot \mathbf{r})W - \frac{H\mathbf{r}}{\zeta} W - D \text{ grad } W \qquad (6.4)$$

where $\underset{\sim}{L}$ is the velocity gradient (here and subsequently, second-order tensors are indicated by a tilde underneath the symbol), H is the spring constant as given by equation (6.3) ($H = 3kT/nb^2$), ζ is a friction constant, and D is a diffusion coefficient. However, the latter two parameters are not independent since they must obey the Fokker–Planck relation

$$\zeta D = kT \qquad (6.5)$$

The three terms which appear in equation (6.4) are the contributions to the flux arising from the three kinds of forces acting on the beads, in the same order as listed above.

The equation of change for the distribution function W is then written in the form of a continuity equation

$$\frac{\partial W}{\partial t} = - \text{div } \mathbf{J} \qquad (6.6)$$

Of course, both the gradient and the divergence operators appearing in equations (6.4) and (6.6) are defined in $\mathbf{r}$-space.

Combining equations (6.4) and (6.6) gives the following differential equation:

$$\frac{\partial W}{\partial t} = - \text{div}(W\underset{\sim}{L} \cdot \mathbf{r}) + \frac{H}{\zeta} \text{div}(W\mathbf{r}) + D\nabla^2 W \qquad (6.7)$$

where ∇^2 denotes the Laplacian operator.

Equation (6.7) should be interpreted in the following way. We consider a "sample" of macromolecules — here modeled as dumbbells — comprising all macromolecules contained in a small volume in the neighborhood of a material point P of the liquid mass. At a given time, say $t = 0$, the distribution W_0 of end-to-end vectors of these molecules is known [e.g., it is the equilibrium Gaussian distribution given by equation (6.1)]. If for all subsequent times the velocity gradient at P is assigned, i.e., $\underset{\sim}{L}(t)$ for $t > 0$ is given, then the solution of equation (6.7) with the initial condition $W = W_0$ provides the distribution function in the neighborhood of P at all times $t > 0$. Actually, also the thermal history $T(t)$ must be known since the parameters H, ζ, and D are temperature dependent.

The distribution function obtained by solving equation (6.7) generally represents more information than actually required. For most purposes, it is sufficient to obtain the *second-order moments* of the distribution, i.e., the ensemble average of the dyad **rr** given by

$$\langle \mathbf{rr} \rangle = \int_{R^3} \mathbf{rr}\, W(\mathbf{r}) d^3\mathbf{r} \tag{6.8}$$

where $d^3\mathbf{r}$ stands for $dxdydz$ and $\int_{R^3}$ means integration over the whole **r**-space.

The equation of change for $\langle \mathbf{rr} \rangle$ is readily obtained by equation (6.7), all terms of which are multiplied by **rr** and integrated over **r**-space:

$$\int_{R^3} \mathbf{rr}\, \frac{\partial W}{\partial t} d^3\mathbf{r} = -\int_{R^3} \mathbf{rr}\, \mathrm{div}(W\underset{\sim}{L} \cdot \mathbf{r}) d^3\mathbf{r} + \frac{H}{\zeta} \int_{R^3} \mathbf{rr}\, \mathrm{div}(W\mathbf{r}) d^3\mathbf{r}$$
$$+ D \int_{R^3} \mathbf{rr}\nabla^2 W d^3\mathbf{r} \tag{6.9}$$

On the left-hand side of equation (6.9), **rr** can be brought inside the time derivative and the order of differentiation with respect to time and integration over **r**-space reversed. The integrals on the right-hand side are manipulated by the use of Green's theorem together with the property of W whereby $r^\alpha W$ goes to zero as r goes to infinity for all values of α. One thus obtains

$$\frac{d}{dt} \langle \mathbf{rr} \rangle = \underset{\sim}{L} \cdot \langle \mathbf{rr} \rangle + \langle \mathbf{rr} \rangle \cdot \underset{\sim}{L}^{\mathrm{T}} - \frac{2H}{\zeta} \langle \mathbf{rr} \rangle + 2D\, \underset{\sim}{1} \tag{6.10}$$

where $\underset{\sim}{L}^{\mathrm{T}}$ is the transpose of $\underset{\sim}{L}$ and $\underset{\sim}{1}$ is the unit tensor.

The advantage of equation (6.10) over equation (6.7) is apparent. Equation (6.10), which is but an ordinary first-order linear differential equation, is finally written as

$$\langle \mathbf{rr} \rangle + \tau \left(\frac{d\langle \mathbf{rr} \rangle}{dt} - \underset{\sim}{L} \cdot \langle \mathbf{rr} \rangle - \langle \mathbf{rr} \rangle \cdot \underset{\sim}{L}^{\mathrm{T}} \right) = \frac{nb^2}{3}\, \underset{\sim}{1} \tag{6.11}$$

where we have used equation (6.5) together with the definition of H. The parameter τ is the *relaxation time* defined as

$$\tau \equiv \frac{\zeta}{2H} = \frac{\zeta nb^2}{6kT} \tag{6.12}$$

If, from a certain time onward, the flow is stopped ($\underset{\sim}{L} = \underset{\sim}{0}$), equation (6.11) reduces to

$$\langle \mathbf{rr} \rangle + \tau \frac{d\langle \mathbf{rr} \rangle}{dt} = \frac{nb^2}{3} \underset{\sim}{1} \tag{6.13}$$

i.e., $\langle \mathbf{rr} \rangle$ relaxes toward the equilibrium isotropic value $(nb^2/3)\underset{\sim}{1}$. Incidentally, this equilibrium value is obviously consistent with equation (6.2).

The rate of approach to equilibrium is regulated by the relaxation time τ. If the temperature is held constant during the relaxation, the decay is exponential. Also τ, as given by equation (6.12), depends on the molecular weight in two ways: directly through n and indirectly through ζ, which obviously must grow as n increases. If the chain is fully "exposed" to the flow of the surrounding liquid, i.e., hydrodynamic interactions among the various parts of the chain are neglected (free-draining case), then ζ is proportional to n and thus τ varies proportionally to the square of the molecular weight. Hydrodynamic interactions, i.e., shielding effects of some parts of the chain on other ones, somewhat weaken this dependence.

We now examine the predictions of equation (6.11) for some categories of flow. Before doing so, it is worth noting that $\langle \mathbf{rr} \rangle$ not only describes the average orientation and elongation of the macromolecules but also gives their contribution to the stress tensor. In fact, consistently with the dumbbell model the stress tensor $\underset{\sim}{\sigma}$ is given by

$$\underset{\sim}{\sigma} = cH\langle \mathbf{rr} \rangle \tag{6.14}$$

where c is the number of macromolecules per unit volume. Equation (6.14) is readily obtained by considering that, for any given value of $\mathbf{r}$, $H\mathbf{r}$ is the elastic force in the dumbbell spring and the probability of intersecting the dumbbell by a crossing plane is proportional to $\mathbf{r}$.

6.3.1. Shear Flow

We consider the case where the fluid is maintained at rest up to $t = 0$ and then sheared isothermally at a constant rate $\dot{\gamma}$ thereafter. The matrix of the velocity gradient $\underset{\sim}{L}$ then has only a single nonzero xy-component $\dot{\gamma}$. Solution of equation (6.11) leads to the viscosity η in the form

$$\eta = ckT\tau = \tfrac{1}{6}c\zeta nb^2 \tag{6.15}$$

By comparing liquid systems obtained by varying the molecular weight M while keeping constant the polymer concentration by weight, equation (6.15) predicts that η should show the same M-dependence as ζ, i.e., it should be proportional to the first power of M if hydrodynamic interactions are negligible, or to a lesser power of M if they are not. This prediction is confirmed experimentally. The intrinsic viscosity of dilute solutions is indeed proportional to M^α with $0.5 < \alpha < 1$. The viscosity of low-molecular-weight melts is approximately proportional to M.

Unlike an ordinary liquid, a polymeric liquid in shear flow also exhibits normal stresses. Because of incompressibility, only normal-stress differences are relevant. In the present case the difference $\sigma_{xx} - \sigma_{yy}$ in the steady state is nonzero and quadratic in $\dot\gamma\tau$.

6.3.2. Elongational Flow

While shear flows are usually encountered when the polymeric liquid moves in pipes, conduits, dies, etc., elongational flows dominate when the polymer is being deformed with free surfaces like those in, e.g., a spinning line or the bubble of a blow extrusion. Elongational flows are characterized by the fact that the Cartesian matrix of the velocity gradient is diagonal.

We shall consider in some detail the case of axially symmetric stretching, where the velocity gradient matrix has diagonal elements $\dot\Gamma$, $-\dot\Gamma/2$, $-\dot\Gamma/2$ with principal stretching rate $\dot\Gamma > 0$. Such uniaxial stretchings are typical of fiber spinning, while biaxial stretchings are encountered in the fabrication and formation of sheets.

For the simplest case where the fluid is kept at rest up to $t = 0$ and stretched isothermally at constant $\dot\Gamma$ thereafter, the solution of equation (6.11) gives

$$\langle x^2 \rangle = \frac{nb^2}{3} \frac{1}{1 - 2\dot\Gamma\tau} \left\{ 1 - 2\dot\Gamma\tau \exp\left[-\frac{t}{\tau}(1 - 2\dot\Gamma\tau) \right] \right\} \tag{6.16}$$

$$\langle y^2 \rangle = \langle z^2 \rangle = \frac{nb^2}{3} \frac{1}{1 + \dot\Gamma\tau} \left\{ 1 + \dot\Gamma\tau \exp\left[-\frac{t}{\tau}(1 + \dot\Gamma\tau) \right] \right\} \tag{6.17}$$

together with the obvious result $\langle xy \rangle = \langle yz \rangle = \langle zx \rangle = 0$.

Equations (6.16) and (6.17) require some discussion. Quantities $\langle y^2 \rangle$ and $\langle z^2 \rangle$ show no special behavior. The effect of the flow decreases their value with respect to the equilibrium value $nb^2/3$. After a transient, they asymptotically approach the steady-state value $nb^2/3(1 + \dot\Gamma\tau)$. Conversely, the behavior of $\langle x^2 \rangle$, which is augmented by the flow, shows a distinct peculiarity. If $\dot\Gamma\tau < 0.5$, the argument of the exponential in equation (6.16) is negative. Thus after a transient, $\langle x^2 \rangle$ reaches a steady-state value equal to $nb^2/3(1 - 2\dot\Gamma\tau)$. The viscosity corresponding to this steady state, called the *elongational viscosity*, is given by

$$\eta_{el} \equiv \frac{\sigma_{xx} - \sigma_{yy}}{\dot\Gamma} = 3ckT\tau \frac{1}{(1 - 2\dot\Gamma\tau)(1 + \dot\Gamma\tau)} \tag{6.18}$$

(It may be noted that if $\dot{\Gamma}\tau \ll 1$, $\eta_{el} = 3\eta$, which is the Trouton result for ordinary liquids.)

However, if $\dot{\Gamma}\tau > 0.5$, $\langle x^2 \rangle$ grows with time without bounds. For $\dot{\Gamma}\tau \gg 1$, equation (6.16) may be approximated by

$$\langle x^2 \rangle \cong \frac{nb^2}{3} \exp(2\dot{\Gamma}t) \tag{6.19}$$

i.e., $\langle x^2 \rangle$ grows exponentially with time. Actually, since $\exp(\dot{\Gamma}t)$ is the stretch ratio of the continuum in the x-direction, the conclusion is drawn that the macromolecules deform affinely with the continuum in such extreme cases.

It is noteworthy that no equivalent case exists for the shear flow. This is due to the *rotational* character of a shear flow that makes the macro-molecules rotate continuously in the liquid. During the rotation the dumbbell length pulsates, i.e., the dumbbell is alternatively stretched and compressed, though it is somewhat elongated on the average with respect to the equilibrium length. On the other hand, in elongational flows no rotational component is present. The dumbbells continuously tend to be aligned and stretched in the direction of the draw.

Also the critical value $\dot{\Gamma}\tau = 0.5$, which separates "weak" from "strong" effects of the flow, can be readily understood by considering the competition between the drag exerted by the solvent on the beads of a dumbbell and the force in the spring. By taking a dumbbell parallel to x, the friction contributes a force given by $\zeta(\dot{\Gamma}x - dx/dt)$ while the force in the spring is Hx. Thus if $\zeta\dot{\Gamma} > H$, dx/dt is positive for all values of x and the dumbbell elongates indefinitely. In the opposite case, namely $\zeta\dot{\Gamma} < H$, dx/dt is negative and the dumbbell would collapse. However the diffusion term, which always acts in the direction of separating the beads, allows an average equilibrium separation to be found in the latter case.

It is finally noted that the indefinite elongation of the dumbbells, when $\dot{\Gamma}\tau > 0.5$, is obviously an exaggeration due to the linearity of the approxi-mate elastic law given by equation (6.3). In reality, the macromolecules cannot go beyond the fully extended length nb. However, depending on the value of n this may be much larger than the rms (root-mean-square) distance of the molecule at equilibrium, which is given by equation (6.2) as $b\sqrt{n}$. When all molecules are fully extended, the elongation viscosity achieves its maximum value

$$\eta_{el,max} = c\zeta n^2 b^2 = 6n\eta \tag{6.20}$$

which is $2n$ times the Trouton value.

6.3.3. Other Flows

Although the shear and elongational flows considered previously are representative of most cases of practical interest, they obviously do not cover all possibilities.

It is noteworthy that equation (6.11) can be formally integrated in the general case. For an arbitrary *isothermal* flow, the integral of equation (6.11) is written as

$$\langle \mathbf{rr} \rangle = \frac{nb^2}{3} \frac{1}{\tau} \int_{-\infty}^{t} \underset{\sim}{C}_{t}^{-1}(t') \exp\left(-\frac{t-t'}{\tau}\right) dt' \qquad (6.21)$$

where $\underset{\sim}{C}_{t}^{-1}(t')$ is the inverse Cauchy–Green deformation tensor, which maps the configuration of the continuum at the present time t into that at the current time t'. This tensor is readily obtained from the velocity gradient $\underset{\sim}{L}(t')$. By the use of a Cartesian coordinate system, if x_i are the coordinates of a material point at time t and ξ_i the coordinates of the same point at time t', the components of $\underset{\sim}{C}_{t}^{-1}(t')$ are obtained as

$$C_{ij}^{-1} = \sum_{\kappa=1}^{3} \frac{\partial x_i}{\partial \xi_\kappa} \frac{\partial x_j}{\partial \xi_\kappa} \qquad (6.22)$$

It should be noted that the integral of equation (6.11) as given by equation (6.21) has not required an initial condition for $\langle \mathbf{rr} \rangle$. The requirement of an initial condition is here replaced by the fact that the integral extends backward in time to $-\infty$, i.e., it contains the full history of deformation up to present time t. This is possible because of the negative exponential kernel in the integral of equation (6.21). The kernel is called the memory function of the material and we are here dealing with a material endowed with a fading memory. The latter expression has the following meaning. Given two deformation histories that differ only in the distant past (i.e., for $t - t' \gg \tau$), their effect on the present value of $\langle \mathbf{rr} \rangle$ is the same. Only differences in the recent past — recent with respect to the relaxation time — may indeed be relevant.

The integration of equation (6.11) under nonisothermal conditions can also be performed in general. Both parameters appearing in equation (6.11), τ and nb^2, depend on temperature. The latter, i.e., the equilibrium square end-to-end distance, depends on temperature rather weakly. However the relaxation time, being proportional to the viscosity of the medium, may depend strongly on temperature.

By setting $\mu = nb^2/3$ for brevity, the nonisothermal integral is given by

$$\langle \mathbf{rr} \rangle = \int_{-\infty}^{t} \frac{\mu(t')}{\tau(t')} \underset{\sim}{C}_{t}^{-1}(t') \exp\left[-\int_{t'}^{t} \frac{d\alpha}{\tau(\alpha)}\right] dt' \qquad (6.23)$$

which obviously requires knowledge of the *thermal history* together with that of the deformation.

We conclude this section by remarking that the usefulness of the linearly elastic dumbbell considered here consists essentially in the simplicity of the results. Many results are in qualitative agreement with experimental indications and predict correct orders of magnitude. However, the limitations inherent in the use of this model are readily apparent.

The elasticity of the entropic "spring" is linear only up to moderately large deformations. Throughout the deformation range, i.e., up to the fully extended chain, an inverse Langevin function should be used instead. Furthermore, the process by which a flexible macromolecule gets extended actually occurs through rotations around bonds (in a carbon backbone, say, by switching from gauche to trans conformations). These rotations occur by overcoming energy barriers so that, if the macromolecule is forced to do so, a dissipation process will result. The effect is usually described in terms of an "internal viscosity," which relates the resisting dissipative force to the rate of change of the end-to-end distance.

All these effects can be introduced in the dumbbell model by suitably modifying the behavior of the "connector" between the two beads. Better approximate results are thus obtained, obviously at the expense of greater mathematical complexity. Particularly, general solutions for these non-linear dumbbells cannot be found and the various cases of possible interest require separate treatment.

For more detailed treatment of the dumbbell model we refer to Bird *et al.*[2] Liquids with memory are treated from the continuum mechanics point of view by, e.g., Astarita and Marrucci.[3] Detailed predictions of a constitutive equation for the stress tensor which, though derived differently, exactly coincides with equation (6.21) are considered at length by Lodge.[4]

6.4. Multiplicity of Friction Points: Rouse–Zimm Model

A natural extension of the dumbbell model that preserves linearity is that obtained by considering a multiplicity of friction beads distributed along the chain length. The model is depicted in Figure 6.5, where again

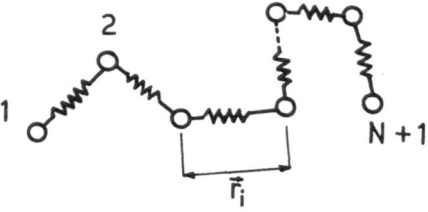

Figure 6.5. The Rouse–Zimm model. Many friction beads along the chain are considered.

the connectors between consecutive beads are made up of springs obeying a linear elasticity law similar to equation (6.3).

If the (arbitrary) number of beads is $N + 1$, the function describing the distribution of possible conformations is now defined over the $3N$-dimensional space of the end-to-end subchain (or connector) vectors, $\mathbf{r}_i$ $(i = 1, \ldots, N)$. At equilibrium, the distribution is Gaussian and one finds

$$\langle r_i^2 \rangle = n_i b^2 \equiv 3\mu_i$$

$$\langle r^2 \rangle = \sum_{i=1}^{N} \langle r_i^2 \rangle = 3N\mu \tag{6.24}$$

where n_i is the number of monomers in the ith subchain and, of course, the last equality holds only if all subchains have the same value of $\mu_i = \mu$ corresponding to $n_i = n/N$. This is the case considered below.

In the Rouse version of the theory,[5] the hydrodynamic interactions among the beads are ignored. The physical situation is similar to that of the dumbbell model with the important difference that on all beads, with the exception of the end ones, two elastic connectors act simultaneously. The corresponding mathematics does not allow one to treat each subchain independently of the others.

The problem is solved by performing a change to so-called "normal" coordinates, which introduces a spectrum of relaxation times given in the Rouse case by

$$\tau_i = \frac{1}{i^2} \tau_{\max} \qquad (i = 1, \ldots, N) \tag{6.25}$$

where $\tau_{\max} = n^2 b^2 \zeta_0 / 6\pi^2 kT$, which plays the same role as τ in the dumbbell model in determining whether or not the flow has a significant influence on orientation and stretching of molecules. The arbitrary subdivision of the chain, namely the value of N, does not play any significant role provided N is not too small. Knowing the spectrum of relaxation times, the stress tensor can be calculated from

$$\underset{\sim}{\sigma} = ckT \int_{-\infty}^{t} \underset{\sim}{C}_i^{-1}(t') \sum_{i=1}^{N} \frac{1}{\tau_i} \exp\left(-\frac{t - t'}{\tau_i}\right) dt' \tag{6.26}$$

In the theory due to Zimm,[6] the influence of hydrodynamic interactions is accounted for in an approximate way, i.e., by preaveraging their effect on the basis of the equilibrium distribution. The mathematics is the same as in the Rouse case with the only difference that the matrix to be diagonalized is somewhat more complicated. Thus a different set of eigenvalues is obtained that generates a different relaxation spectrum.

All the results so far reported refer to a sample of chains all with the same molecular weight. Real polymers are polydisperse in most cases, i.e., they have a distribution of molecular weights. Equation (6.26) can be readily extended to include this case. Again, what will change is the relaxation spectrum, which will become broader and more "continuous" than in the monodisperse case. In general terms, equation (6.26) may be written as

$$\underset{\sim}{\sigma} = \int_{-\infty}^{t} \underset{\sim}{C_t^{-1}}(t')f(t-t')dt' \tag{6.27}$$

where $f(t)$ is the *memory function*, related to the relaxation spectrum by

$$f(t) = \sum_i \frac{G_i}{\tau_i} \exp\left(-\frac{t}{\tau_i}\right) \tag{6.28}$$

or

$$f(t) = \int_0^\infty \frac{H(\tau)}{\tau^2} \exp\left(-\frac{t}{\tau}\right) d\tau = \int_{-\infty}^\infty \frac{H(\tau)}{\tau} \exp\left(-\frac{t}{\tau}\right) d\ln\tau \tag{6.29}$$

depending on whether the spectrum is discrete or continuous, respectively. In the former case, G_i (with dimensions of a modulus) is the "weight" associated to the ith relaxation time τ_i. In the case of a Rouse spectrum, all relaxation times have the same weight, given by ckT. In the continuous case, $H(\tau)$ represents the distribution of relaxation times and is called the "spectrum." Since the spectrum usually extends in time over many decades, a logarithmic representation is often preferred. The continuous equivalent of a Rouse spectrum is of the form

$$H(\tau) = \text{const} \cdot \tau^{-1/2} \tag{6.30}$$

The relaxation spectrum of a polymeric substance is commonly determined by means of experiments of *linear viscoelasticity*. These are experiments where either the deformation or the deformation rate are sufficiently small at all times. In such a case, the relationship between the stress and the history of motion becomes

$$\underset{\sim}{\sigma} = \int_{-\infty}^{t} 2\underset{\sim}{D}(t')G(t-t')dt' \tag{6.31}$$

where $\underset{\sim}{D}$ is the symmetric part of the velocity gradient $(\underset{\sim}{L} + \underset{\sim}{L}^T)/2$ and G is the *relaxation modulus*. In a stress relaxation experiment performed by imposing a sudden strain at $t = 0$, $G(t)$ gives the relaxing stress per unit strain.

The relationship between $G(t)$ and $H(\tau)$ is as follows:

$$G(t) = \int_{-\infty}^\infty H(\tau)\exp\left(-\frac{t}{\tau}\right) d\ln\tau \tag{6.32}$$

Equation (6.31) describes the linear superposition principle named after Boltzmann. In fact, $2D(t')dt'$ represents the increment in strain at time t'; the stress generated by the strain increment relaxes up to present time t according to $G(t - t')$. The present stress is the sum of all such contributions. For a thorough treatment of linear viscoelasticity, we refer to the book by Ferry.[7]

It should be noted that equation (6.27) properly degenerates into linear viscoelasticity for the case of small deformations or small deformation rates. To formally obtain equation (6.31) from equation (6.27) in such a limiting case, an integration by parts is required that makes use of the relationship between $G(t)$ and $f(t)$, i.e. [cf equations (6.29) and (6.32)],

$$f(t) = - dG(t)/dt \qquad (6.33)$$

In a way, equation (6.27) represents a nonlinear superposition principle, which applies to polymeric liquids conforming to the Rouse–Zimm theory, i.e., below the critical values of molecular weight and/or concentration. At large deformations, linear elasticity looses any meaning since the deformation measure is not defined uniquely.[3] The nonlinearity of equation (6.27) resides in the fact that C^{-1} is generally not proportional to the infinitesimal strain tensor of Hookean elasticity, yet the *superposition in time* is still obeyed by equation (6.27). Indeed, any constitutive equation that can be written as an integral over time of the product of a history of deformation times a memory function *independent of deformation* gives rise to a superposition rule. Other examples will be encountered in the next section.

It should also be noted that, since in the Rouse theory all relaxation times are proportional to τ_{max} via a numerical constant, they all change by the same factor as a consequence of a change in temperature. Because the moduli are proportional to absolute temperature only, this implies that the whole spectrum, when plotted on a log scale, shifts almost horizontally by changing the temperature. This result forms the basis of the so-called time–temperature superposition principle, which turns out to be of wide applicability. The principle can be stated in the form that a change in temperature is tantamount to an appropriate change in the time scale of all rates and time-dependent phenomena.

Before concluding this section, we emphasize that the Rouse–Zimm theory remains approximate, more or less for the same reasons as already discussed in relation to the linear dumbbell. The nonlinear entropic elasticity at high chain extension as well as the effects of internal viscosity are here similarly ignored. The approximate nature of the Rouse–Zimm theory is best appreciated by considering that, as for the dumbbell model, the shear viscosity is predicted to be independent of shearing rate.

Although this is approximately true for dilute solutions and relatively low-molecular weight melts, deviations are found at large values of the shear rate.

6.5. Concentrated Systems

High-molecular-weight polymers in concentrated solutions and melts behave in a manner significantly different from that described by the models so far considered. A well-known peculiar feature of concentrated systems is the molecular-weight dependence of the shear viscosity. While at low M-values the viscosity is approximately proportional to M, as predicted by the Rouse and dumbbell models, at large M-values the dependence becomes much stronger and can be empirically described by the law $\eta \propto M^{3.4}$. The transition between the two kinds of behavior is quite abrupt and occurs in the neighborhood of a molecular weight M_c which is called "critical."

Also, the shear sensitivity of the viscosity becomes much larger for $M > M_c$. By increasing the shear rate, the viscosity may drop by several orders of magnitude in some cases.

It must be said immediately that concentrated system theories for high-molecular-weight polymers are in a developing stage and the present situation is still far from satisfactory. However, significant progress has nonetheless proved possible.

Attempts to describe the behavior of these systems have long been made by using ideas borrowed from the Rouse model on the one hand, and from rubber-elasticity theory on the other. It has been recognized that concentrated systems made up of very long and flexible chains are so thoroughly entangled as to resemble networks. The role played by the cross-links in a permanent rubber network would here be replaced by that of the entanglements, albeit on a temporary basis, i.e., for deformation processes fast enough not to allow effective chain disentanglement to occur. From the network viewpoint, the basic unit of the system is not the whole chain but rather the average subchain, which forms between two consecutive entanglements. On the other hand, the entire chain also preserves its identity so that its length and the associated friction properties must play a role. From the latter standpoint, attempts were made to modify the Rouse model to account for the effect of entanglements in enhancing the friction.

It is noteworthy that impermanent network models also lead to equations similar to those obtained for the dumbbell and Rouse models. In the theory by Lodge,[4] an impermanent network of subchains is considered. The subchains periodically detach from the network to rejoin it again subsequently. As long as they remain attached to the network, their end-to-end vector **r** deforms affinely with the continuum, while when they join the network their **r**-values are distributed according to the equilibrium Gaussian. By calling τ the average lifetime of a subchain in the network and assuming that τ is a constant, equation (6.21) is obtained exactly. Similarly, if subchains having different lifetimes τ_i are considered, equation (6.26) is obtained instead.

These results are not particularly encouraging, since they imply — among other things — that the shear viscosity should be independent of the shear rate, a prediction largely disproved by the actual behavior of concentrated long-chain systems. Moreover, the theory is unable to make predictions about the values of the relaxation times τ_i, which are thus necessarily taken as *a priori* parameters.

Various modifications of Lodge's model were proposed to overcome the difficulty of a constant viscosity and of other major discrepancies. In one way or the other, they assume the relaxation times to depend on the chain extension, or the velocity gradient, or the entanglement density. In the latter case, an equation of change for this density, somehow related to the history of deformation, is required. All these attempts lead to equations of the form of equation (6.27) where, however, the memory function also depends on some scalar measure of the history of deformation.

Though some of these proposals compare favorably with many experimental observations, they remain empirical in many respects. Furthermore, some subtle fundamental difficulties remain unsolved. These are rooted in the tensor $\underset{\sim}{C}^{-1}$, which is anyhow present in equations based on classical rubberlike models. We mention a typical instance.

When dealing with the predictions of the dumbbell model for a shear flow we found that a nonzero normal-stress difference $\sigma_{xx} - \sigma_{yy}$ exists. This is called the *first* normal-stress difference. The *second* normal-stress difference, $\sigma_{yy} - \sigma_{zz}$, was found to be zero. Actually, the experimental results show that a nonzero second normal-stress difference also exists, though smaller than the first. Now, no matter how the memory function is modified in equation (6.27), as long as the tensorial part of the equation is given by $\underset{\sim}{C}^{-1}$, the second normal-stress difference in a shear flow is predicted to be zero.

The attempts to tackle the problem from the viewpoint of the Rouse theory must also be regarded as inadequate. The arbitrary assumption is made that the entanglements enhance the friction at a few spots along the chain, which obviously results in a spectrum having a set of longer relaxation times. Yet the modified Rouse theory does not predict correctly a number of linear viscoelastic properties. For example, the steady-state compliance J_e^0 is predicted to be proportional to the molecular weight while the experimental results show that, at least for linear polymers, J_e^0 becomes virtually independent of M at large values of M.[7] Furthermore, as previously shown, the Rouse theory predicts a C^{-1} tensorial dependence at large deformations as in the rubberlike models.

A completely different approach was taken by Doi and Edwards.[8] Already Edwards[9] had envisaged that a long chain belonging to a concentrated system is topologically constrained by neighboring chains in such a way that sideway thermal motions are limited within relatively short distances. Later, de Gennes[10] developed this idea into a workable mathematics and coined the descriptive name of "reptation" to indicate the motion of a chain, laterally constrained, along its length. De Gennes obtained two important results. He showed that rearrangements of the chain *within* the existing constraints obey a Rouse-type mathematics and thus occur in a time proportional to M^2. However, the time required for the chain to reptate out of the existing constraints is proportional to M^3. Now, when the polymeric system is deformed, so are the constraints and the enclosed chain is no longer at equilibrium. If the only way for the chain to return to complete equilibrium is to diffuse out of the deformed constraints, the longest relaxation time will become proportional to M^3. As shown by Doi,[11] this result can be transferred to the viscosity, which also comes out proportional to M^3 in such a case. Although the experimental results actually indicate an $M^{3.4}$ dependence, the agreement is close enough to encourage further studies along these lines.

In a systematic work by Doi and Edwards,[8] the model is further developed into a constitutive equation for the stress tensor. The highlights of the development are as follows.

Each chain is assumed to be enclosed in a "tube," which is taken to represent the topological hindrance of the surrounding chains (cf Figure 6.6). The tube radius is a parameter of the model and is indicated by the symbol a. The centerline of the tube is called the *primitive chain*. Under equilibrium conditions, the primitive chain can itself be considered as a

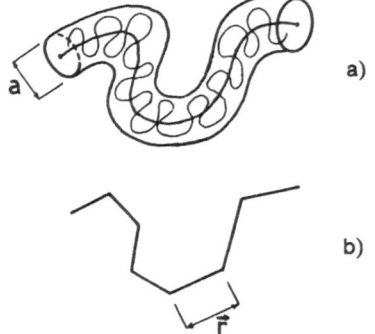

a)

b)

Figure 6.6. The "tube" model. (a) The polymeric chain together with its tube of topological constraints. The tube center-line is the primitive chain. (b) The primitive chain in discretized form.

random coil made up of segments of length a. The number N_0 of these segments is related to the total number of monomers in the chain n_t by

$$N_0 a^2 = n_t b^2 \qquad (6.34)$$

Still at equilibrium, each segment of the primitive chain contains a number n_0 of monomers given by

$$n_0 = \frac{n_t}{N_0} = \frac{a^2}{b^2} \qquad (6.35)$$

Throughout the analysis, the stress tensor is calculated by considering the entropic forces in the segments of the primitive chain as given by the Gaussian assumption, i.e., by using equation (6.3) with $\mathbf{r}$ the end-to-end vector of the primitive chain segment and n the corresponding number of monomers. By calling c the polymer concentration, the stress tensor is calculated through the average:

$$\underset{\sim}{\sigma} = \frac{3kTcN}{b^2} \left\langle \frac{\mathbf{rr}}{n} \right\rangle \qquad (6.36)$$

As discussed below, n and N do not necessarily coincide with the equilibrium values n_0 and N_0.

The other general assumption is that the segments of the primitive chains deform affinely with the continuum. If $\underset{\sim}{E}$ is the deformation gradient, which maps the equilibrium configuration into the deformed one, the "affine" assumption implies

$$\mathbf{r} = a\underset{\sim}{E} \cdot \mathbf{u} \qquad (6.37)$$

where $\mathbf{u}$ is a unit vector and the average in equation (6.36) is calculated by assuming that vectors $\mathbf{u}$ are randomly oriented.

First consider the case of a stress relaxation experiment, where an impulsive deformation is imposed at $t = 0$ to the polymeric liquid previously unperturbed. Just after the deformation has been applied, n and N keep their equilibrium values n_0 and N_0. The corresponding stress is readily obtained from equations (6.36) and (6.37) and is identical to that of an ideal rubber network made up of cN_0 active chains:

$$\underset{\sim}{\sigma} = 3kTcN_0 \langle (\underset{\sim}{E} \cdot \mathbf{u})(\underset{\sim}{E} \cdot \mathbf{u}) \rangle \qquad (6.38)$$

The subsequent stress relaxation results from the combination of two processes, which have a different time scale and thus essentially occur one after the other.

The first relaxation process, i.e., the faster one, is a redistribution of monomers along the primitive chain and takes place while the primitive chain itself remains fixed in the deformed conformation. The second relaxation process, much slower than the first, brings the primitive chain back to equilibrium through a reptation of the whole chain out of the deformed constraints.

An important result by Doi and Edwards, relegated by them to an Appendix,[8] gives the equilibrium line-density of monomers for a chain constrained in a tube. If l is the tube length and a is the tube radius, within a numerical constant this result is written as

$$\frac{n}{l} = \frac{a}{b^2} \tag{6.39}$$

[One may notice, incidentally, that equations (6.34) and (6.35) are consistent with equation (6.39).]

As a consequence of the deformation, the monomer density along the primitive chain is altered. In fact it is decreased on the average, since the segments of the primitive chain elongate on the average for all deformations that preserve the volume. The first relaxation process takes care of this effect. It corresponds to the monomer density going back to the equilibrium value given by equation (6.39).

During this relaxation, some of the segments of the primitive chain, those close to the chain ends, vanish. At the end of the first relaxation process, the situation is as follows. Each segment of the primitive chain that has not vanished contains a number of monomers n, which is proportional to the end-to-end segment length in accordance with equation (6.39):

$$n = \frac{a}{b^2} r \tag{6.40}$$

The number of primitive chain segments has correspondingly become

$$N \simeq \frac{n_t}{\langle n \rangle} = \frac{n_t b^2}{a} \frac{1}{\langle r \rangle} = N_0 \frac{a}{\langle r \rangle} \tag{6.41}$$

Substituting equations (6.40) and (6.41) into equation (6.36) gives the stress as

$$\underset{\sim}{\sigma} = 3kTcN_0 \frac{1}{\langle r \rangle} \left\langle \frac{\mathbf{rr}}{r} \right\rangle = 3kTcN_0 \frac{1}{\langle |\underset{\sim}{E} \cdot \mathbf{u}| \rangle} \left\langle \frac{(\underset{\sim}{E} \cdot \mathbf{u})(\underset{\sim}{E} \cdot \mathbf{u})}{|\underset{\sim}{E} \cdot \mathbf{u}|} \right\rangle \equiv G_0 \underset{\sim}{Q}(\underset{\sim}{E}) \tag{6.42}$$

where G_0 is the constant factor $3kTcN_0$ and $\underset{\sim}{Q}(\underset{\sim}{E})$ describes the dependence on deformation. Of course, the stress as given by equation (6.42) is smaller in magnitude than that given by equation (6.38) though individual components may be larger. The change of the stress in time follows a Rouse-like behavior with a largest relaxation time proportional to M^2.

The second relaxation process corresponds to a longitudinal diffusion of the chain along the tube. The chain can move in either direction but, though moving back and forth, it will eventually disengage itself from the deformed tube. During this process the deformed tube can be considered as fixed. Indeed, though the tube "wall" is made up of other chains, which are themselves diffusing, it can be shown that its relaxation time — resulting from the cooperative motion of many chains — would be much larger than that of a single chain.[12] The deformed tube progressively vanishes, however, starting from the ends, which are alternatively abandoned by the chain moving back and forth. The deformed tube is progressively replaced by a new tube, which has an equilibrium random conformation.

At any instant of time, those parts of the chain that belong to a new equilibrium tube do not contribute to the deviatoric part of the stress tensor. Those entrapped in the surviving part of the deformed tube contribute according to equation (6.42). Thus, during the second relaxation process, the stress tensor is given by

$$\underset{\sim}{\sigma} = G_0 \underset{\sim}{Q}(\underset{\sim}{E})\mu(t) \tag{6.43}$$

where $\mu(t)$ is the average fraction of deformed tube that survives at time t.

By solving the diffusion problem, $\mu(t)$ is obtained as

$$\mu(t) = \sum_{p \text{ odd}} \frac{8}{p^2 \pi^2} \exp\left(-p^2 \frac{t}{T_d}\right) \tag{6.44}$$

where

$$T_d = \frac{L^2}{\pi^2 D} \tag{6.45}$$

with L $(L = aN_0)$ the equilibrium length of the primitive chain and D the longitudinal diffusion coefficient of the chain.

Since L is proportional to M, and D to M^{-1}, the relaxation time T_d is proportional to M^3. If M is large, the first and second relaxation processes are well separated in time and the first one may be assumed instantaneous at least as long as one is not interested in what occurs at short times but rather at values of t of order T_d. Further, notice that equation (6.43) gives the relaxing stress as a product of two terms, which separately depend on deformation and time. This is analogous to what is found in the Rouse–Zimm theory and in the original Lodge rubberlike model. However, the tensor $\underset{\sim}{C}^{-1}$, which emerges from either theories, is here replaced by a new deformation tensor $\underset{\sim}{Q}(\underset{\sim}{E})$.

Further development of the theory from the relatively simple case of a stress relaxation to that of a general deformation history is also achieved by Doi and Edwards.[8] They use complex and elegant mathematical machinery, which is not repeated here. Rather, a simple approach leading to the same result is considered.

We take the time–deformation separation expressed by equation (6.43) as the basis of a nonlinear superposition principle analogous to that discussed in the previous section. For an arbitrary deformation history

$E_t(t')$, we consider the time-dependent tensor $Q_t(t')$, which depends on $\tilde{E}_t(t')$ just as Q depends on E in equation (6.42). The superposition principle is then written as

$$\sigma = G_0 \int_{-\infty}^{t} \dot{Q}_t(t')\mu(t-t')dt' \tag{6.46}$$

where $\dot{Q}$ is the derivative of Q with respect to t'.

The meaning of equation (6.46) is readily explained: $\dot{Q}dt'$ is the increment of Q that takes place in the time interval t', $t' + dt'$ and $G_0\dot{Q}dt'$ is the corresponding increment in stress; $G_0\dot{Q}\mu(t-t')dt'$ is what survives of that stress increment at the present time t according to equation (6.43). An integration by parts then gives

$$\sigma = G_0 \int_{-\infty}^{t} Q_t(t')\mu'(t-t')dt' \tag{6.47}$$

where

$$\mu'(t) = -\frac{d}{dt}\mu(t) = \frac{8}{\pi^2 T_d} \sum_{p\,odd} \exp\left(-p^2\frac{t}{T_d}\right) \tag{6.48}$$

Equation (6.47), equivalent to that by Doi and Edwards, is the constitutive equation for the stress tensor. As compared to that obtained from both the Rouse–Zimm and Lodge theories, it contains the important difference represented by tensor Q in place of C^{-1}. Furthermore, it makes definite predictions about the relaxation spectrum.

Doi and Edwards[8] have shown that tensor Q provides for various rheological predictions, which are qualitatively in agreement with the experimental observations. Most relevant are the predictions for a shear flow where it is found that:

1. The viscosity is a decreasing function of the shear rate.
2. A nonzero second normal-stress difference, as well as a first one, exists and is of the correct sign and order of magnitude.
3. If the shear rate is large, the transient shear stress following start-up of the flow passes through a maximum (overshoot).

The predictions for elongational flows are perhaps a little less satisfactory. It is found that η_{el} never exceeds the Trouton value $3\eta_0$, where η_0 is the zero-shear viscosity. Although in polymer melts, unlike solutions, values of the elongational viscosity much larger than the Trouton values are never observed, values of η_{el} somewhat in excess of $3\eta_0$ have been reported. We shall reconsider this point a little later.

As for the predictions that depend on the spectrum, it is satisfactory that the steady-state compliance J_e^0 is found to be independent of molecular weight. Of course, there remains the disturbing result whereby the zero-shear viscosity η_0 is predicted to be proportional to M^3 instead of $M^{3.4}$ as observed experimentally.

In the above discussion, the assumption that the tube diameter remains constant throughout the deformation process is involved. It has been pointed out that since the tube represents the topological constraints that other chains exert on the given one, the assumption that the tube deforms affinely in the longitudinal direction [cf equation (6.37)] while it remains unchanged in the lateral ones [cf equation (6.40)] presents a problem.

Marrucci and de Cindio[13] performed stress relaxation experiments on molten PMMA (polymethylmethacrylate) at large extensional deformations and found an intermediate behavior between that predicted by tensor $\underset{\sim}{C}^{-1}$, which overestimates the deformation dependence, and that predicted by tensor $\underset{\sim}{Q}$, which underestimates it. Better agreement was found by assuming that the constraining tubes deform affinely both in the longitudinal and in the radial directions. This assumption leads to a tensor $\underset{\sim}{Q}'$ that, while maintaining the nice properties of tensor $\underset{\sim}{Q}$ in a shear flow, gives intermediate predictions between $\underset{\sim}{C}^{-1}$ and $\underset{\sim}{Q}$ in elongation.

Marrucci and Hermans[14] studied the problem of the tube deformation in the radial direction in a more systematic way. First of all, they extended the thermodynamic treatment of a chain constrained in a tube and found, beyond the equilibrium line-density of monomers given by equation (6.39), the expression for the *free energy* of a chain so constrained. From the free energy, one derives not only an entropic chain tension in the longitudinal direction, as already accounted for, but also a chain *pressure* on the tube wall. Thus, when a deformation is applied, the contraction of the tube cross section that would result by an affine deformation is opposed by the corresponding increase in the chain pressure. A balance of these counteracting effects is found at an appropriate value of the tube diameter, which depends on deformation as well as on a constitutive parameter of the polymeric liquid.

The final result is a constitutive equation like equation (6.47) where, however, tensor $\underset{\sim}{Q}$ is replaced by $\underset{\sim}{Q}^*$:

$$\underset{\sim}{Q}^*(\underset{\sim}{E}) = \frac{1}{\langle|\underset{\sim}{E}\cdot\mathbf{u}|\alpha\rangle}\left\langle\frac{(\underset{\sim}{E}\cdot\mathbf{u})(\underset{\sim}{E}\cdot\mathbf{u})}{|\underset{\sim}{E}\cdot\mathbf{u}|\alpha}\right\rangle \tag{6.49}$$

where α depends on $|\underset{\sim}{E}\cdot\mathbf{u}|$ through the equation

$$\frac{1}{\alpha^3} - 1 = B\left(1 - \frac{1}{\alpha}|\underset{\sim}{E}\cdot\mathbf{u}|^{-1/2}\right) \tag{6.50}$$

In equation (6.50), B is the constitutive parameter related in a specific way to the monomer dimensions, to the equilibrium tube diameter, and to the polymer concentration.

An interesting feature of tensor Q^* is that of predicting an elongational viscosity that, by increasing the stretching rate, first increases above the Trouton value and then drops again. This behavior has been experimentally observed. Furthermore, equation (6.49) may discriminate among the somewhat different behaviors of various polymeric liquids via the effect of the parameter B.

It is clear from the above that the approach of Doi and Edwards is important for describing the intricate behavior of concentrated polymeric systems, especially for what concerns the response at large deformations as encountered in practice.

6.6. Effects of Flow on Crystallization

The models described so far are used in the analysis of polymer processing, especially for what goes under the general name of "processability." This term refers to various difficulties that may be encountered in processing liquid polymers. The difficulties may arise because the liquid is much too viscous, or else too little viscous, or maybe too "elastic," the latter term usually referring to the normal stresses arising in a shear flow. Depending on processing conditions and properties of the polymer, flow instabilities may arise which disrupt the steady-state operation and cause irregular manufactured items to be produced. Many of these effects have been studied with the help of constitutive equations similar to those described in the previous sections.

A further possible use of the molecular models considered previously is that of helping to understand the variety of crystalline morphologies obtained in polymeric manufactured items, depending on processing conditions, and possibly to determine those conditions that would produce a desired result. This application is far less developed and the following brief outline should be interpreted mainly as a research proposal.

Let us start with some rather obvious remarks about the crystallization entropy of a polymeric substance. For a pure substance in general, the melting (and crystallization) temperature is given by

$$T_m = T_c = \frac{\Delta H}{\Delta S} \tag{6.51}$$

where ΔH and ΔS are the enthalpy and entropy changes that take place in the melting process. An actual crystallization process will occur at a temperature T lower than T_c and the crystallization rate grows by increasing the subcooling $T_c - T$ unless T becomes so low as to drastically reduce the molecular mobility.

For a flexible polymer, the greatest part of ΔS is made up of the conformational entropy of the liquid polymer that is lost in the crystal. For example, for polyethylene it is found experimentally that $\Delta H \cong 4100$ J/mol CH_2 and $T_m = 414$ °K, thus from equation (6.51), $\Delta S \cong 9.9$ J/°K mol CH_2. On the other hand, since in the amorphous phase each CH_2 has three conformational states (one trans + two gauche), one may roughly calculate the conformation entropy S_c as

$$S_c = R \ln 3 = 9.12 \text{ J/°K mol } CH_2 \qquad (6.52)$$

More sophisticated calculations give a smaller value, $S_c = 7.41$ J/°K mol CH_2, which is anyhow 75% of the crystallization entropy change.

These numbers give an idea of the dramatic effects that might result from a flow-induced molecular orientation in the liquid phase. If the conformational entropy is reduced by say 10%, the crystallization temperature is increased by as much as 40 °C and, at any given working temperature, the effective subcooling is also increased by the same amount.

At this point, we need to establish the change in conformational entropy brought about by a flow process. In this regard, we can borrow well-known concepts from the classical theory of rubber elasticity. Thus, for the dumbbell model, we may use equation (6.1), which gives the *a priori* probability of finding a given end-to-end vector **r**.

For a collection of c chains, which change their average end-to-end distance as a consequence of flow, we write the corresponding free-energy change as

$$\Delta A = -T\Delta S \cong \tfrac{3}{2}ckT\left(\frac{\langle r^2 \rangle}{nb^2} - 1\right) \qquad (6.53)$$

Equation (6.53) establishes the required link between the entropy change and the history of deformation and temperature (thermomechanical history) via $\langle r^2 \rangle$. The latter is readily obtained from the constitutive equation for the average dyad $\langle \mathbf{rr} \rangle$, namely equation (6.11) or equation (6.23).

Since $\langle \mathbf{rr} \rangle$ is proportional to $\underset{\sim}{\sigma}$ (cf equation (6.14)), it may be useful to establish a direct link between the entropy change and the stress tensor. To this end, we suitably redefine the stress tensor as given by equation (6.14) by adding to it an isotropic term, which is permissible in an incompressible liquid:

$$\underset{\sim}{\sigma} = cH\langle \mathbf{rr} \rangle - ckT\underset{\sim}{1} \qquad (6.54)$$

By recalling the definition of the spring constant H, equations (6.53) and (6.54) are then combined to give

$$\Delta A = \tfrac{1}{2}\operatorname{tr}\underset{\sim}{\sigma} \qquad (6.55)$$

where tr is the trace operator.

Equation (6.55) may not seem particularly useful, since $\operatorname{tr}\underset{\sim}{\sigma}$ cannot be measured in an incompressible liquid. However, one should not forget that

σ as defined by equation (6.54) obeys a constitutive equation, which allows one to make definite predictions.

For example, the tensile stress arising in elongational flow at high gradients is soon dominated by the $\langle x^2 \rangle$ component (cf the discussion in Section 6.3), which becomes much larger than $\langle y^2 \rangle = \langle z^2 \rangle$. By calling σ the tensile stress, equation (6.55) becomes in such a case

$$\Delta A = - T\Delta S \cong \tfrac{1}{2}\sigma \qquad (6.56)$$

where now σ is a measurable quantity. If we are dealing, e.g., with fiber spinning, σ is the tensile force in the spinning line divided by the area of the filament cross section.

Similarly, equation (6.55) tells us that, in a shear flow, what determines the entropy change is not the tangential shear stress but rather the normal stresses. It may be recalled that these are quadratic in the shear rate (cf Section 6.3) and thus grow rapidly at large $\dot{\gamma}$.

Equation (6.56), though approximate in various respects, allows one to estimate the order of magnitude of a tensile stress, which induces significant effects on the crystallization temperature. From the discussion above, we take 1 J/°K per mole of monomer as a quite significant change in conformational entropy. By assuming that equation (6.56) applies to a polymer melt ($T = 500$ °K) with a density of 10^3 kg/m^3 made up of monomers of molecular mass $M = 100$, we obtain $\sigma = 10^7$ Pa.

Equation (6.55) was obtained earlier by Marrucci.[15] Subsequently Sarti and Marrucci[16] derived a somewhat improved expression and extended the free-energy calculation to the Rouse–Zimm model. However, the dumbbell and Rouse–Zimm models strictly belong to dilute solutions. The case of concentrated entangled systems has not been worked out at the time of writing. The free-energy expression derived by Marrucci and Hermans[14] should provide a basis for further work.

Various predictions of dilute-solution molecular theories find confirmation in experiments of flow-induced polymer cystallization from dilute solutions. From the studies of Keller and co-workers,[17] Pennings et al.,[18] McHugh,[19] etc., the following general results are obtained:

1. For subcoolings insufficient to induce crystallization in the quiescent solution, a rapid nucleation of crystals takes place if a flow is established. However, the flow must be elongational in character rather than shear, at least at some points of the flow field. The greater effectiveness of an elongational flow with respect to a shear flow in extending the macromolecules in dilute solutions has been discussed in connection with the dumbbell model. For crystal growth (secondary nucleations), a shear flow may be sufficient, provided it is strong enough.

2. The crystalline morphology is fibrous in its general character. This

is obviously related to the uncoiling of the chains brought about by the flow. The details of the morphology are quite complex, however, and include lamellar structures. The structural details are variously influenced by the particular conditions of the experiment.

3. Very large values of the molecular weight are required to obtain significant effects. This is related to the influence of M on relaxation time and consequently on the effectiveness of the flow. We recall that $\tau\dot{\Gamma}$ or else $\tau\dot{\gamma}$ must be large in order that the molecules may orient.

4. Intense flows must be maintained for some time to observe effects. Typical is the growth of crystals in "stagnation" regions of the flow field, where gradients are large though the translation velocity is almost nil. This is related to the fact that extension of a chain not only requires a strong gradient but also takes some time, which is longer the longer the chain. The transient part of equations like equation (6.16) describes this effect.

In molten polymers, flow-induced crystallization shows similar aspects as well as different ones. Like in solutions, the crystallites are fibrous rather than lamellar. Yet as shown, e.g., by Yeh[20], they are fairly small, of order 100 Å, not only in the transversal direction but also in the longitudinal one, i.e., along the c-axis of the elementary cell. Shear as well as elongational flows are similarly effective. The overall deformation required to induce crystallization is relatively small, provided the deformation rate is sufficiently large. The results of Lagasse and Maxwell[21] on linear polyethylene indicate a value of 5 to 7 shear units. This suggests that whole molecules need not be stretched but only relatively small segments of them. These are reminiscent of the primitive chain segments in the model by Doi and Edwards.

Finally, it is noted that in many cases the crystallization process is carried along in subsequent steps. Typical is the fabrication of a biaxially oriented sheet of PET (polyethylenterephthalate), which is first produced as an amorphous sheet by rapidly quenching the melt after the extrusion. The amorphous sheet is then heated just above the glass transition temperature and stretched in one direction. This produces some crystallinity. Subsequently, it is stretched at a right angle and further crystallinity is added. The final product is then obtained by a constrained annealing, i.e., by maintaining the sheet at a given temperature for some time without allowing it to retract.

For the greatest part of this process two phases are present: the crystalline and the amorphous. The system may be considered as a very viscous liquid into which a solid phase is dispersed. The rheological modeling of such a heterogeneous system, where a phase change occurs progressively, does not appear to be an easy problem to solve.

6.7. Recent Developments (Note Added in Proof)

From the time when the contents of this chapter were used for a short course (Spring 1980), further progress in the understanding of the complex behavior of concentrated polymeric liquids has been made, based mainly on the "tube" model of Doi and Edwards discussed in Section 6.5. By way of example, we note briefly the following contributions:

The discrepancy between the experimental and theoretical dependencies of the viscosity on molecular weight (3.4-vs-3 power law) has been attributed by Doi to longitudinal fluctuations of the chain in the tube. Approximate calculations of the effect of these fluctuations substantiate this interpretation [M. Doi, *J. Polym. Sci.: Polym. Phys. Ed.* **21**, 667 (1983); *J. Polym. Sci.: Polym. Lett. Ed.* **19**, 265 (1981)].

Relaxation due to longitudinal fluctuations in the constraining tube is the dominant mode in branched, star-shaped polymers. A quantitative model has been developed [M. Doi and N. Y. Kuzuu, *J. Polym. Sci.: Polym. Lett. Ed.* **18**, 775 (1980)] which compares favorably with several data by Graessley and coworkers on carefully prepared star-shaped polymers [W. W. Graessley and J. Roovers, *Macromolecules* **12**, 959 (1979); V. R. Raju, E. V. Menezes, G. Marin, W. W. Graessley and L. J. Fetters, *Macromolecules* **14**, 1668 (1981)].

Expressions for the change in free energy of concentrated polymers due to deformation and flow have been developed [G. Marrucci and N. Grizzuti, *J. Rheol.* **27**, 433 (1983)]. They were used to interpret certain anomalous data of nonlinear stress relaxation in terms of a deformation instability.

Finally, it should be noted that there is a growing interest in rigid and semirigid polymers that may exist in the liquid crystalline state. Many experimental and theoretical works have been contributed recently in this field. The influence of flow on molecular orientations appears to be particularly important for these polymers.

References and Bibliography for Part I

Some of the Underlying Bibliography (Especially Chapters 1–5)

General Texts on Polymer Science

P. J. Flory, *Principles of Polymer Chemistry*, Cornell University Press, Ithaca (1953).

F. Billmeyer, *Textbook of Polymer Science*, Interscience, New York, London (1962).

M. Gordon, *High Polymers; Structure and Physical Properties*, Iliffe Books, London (1963).

T. Alfrey and E. F. Gurnee, *Organic Polymers*, Prentice-Hall Inc., Englewood Cliffs, NJ (1967).

L. R. G. Treloar, *Introduction to Polymer Science*, The Wykeham Science Series, London (1970).

On Crystallography

L. E. Alexander, X-ray Diffraction Methods in Polymer Science, Wiley–Interscience, New York (1969).

H. Tadokoro, *Structure of Crystalline Polymers*, John Wiley, New York (1979).

On Crystallization, Morphology, and Solid State

P. H. Geil, *Polymer Single Crystals*, Interscience Publ., New York, London (1963).

L. Mandelkern, *Crystallization of Polymers*, McGraw-Hill, New York (1964).

A. Keller, *Polymer Crystals, Rep. Progr. Phys.* **31** (Part 2), 623–704 (1968).

A. Keller, Solution grown polymer crystals, *Kolloid-Z. Z. Polym.* **231**, 386 (1969).

A. Keller, *Morphology of Lamellar Polymer Crystals* in: *International Review of Science; Macromolecular Science. Physical Chemistry*, Series 1, v. 8, p. 105–159, MTP, Butterworths, London (1972).

B. Wunderlich, *Macromolecular Physics*, Volumes 1, 2, 3, Academic Press, New York, London (1973, 1976, 1980).

J. D. Hoffman, G. T. Davis, and J. I. Lauritzen, in: *Treatise on Solid State Chemistry*, (B. Hannay, ed.), Vol. 3, Ch. 7, Plenum Press, New York (1976).

F. Khoury and E. Passaglia, in: *Treatise on Solid State Chemistry* (B. Hannay, ed.), Vol. 3, Ch. 6, Plenum Press, New York (1976).

J. H. Magill, Morphogenesis of solid polymer microstructures, in: *Treatise on Materials Science and Technology*, Vol. 10, Part A, pp. 3–341, Academic Press, New York, London (1977).

Organization of Macromolecules in the Condensed Phase, *Faraday Discussion*, No. 68 (1979).

I. Hall, ed., *Structure of Crystalline Polymers*, Applied Science Publ., London (1984).

References to Sources of Illustrations in Chapters 1–5

1. G. Natta and P. Corradini, *Nuovo Cimento, Suppl.* **XV**, serie X, No. 1, p. 6 (1960).
2. A. V. Tobolsky, *Property and Structure of Polymers*, Wiley, New York (1980).
3. C. W. Bunn, in: *Fibres from Synthetic Polymers* (R. Hill, ed.), Elsevier, Amsterdam (1953).
4. P. H. Geil, *Polymer Single Crystals*, Interscience, New York (1963).
5. G. Natta and P. Corradini, *Nuovo Cimento, Suppl.* **XV**, serie X, No. 1 (1960).
6. A. Keller, Growth and perfection of crystals, *Proc. Int. Conf. on Crystal Growth, Cooperstown*, Wiley, New York.
7. T. Kawai and A. Keller, *Phil. Mag.* **8**, 1203 (1963).
8. J. D. Hoffman, G. T. Davies, and J. I. Lauritzen, in: *Treatise On Solid State Chemistry* (N. B. Hannay, ed.), Vol. 3, Chap. 7, Plenum Press, New York (1976).
9. D. C. Bassett, F. C. Frank, and A. Keller, *Phil. Mag.* **8**, 1739 (1963).
10. A. Keller and D. C. Bassett, *J. Roy. Micr. Soc.* **49**, 243 (1960).
11. P. H. Lindenmeyer, *J. Polym. Sci. C* **1**, 1 (1963).
12. A. Keller (with contribution by S. Mitsuhashi), *Kolloid-Z. Z. Polym.* **219**, 118 (1967).
13. S. Mitsuhashi and A. Keller, *Polymer* **2**, 109 (1961).
14. M. M. Winram, D. T. Grubb, and A. Keller, *J. Mater. Sci.* **13**, 791 (1978).
15. E. W. Fischer, *Kolloid-Z. Z. Polym.* **231**, 458 (1969).
16. B. Wunderlich, R. B. Prime, and L. Melillo, J. Polymer Sci. A-2, **7**, 209 (1969).
17. *Faraday Discussions*, No. 68, p. 145 (1979).
18. M. J. Hill, P. J. Barham, and A. Keller, *Colloid Polym. Sci.* **258**, 1023 (1980).
19. A. Keller and P. J. Barham, *Plast. Rubber Int.* **6**, 19 (1981).
20. A. J. Pennings, *J. Polym. Sci., Polym. Symp.* **59**, 55 (1977).
21. A. Keller and M. J. Machin, in: *Deformation and Flow* (R. L. Wetton and R. W. Whorlow eds.), Macmillan, London (1968).
22. P. J. Barham, L. Otun, A. Keller and P. Holmes, *J. Material Sci.* in press.
23. A. Keller in: *Structure-Property, Relationships of Polymeric Solids*, ed., A. Hiltner, Plenum Press (1983).

Specialist References for Chapter 6

1. W. Kuhn, *Kolloid Z.* **76**, 258 (1936); W. Kuhn and F. Grun, *Kolloid Z.* **101**, 248 (1942).
2. R. B. Bird, R. C. Armstrong, O. Hassager, and C. F. Curtiss, *Dynamics of Polymeric Liquids*, J. Wiley and Sons, New York (1977).
3. G. Astarita and G. Marrucci, *Principles of Non-Newtonian Fluid Mechanics*, McGraw-Hill, New York (1974).
4. A. S. Lodge, *Elastic Liquids*, Academic Press, New York (1964).
5. P. E. Rouse, *J. Chem. Phys.* **21**, 1272 (1953).
6. B. H. Zimm, *J. Chem. Phys.* **24**, 269 (1956).
7. J. D. Ferry, *Viscoelastic Properties of Polymers*, J. Wiley and Sons, New York (1970).
8. M. Doi and S. F. Edwards, *J. Chem. Soc., Faraday Trans. 2*, **74**, 1789 (1978); **74**, 1802 (1978); **74**, 1818 (1978); **75**, 38 (1979).
9. S. F. Edwards, *Proc. Phys. Soc.* **92**, 9 (1967).
10. P. G. de Gennes, *J. Chem. Phys.* **55**, 572 (1971).
11. M. Doi, *Chem. Phys. Lett.* **26**, 269 (1974).
12. J. Klein, *Macromolecules* **11**, 852 (1978).
13. G. Marrucci and B. de Cindio, *Rheol. Acta* **19**, 68 (1980).
14. G. Marrucci and J. J. Hermans, *Macromolecules* **13**, 380 (1980).
15. G. Marrucci, *Trans. Soc. Rheol.* **16**, 321 (1972).
16. G. C. Sarti and G. Marrucci, *Chem. Eng. Sci.* **28**, 1053 (1973).
17. D. P. Pope and A. Keller, *Coll. Polym. Sci.* **255**, 633 (1977), and references cited therein.
18. A. Zwijnenburg and A. J. Pennings, *Coll. Polym. Sci.* **254**, 868 (1976), and references cited therein.
19. A. J. McHugh, P. Vaughn, and E. Ejike, *Polym. Eng. Sci.* **18**, 443 (1978).
20. G. S. Y. Yeh, *Polym. Eng. Sci.* **16**, 138 (1976).
21. R. R. Lagasse and B. Maxwell, *Polym. Eng. Sci.* **16**, 189 (1976).

II

Liquid Crystals

Structural Classification of Thermotropic Liquid Crystals

S. Chandrasekhar

7.1. Introduction

Most crystals transform directly into the liquid phase, so that the long-range translational order as well as the long-range orientational order of the molecules are destroyed simultaneously. However, if the constituent molecules have pronounced anisotropy of shape, the disappearance in one, two, or three dimensions of the long-range translational periodicity in the crystal may precede the collapse of the long-range orientational order; the intermediate phases are then referred to as *liquid crystals*.

Transitions to the intermediate (or meso-) phases may be brought about in two different ways, one by purely thermal processes and the other by the influence of solvents. Mesophases obtained by the first method are called *thermotropic liquid crystals* and are exhibited by a large number of pure compounds having molecular weights in the range of about 250–500 (or by mixtures of such compounds). Liquid crystallinity induced by the second method is called *lyotropic mesomorphism*; the mesophase then necessarily consists of two or more components, one of which is often an amphiphilic molecule. Liquid crystals are also formed by certain macromolecules (e.g., long-chain polymers), usually in solution but sometimes even in the pure state. We shall not however be concerned here with lyotropic and macromolecular systems but shall confine attention to the

S. Chandrasekhar · Raman Research Institute, Bangalore, India.

structure and physical properties of thermotropic liquid crystals composed of relatively simple, pure compounds. The emphasis will be on providing a broad perspective of the remarkable properties of these materials rather than a detailed and rigorous mathematical treatment of selected topics. For further study the interested reader may consult elsewhere.[1–6]

7.2. Rod-like Molecules

Almost all pure compounds exhibiting thermotropic mesomorphism have one distinctive feature in common, namely the rod-like shape of the molecule. Mesophases of rod-like molecules are classified broadly into three types: nematic, cholesteric, and smectic.

The nematic liquid crystal has a high degree of long-range orientational order but no long-range translational order (Figure 7.1a). The preferred axis of orientation, referred to as the director and represented by an apolar unit vector $\mathbf{n}$, usually varies from point to point in the medium, but a monodomain sample is optically uniaxial and strongly birefringent. Several other properties, such as the diamagnetic susceptibility and dielectric constant, are also anisotropic. The anisotropy is a function of the degree of orientational order of the molecules, which decreases appreciably with rise in temperature and drops abruptly to zero at the nematic-isotropic point in a weak first-order transition.

The cholesteric is also a nematic type of liquid crystal except that it is composed of optically active molecules. As a result the structure acquires a spontaneous twist about an axis perpendicular to the director (Figure 7.1b). Optically inactive molecules or racemic mixtures result in a helix of infinite pitch, which corresponds to the true nematic.

Smectic liquid crystals have stratified structures, but a variety of molecular arrangements are found within each stratification. In smectic A (S_A) the mean orientation of the molecular long axes is normal to the layers, and the molecular centers in each layer are irregularly spaced in a "liquid-like" fashion (Figure 7.1c). Smectic C (S_C) is a tilted form of smectic A, i.e., the molecules are inclined with respect to the layer normal (Figure 7.1d). As many as a dozen other smectic modifications have been found,[7] but the detailed molecular arrangements are not known with certainty for all of them. The broad structural features of the phases so far identified are summarized in Table 7.1. Recent high-resolution X-ray studies have established that several of these phases possess three-dimensional (3D) positional order, though with extremely weak interlayer forces. These highly ordered smectics offer convenient systems for verifying important ideas on 2D melting and on possible mechanisms of unregistering of layers in 3D systems.

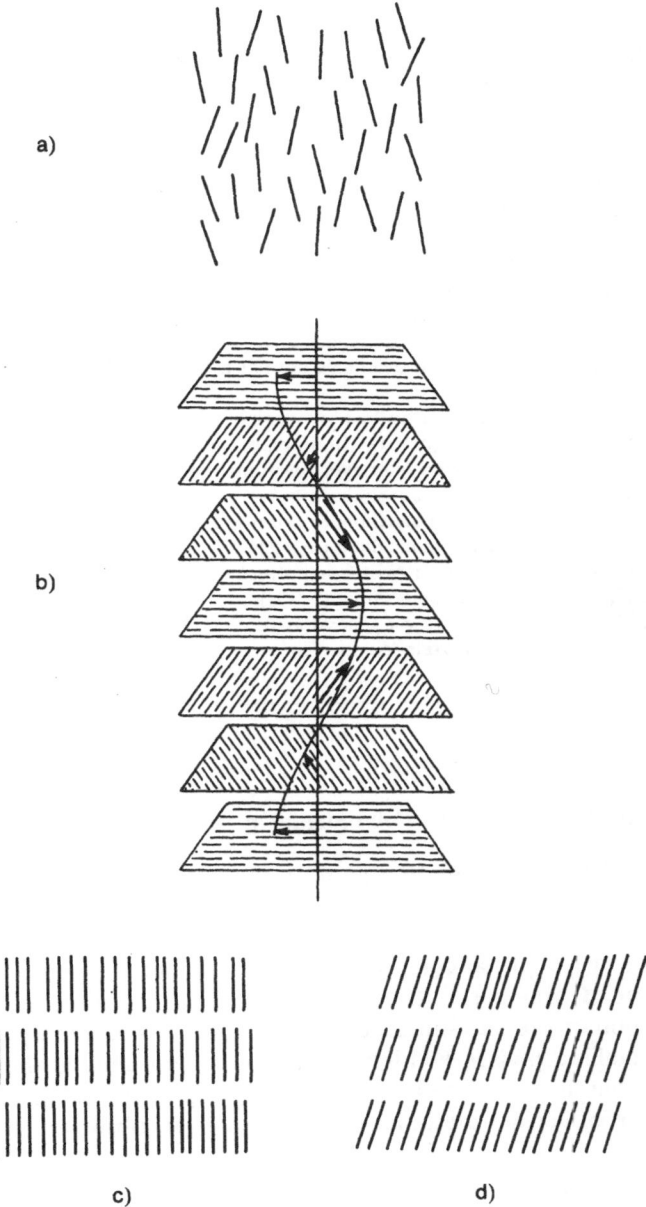

Figure 7.1. Schematic representation of molecular arrangements in (a) nematic, (b) cholesteric, (c) smectic A, and (d) smectic C phases.

Table 7.1. Structural Classification of Thermotropic Liquid Crystals of Rod-like Molecules

Nematic (N)	Long-range orientational order, but no long-range translational order (Figure 7.1a)
Cholesteric (Ch)	Chiral nematic (Figure 7.1b)
Smectic A (S_A)	Liquid-like layers with upright molecules (Figure 7.1c)
B (S_B)	Two distinct types of S_B have been identified: (i) 3D crystal, hexagonal lattice, upright molecules (ii) stack of interacting "hexatic" layers with in-plane short-range positional correlation and long-range 3D six fold "bond-orientational" order
C (S_C)	Tilted form of S_A (Figure 7.1d)
C* (S_{C^*})	Chiral S_C with twist axis normal to the layers (Figure 10.5)
D (S_D)	Cubic
E (S_E)	3D crystal, orthorhombic, upright molecules
F (S_F)	Monoclinic ($a > b$) with in-plane short-range positional correlation and weak or no interlayer positional correlation (tilted hexatic?)
G (S_G)	3D crystal, monoclinic ($a > b$)
G' ($S_{G'}$)	3D crystal, monoclinic ($b > a$)
H (S_H)	3D crystal, monoclinic ($a > b$)
H*($S_{H'}$)	Chiral S_H with twist axis normal to the layers
H' ($S_{H'}$)	3D crystal, monoclinic ($b > a$)
I (S_I)	Monoclinic ($b > a$), possibly hexatic with slightly greater in-plane positional correlation than S_F
I*($S_{I'}$)	Chiral S_I

7.2.1. Effect of Pressure on Polymorphism

Examples of typical mesogenic compounds and the sequence of transitions occurring in them may be found in any of the standard references on the subject and need not be repeated here. The polymorphism of some of these compounds can be modified significantly by the application of pressure.[8] The most spectacular of these effects is the phenomenon of pressure-induced mesomorphism, i.e., the appearance of liquid-crystalline phases at high pressures in compounds that are non-mesomorphic at atmospheric pressure. The opposite effect, namely the suppression of certain mesophases, has also been observed.

7.2.2. The Reentrant Phenomenon

As a rule the smectic phases are more ordered than the nematic phase and when a compound shows both types of mesophase, the nematic occurs

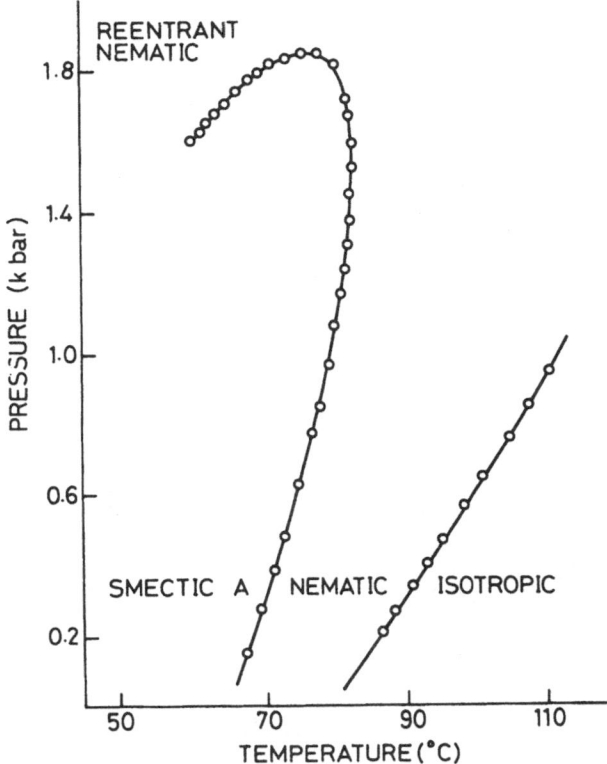

Figure 7.2. Experimental phase diagram of 8 OCB on cooling from the isotropic phase (adapted from Cladis *et al.*[9]).

at a higher temperature. Exceptions to this rule have been discovered in certain compounds with the highly polar cyano (i.e., C≡N) end group. The first observations were in binary mixtures: the sequence of transitions on cooling was

$$\text{Isotropic} \rightarrow N \rightarrow S_A \rightarrow N$$

Later, a similar effect was found in a pure compound at elevated pressures[9] (Figure 7.2). The second (lower temperature) nematic phase is referred to as the reentrant nematic phase. Reentrant phases have been found in pure compounds even at atmospheric pressure. Two examples are given below along with the sequence of transitions on cooling:

a)

b)

Figure 7.3. Mesophases of disk-like molecules. Schematic representation of (a) columnar and (b) nematic-like structures.

4-Cyanophenyl-3-methyl-4(4-n-undecylbenzoyloxy)benzoate(11 CPMBB):

$$I \xrightarrow{152.5\,°C} N \xrightarrow{127\,°C} S_A \xrightarrow{78.5\,°C} N \xrightarrow{65\,°C} \text{Solid}$$

Octyloxy-4′-benzoyloxy-4-cyanostilbenzene (T8):

$$I \xrightarrow{283\,°C} N \xrightarrow{247\,°C} S_A \xrightarrow{138\,°C} N \xrightarrow{96\,°C} S_A \xrightarrow{94.5\,°C} \text{Solid}$$

It may be mentioned that from the phenomenological viewpoint the appearance of reentrant phases is of considerable interest in the physics of condensed matter in general. The case of T8 given above with transitions $N–S_A–N–S_A$ is the first example of a system with two reentrant phases.

7.3. Disk-like Molecules

The first observations of liquid-crystalline behavior in pure compounds of disk-like molecules were reported in 1977.[10] These compounds form an entirely new class of thermotropic liquid crystal, the structures of which are illustrated in Figure 7.3a. F. C. Frank has suggested that this type of mesophase may be called "canonic" ($\chi\alpha\nu\omega\nu$ = rod); W. Helfrich was the first to use the term "columnar"; the word "discotic" was proposed by J. Billard to describe the disk-like molecules as well as the mesophases formed by them. The simplest columnar phase is the one in which the disks are aperiodically stacked in columns, the different columns forming a hexagonal array. The structure therefore has translational periodicity in two dimensions and "liquid-like" disorder in the third. This basic structure has been confirmed in all the discotic systems studied so far but a number of variants have been found, such as a rectangular array of columns and a tilted columnar arrangement. The hexa-n-alkoxy or alkyl benzoates of triphenylene show an interesting type of polymorphism: a transition occurs from a columnar to a nematic-like phase (Figure 7.3b).

Mesophases of plate-like molecules are known to occur during the carbonization of graphitizable substances such as petroleum and coal-tar pitches. However these are rather complex systems consisting of large molecules having a range of molecular weights, typically around 2000, and will not be considered here.

To recapitulate, therefore, the columnar or canonic phase, the smectic phase (in its simplest form), and the nematic phase represent orientationally ordered states of matter in which the melting of the spatial periodicity has taken place in one, two, and three dimensions, respectively.

Nematic Liquid Crystals

S. Chandrasekhar

8.1. Elastic Properties

8.1.1. Basic Equations

The nematic liquid crystal has "curvature elasticity," i.e., elasticity associated with orientational deformations of the director. The three principal types of deformation are splay, twist, and bend as illustrated schematically in Figure 8.1.

Consider first the case of pure splay (Figure 8.1a). The elastic free energy per unit volume is $\frac{1}{2}k_{11}(d\phi/dx)^2$, where ϕ is the tilt of the director and k_{11} the splay elastic constant. Remembering that the director is assumed to be of unit magnitude, $\phi = n_x$ and the free energy density may be written as

$$F = \tfrac{1}{2}k_{11}\left(\frac{\partial n_x}{\partial x}\right)^2 \tag{8.1}$$

Similarly for pure twist (Figure 8.1b)

$$F = \tfrac{1}{2}k_{22}\left(\frac{\partial n_x}{\partial y}\right)^2 \tag{8.2}$$

and for pure bend (Figure 8.1c)

$$F = \tfrac{1}{2}k_{33}\left(\frac{\partial n_x}{\partial z}\right)^2 \tag{8.3}$$

S. Chandrasekhar · Raman Research Institute, Bangalore, India.

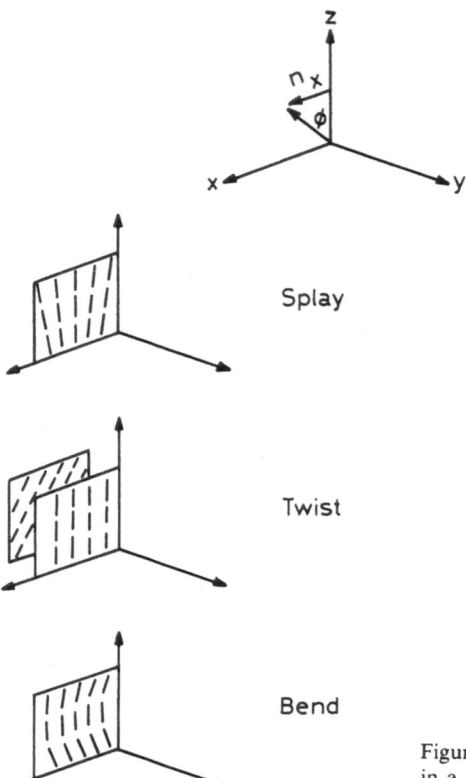

Figure 8.1. Three principal types of deformation in a nematic liquid crystal.

More generally, one may write

$$F = \tfrac{1}{2}k_{11}(\nabla \cdot \mathbf{n})^2 + \tfrac{1}{2}k_{22}(\mathbf{n} \cdot \nabla \times \mathbf{n})^2 + \tfrac{1}{2}k_{33}(\mathbf{n} \times \nabla \times \mathbf{n})^2 \qquad (8.4)$$

The elastic constants k_{11}, k_{22}, and k_{33} (often referred to as Frank constants) are usually of the order of 10^{-6}–10^{-7} dyn.

In cholesterics there is an additional term $k_2(\mathbf{n} \cdot \nabla \times \mathbf{n})$ in the free energy to allow for the intrinsic twist in the structure. The twist per unit length is then given by $q = k_2/k_{22}$.

8.1.2. Determination of the Elastic Constants: The Freedericksz Effect

The most direct, and probably the most accurate, method of measuring the elastic constants is by studying the distortions due to an external

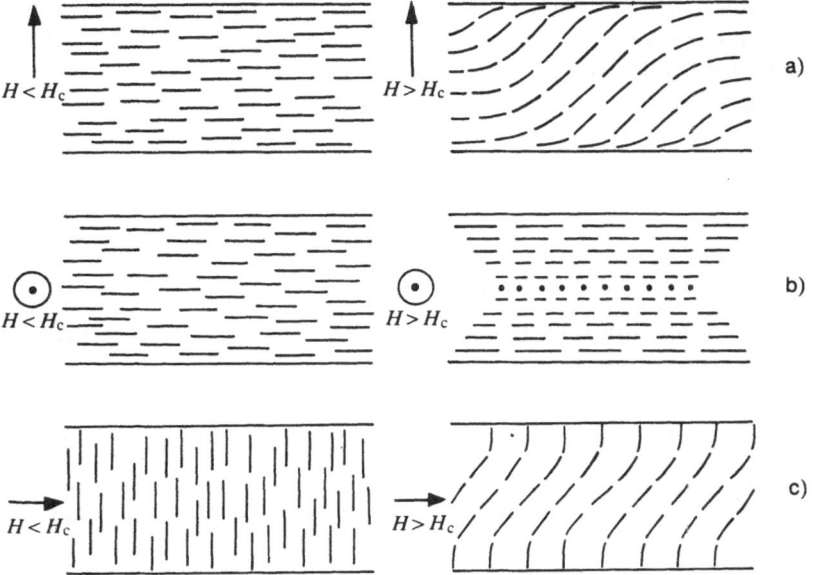

Figure 8.2. Freedericksz geometries for measuring the three elastic constants of a nematic.

magnetic field. The geometry has to be so chosen that the orienting effect of the field conflicts with the orientations imposed by the surfaces with which the liquid crystal is in contact. The three principal experimental geometries are illustrated in Figure 8.2. Above a certain critical field, a distortion sets in that can be detected optically. The threshold field is given by

$$H_c = \frac{\pi}{d} \left(\frac{k_{ii}}{\chi_a}\right)^{1/2} \tag{8.5}$$

where $\chi_a = \chi_{\parallel} - \chi_{-}$, $\chi_{\parallel}$ and χ_{-} being the principal diamagnetic susceptibilities per unit volume along and perpendicular to the director axis.

To explain the mechanism of the Freedericksz "transition," let us consider the twist geometry (Figure 8.2b). The free-energy density in the presence of the field is

$$F = \tfrac{1}{2}k_{22}\left(\frac{d\theta}{dz}\right)^2 - \tfrac{1}{2}\chi_a H^2 \sin^2 \theta \tag{8.6}$$

and its minimization yields

$$k_{22}\frac{d^2\theta}{dz^2} + \chi_a H^2 \sin \theta \cos \theta = 0 \tag{8.7}$$

For small deformations, we may put $\theta = \theta_m \cos qz$, neglecting higher harmonics, where $q = \pi/d$, d being the thickness of the sample. We are assuming that the director is firmly anchored to the wall by prior treatment of the glass surfaces. We then have

$$(\chi_a H^2 - k_{22} q^2)\theta = 0$$

so that

$$H_c = \frac{\pi}{d}\left(\frac{k_{22}}{\chi_a}\right)^{1/2}$$

If we take powers of θ up to the fourth in the expression for F, the energy per unit area is

$$E = \int_{-d/2}^{d/2} F\,dz = \frac{\frac{1}{2}\pi^2 k_{22}}{d}\left[\left(1 - \frac{H^2}{H_c^2}\right)\frac{\theta_m^2}{2} + \frac{H^2}{H_c^2}\frac{\theta_m^4}{8}\right]$$

A plot of E vs θ_m for different values of H/H_c is reminiscent of the free-energy curves depicting a second-order phase transition. For $H < H_c$, E is a minimum at $\theta_m = 0$, corresponding to the undistorted state. For $H > H_c$ the distorted state is the stable one. The threshold field for the deformation to occur is $H = H_c$, which at once gives the appropriate elastic constant of the material.

It is obvious that curvature elasticity is a consequence of the orientational order of the molecules. The orientational order decreases with rise of temperature and hence the elastic constants also decrease very rapidly. It turns out that in ordinary nematics $k_{ii} \propto s^2$ approximately, where s is the orientational order parameter. Also $k_{33} > k_{11} > k_{22}$, as expected from simple geometric considerations.

8.1.3. Orientational Fluctuations and Light Scattering

Nematics scatter light very strongly, so much so that even a relatively thin monodomain film appears quite turbid. The scattering arises from the orientational fluctuations of the director. The rigorous theory of the phenomenon is rather elaborate, but we shall present here some elementary arguments that will illustrate the important consequences of the theory.

In ordinary liquids the scattering arises from density fluctuations, which give rise to fluctuations in the refractive index. The scattering cross-section per unit solid angle is given by

$$\frac{d\sigma'}{d\Omega} = \frac{\pi^2}{\lambda^4}\left\langle\left(\rho\frac{\partial\varepsilon}{\partial\rho}\right)^2\right\rangle \beta k_B T \tag{8.8}$$

(ignoring the polarization factor), where ρ is the density, ε the dielectric constant at optical frequencies, and β the isothermal compressibility.

In the case of nematics, the angular fluctuations of the director (or the local dielectric ellipsoid) are responsible for the scattering. The variation of the dielectric constant with angle is given by

$$\frac{\cos^2 \theta}{\varepsilon_{\parallel}} + \frac{\sin^2 \theta}{\varepsilon_{\perp}} = \frac{1}{\varepsilon}$$

where $\varepsilon_{\parallel}$ and $\varepsilon_{\perp}$ are the principal values along and perpendicular to the director axis. Therefore

$$\frac{\sin 2\theta \varepsilon_a}{\varepsilon^2} \delta\theta \simeq \frac{-\delta\varepsilon}{\varepsilon^2}$$

where ε_a is the dielectric anisotropy $\varepsilon_{\parallel} - \varepsilon_{\perp}$. Now $\delta\theta = \delta n$, so that as an order-of-magnitude estimate, we may put

$$\varepsilon_a^2 \langle \delta n^2 \rangle \sim \langle \delta\varepsilon^2 \rangle$$

This expression gives the director fluctuations in terms of the dielectric-constant fluctuations. Writing the director fluctuation as a wave and using (8.1), the free energy of elastic deformation associated with this mode is $\frac{1}{2}kq^2 \langle \delta n^2 \rangle$, where k is an appropriate elastic constant. From the equipartition theorem, this energy may be equated to $\frac{1}{2}k_B T$ and hence

$$\frac{d\sigma}{d\Omega} = \frac{\pi^2}{\lambda^4} \varepsilon_a^2 \frac{k_B T}{kq^2} \tag{8.9}$$

Taking $\rho \partial\varepsilon/\partial\rho \sim \varepsilon_a$, we have $d\sigma'/d\sigma \sim \beta kq^2$. Typically $\beta \sim 10^{-11}$ cm^2 dyn^{-1}, $k \sim 10^{-6}$ dyn and the scattering vector $q \sim 10^5$ cm^{-1}, so that $d\sigma'/d\sigma \sim 10^{-7}$. Thus the orientational fluctuations make the predominant contribution, as first shown by de Gennes.[11]

In addition, we may draw the following conclusions:

1. From (8.9) it is seen that with decrease of scattering angle ($q \to 0$), the intensity of scattering should increase enormously.
2. Since $k \propto s^2$ and $\varepsilon_a \propto s$, $d\sigma/d\Omega$ should be practically independent of temperature.
3. For simplicity, we have throughout ignored the polarization factors, but it is easy to see that when the scattering arises from the tilt of the director, the scattered light should be strongly depolarized.

All these conclusions are in accord with experiment.

In principle, a measurement of the intensity of light scattering and its angular dependence should yield the elastic constants of the material.

8.1.4. Disclinations

We shall now discuss some applications of the continuum theory of elasticity to a study of disclinations, in particular to evaluating their energies and interactions. First let us recall briefly what disclinations look like when viewed through a microscope. Between crossed polarizers, a nematic film of thickness about 10 μm shows a number of dark brushes originating from points. These are due to line singularities (or disclinations) perpendicular to the layer. Some points have four dark brushes while others have only two. On rotating the crossed polarizers, the positions of the points remain unchanged but the brushes themselves rotate continuously. The sense of rotation may either be the same sense as that of the polarizers (positive disclinations) or opposite (negative disclinations). The rate of rotation is about equal to that of the polarizers when the disclination has four brushes and is twice as fast when it has only two. The strength of the disclination is defined as s = (number of brushes)/4. So far, $s = +\frac{1}{2}, -\frac{1}{2}, +1$, and -1 have been observed. Neighboring disclinations connected by brushes are of opposite signs and the sum of the strengths of all disclinations in the sample tends to zero. Normally, the disclinations remain static in the field of view but as the temperature approaches the nematic-isotropic transition point they tend to become mobile and disclinations of opposite signs are seen to attract one another and coalesce. They may then disappear altogether ($s_1 + s_2 = 0$) or form a new singularity ($s_1 + s_2 = s'$).

We shall now interpret these facts in terms of the continuum theory.[12,13] We consider a planar sample in which the director orientation is parallel to the glass surfaces and not a function of z, the normal to the film. (This assumption will not, of course, be valid very close to the singularities.) Under these circumstances, only two elastic constants come into play, k_{11} and k_{33}; we shall suppose $k_{11} = k_{33} = k$. Let us choose a cylindrical coordinate system (r, α) and seek solutions in which the director orientation ψ is independent of r (Figure 8.3). The elastic free energy is then

$$F = \frac{\frac{1}{2}k}{r^2} \left(\frac{\partial \psi}{\partial \alpha}\right)^2 \tag{8.10}$$

The condition for its minimization gives ψ = const, which describes the uniformly oriented nematic sample, or

$$\psi = s\alpha + c \tag{8.11}$$

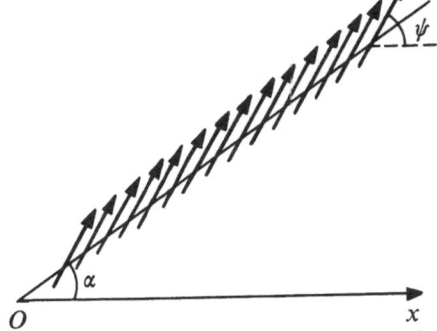

Figure 8.3. Director orientation (indicated by arrows) along a polar line making an angle α. Incident light that is linearly polarized at angle ψ or $\psi \pm \pi/2$ will be extinguished by a crossed analyzer and will give rise to a dark brush at an angle α.

where c is a constant. In the nematic, the orientational order is taken to be apolar and hence

$$s = \pm\tfrac{1}{2}, \pm 1, \pm\tfrac{3}{2}, \ldots \qquad \text{with } 0 < c < \pi$$

If the incident light is polarized at an angle ψ with respect to the x-axis, it is seen from Figure 8.3 that the polarization will be unchanged at all points on the polar line α and hence will not be transmitted by the crossed analyzer. This will result in a black brush at an angle α. A similar situation arises when ψ changes by $\pi/2$. The angle between two successive dark brushes is therefore $\Delta\alpha = \Delta\psi/s = \pi/2s$. Thus the number of dark brushes per singularity is $2\pi/\Delta\alpha = 4|s|$. Also, if the polarizers are turned through an angle ω, the brushes rotate by an angle ω/s. Consequently the rate of rotation of the brushes of the two-brush disclination ($s = \pm\tfrac{1}{2}$) is twice as fast as that of the four-brush disclination ($s = \pm 1$). Observations in polarized light therefore enable one to determine s in both sign and magnitude.

The molecular orientation in the neighborhood of a disclination is shown in Figure 8.4 for a few values of s. The curves represent the projection of the director field in the xy-plane. For $s \neq 1$, a change in c merely causes a rotation of the figure by $c/(1-s)$, while for $s = 1$ the pattern itself is changed.

When two neighboring disclinations are of equal and opposite strengths the brushes connecting them are observed to be circular. By superposition of solutions of the type (8.11) and putting $s_1 = -s_2 = s$, we have

$$\psi = \psi_1 + \psi_2 = s\beta + \text{const}$$

where $\beta = \alpha_1 - \alpha_2$. Thus curves of constant ψ will be arcs of circles passing through the two disclinations.

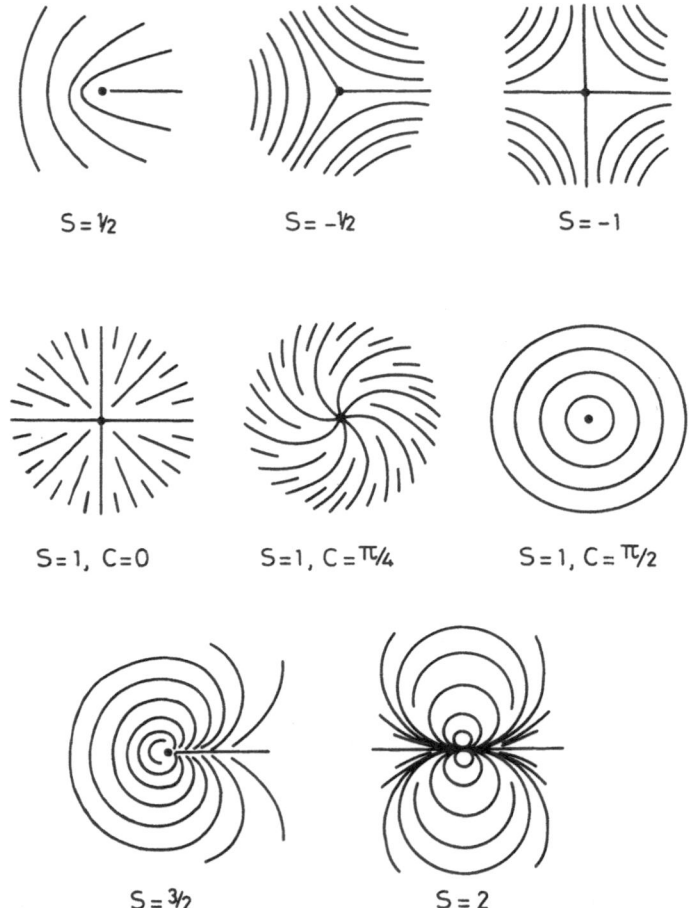

Figure 8.4. Director field in the neighborhood of a disclination (after Frank[12]).

The deformation energy of an isolated disclination in a circular layer of radius R and of unit thickness is

$$W = \int_0^R \int_0^{2\pi} \frac{1}{2} \frac{k}{r^2} \left(\frac{d\psi}{d\alpha}\right)^2 r\,dr\,d\alpha$$

This expression is not valid too close to the center of a disclination; we thus postulate a cut-off radius r_c around the disclination and integrate for distances greater than r_c. Remembering that $d\psi/d\alpha = s$, we obtain

$$W = \pi k s^2 \ln(R/r_c) \qquad (8.12)$$

As $R \to \infty$, $W \to \infty$, i.e., an isolated disclination in an infinitely extended layer has infinite energy. However, such a situation does not arise in practice due to the presence of pairs of disclinations of opposite signs.

The interaction between disclinations may be calculated by using the superposition principle ($\psi = \psi_1 + \psi_2$). Proceeding as before, we obtain for a pair of disclinations, separated by a distance r_{12},

$$W = \pi k (s_1 + s_2)^2 \ln(R/r_c) - 2\pi k s_1 s_2 \ln(r_{12}/r_c) \qquad (8.13)$$

We have assumed here that $r_c < r_{12} < R$. The force between two singularities is therefore $-2\pi k s_1 s_2 / r_{12}$. Accordingly singularities of opposite signs attract one another and those of like signs repel. The force is inversely proportional to the distance, similar to the case of two current-carrying conductors.

8.1.4.1. Interaction Between a Cavity (Air Bubble) and a Singularity

It has been observed that an air bubble in a nematic sample interacts with a singularity.[14] For simplicity, we shall consider the two-dimensional problem of a disclination of strength s placed at A at distance D from the center of a cylindrical cavity of radius R (i.e., $r = D$ and $\alpha = 0$) such that the line of the singularity is parallel to the axis of the cylinder (Figure 8.5). Let a disclination of strength $(1 - s)$ be at the center of the cylinder and one of strength s at A' distance D' such that $DD' = R^2$. The net orientation at any point (x, y) on the cylinder is then[15]

$$\psi = (1 - s)\tan^{-1}(y/x) + s \tan^{-1}(y/x) + c$$
$$= \alpha + c$$

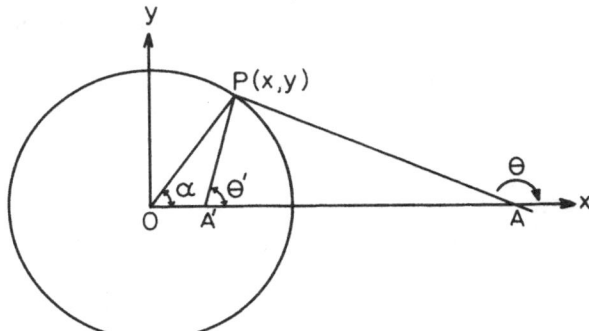

Figure 8.5. A line singularity of strength s located at A at distance D from the center O of a cylindrical cavity of radius R.

which implies a uniform alignment of the director on the surface of the cylinder. From surface-scattering and surface-tension studies[2] there is evidence that the molecules at the free surface of a nematic are aligned at a constant angle with respect to the boundary. The angle may vary from substance to substance — it may be normal, tangential, or tilted depending on the chemical constitution of the molecule — but for a given substance it has a certain definite value. Thus the solution we have obtained describes a realistic situation. At far-off points, the cylinder behaves as a $+1$ disclination located at its center. Consequently a negative disclination $(s < 0)$ is attracted by the cylinder when it is at large distances. When it moves closer than a distance $D_0 = R(1 - s)^{1/2}$ it is repelled. At D_0 it forms a "dipole disclination" with the cylinder. Such dipoles have been observed experimentally.[14] A positive disclination $(s > 0)$ on the other hand is always repelled by the cylinder at all distances. These ideas can be extended qualitatively to singular points and spherical cavities and predictions are in general agreement with observations.

It is seen that this theory bears a close resemblance to problems in electrostatics. Indeed, image forces come into play when a disclination approaches a boundary. The force can be attractive or repulsive depending on the nature of the boundary condition.

8.1.4.2. Effect of Elastic Anisotropy

We have so far assumed $k_{11} = k_{33} = k$. When $k_{11} \neq k_{33}$, it is obvious that there should be some changes in the director patterns (Figure 8.4) and also in the radial force of interaction between disclinations.

In addition, the elastic anisotropy gives rise[16] to a new type of force, which is not expected from the simpler theory. It turns out that energy is now required for one defect to move around another, thus resulting in angular forces. The physical basis for this effect is readily understood by reference to Figure 8.6, which shows director patterns for two pairs of unlike defects $(+1, -1)$ and $(+\frac{1}{2}, -\frac{1}{2})$ in two situations. It is seen that there are significant differences in the patterns depending on whether the director at large distances is parallel or perpendicular to the line joining the defects. In the case of $(+\frac{1}{2}, -\frac{1}{2})$ the central region is predominantly splay for one configuration and predominantly bend for the other. On the other hand, in the case of $(+1, -1)$ a structure that is mainly splay goes over to one that is mainly bend. In addition, the $+1$ defect itself has changed from a radial to a circular pattern. Thus depending on the sign of the elastic anisotropy, one or the other configuration will be energetically favored. In other words, given a boundary condition, angular forces should come into play and the disclination pair will be allowed only one configuration (i.e., only one value of the constant c).

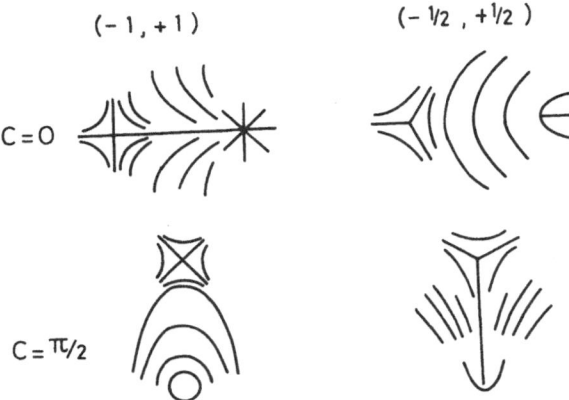

Figure 8.6. Director patterns for $(-1, +1)$ and $(-\frac{1}{2}, \frac{1}{2})$ pairs in two situations, $c = 0$ and $c = \pi/2$ (after Ranganath[16]).

This result has interesting consequences in regard to disclinations in long-pitched cholesterics, in which one may assume each section or layer to be nematic-like. Since each layer is allowed only one particular value of c, and the layers themselves rotate about the helical axis, pairs of disclinations should wind round each other to form helixes. This is indeed found to be the case experimentally. For example, pairs of like disclinations form double helices. The $(+1, -1)$ pair has a somewhat special configuration. It turns out that the $+1$ singularity has an energy of $3\pi k$ and the -1 an energy of πk per unit length. Thus one would expect the -1 to twist around three times as much as $+1$. A nearly straight $+1$ with a -1 going round it helically with a pitch equal to the diameter almost satisfies this condition, and this is precisely what has been observed by Rault.[17]

8.2. Viscous Properties

8.2.1. Experimental Determination of the Viscosity Coefficients

In a nematic liquid crystal, the translational motion of the fluid is, in general, coupled with the orientational motion of the director, and because of this its flow properties are remarkably different from those of an ordinary liquid. An excellent discussion of the basic conservation laws and constitutive relations describing these properties may be found in a review article by Leslie.[18] For a concise account of the most important equations of the theory and their applications, the reader is referred to Chapter 3 of

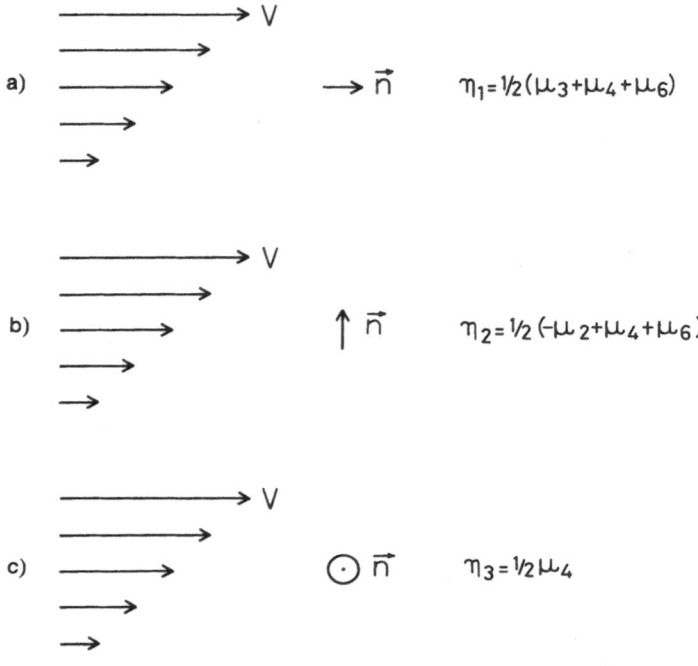

Figure 8.7. The three principal geometries of Miesowicz's experiment.

my book,[2] since we shall throughout adopt the same symbols and definitions as in that book.

The nematic has six viscosity coefficients, $\mu_1, \ldots, \mu_6$, but assuming Onsager's reciprocal relations for irreversible processes it turns out that

$$\mu_2 + \mu_3 = \mu_6 - \mu_5 \tag{8.14}$$

and hence the number of independent coefficients reduces to five. We shall now describe two experiments, which enable a direct determination of all the five coefficients.

In the first experiment, the nematic is oriented in a strong magnetic field and the apparent viscosity is measured in the following three geometries[19] (Figure 8.7): (a) **n** parallel to the flow, (b) **n** parallel to the velocity gradient, and (c) **n** perpendicular to the flow and to the velocity gradient. The apparent viscosity for any geometry is

$$\eta = \frac{\text{shear stress}}{\text{velocity gradient}}$$

For the three cases, this yields[2]

$$\eta_1 = \tfrac{1}{2}(\mu_3 + \mu_4 + \mu_6)$$

$$\eta_2 = \tfrac{1}{2}(-\mu_2 + \mu_4 + \mu_5)$$

$$\eta_3 = \tfrac{1}{2}\mu_4$$

The method can be extended to arbitrary directions of the magnetic field[20] and allows other combinations of μ's to be measured.

In the second experiment, a tube containing the nematic is suspended in a uniform magnetic field H acting in a horizontal plane and is spun at a constant angular velocity Ω about a vertical axis.[21] Below a critical angular velocity Ω_c, there is a steady-state solution[2]

$$\sin 2\phi = \Omega/\Omega_c$$

where

$$\Omega_c = -\chi_a H^2/2\lambda_1 \quad \text{and} \quad \lambda_1 = \mu_2 - \mu_3$$

In other words, the director makes a constant angle ϕ with respect to the magnetic field. This offers a direct method of determining $-\lambda_1$, which may be called the *twist viscosity coefficient*, i.e., the viscosity associated with the rotation of the director without material flow.

Other indirect methods of measuring the viscosity coefficients are available. The most important of these is by the application of laser beat spectroscopy to analyze the frequency spectrum of the scattered light. The integrated intensity gives the elastic constants (as we have seen earlier) and the half-width the viscosity coefficients. By selecting appropriate geometries, four coefficients $\mu_2-\mu_5$ have been determined. Reflection of ultrasonic shear waves have also been used to determine the viscosity coefficients.

In 10^{-2} poise, for the experimental values for p-azoxyanisole (PAA) are given by[22]

$$\mu_1 = -3.8 \qquad \mu_4 = 6.8$$

$$\mu_2 = -6.8 \qquad \mu_5 = 4.8$$

$$\mu_3 = 0.0 \qquad \mu_6 = -2.0$$

8.2.2. Viscous Torques

In order to appreciate the significance of the coefficients $\mu_1-\mu_6$, we shall examine their individual contributions to the viscous torque arising from fluid motion. The general expression for the viscous stress is

$$t_{ji} = \mu_1 n_k n_m d_{km} n_i n_j + \mu_2 n_j N_i + \mu_3 n_i N_j + \mu_4 d_{ji}$$
$$+ \mu_5 n_j n_k d_{ki} + \mu_6 n_i n_k d_{kj} \tag{8.15}$$

where

$$d_{ij} = \tfrac{1}{2}(v_{i,j} + v_{j,i}) \tag{8.16}$$

$$N_i = \dot{n}_i - w_{ik} n_k \tag{8.17}$$

$$w_{ij} = \tfrac{1}{2}(v_{i,j} - v_{j,i}) \tag{8.18}$$

In the above equations, we use the cartesian tensor notation, repeated tensor indexes being subject to the usual summation convention; the comma denotes partial differentiation with respect to spatial coordinates and the superposed dot a material time derivative. Let us choose a right-handed cartesian coordinate system with the flow along x, the velocity gradient along y, and the director confined to the xy-plane. In such a case, $v = v_x$, $v_{i,j} = v_{x,y}$, and if θ is the angle made by the director with x, $n_x = \cos\theta$, $n_y = \sin\theta$, and $n_z = 0$. Therefore, from (8.15), the respective contributions of $\mu_1, \mu_2, \ldots, \mu_6$ to the viscous torque Γ_z are

1. $\Gamma_z(\mu_1) = t_{xy}(\mu_1) - t_{yx}(\mu_1) = 0$
2. $\Gamma_z(\mu_2) = \tfrac{1}{2}\mu_2 v_{x,y} + \mu_2 \dot{\theta}$
3. $\Gamma_z(\mu_3) = -\tfrac{1}{2}\mu_3 v_{x,y} - \mu_3 \dot{\theta}$
4. $\Gamma_z(\mu_4) = 0$

It is seen from (8.15) that μ_4 is uncoupled with the director; thus μ_4 may be regarded as the ordinary hydrodynamic viscosity of the fluid.

5. $\Gamma_z(\mu_5) = \tfrac{1}{2}\mu_5 \cos 2\theta v_{x,y}$
6. $\Gamma_z(\mu_6) = -\tfrac{1}{2}\mu_6 \cos 2\theta v_{x,y}$

The total torque is

$$\Gamma_z = \tfrac{1}{2} v_{x,y}[\mu_2 - \mu_3 + (\mu_5 - \mu_6)\cos 2\theta] + \dot{\theta}(\mu_2 - \mu_3)$$

or, making use of (8.14),

$$\Gamma_z = v_{x,y}(\mu_2 \sin^2\theta - \mu_3 \cos^2\theta) + \dot{\theta}(\mu_2 - \mu_3)$$

In the steady state ($\dot{\theta} = 0$), Γ_z vanishes when the director orientation assumes an equilibrium value θ_0 given by

$$\tan^2\theta_0 = \mu_3/\mu_2$$

It turns out that both μ_2 and μ_3 are negative in most ordinary nematics and θ_0 is usually a small angle. For example, in 4-methoxybenzylidene-4'-butyl aniline (MBBA), $\theta_0 \simeq 7°$, and for PAA, $\theta_0 \simeq 0$. This equilibrium orientation θ_0 is attained in practice in relatively thick samples at high flow rates, so that the aligning effects of the walls have a negligible influence. In thin samples and at moderate flow rates, the elastic terms cannot be neglected and the velocity and orientation profiles can be evaluated only by numerical techniques taking explicit account of the director orientation at the boundaries.[2]

In nematics that show a smectic phase at lower temperatures (such as HBAB) it is found that $\mu_3 > 0$ at temperatures close to the A–N transition point.[20] Under these circumstances the contributions of μ_2 and μ_3 to Γ_z reinforce each other and there is no equilibrium value of θ_0. Thus in the absence of an orienting effect due to either the walls or a strong external field, the flow becomes unstable.

Clearly therefore the behavior of a nematic is markedly different from that of an isotropic liquid composed of anisotropic molecules, for which the flow-induced orientation is always 45°.

8.2.3. Orientational Relaxation

We have already discussed the static theory of the Freedericksz effect (see Section 8.1.2); we shall now examine its dynamical behavior. Equation (8.7) for the twist geometry is modified to

$$k_{22} \frac{d^2\theta}{d^2z} + \chi_a H^2 \sin\theta \cos\theta = -\lambda_1 \frac{d\theta}{dt}$$

If the field is switched off suddenly from $H > H_c$ to zero, the decay rate is

$$\tau^{-1} = -\frac{k_{22}}{\lambda_1} \frac{\pi^2}{d^2}$$

which is typically about 25 ms for a film of thickness 10 μm. This gives an idea of the switch-off time in nematic liquid crystal devices.

The case of twist relaxation is particularly simple, because the orientational motion of the director does not give rise to translational motion of the centers of gravity of the molecules. On the other hand, relaxation from a homeotropic to a planar configuration is more complicated as it results in a new effect, namely hydrodynamic flow induced by orientational deformation. Consider, for example, the geometry of Figure 8.2a of the Freedericksz experiment. The velocity and orientational profiles at different instants of time after the field is switched off from $H \gg H_c$ to zero can be determined numerically. The velocity is zero at the boundaries ($z = \pm d/2$) and at the middle of the sample ($z = 0$); at intermediate regions ($z \sim \pm d/4$) the fluid first moves one way, then with passage of time reverses direction, and finally settles down to the equilibrium zero value. The director orientation in the middle of the sample initially tips over to an angle greater than $\pi/2$ and then gradually relaxes to $\theta = 0$. These are referred to as the backflow and kickback effects, respectively. The explanation of these effects will be clear from Figure 8.8. The initial orientation profile of the director is plotted on the left. It is obvious that the elastic torque will be greatest around $z \sim \pm d/4$ where the curvature is greatest; this elastic torque is balanced by the magnetic torque. When the field is switched off the unbalanced elastic torque causes a clockwise rotation of the director in the region close to the boundaries (indicated in Figure 8.8b). As a result of the coupling between the director rotation and hydrodynamic motion, fluid flow is induced as indicated by the long arrows. This fluid motion in turn results in a counterclockwise torque on the director in the middle region ($z \sim 0$). This overcomes the elastic torque, which is weak in this region, and gives rise to a counterclockwise rotation of the director. The director therefore tilts over to an angle greater than $\pi/2$; when it relaxes back to $\theta = 0$, it produces a small amount of fluid motion in the reverse direction. Finally, at large times the system settles down to the undistorted equilibrium configuration.

8.3. Nematic-Isotropic Transition

8.3.1. Molecular Theories of the Nematic Phase

A microscopic approach that has been very useful in describing the long-range orientational order and some of the related properties of the nematic phase is that due to Maier and Saupe. It is based on the mean field approximation, the single-particle potential being taken in the form

$$u_i = -AsP_2(\cos \theta_i)$$

where A is a function of volume, $s = \langle P_2(\cos \theta) \rangle$ is the orientational order

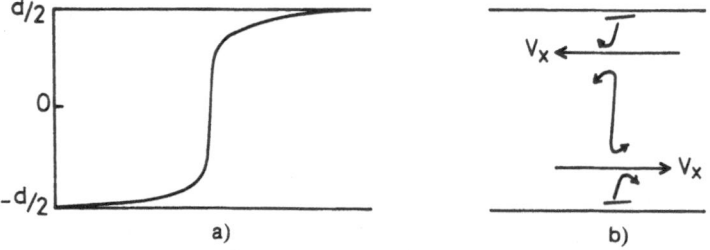

Figure 8.8. Interpretation of backflow and kickback effects.

parameter, $P_2(\cos \theta)$ is the Legendre polynomial of the second order, and θ is the angle that the long axis of the molecule makes with **n**. Such a potential is consistent with the fact that **n** is apolar. In their original presentation, Maier and Saupe regarded the dipole–dipole part of the anisotropic dispersion forces to be responsible for nematic stability and hence they assumed $A \propto V^{-2}$, V being the molar volume. However, experimental data on the pressure dependence of the order parameter indicate that $A \propto V^{-m}$ where $m = 4$, or in some cases even greater than 4. On the other hand, an argument due to Cotter suggests that thermodynamic consistency of the mean field theory requires that $A \propto V^{-1}$. For the present purpose we shall ignore the volume dependence of the potential, as it will not affect the major conclusions of the theory. The free energy due to order is

$$F = N\left[\tfrac{1}{2}As(s+1) - k_B T \ln \int \exp \left(\frac{3}{2}\frac{As}{k_B T} \cos^2 \theta_i d(\cos \theta_i)\right)\right]$$

The condition for transition from the nematic to the isotropic phase is $F = 0$, which gives $A/k_B T_{NI} = 4.541$, or $s \simeq 0.43$ at T_{NI}. The theoretical plot of s vs T/T_{NI} is a universal curve for all substances. This is in fair agreement with the data for a number of compounds, but there are systematic deviations.

A difficulty with this approach becomes apparent when we evaluate the latent heat of transition ΔH. The theoretical value of ΔH turns out to be much too high, usually by a factor of 2 or 3. Large discrepancies are also found in the specific heat C_v and isothermal compressibility β.

There is a further serious difficulty. Until 1973, only $\langle P_2 \rangle$ was accessible experimentally, but now a technique for measuring both $\langle P_2 \rangle$ and $\langle P_4 \rangle$ is available.[23] This involves polarized Raman scattering measurements using aligned samples. A number of cyano-compounds have been investigated by this method using the C≡N stretch vibration for the Raman measurements. It turns out that in every case studied so far $\langle P_4 \rangle$ is negative for at least a part of the nematic range. While the observed $\langle P_2 \rangle$ can be quantitatively accounted for by the introduction of a higher-order term involving $\langle P_4 \rangle$ in the Maier–Saupe potential, the observed negative $\langle P_4 \rangle$ cannot be accounted for theoretically. Another difficulty is that according to Cotter's thermodynamic consistency condition, $\gamma = (\partial T/\partial V)_s = 1$, while pressure studies show that γ may be as high as 7 in certain compounds. The principal drawback of the Maier–Saupe theory is that it does not take explicit account of the geometrical anistropy of the molecule, which as we know is an essential requirement for mesomorphism to occur. A more promising approach is to treat the system as an assembly of hard rods with a superimposed attractive potential.[24]

8.3.2. Short-Range Order Effects in the Isotropic Phase: The Landau–de Gennes Model

The isotropic phase of a nematic shows certain anomalous properties that can be attributed to a persistence of nematic-like short-range order above the transition point. For example, the magnetic birefringence close to the transition may be as much as 100 times higher than in ordinary organic liquids. Foex noted in 1933 that the behavior is closely analogous to that of a ferromagnetic material above the Curie temperature. A phenomenological description of these effects has been proposed by de Gennes[11] on the basis of the Landau theory of phase transitions.

Consider an expansion of the excess free energy in terms of a scalar order parameter s:

$$F = \tfrac{1}{2}As^2 - \tfrac{1}{3}Bs^3 + \tfrac{1}{4}Cs^4 + \cdots$$

where $B > 0$ and $C > 0$. Such an expression leads to a first-order transition. For a second-order transition, $B = 0$ and A may be taken to be of the form $a(T - T^*)$, where T^* is the second-order transition temperature. The N–I transition is weakly first order and therefore de Gennes suggested that the same form of $A(T)$ may be retained but with T^* representing a hypothetical second-order transition point slightly below T_{NI}. The term of order s^3 is not precluded by symmetry, for s and $-s$ represent two entirely different kinds of molecular arrangements, which are not symmetry related and do not have equal free energies. In the former case, the molecules are more nearly parallel to the unique axis, while in the latter they are nearly perpendicular to it. This model is conveniently applied to the weakly ordered isotropic phase.

8.3.2.1. Magnetic and Electric Birefringence

The free energy per mole of the isotropic phase in the presence of an external magnetic field may be written as

$$F = \tfrac{1}{2}a(T - T^*)s^2 - \tfrac{1}{3}Bs^3 + \tfrac{1}{4}Cs^4 \cdots - \tfrac{1}{3}N\chi_a H^2 s$$

where χ_a is the anisotropy of diamagnetic susceptibility of the molecule and N is Avogadro's number. The magnetically induced order is extremely weak ($\sim 10^{-5}$) so that we may neglect cubic and higher powers of s. The condition $\partial F/\partial s = 0$ then leads to the result

$$s_H = \frac{1}{3}\frac{N\chi_a H^2}{a(T - T^*)}$$

Thus the magnetic birefringence, which is proportional to s_H, varies as

$(T - T^*)^{-1}$. This is borne out by experiments, with $T_{NI} - T^*$ usually of the order of 1 °K.

The case of electric birefringence is slightly more complicated because the orientational energy due to an electric field consists of two contributions: (1) that due to the anisotropy of the low frequency polarizability and (2) that due to the permanent dipole moment. However, if the molecule has a strong dipole moment parallel to its length, $1/s_E$ varies linearly with temperature exactly as in the magnetic case and this has been confirmed experimentally.[2]

8.3.2.2. Light Scattering

The intensity of light scattering from the isotropic phase increases enormously as the temperature approaches T_{NI}.[11] We shall first give a simple physical interpretation of the origin of this effect. In the absence of an externally applied field, the mean value of the order parameter in the isotropic phase is zero, but fluctuations can occur about this mean value. We may expand the order-parameter fluctuations as a Fourier series, and if $\langle s^2 \rangle_q$ is the mean-square fluctuation corresponding to a given wave vector q, then by the equipartition theorem $F = \frac{1}{2} a (T - T^*)\langle s^2 \rangle_q = \frac{1}{2} k_B T$. Now the order-parameter fluctuations are directly proportional to the dielectric-constant fluctuations and thus the scattered intensity varies essentially as $(T - T^*)^{-1}$.

To discuss the effect in somewhat greater detail we shall write the excess free energy in the more general form

$$F = \tfrac{1}{2} A s_{\alpha\beta} s_{\beta\alpha} - \tfrac{1}{3} B s_{\alpha\beta} s_{\beta\gamma} s_{\gamma\alpha} + \tfrac{1}{2} L_1 \partial_\alpha s_{\beta\alpha} \partial_\alpha s_{\beta\gamma} + \tfrac{1}{2} L_2 \partial_\alpha s_{\alpha\gamma} \partial_\beta s_{\beta\gamma}$$

where $s_{\alpha\beta}$ is the order-parameter tensor for molecules of arbitrary shape, with $\alpha = \beta = 3$ corresponding to the long-molecular axis; the terms involving $\partial_\alpha = \partial / \partial x_\alpha$ describe gradients in the order parameter. The coherence length ξ of the order-parameter fluctuations, which may be viewed as a measure of the average size of an ordered region, increases rapidly as the temperature approaches the transition point. In fact, two coherence lengths may be defined:

For gradients along the local director axis

$$\xi_1^2(T) = \frac{L_1 + (2L_2/3)}{A(T)}$$

For gradients in the perpendicular direction

$$\xi_t^2(T) = \frac{L_1 + (L_2/6)}{A(T)}$$

Both ξ_1 and ξ_t diverge as $T \to T^*$, a standard result in the theory of phase transitions. Expanding the order-parameter fluctuations as a Fourier series, applying the equipartition theorem, and expressing the order-parameter fluctuations in terms of the corresponding dielectric-constant fluctuations $\delta\varepsilon$, we obtain, for the case in which the incident and scattered light are both polarized perpendicular to the scattering plane,[25]

$$\langle \delta\varepsilon^2_{VV}(q) \rangle = \frac{4}{3}\left[\frac{2k_{\mathrm{B}}T(\Delta\varepsilon)^2}{9a(T-T^*)}\right][1 - \xi_1^2 q^2 - (\xi_2^2 q^2/6)]$$

where $\Delta\varepsilon$ is the anisotropy of the dielectric constant (for optical frequencies) for perfectly parallel alignment of the molecules, and $\xi_1 = L_1/A(T)$ and $\xi_2 = L_2/A(T)$. Similarly, for incident light polarized normal to the plane of scattering and the scattered light polarized parallel to it,

$$\langle \delta\varepsilon^2_{VH}(q) \rangle = \left[\frac{2k_{\mathrm{B}}T(\Delta\varepsilon)^2}{9a(T-T^*)}\right](1 - \xi_1^2 q^2 - \tfrac{1}{2}\xi_2^2 q^2 \cos^2(\theta/2))$$

where θ is the scattering angle. An analogous expression may be written for the HH geometry also. The intensity of scattering I is proportional to $\langle \delta\varepsilon^2 \rangle$. Studies on the angular dependence of the scattering show that ξ is small compared with the wavelength of light, although not negligibly so. If we assume to a first approximation that $q\xi \ll 1$, we note that $I \propto (T-T^*)^{-1}$ and $I_{VV}/I_{VH} = 4/3$. Both these predictions have been confirmed experimentally.[25] Further, $I_{VH}(\theta) - I_{VH}(180 - \theta)$ yields ξ_1 (the ξ_2-term cancels out from this difference since $q = 4\pi n \sin(\theta/2)/\lambda$) and it is easily shown that $I_{VH} - I_{HH}$ yields ξ_2. Studies on MBBA have verified that the temperature dependence of the coherence lengths is in accord with the theory and also that there is a slight anisotropy in the shape of the correlated region.

8.3.2.3. Pretransition Effects in the Dynamic Properties

In developing the hydrodynamic theory of truly isotropic liquds, one treats the velocity gradient tensor as a "flux" and the corresponding viscous stress tensor $t_{\alpha\beta}$ as the conjugate "force." In the isotropic phase of a nematic, there is a fluctuating local order parameter, and therefore there is an additional flux given by

$$R_{\alpha\beta} = \frac{\delta s_{\alpha\beta}}{\delta t} \simeq \frac{\partial s_{\alpha\beta}}{\partial t}$$

the conjugate force being

$$\phi_{\alpha\beta} = -\frac{\partial F}{\partial s_{\alpha\beta}} = -As_{\alpha\beta}$$

With the strain tensor $d_{\alpha\beta}$ defined in equation (8.16), we have the following equations coupling the fluxes and forces:

$$\begin{bmatrix} t_{\alpha\beta} \\ \phi_{\alpha\beta} \end{bmatrix} = \begin{bmatrix} \eta & \mu \\ \mu' & \nu \end{bmatrix} \begin{bmatrix} d_{\alpha\beta} \\ R_{\alpha\beta} \end{bmatrix} \qquad (8.19)$$

where μ, η, and ν are viscosity coefficients[11]; further assuming Onsager's reciprocal relations, we have $\mu = \mu'$.

We shall now discuss the application of these equations to two other properties which exhibit pretransition anomalies.

  a. *Flow Birefringence.* Consider a shear flow along x with a velocity gradient dv/dz. The flow induces a birefringence proportional to the velocity gradient with the principal axes of the index ellipsoid inclined at 45° to the x, z-axes. In the steady state $R_{\alpha\beta} = 0$, $\phi_{xz} = \frac{1}{2}\mu(dv/dz)$, and hence

$$s_{xz} = -\frac{\mu}{2a(T-T^*)}\frac{dv}{dz}$$

Now, the flow birefringence is proportional to s_{xz} and therefore shows a $(T-T^*)^{-1}$ dependence. This has been confirmed experimentally in the case of MBBA.

  b. *Spectrum of Scattered Light.* For small q one may neglect the velocity terms in equation (8.19) and write

$$\frac{\partial s_{\alpha\beta}}{\partial t} = -\Gamma(T)s_{\alpha\beta}$$

where

$$\Gamma(T) = \frac{a(T-T^*)}{\nu}$$

$\Gamma(T)$ can be determined from the line width of the scattered spectrum. However, since the viscosity coefficient ν is itself temperature-dependent, the measurements have to be very precise.[26] Assuming that ν is proportional to the shear viscosity, it has been found that the experimental variation of $\Gamma(T)$ for MBBA is in agreement with the theory.

Despite the apparent success of the Landau–de Gennes model, some important questions have been raised. Is the N–I transition in fact near-tricritical rather than classical mean field? Is the model applicable to a transition, which is not truly second order? These doubts are certainly pertinent, but more experimental data are needed — very precise data concerning the specific heat, order parameter, etc., in the vicinity of the transition and the critical exponents associated with them.

8.3.3. Near-Neighbor Correlations

The phenomenological model just discussed will now be compared with the molecular statistical theory of Maier and Saupe. The free energy of the weakly ordered isotropic phase in the presence of an external magnetic field is, according to the mean field theory,

$$F = N\left\{\tfrac{1}{2}As(s+1) - k_B T \ln \int_0^{\pi/2} \exp\left[\frac{3}{2k_B T}(As + \tfrac{1}{3}\chi_a H^2)\right]\cos^2\theta \sin\theta d\theta\right\}$$

Expanding and integrating one finds

$$F = Nk_B T\left[\frac{As^2}{2k_B T^2}(T - T^*) - 0.0762\frac{A^3 s^3}{8k_B^3 T^3} + 0.0122\frac{A^4 s^4}{16k_B^4 T^4}\right] - \tfrac{1}{3}N\chi_a H^2 s$$

where $T^* = A/5k_B$. This expression is identical in form to the free-energy expansion of the Landau model. However, it does not yield a satisfactory value of T^*. Since $A/k_B T_{NI} = 4.54$, $T^*/T_{NI} = 0.908$. For PAA, $T_{NI} = 408$ °K so that $T_{NI} - T^* \simeq 40$ °K, while empirically $T_{NI} - T^* \simeq 1$ °K.

Clearly near-neighbor correlations have to be allowed for in the molecular statistical approach to give an improved description of the pretransition effects. A theory of this type has been developed,[2] based on the method proposed originally by Bethe for treating order–disorder effects in binary alloys. We suppose that every molecule is surrounded by z-neighbors ($z \geq 3$) and that no two of the z-neighbors are nearest neighbors with respect to each other. Let the pair potential between the central molecule 0 and one of its neighbors j be $E(\theta_{0j})$, where θ_{0j} is a function of the usual spherical coordinates θ_0, ϕ_0, θ_j, ϕ_j, and let every outer-shell molecule j be coupled with the remaining (external) molecules of the uniaxial medium by an interaction potential $V(\theta_j)$.

For thermodynamic equilibrium of the system we make use of the condition that the probability of central molecule assuming a certain orientation θ_1, ϕ_1 should be the same as that for an outer-shell molecule assuming exactly the same orientation. If we take

$$E(\theta_{0j}) = -B^* P_2(\cos\theta_{0j}) \tag{8.20}$$

$$V(\theta_j) = -BP_2(\cos\theta_j) - CP_4(\cos\theta_j) \tag{8.21}$$

solutions can be found that satisfy the consistency relation to very good accuracy and the thermodynamic properties of the system can be evaluated. The value of $(T_{NI} - T^*)/T_{NI}$ is now significantly better than the mean field value, though still rather high as compared with experiment. Agreement with the experimental value improves with decreasing z.

8.3.3.1. Antiparallel Correlation in Polar Liquid Crystals

The absence of ferroelectricity in the nematic phase shows that there is equal probability of the dipoles pointing in either direction, and hence it was generally assumed that the permanent dipolar contribution to the orientational order is negligibly small. However, from simple energy considerations it is evident that interactions between neighboring dipoles can by no means be neglected, particularly in strongly polar materials. This effect can be treated again by the Bethe approximation, by modifying the pair potential (8.20) to

$$E(\theta_{0j}) = A^* P_1(\cos \theta_{0j}) - B^* P_2(\cos \theta_{0j})$$

which favors an antiparallel arrangement of the permanent dipoles.

An important consequence of the theory is that the mean dielectric constant $\bar{\varepsilon} = \frac{1}{3}(\varepsilon_{\parallel} + 2\varepsilon_{\perp})$ should increase by a few percent on going from the nematic to the isotropic phase because of the diminution in the antiparallel short-range order $\langle P_1(\cos \theta_{0j}) \rangle$. This has been verified to be the case experimentally in a number of strongly positive materials. Additional evidence of this effect has been obtained by measuring the molecular dipole moment of 5CB dissolved in a nonpolar solvent (namely benzene) as a function of concentration. There is a marked decrease in the effective dipole moment per molecule with increasing concentration proving that the near-neighbor correlations are indeed antiparallel. Neutron diffraction studies of isotopically substituted cyano-compounds appear to provide supporting evidence that neighboring molecules adopt an overlapping head-to-tail arrangement.

In summary, therefore, the application of the Bethe approximation has been useful in bringing out some qualitative features of short-range order in nematics. For example, the concept of antiparallel correlations has important consequences in the structure of bilayer smectics and the reentrant phenomenon, as we shall see later. However, this approximation certainly cannot be regarded as adequate from a quantitative point of view.

Cholesteric Liquid Crystals

S. Chandrasekhar

9.1. Optical Properties

For light propagating along its optic axis, the cholesteric liquid crystal possesses a very high optical rotatory power, usually a few orders of magnitude larger than that of an ordinary optically active liquid. When the wavelength of the light in the medium is cqual to the pitch P of the helical structure, Bragg reflection takes place. The reflected light is strongly circularly polarized; one circular component is almost totally reflected over a spectral range of some 100 Å while the other passes through practically unchanged. In the neighborhood of the region of reflection, the rotatory dispersion is anomalous and the sign of the rotation is opposite on opposite sides of the reflected band. The behavior is not unlike that of an optically active molecule in the vicinity of an absorption, except that in this case the anomalous rotatory dispersion is present even when the molecules are nonabsorbing.

These properties are quite well understood. The high rotatory power does not arise from the chirality of the individual molecules but rather from their helical arrangement as represented in Figure 7.1b. The structure may be regarded as a spiraling dielectric ellipsoid whose principal axis Oc is always parallel to z, the axis of twist; the other two principal axes Oa and Ob (with principal values ε_a and ε_b) spiral around z with a twist angle $q = 2\pi/P$ per unit length. Rigorous solutions of the wave equation for propagation along the optic axis are known since the work of Mauguin, Oseen, and de Vries (for a full discussion of the theory and its applications,

S. Chandrasekhar · Raman Researsch Institute, Bangalore, India.

see elsewhere[2]). The normal waves may be written as

$$u_1 = \begin{bmatrix} \exp(iK_1 z) \\ d \exp i(K_1 - 2q)z \end{bmatrix} \tag{9.1}$$

$$u_2 = \begin{bmatrix} f \exp i(K_2 + 2q)z \\ \exp(iK_2 z) \end{bmatrix} \tag{9.2}$$

where

$$K_1 = q + [K_m^2 + q^2 - (4K_m^2 q^2 + \alpha^2 K^4)^{1/2}]^{1/2} \tag{9.3}$$

$$K_2 = -q + [K_m^2 + q^2 + (4K_m^2 q^2 + \alpha^2 K^4)^{1/2}]^{1/2} \tag{9.4}$$

$$d = \frac{K_1^2 - K_m^2}{\alpha K^2} \qquad f = \frac{K_2^2 - K_m^2}{\alpha K^2} \tag{9.5}$$

$$\alpha = \tfrac{1}{2}(\varepsilon_a - \varepsilon_b) = \tfrac{1}{2}(n_a + n_b)(n_a - n_b) = n\delta n \qquad K = 2\pi/\lambda \qquad K_m = 2\pi n/\lambda$$

where λ is the wavelength in vacuo. Each normal wave is made up of two circularly polarized components with wave vectors differing by $2q$. This is a consequence of the Bragg reflection. The superposition at any point in the medium of the two components of each normal wave results in general in an elliptic vibration, the ellipticity of which is determined by the coefficient d or f. The wavelength dependence of d and f is shown in Figure 9.1a. It is seen that d varies significantly with λ in the vicinity of the reflection band; it follows that the u_1-wave is elliptically polarized outside the region of total reflection and linearly polarized inside it (Figure 9.1b). In contrast, f is throughout very small and exhibits no anomaly around the reflection band, so that the u_2-wave is almost circular at all wavelengths. We therefore have here a remarkable instance of an optical system in which the normal waves are nonorthogonal, i.e., they bear no simple geometrical relationship to each other and cannot be represented by diametrically opposite points on the Poincaré sphere.

Figure 9.2 gives the reflection curve and Figure 9.3 a plot of the wave vectors K_1 and K_2 as functions of λ. The real part of K_1 shows a gap within the reflection band — analogous to the familiar band gap in solid-state physics — while the imaginary part grows rapidly in the same region. On the other hand, K_2 exhibits normal behavior throughout. The rotatory power $\rho = \tfrac{1}{2}(K_1 - K_2)$ is therefore anomalous around the region of reflection (Figure 9.4). If we introduce the approximation that $(K_1 - K_2)/q \ll 1$, or that the rotation per pitch is small compared to π, which is certainly valid in cholesterics, then

$$\rho = \frac{x - (x^2 - \alpha^2 K^4)^{1/2}}{4q} \tag{9.6}$$

where $x = K_m^2 - q^2$. When $x^2 < \alpha^2 K^4$, ρ becomes a complex quantity: the real part gives the rotatory power and the imaginary part the circular dichroism. The attenuation of the wave does not arise from ordinary absorption but from the reflection of one of the components. The reflection band is centered at $x = 0$, i.e., $K_m = q$ or $\lambda_0 = nP$, where λ_0 is the wavelength in vacuo. The spectral width of the reflection is $\Delta\lambda = P\delta n$.

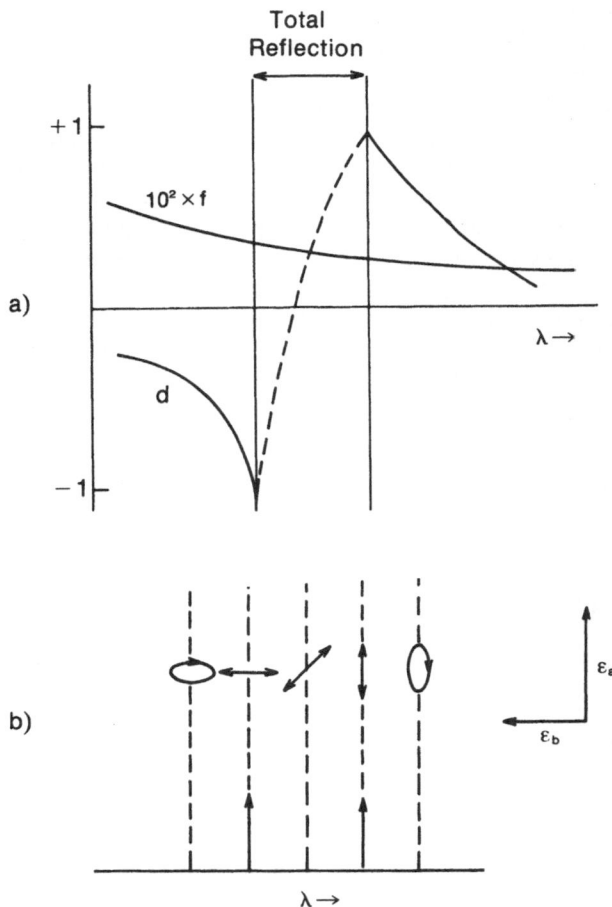

Figure 9.1. a. Coefficients d and f vs wavelength. The dashed portion of the curve for d is meant to indicate that $|d|$ remains 1 within the region of total reflection while the phase varies from 180° to 0°; in other words, d moves along the semicircle of unit radius in the upper half of the complex plane, varying from -1 to $+1$ as λ increases from one edge of the reflection band to the other. b. Polarization of the normal wave u_i vs λ.

When $x^2 \gg \alpha^2 K^4$, which is not valid close to or inside the reflection band,

$$\rho = -\alpha^2 K^{-4}/8qx = \frac{-\pi(\delta n)^2 P}{4\lambda^2(1 - \lambda^2/\lambda_0^2)} \qquad (9.7)$$

This is often referred to as the de Vries equation. The sign of the rotatory power reverses on crossing the reflection band. When $\lambda \ll \lambda_0$, equation (9.7) reduces to equation (9.8), which will be discussed presently, and when $\lambda \gg \lambda_0$, ρ tends asymptotically to zero. These predictions are in conformity with observations.

The relationship between rotatory dispersion, reflection, and circular dichroism may also be interpreted in a simple and elegant fashion in analogy with X-ray diffraction theory. The reflection curve and the amplitude attenuation factor, $\exp(-\xi)$, for circularly polarized light can be derived by setting up "difference equations" similar to those formulated by Darwin in his dynamical theory of X-ray diffraction from perfect crystals. Within the range of reflection ξ is real, primary extinction occurs, and the medium is highly circularly dichroic. Outside the range of reflection, ξ is imaginary and is of opposite signs on opposite sides of the reflection band, and therefore there is a reversal of the sign of the rotation on crossing the band. The predictions of this simple approach are in good agreement with those of the rigorous theory (Figures 9.2–9.4).

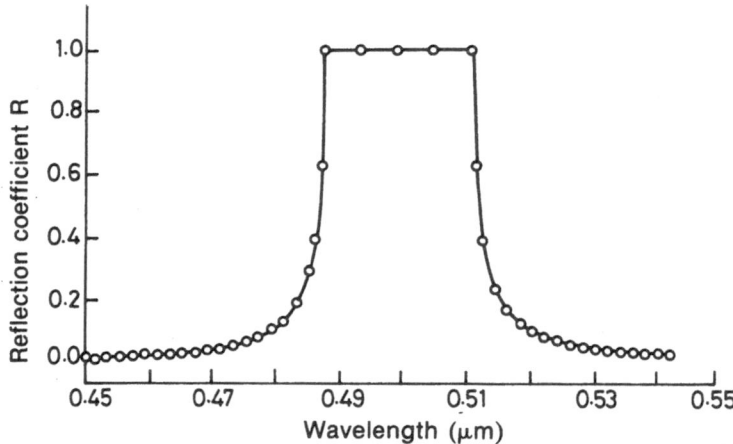

Figure 9.2. Reflection coefficient R at normal incidence plotted against wavelength for a semi-infinite nonabsorbing cholesteric. Circles represent values computed from the exact theory assuming that the medium external to the cholesteric (such as glass) has a refractive index of 1.5; the curve is derived from a modified form of the dynamical theory of X-ray diffraction. The parameters used in the calculations are $n = 1.5$, $\delta n = 0.07$, and $\lambda_0 = nP = 0.5 \ \mu m$.

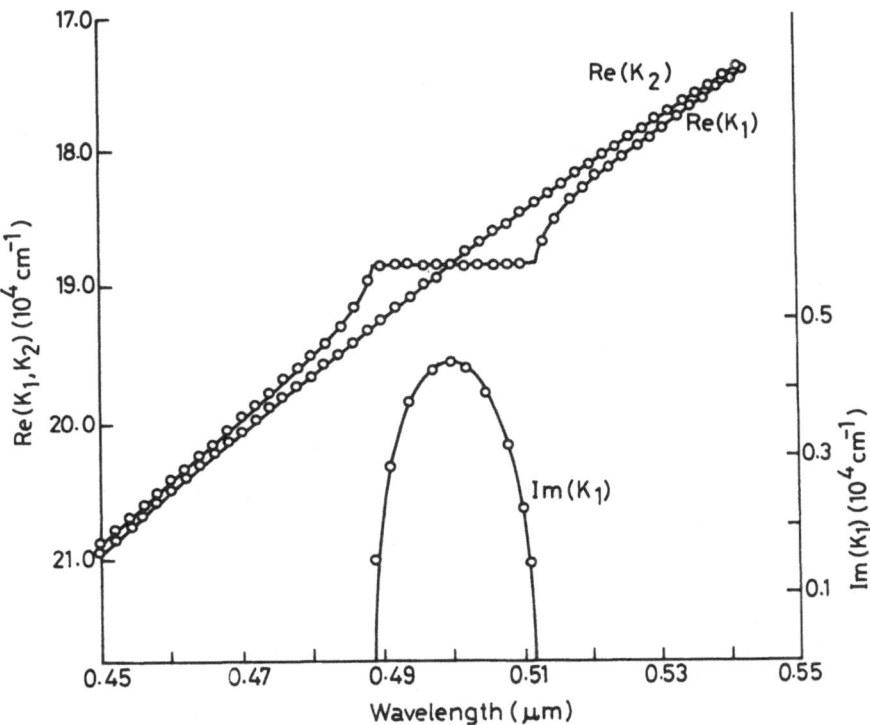

Figure 9.3. Wave vectors K_1 and K_2 of the normal waves as functions of wavelength in a semi-infinite nonabsorbing cholesteric. (For explanation of symbols see Figure 9.2.)

An interesting consequence of the dynamical theory of X-ray diffraction is that there should be an anomalous increase in the transmitted X-ray intensity when an absorbing crystal is set for Bragg reflection. This is known as the Borrmann effect. A similar effect occurs in absorbing cholesteric liquid crystals in the vicinity of the reflection band, except that, in contrast to the X-ray case, the polarization of the wave field and the linear dichroism of the molecules play an essential part. The existence of the effect was established by studies on cholesteryl nonanoate mixed with small quantities of PAA, which has a strong linearly dichroic band at about 3500 Å. The PAA molecules arrange themselves in a helical fashion in the cholesteric structure so that locally the medium becomes linearly dichroic in addition to being linearly birefringent. When the reflection band is adjusted to overlap with the strongly dichroic band of the solute molecules, the transmission spectrum exhibits the features predicted by theory.

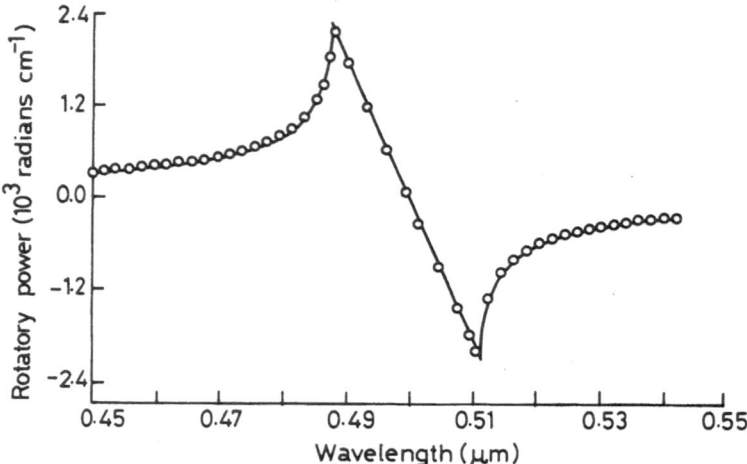

Figure 9.4. Rotatory power vs wavelength for a semi-infinite nonabsorbing cholesteric. (For explanation of symbols see Figure 9.2.)

The physical origin of the anomalous increase in transmission may be explained as follows.[27] We have noted earlier that inside the spectral range of reflection, a superposition of the two components of the normal wave u_1 gives rise to a linearly polarized vibration at any point. As z is varied, the azimuth of this linear vibration follows the director axis, making a constant angle with it. At one edge of the reflection band the polarization is along the director, while at the other edge it is at right angles (Figure 9.1b). Now, if the molecules are linearly dichroic with greater absorption for polarization along the director axis, it is evident that the wave will be attenuated to a greater extent on one side of the reflection band than on the other. This attenuation will, of course, be over and above that due to primary extinction. Consequently there will be an enhanced transmission near one edge of the reflection band; this is the Borrmann effect.

When $P \gg \lambda$, reflection and interference effects are small and the calculations become simpler. The optical behavior in this case may be divided into three distinct regimes:

1. When $\frac{1}{2}P\delta n \ll \lambda$, i.e., when the pitch is not too large, the cholesteric behaves approximately like a pure rotator. The rotatory power is given by

$$\rho = -\pi(\delta n)^2 P/4\lambda^2 \tag{9.8}$$

which is the de Vries equation (9.7) for $\lambda \ll \lambda_0$. The negative sign in (9.8) indicates that the sense of the rotation is opposite to that of the helical twist of the structure. Typically, $\delta n \simeq 0.05$ for a choies-teric; taking $P = 5$ μm and $\lambda = 0.5$ μm, $\rho \simeq 2000°/mm$. The rotat-ory power increases with increasing pitch.

2. When $\frac{1}{2}P\delta n$ is comparable to or greater than λ, the medium is equivalent to a combination of a rotator and a birefringent plate.

3. When $\frac{1}{2}P\delta n \gg \lambda$, i.e., when the pitch is extremely large, the wave at any point in the medium can be resolved into two linear vibrations polarized along the local principal axes. The polarization directions of these two vibrations rotate with the local principal axes as they travel along the axis of twist, the phase difference between them being the same as in the untwisted nematic medium. Hence in this case the optical rotation is in the same sense as that of the helical structure. It is this property that is employed in the so-called "twisted nematic" device. Clearly, therefore, on passing from regime (1) or (2) to (3), the rotatory power drops to zero and then reverses sign. When $P = \infty$, the structure becomes nematic and ρ again drops to zero.

We have so far considered only the semi-infinite medium. Rigorous calculations have been conducted for thin films, both nonabsorbing and absorbing, for $\lambda \sim \lambda_0$ as well as $\lambda \ll \lambda_0$. The interesting fact emerges that in both regimes the optical rotation and circular dichroism per unit thickness are functions of the thickness of the film and of the azimuth of the plane of polarization of the incident light. This emphasizes the difference between the properties of cholesterics and those of ordinary optically active substances. However, the dependence on thickness and azimuth is signifi-cant only for very thin films and becomes practically negligible when the sample thickness exceeds a few tens of pitch lengths.

The theory of propagation inclined to the optic axis is very much more complicated and only approximate analytical solutions have so far been obtained. The problem has been solved by numerical techniques and the calculations are in good agreement with observations.

9.2. Flow Properties

The cholesteric liquid crystal is highly non-Newtonian in its flow properties, its viscosity increasing by about a million times as the shear rate drops to a low value. Helfrich[28] proposed a simple physical mechanism to account for the very high apparent viscosity η_{app} at low shear rates. He suggested that flow takes place along the helical axis without the helical structure itself moving. Assuming that the velocity profile is flat, the energy gained by the translational motion of the fluid in the pressure gradient should be equal to that dissipated by the rotational motion of the director. Now the viscous torque $= -\lambda_1\dot{n} = -\lambda_1 qv$, where $q = 2\pi/P$ is the twist per unit length and v is the velocity. Therefore

$$-\lambda_1(qv)^2 = \frac{dp}{dz}\, v$$

The quantity of fluid flowing per second is

$$Q = \pi R^2 v = -\frac{\pi R^2 (dp/dz)}{\lambda_1 q^2}$$

or applying Poiseuille's law

$$\eta_{app} = -\lambda_1 \frac{q^2 R^2}{8} \tag{9.10}$$

Typically $R \sim 500$ μm, $P = 2\pi/q \sim 1$ μm so that $\eta_{app} \sim -10^6\lambda_1$, which accounts for the very high apparent viscosity at low pressure gradients. The essential features of the Helfrich model can be justified in terms of the continuum theory.

For shear flow normal to the helical axis, the cholesteric is expected to behave more or less like a nematic. The basic equations for this geometry were set up by Leslie (see elsewhere[2]) and the most complete calculations are due to Kini.[29] An interesting prediction is that for a given shear rate and sample thickness the apparent viscosity vs pitch should show oscillatory behavior, the effect being more pronounced when the pitch becomes comparable to the sample thickness. This prediction was verified experimentally in a 1.65: 1 by weight mixture of right-handed cholesteryl chloride and left-handed cholesteryl myristate.[30] Such a mixture has a helical structure, but the pitch is extremely sensitive to temperature. The measured value of η_{app} showed an oscillatory temperature dependence. The reason for this effect is quite easily understood: the greater the proportion of the molecules lying along the flow direction, the smaller will be the apparent viscosity. This proportion evidently depends on the ratio of the pitch-to-sample thickness and when this ratio is varied there should be variations of η_{app}.

Smectic Liquid Crystals

S. Chandrasekhar

10.1. Smectic A

10.1.1. Continuum Theory of Smectic A

Consider a smectic A structure in which each layer is in effect a two-dimensional fluid with its director normal to the layers. If we assume the layers to be incompressible, the integral

$$\frac{1}{a} \int_P^O \mathbf{n} \cdot d\mathbf{r}$$

represents the number of layers crossed on going from P to Q, where a is the layer thickness. In a disclination-free sample this number should be independent of the path chosen so that $\nabla \times \mathbf{n} = 0$. Hence $\mathbf{n} \cdot \nabla \times \mathbf{n} = 0$ and $\mathbf{n} \times \nabla \times \mathbf{n} = 0$. In other words both twist and bend distortions are absent and only the splay term remains in the Frank free-energy expression.

Thus the stratified structure of smectic A imposes certain restrictions on the types of deformation that can take place in it. A compression of the layers requires considerable energy and therefore only those deformations that tend to preserve the interlayer spacing are easily possible. It is obvious that the most favorable distortion is bending of the layers, since this involves only splay and does not affect the layer thickness (Figure 10.1). Such deformations can be observed optically. For example, when there is a center of attachment at the glass surface the molecules adopt a radiating or

S. Chandrasekhar · Raman Research Institute, Bangalore, India.

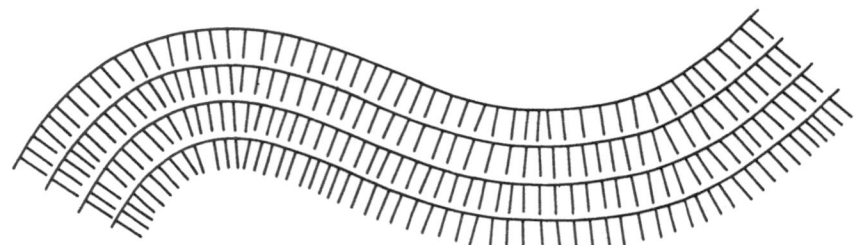

Figure 10.1. Flexibility of smectic A layers: the interlayer spacing is preserved.

fan-like arrangement and the layers form a family of equispaced surfaces normal to the molecular directions. Under a polarizing microscope these give rise to beautiful optical patterns, referred to as focal conic textures.

A more complete description of smectic A must account for layer compressibility, though, of course, the elastic constant B for compression may be expected to the quite large. For small displacements of layers normal to their planes, the free energy in the presence of a magnetic field along z takes the form[1]

$$F = \tfrac{1}{2}B\left(\frac{\partial u}{\partial z}\right)^2 + \tfrac{1}{2}\chi_a H^2\left[\left(\frac{\partial u}{\partial x}\right)^2 + \left(\frac{\partial u}{\partial y}\right)^2\right] + \tfrac{1}{2}k_{11}\left(\frac{\partial^2 u}{\partial x^2} + \frac{\partial^2 u}{\partial y^2}\right)^2 \qquad (10.1)$$

Here the physically reasonable assumption is made that twist and bend distortions are not allowed, though $\nabla \times \mathbf{n}$ does not strictly vanish when the layers are taken to be compressible. Writing the free energy in terms of Fourier components of u and applying the equipartition theorem, we have

$$\langle |u_q|^2\rangle = \frac{k_B T}{Bq_z^2 + k_{11}(q_\perp^2 + \xi^{-2})q_\perp^2}$$

where $\xi = (k_{11}/\chi_a)^{1/2}H^{-1}$. Therefore the mean-square fluctuation is

$$\langle u^2(\mathbf{r})\rangle = \frac{k_B T}{16\pi(Bk_{11})^{1/2}}\ln(\xi/a)$$

where a is the layer spacing. As $H \to 0$, $\langle u^2\rangle \to \infty$, implying that such a structure cannot be stable. This is the well-known Peierls–Landau instability of an unbounded one-dimensional solid. For a sample of dimension L

along z,

$$\langle u^2(\mathbf{r}) \rangle = \frac{k_B T}{16\pi(Bk_{11})^{1/2}} \ln(L/a)$$

in the absence of a magnetic field. Taking $B \sim 10^8$ dyn/cm^2, $k \sim 10^{-6}$ dyn, $a = 30$ Å, and $L = 1$ cm, $\langle u^2(\mathbf{r}) \rangle^{1/2} \sim 4$ Å, which is small compared to a. The model can therefore justifiably be applied to finite samples, which in any case are known to be quite stable and undergo sharp transitions to the nematic or isotropic phase.

10.1.1.1. Fluctuations and Rayleigh Scattering

The intensity of light scattering is proportional to the mean-square fluctuation of the director. Hence for $H = 0$

$$\langle |\delta n|^2 \rangle = \langle |u_q q_\perp|^2 \rangle = \frac{k_B T}{B q_z^2 q_\perp^{-2} + k_{11} q_\perp^2}$$

When $q_z = 0$, which corresponds to an "undulation" of the layers (Figure 10.1),

$$\langle |\delta n|^2 \rangle = \frac{k_B T}{k_{11} q_\perp^2}$$

which is a large quantity since it involves only the nematic splay constant k_{11}. On the other hand, when q_z and $q_\perp$ are comparable

$$\langle |\delta n|^2 \rangle = \frac{k_B T}{B(q_z/q_\perp)^2}$$

which is quite small. Consequently, in certain geometries smectic A does not appear turbid.

The most important contribution to light scattering in a perfect sample of smectic A comes from the undulation mode with the wave vector parallel to the layers ($q_z = 0$). In this mode the layers are deformed, but there is no change in layer spacing. Also, it is a highly damped mode with relaxation rate

$$\tau^{-1} = \frac{k_{11} q^2}{\eta} \sim 10^3 \text{ s}^{-1}$$

where η is an effective viscosity coefficient. As expected k_{11}/η is comparable to that of a nematic.

10.1.1.2. Ultrasonic Propagation and Brillouin Scattering

To discuss ultrasonic propagation we include the volume compression θ and write the free-energy density as

$$F = \tfrac{1}{2}A\theta^2 + \tfrac{1}{2}B_0\left(\frac{\partial u}{\partial z}\right)^2 + C_0\theta\frac{\partial u}{\partial z} + \text{higher terms}$$

Ignoring viscous effects and interlayer permeation, expressions can be derived for the velocities of propagation. It turns out that there are two acoustic modes for any arbitrary direction of the wave vector. One is the usual longitudinal mode whose velocity

$$c_1 \simeq (A_0/\rho)^{1/2}$$

is practically independent of the direction of propagation. The other mode has a much lower velocity, which is strongly orientation dependent:

$$c_2 \simeq (B_0/\rho)^{1/2}\sin\phi\cos\phi$$

where ϕ is the angle between the wave vector and its projection on the layer. This mode is associated with changes in the layer spacing without appreciable changes in density. As we shall see later, this corresponds to fluctuations in the phase of a complex order parameter and for this reason may be compared with the phonon branch in superfluids known as "second sound."

The two phonon branches in S_A have been clearly identified from Brillouin scattering experiments. Second sound is a critical mode, its velocity (for arbitrary ϕ) going to zero as the temperature approaches the A–N transition.

10.1.2. The Smectic A–Nematic Transition

10.1.2.1. McMillan's Molecular Model

The smectic A phase has one-dimensional translational order in addition to the usual orientational order common to all liquid crystals. McMillan[31] extended the Maier–Saupe theory to include an additional order parameter for characterizing the translational periodicity of such a layered structure. The anisotropic part of the pair potential may be conveniently taken in the form

$$V_{12}(r_{12}, \cos\theta_{12}) = -(V_0/Nr_0\pi^{3/2})\exp[-(r_{12}/r_0)^2]\frac{(3\cos^2\theta_{12}-1)}{2}$$

where the exponential term reflects the short-range character of the interaction, r_{12} is the distance between the molecular centers, and r_0 is of the order of the length of the rigid part of the molecules. If the layer thickness is a, then, retaining only the leading term in the Fourier expansion, the single-particle potential may be written as

$$V_1(z, \cos \theta) = - V_0[s + \sigma\alpha \cos(2\pi z/a)]\tfrac{1}{2}(3 \cos^2 \theta - 1)$$

where $\alpha = 2 \exp[-(\pi r_0/a)^2]$, s and σ are order parameters to be defined presently. With this form of the potential, the energy is a minimum when the molecule is in the smectic layer with its axis along z.

The single-particle distribution function is then

$$f_1(z, \cos \theta) = \exp[- V_1(z, \cos \theta)/k_B T]$$

and self-consistency requires that

$$s = \left\langle \frac{3 \cos^2 \theta - 1}{2} \right\rangle$$

$$\sigma = \left\langle \cos(2\pi z/a) \frac{(3 \cos^2 \theta - 1)}{2} \right\rangle$$

where the angle brackets denote statistical averages over the distribution f_1; s is the usual orientational parameter of the Maier–Saupe theory and σ is a new order parameter, which is a measure of the amplitude of the density wave describing the layered structure. The last two equations can be solved numerically to obtain the following types of solutions:

1. $\sigma = s = 0$ (isotropic phase)
2. $\sigma = 0$, $s \neq 0$ (nematic phase)
3. $\sigma \neq 0$, $s \neq 0$ (smectic phase)

The free energy of the system can be calculated in the usual manner. The two parameters characterizing the material are V_0, which determines the nematic-isotropic transition temperature, and α, a dimensionless interaction strength, which can vary between 0 and 2 and increases with increasing chain length of the alkyl tails of the molecule.

For $\alpha > 0.98$, smectic A transforms directly into the isotropic phase, while for $\alpha < 0.98$ there is a smectic A–nematic (A–N) transition followed by a nematic-isotropic transition at a higher temperature. For $\alpha < 0.70$ the model predicts a second-order A–N transition. The particular value of α ($=(0.70)$) at which the line of first-order A–N transition points goes over to a line of second-order transition points corresponds to a tricritical point.

In later work McMillan developed a refined model in which he introduced a partial decoupling between translational and orientational ordering. The calculations now involve three model-potential parameters, which are fixed by requiring the theory to fit T_{AN}, T_{NI}, and S_{AN}, the entropy of the A–N transition. The results are essentially the same as with the simpler potential but there are quantitative improvements.

A direct method of studying the translational order (or the amplitude of the density wave) is by measuring the intensity of the Bragg scattering from the smectic planes. McMillan himself carried out precise X-ray intensity measurements, which confirmed the trends predicted by his theory; in point of fact there was excellent quantitative agreement with his refined model. A noteworthy result that emerged from his experiments was that the X-ray intensities reveal an appreciable pretransitional smectic-like behavior in the cholesteric (nematic) phase. The existence of smectic-like clusters in the nematic phase was known from earlier X-ray work and had been designated as "cybotactic" groups. The occurrence of this type of short-range order cannot, of course, be accounted for by the simple molecular theory that we have just discussed. This aspect of the problem will be dealt with in the next section.

The orientational order parameters in the smectic and nematic phases, studied by magnetic resonance and other techniques, also follow the predicted type of behavior as the length of the alkyl end-chain is increased. In particular, a continuous change of s at T_{AN}, as expected of a second-order (or nearly second-order) transition, has been found (within experimental limits) in several compounds.

10.1.2.2. Phenomenological Theory of the A–N Transition

We now proceed to consider a Landau type of phenomenological description of the fluctuations attending the A–N transition. This approach to the problem is due to de Gennes and to McMillan, both of whom recognized the analogy with similar phenomena in superfluids.

We shall suppose for the present that A–N transition is of second order, which as we have seen is a possibility predicted by McMillan's simple microscopic theory. We start with the density wave in the smectic phase

$$\rho(z) = \rho_0[1 + c\,|\psi|\cos(q_s z - \phi)]$$

where ρ_0 is the mean density, $|\psi|$ the amplitude, and $q_s = 2\pi/a$ the wave vector of the density wave, a the interlayer spacing, and ϕ a phase factor that gives the position of the layers, and c a constant. Thus the smectic order can be fully specified by the complex parameter

$$\psi = |\psi|\exp(i\phi)$$

Near the transition, the free energy may be expanded in powers of $|\psi|$ and its gradients. For a fixed orientation of the director

$$F_s = \alpha |\psi|^2 + \frac{\beta}{2}|\psi|^4 + \frac{1}{2M_V}\left(\frac{\partial\psi}{\partial z}\right)^2 + \frac{1}{2M_T}\left[\left(\frac{\partial\psi}{\partial x}\right)^2 + \left(\frac{\partial\psi}{\partial y}\right)^2\right] \quad (10.2)$$

From symmetry considerations it is clear than only even powers of $|\psi|$ may be included. In the mean field approximation we may set $\alpha = \alpha_0(T - T^*)$.

If the director orientation is not fixed, it is the relative tilt between the layers and the director that should be considered and therefore (10.2) takes the generalized form

$$F = \alpha |\psi|^2 + \frac{\beta}{2}|\psi|^4 + \frac{1}{2M_V}\left(\frac{\partial\psi}{\partial z}\right)^2 + \frac{1}{2M_T}|\nabla_T - iq_s\delta n\psi|^2 \quad (10.3)$$

where ∇_T is the gradient operator in the plane of the layers. This equation is reminiscent of the Landau–Ginzburg expression for the free energy of superconductors; $\mathbf{n}$ corresponds to the vector potential $\mathbf{A}$, $\nabla \times \mathbf{A}$ being the local magnetic field.

Above T_{AN}, we may ignore the term involving the fourth power in (10.3) and from the equipartition theorem obtain in the usual manner

$$\langle|\psi(q)|^2\rangle = k_B T/\{\alpha + q_z^2/2M_V + [(q_x^2 + q_y^2)/2M_T]\}$$

from which we may define the coherence lengths

$$\xi_V^2 = \frac{1}{2M_V\alpha} \quad \text{and} \quad \xi_T^2 = \frac{1}{2M_T\alpha}$$

Since $\alpha \rightarrow 0$ as $T \rightarrow T^*$, $\langle|\psi(q)|^2\rangle$, ξ_V, and ξ_T diverge. The variation of $\langle|\psi(q)|^2\rangle$ can be seen directly in the intensity of the Bragg scattering.

Because of the smectic-like clusters developing in the nematic phase, the twist and bend elastic constants increase rapidly as the temperature approaches the A–N transition point. The additional contribution due to the smectic-like fluctuations is proportional to ξ. The critical divergence of the bend constant for one compound[32] is shown in Figure 10.2. The splay constant, on the other hand, does not exhibit such behavior. (A consequence of this is that the intensity of light scattering arising from a mode involving the twist or bend fluctuation will no longer be temperature independent; see Section 8.1.3.) Similarly, if the cholesteric shows a smectic A phase at a lower temperature, the pitch $P = 2\pi k_{22}/k_2$ diverges as the temperature approaches the cholesteric–smectic A transition point. It is this high temperature dependence of the pitch that is made use of in

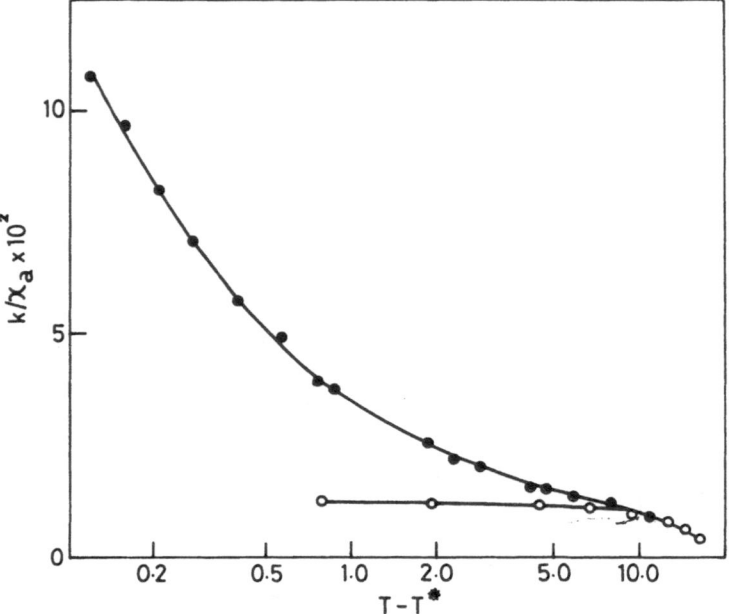

Figure 10.2. Variation of k_{11} (open circles) and k_{33} (closed circles) in the nematic phase of trans-p-n-decyloxy-d-methylcyanophenyl cinnamate, 10 OMCPC (after Karat and Madhusudana[32]).

thermography, i.e., in the mapping of surface temperatures. Some viscosity coefficients should also diverge because of an additional contribution to the torque due to the smectic-like regions. In the mean field approximation

$$\xi \propto (T - T^*)^{-1/2}$$

while, if one invokes the analogy with superfluids,

$$\xi \propto (T - T^*)^{-2/3}$$

Based on their very precise high-resolution X-ray and light-scattering studies of the A–N transition, Litster et al.[33] conclude that the pretransitional behavior in the nematic phase is essentially consistent with the He^4 analogue proposed by de Gennes; ξ is generally of the order of 10^2–10^3 Å at $(T - T_{AN}) \simeq 10^{-3}$ °K.

10.1.3. Bilayer Smectic A and the Reentrant Phenomenon

In Section 8.3.3 we discussed the concept of antiparallel near-neighbor correlations in strongly polar materials and presented experimental evidence in favor of it. A consequence of this type of correlation is that the S_A phases of these materials often consist of "bilayers," the molecules arranged in an antiparallel, overlapping interdigitated structure with a layer spacing of about 1.4 times the molecular length (Figure 10.3). X-ray studies on some cyano compounds that exhibit re-entrant phases (see Section 7.2) reveal unusual trends in the dependence of the bilayer spacings and molecular associations on pressure or temperature.

For example, in the case of 8 OCB (see Figure 7.2), the spacing decreases at first more or less linearly with increase of pressure till about 1.4 kbar, after which it increases.[34]

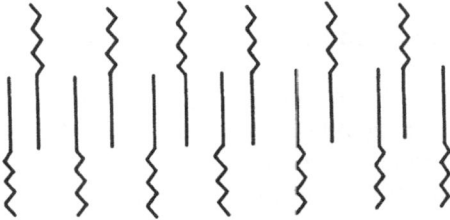

Figure 10.3. Schematic representation of the antiparallel, interdigitated arrangement of the molecules in bilayer smectic A.

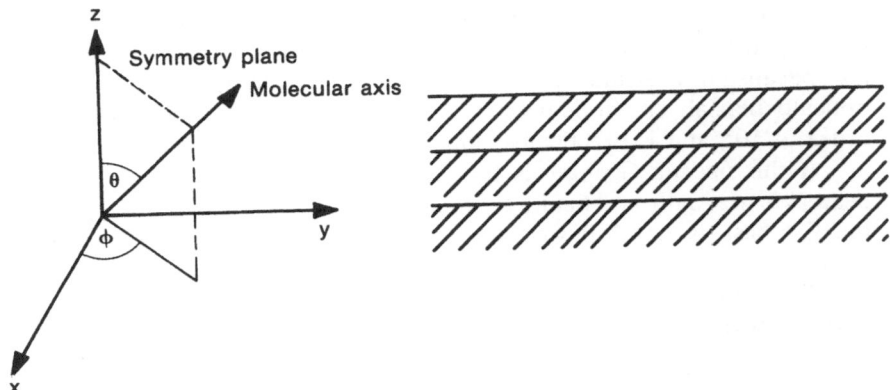

Figure 10.4. Tilt angle θ, azimuthal angle ϕ, and symmetry plane in the smectic C structure.

A somewhat similar type of behavior is seen in the S_A phase of 11 CPMBB, which is a room-pressure reentrant nematogen. In this case, on cooling the sample the S_A layer spacing decreases at first and then expands as the temperature approaches the S_A reentrant nematic transition point.[35] Dielectric studies on the same compound also reveal a structural rearrangement in the S_A phase as a precursor to the formation of the reentrant nematic. Moreover, the activation energy turns out to be much higher for the reentrant nematic than for the normal nematic, proving that the molecular associations in these two phases are significantly different.

The structural changes in "T8," which shows two reentrant phases at atmospheric pressure, are much more drastic and quite different from the cases discussed above.[36] The higher-temperature S_A phase has a bilayer structure, while the lower-temperature reentrant S_A phase has a monolayer structure.

It is clear that the extent of the interdigitation of the molecules is an additional parameter that has to be taken into account in any molecular theory seeking to explain the occurrence of the re-entrant phases.

10.2. Smectic C

In smectic C, the molecules are disordered within the layers (as in smectic A) but inclined with respect to the layer normal (Figure 7.1d). The orientation of the director in such a structure is specified by two angles, the tilt angle θ and the azimuthal angle ϕ (Figure 10.4). The former is coupled with the layer thickness while the latter is not. Therefore, at any given temperature, the amplitude of the θ oscillations of the director are small compared with those of the ϕ oscillations, with the result that the uniaxial symmetry about the mean molecular direction disappears. There exists now a plane of symmetry as indicated in the figure. Because of this, and also because of possible anisotropic polarization-field effects, smectic C is optically biaxial. The optic biaxial angle $2V$ is generally quite small, of the order of $10°$, and is practically independent of temperature.

If the smectic C phase is followed by an A phase, the tilt angle θ decreases gradually and finally becomes zero at the C–A transition point. If the C phase is not followed by an A phase, the tilt angle is almost temperature independent and usually about $45°$.

To discuss the C–A transition, it is obvious that the order parameter requires two components, θ and ϕ, and may therefore be written as

$$\chi = \theta \exp(i\phi)$$

This again brings out the analogy with superfluids. The free energy may be written as before in the form

$$F = e(T)|\chi|^2 + \tfrac{1}{2}f(T)|\chi|^4 + \text{gradient terms}$$

Using scaling arguments, de Gennes has predicted that in the ordered phase, $T < T_{CA}^*$,

$$\theta = |\chi| \sim (\Delta T)^\beta$$

where $\Delta T = T_{CA}^* - T$, T_{CA}^* being the second-order transition point and $\beta \sim 0.35$. In the mean field approximation, $\beta \sim 0.5$. However, experimentally it is not yet certain which of these two values of β holds for this transition.

The C–N transition is rather more complicated than the A–N or C–A transitions, because it involves layer formation and also a director tilt with respect to the layer normal. Landau theories of this transition have been proposed by de Gennes[37] and by McMillan.[38] A pretransition S_C-type short ordering occurs in the nematic phase — forming what are called skew-cybotactic groups — and all three elastic constants diverge near the transition (while only the twist and bend are renormalized near an A–N transition). The increase in k_{11} and k_{33} near the C–N transition has been observed, but no quantititative estimates of the critical exponents have yet been made. Some viscosity coefficients of the nematic phase, such as λ_1, become renormalized near the C–N transition. However, remarkably, in the S_C phase itself λ_1 is not renormalized. Physically this is easily understood: once the long-range translational order has set in, the twist viscosity is low because fluid motion can take place parallel to the layers.

S_C bears some interesting similarities to the nematic liquid crystal, as was first pointed out by Saupe. Suppose one defines a unit vector **u** to represent the preferred orientation of the projection of the molecules on the basal (xy) plane; then it is clear that **u** can be compared to the nematic director **n** in a homogeneously aligned sample. For example, the orientational fluctuations of **u** can be large and the smectic C appears quite turbid. Also, as in the case of a nematic, smectic C exhibits schlieren textures, except that as **u** and $-$**u** correspond to tilts in opposite directions, they are not equivalent and only disclinations with integral values of s are possible (see Section 8.1.4).

Another point of similarity with the nematic is that S_C can be easily twisted by the addition of optically active molecules. Pure chiral compounds showing a twisted smectic C — referred to as smectic C* — have also been discovered. The structure is helicoidal with a "precession" of the director about the layer normal (Figure 10.5). Smectic C* can exhibit ferroelectricity, as we shall see in a later section.

Figure 10.5. Schematic diagram of the structure of twisted smectic C.

10.3. Flexoelectricity in Liquid Crystals

If the molecule is polar in shape, in addition to having a permanent dipole moment, then a splay or bend deformation will polarize the material, or, conversely, an electric field will induce a deformation (Figure 10.6). In a first-order theory, the polarization P will be proportional to the distortion:

$$P = e_1[\mathbf{n}(\nabla \cdot \mathbf{n})] + e_3(\mathbf{n} \times \nabla \times \mathbf{n})$$

where e_1 and e_3 are the flexoelectric coefficients. The effect, which is the analogue of piezoelectricity in solids, was first suggested by Meyer. Prost and Marcerou pointed out that even quadrupole moments are sufficient to give flexoelectricity.

Experimentally, the existence of the effect in both nematic and smectic A liquid crystals was convincingly demonstrated by Prost and Pershan.[39] A homeotropically aligned sample was taken between two glass plates and a periodic electrostatic potential was applied by means of interdigitated electrodes coated on one of the plates, as shown in Figure 10.7. The flexoelectric effect being linearly proportional to the applied voltage, the

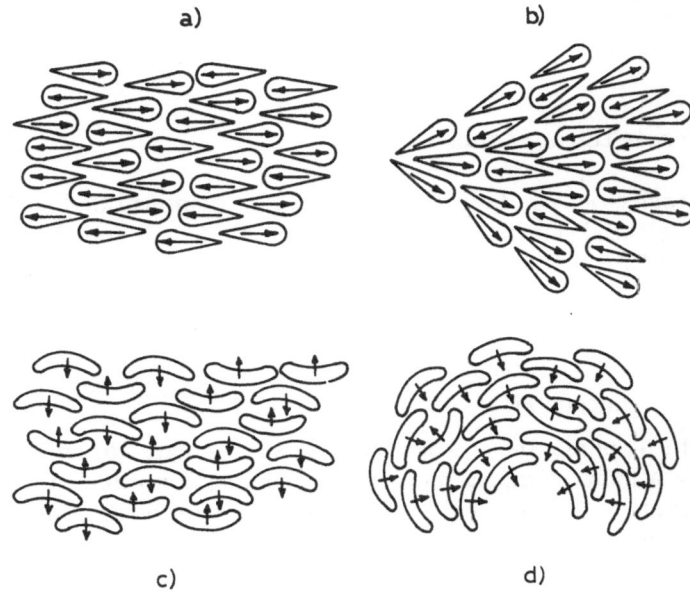

Figure 10.6. Origin of flexoelectricity: the nematic medium composed of polar molecules is nonpolar in the undeformed state, (a) and (c), but polar under splay (b) or bend (d).

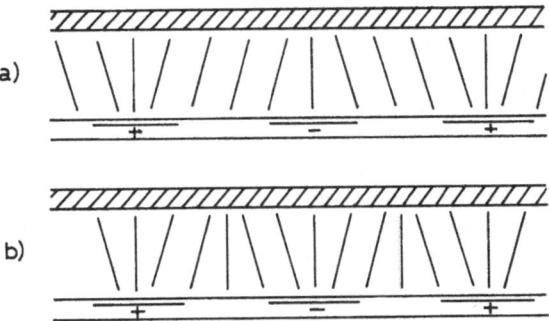

Figure 10.7. A periodic electrostatic potential applied to a homeotropically aligned sample gives rise to (a) a flexoelectric distortion having periodicity $2d$, where d is the spacing between the electrodes, and (b) a dielectric distortion having periodicity d (after Prost and Pershan[39]).

resulting distortion has periodicity $2d$, where d is the spacing between neighboring electrodes. On the other hand, the distortion due to dielectric alignment, which is proportional to the square of the applied voltage, has periodicity d. By optical diffraction it was possible to distinguish between these two types of distortion. It was also shown that the diffracted intensity due to the flexoelectric distortion increases linearly with V, as expected. The flexoelectric coefficient was found to be approximately 10^{-4} esu/cm.

10.4. Ferroelectric Liquid Crystals

Ferroelectricity in liquid crystals was first demonstrated by Meyer *et al.*[40] in the chiral smectic C and H phases (i.e., the C* and H* phases) of DOBAMBC, which shows the following transitions:

$$\text{crystal} \xrightarrow{76\,°C} \text{smectic C*} \xleftrightarrow{95\,°C} \text{smectic A} \xleftrightarrow{117\,°C} \text{isotropic}$$

with a transition at $63\,°C$ to smectic H*.

In the S_A phase, the rod-like molecules (which are not only chiral but also have a nonzero dipole moment) are arranged normal to the layers. Since there is no "head-to-tail" ordering (the director being apolar) there is no polarization normal to the layers. Further, since the molecules are rotating about their long axes, the transverse component of the dipole moment is averaged out and there is no net polarization parallel to the layers.

In the smectic C* phase, the molecules are tilted and their rotation about their long axes is biased. The symmetry plane of the ordinary smectic C structure (see Figure 10.4) is now absent because the molecules are chiral. The only symmetry element left is a twofold rotation axis parallel to the layers and normal to the long molecular axis. This allows the existence of a permanent dipole moment parallel to this axis.

Thus in smectic C* each layer is spontaneously polarized. The tilt and the polarization directions rotate from one layer to the next (see Figure 10.5). When an electric field E is applied normal to the helical axis, the helix becomes distorted (Figure 10.8); above a critical field E_c it is completely unwound and the sample is poled with the molecules tilted along a preferred direction normal to E. Hysteresis loops have been observed, whose strength decreases as the temperature is increased and vanishes at the C*–A transition point. Since this unwinding involves the rotation of the tilt direction, the response is damped by the rotational viscosity of the fluid. Therefore E_c increases with frequency. For the same reason the dielectric constant exhibits a relaxation in the few hundred hertz range.[41]

Not surprisingly, the value of the spontaneous polarization in these ferroelectrics is very small, about 3×10^{-8} C/cm^2 as compared with 5×10^{-6} C/cm^2 for KH$_2$PO$_4$. In terms of the dipole moment this turns out to be only approximately 0.25 debye/molecule.

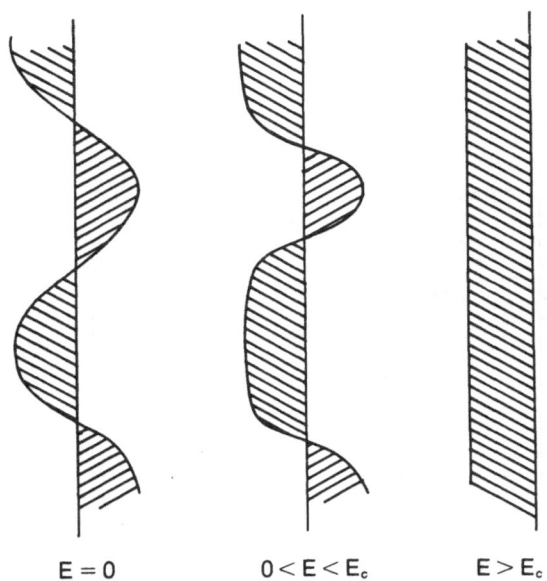

E = 0   0 < E < E_c   E > E_c

Figure 10.8. Unwinding of the smectic C* structure when an electric field is applied normal to the helical axis. For $E > E_c$ the helix is completely unwound.

The small difference (about 1 °C) in the A–C transition temperatures between the chiral and racemic forms of the same compound shows that the transition is driven by the intermolecular forces producing the tilted smectic phase and not by the (dipole–dipole) ferroelectric coupling. As a result, liquid-crystalline ferroelectrics can be classified as "improper" ferroelectrics.

In the few materials studied so far the A–C transition is apparently second order. Both the tilt angle θ and polarization go to zero continuously as the temperature is raised to the transition temperature T_c[(41)] (Figure 10.9). Actually the A–C transition point is itself the Curie point. To investigate some properties in the vicinity of the Curie point, let us write down a mean-field expression for the free energy of the material (assuming the ferroelectric to be an improper one in which θ and P have the same symmetry):

$$F = F_0 + \tfrac{1}{2}A\theta^2 + \cdots + \tfrac{1}{2}\chi^{-1}P^2 - PE - \frac{1}{8\pi}\varepsilon^0 E^2 - t\theta P \qquad (10.4)$$

Here we consider a spatially homogeneous case (i.e., the helix completely unwound, $q = 0$) and ignore the possible effects of the spontaneous helical torsion that appears below T_c. The first two terms are the leading terms of

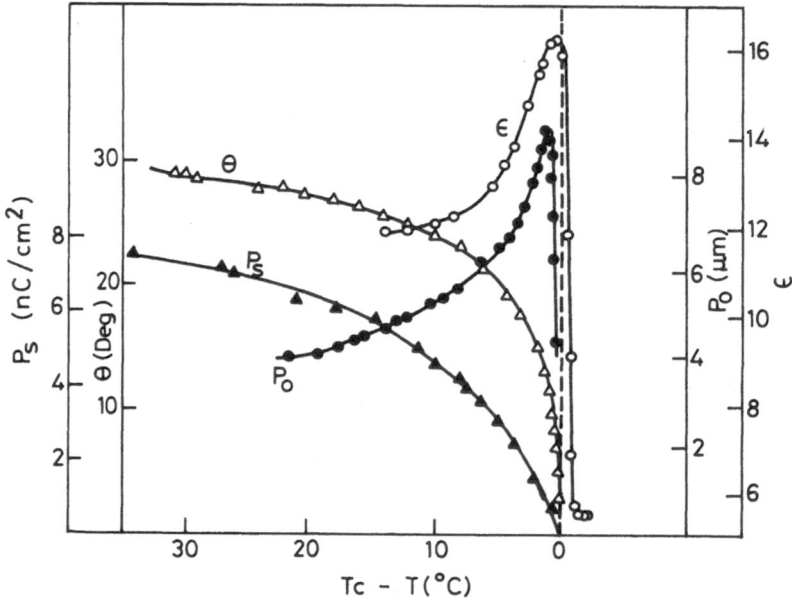

Figure 10.9. Variation in pitch P_0, tilt angle θ, dielectric constant ε, and polarization P_s in the smectic C* phase of DOBAMBC (after Ostrovskii *et al.*[41]).

the usual Landau series expansion in terms of the primary order parameter θ. The next three terms describe the electrostatic free energy and the last term describes the coupling between P and θ. By minimizing F with respect to P and θ we obtain

$$P = \chi(E + t\theta) \qquad (10.5)$$

where

$$\theta = \frac{t\chi E}{A'} \qquad (10.6)$$

and hence

$$F = F_0 - \frac{1}{8\pi}\left[\varepsilon^0 + 4\pi\chi + 4\pi\frac{(\chi t)^2}{A'}\right]E^2 \qquad (10.7)$$

Here $A' = (A - \chi t^2)$ represents the shift in the transition temperature due to the coupling of P and θ. Therefore the coupling between P and θ produces a divergent component in the dielectric constant. All these predictions are in broad agreement with experimental observations (Figure 10.9).

Equation (10.5) suggests that polarization can be induced by a shear stress in the absence of an electric field. There is evidence that shearing of the smectic layers does indeed produce an aligning effect on the molecules and hence polarization in a perpendicular direction.

According to equation (10.6) the field-induced tilt in the smectic A phase should diverge at T_c. This is the "electroclinic" effect observed by Meyer in DOBAMBC. An electric field in the plane of the layers produces a polarization and hence a tilt normal to the field that increases as $T \to T_c$. However, very close to T_c, probably because the order-parameter fluctuations become comparable to the pitch of the helix, neither ε nor the field-induced tilt exhibit a true divergence near T_c.

The Landau theory can be extended to take into account coupling between the polarization and the spontaneous torsion that sets in below T_c. It is expected that for $T < T_c$ there is, in addition to the soft mode, a symmetry-recovering Goldstone mode. Attempts have been made to identify these modes from dielectric-relaxation measurements.

The pyroelectric behavior of these ferroelectric smectics has been studied.[42] The pyroelectric coefficient $\gamma(T) = dP_s/dT$ increases with increase in temperature and near T_c it is approximately $2 \times 10^{-9} \, C \, cm^{-2} \, deg^{-1}$, comparable to the values for solid pyroelectrics. Away from T_c, γ is approximately $2 \times 10^{-11} \, C \, cm^{-2} \, deg^{-1}$. The pyroelectric response to a heat pulse exhibits an exponential decay and the time constant is equal to the relaxation time of the dipoles responsible for the spontaneous polarization; the relaxation frequency is of the order of 10^4–10^5 Hz.

11

Optical Applications of Liquid Crystals

G. Durand

In this chapter, physical principles of industrial applications using liquid crystals are discussed. We mostly concentrate on optical effects that use electrical or thermal fields and do not enter into details of the technology, though the interested reader can find more information elsewhere.[43–45]

11.1. Why Use Liquid Crystals for Optical Applications?

11.1.1. Mechanical Properties of Liquid-Crystal Phases

To understand the texture distortions used in optical applications, one must understand the mechanical properties of liquid-crystal phases from the static and dynamic points of view. We shall now summarize these properties.

11.1.1.1. Elasticity of Liquid-Crystal Textures

The mean orientation of molecules (the "director") in the nematic phase is defined by a unit vector $\mathbf{n}$. An undisturbed state is defined by $\nabla \mathbf{n} = 0$.

When curvature of the texture exists (in the form of "bend," "twist," or "splay"), curvature elastic free energy F is stored in the distortion,

---

G. Durand · Laboratoire de Physique des Solides, Université de Paris-Sud, Orsay, France.

given by

$$F = \int d^3r [\tfrac{1}{2} K (\nabla \mathbf{n})^2]$$

with three different curvature elastic constants $K_{1,2,3} \sim 10^{-6}$ cgs, where 1, 2, 3 correspond to splay, twist, or bend.

In Fourier analysis, the free-energy density $\tfrac{1}{2} K (\nabla \mathbf{n})^2$ becomes $\tfrac{1}{2} K q^2 [\delta n(q)]^2$, where q is the wave vector associated with the observed spatial distortion. The curvature torque, which tends to restore the equilibrium texture, is then $K q^2 \delta n(q)$.

The undisturbed state in the cholesteric phase is a helix of pitch $p = 2\pi/q_0$ in the range 1000 Å–100 μm. The elastic stored energy is simply $\tfrac{1}{2} K_2 (\mathbf{n} . \operatorname{curl} \mathbf{n} + q_0)^2$ + nematic-like terms in K_1 and K_3, which expresses the equilibrium helical state $q = q_0$.

In addition to nematic-like terms associated with curvature distortion, one must add for a smectic A phase the one-dimensional solid-like elasticity associated with layer compression. Calling u the normal layer displacement along $\mathbf{z}$, the corresponding free-energy density is $\tfrac{1}{2} B (\partial u/\partial z)^2$ where B is a solid-like elastic constant equal to approximately 10^8 cgs. To express the stability of the molecular orientation inside the layers, one introduces yet another elastic constant $B_\perp$ (approximately 10^7 cgs). The respective free-energy density is $\tfrac{1}{2} B_\perp \theta^2$, where θ is the tilt angle of molecules relative to the layer normal. For compounds that undergo an A $\rightarrow$ C phase transition, $B_\perp$ is expected to vanish $\propto (T - T_c)$, where T_c is the transition temperature. In the presence of smectic solid-like distortions, the restoring force and torque density are $B \partial u/\partial z$ and $B_\perp \theta$, respectively. Distortions of smectic texture are generally a mixture of curvature distortions of the director and of layer compression.

11.1.1.2. Dynamical Response

The dynamics of liquid-crystal distortions is based on the idea that inertia effects are negligible. In a nematic, for instance, the return to equilibrium from a distorted texture is simply expressed by the balance between elastic and viscous torques. Calling $\delta \dot{n}$ the angular velocity of the director, this results in $K q^2 \delta n = \eta \delta \dot{n}$ where η (approximately 0.1 cgs) is some viscosity. A curvature distortion relaxes down to zero, with characteristic frequency $1/\tau = K q^2/\eta$, typically in the range of 1 kHz for a 1-μm wavelength. One verifies easily that inertia associated with flow induced by the rotation of $\mathbf{n}$ is negligible, by noting that the damping frequency $\eta q^2/\rho$ associated with decay of the corresponding vortex is always much larger than $K q^2/\eta$ (here ρ is the specific mass of the nematic fluid).

The dynamics of cholesterics is a little more complicated. Let us remark simply that, for uniform ($q = 0$) distortion, the helix-restoring frequency is $1/\tau = Kq_0^2/\eta$, also in the kHz range for pitch in the μm range. The dynamics of smectics is complex. We note that some distortions of smectic texture are nematic-like, for instance in smectic C, the one where the molecules are allowed to rotate on a cone of constant θ. The perpendicular distortion that implies a change of θ relaxes more rapidly with frequency $B_\perp/\eta$, approximately equal to 10^8 Hz.

To summarize, liquid-crystal textures are easily distorted. On a molecular scale m, curvature (nematic-like) energy and layer-compression (solid-like) energy are comparable, both involving the same intermolecular van der Waals forces. A distortion on a macroscopic scale a implies a much weaker free energy, in the ratio $(m/a)^2$. The same scaling factor explains why these soft textures relax slowly to the equilibrium texture. The relaxation frequency $Kq^2/\eta \sim K/(a^2\eta)$ is $(a/m)^2$ smaller than the molecular frequency $K/m^2\eta \sim 10^9$ Hz; $(a/m)^2$ can be typically 10^6 Hz.

11.1.2. Coupling to External Fields

One interesting aspect of liqud crystals in optical applications is their easy coupling to an external electric field **E** using some property related to the anisotropy of the liquid-crystal phase. The nematic order is defined by the value of the "tensorial" order parameter $S = \frac{1}{2}\langle 3\cos^2\theta - 1\rangle$, where θ is the angle of one individual (rod-like) molecule with the mean director **n**. All tensorial quantities are proportional to S, which varies with temperature from $S = 0.3$ near isotropic transition to $S = 0.7$ or 0.8 near smectic transition.

The most widely used coupling between the texture orientation and **E** originates from the *dielectric anisotropy* of the phases.

In a nematic, the dielectric-constant tensor $\bar{\bar{\varepsilon}}$ has uniaxial symmetry about **n**. The anisotropy $\varepsilon_a = \varepsilon_\parallel - \varepsilon_\perp$, namely the difference between the longitudinal and transverse eigenvalues of $\bar{\bar{\varepsilon}}$, can be positive or negative depending on the chosen compound. For example,

Depending on the sign of ε_a, the molecules tend to align parallel ($\varepsilon_a > 0$) or perpendicular ($\varepsilon_a < 0$) to **E**. This is shown by the free energy f from

$$\mathbf{D} = \varepsilon_\perp \mathbf{E} + \varepsilon_a (\mathbf{n} \cdot \mathbf{E})\mathbf{n} \qquad \text{and} \qquad f = \frac{1}{2}\frac{1}{4\pi}\mathbf{E} \cdot \mathbf{D} \rightarrow \frac{1}{2}\frac{\varepsilon_a}{4\pi}(\mathbf{n} \cdot \mathbf{E})^2$$

For a small-angle distortion θ with respect to equilibrium, f becomes $\frac{1}{2}(\varepsilon_a/4\pi)E^2\theta^2$ with restoring torque $(\varepsilon_a/4\pi)E^2\theta$.

11.1.2.1. Flexoelectric Coupling

In presence of bend or splay, a bulk polarization **p** can appear in the form

$$\mathbf{p} = e_1\mathbf{n}(\mathrm{div}\,\mathbf{n}) + e_3(\mathrm{curl}\,\mathbf{n}) \times \mathbf{n} \qquad (11.1)$$

p couples to **E** to minimize the free energy $-\mathbf{p} \cdot \mathbf{E}$ so that a field **E** can induce curvature distortion.

The effect was first predicted by considering the lack of balance between longitudinal dipoles for pear-shaped molecules, e.g., in the presence of splay. In fact, it seems that the effect is mostly due to the quadrupolar moment $\bar{\bar{Q}}$ of liquid-crystal molecules, through the term $\mathbf{p} = -\bar{\nabla}\bar{\bar{Q}}$. The coupling coefficients $e_{1,3}$ are of the order of 10^{-4} cgs. Flexoelectric coupling has been tried to make optical modulators, with smectic materials. In a homogeneous field, weak anchoring conditions are necessary to achieve the distortion. The difficulty in mastering these anchoring conditions may explain why flexocoupling has not been used very much for applications.

The chiral C* phase is the only true liquid-crystal phase that is ferroelectric, resulting in *ferroelectric coupling* with **E**. The spontaneous polarization **p** is field independent, so that the coupling $-\mathbf{E} \cdot \mathbf{p}$ is linear in E, in contrast to dielectric anisotropy coupling that is quadratic in E. This effect has been recently proposed for rapid optical switching (see Section 11.5 on smectics).

11.1.3. Coupling to Light

Most applications use the optical anisotropy of the liquid-crystal texture. In the optical-frequency range, the only cause of anisotropy in $\bar{\bar{\varepsilon}}$ is electronic polarizability, always larger along the rod-like molecular axis (the opposite is the case for "discotics"). All systems of rod-like molecules ("calamitic" compounds) can be assumed uniaxial, with ordinary and extraordinary indexes $n_o \sim 1.5$ and $n_e \sim 1$. This remains almost true in the uniaxial smectic A phase. Although biaxial, the smectic C and C* phases can be considered practically as tilted uniaxial systems.

The optical anisotropy is used in two ways. Sometimes one used the "wave-guide" property of a twisted texture, as in cholesterics or in the twisted nematic display. This regime will be discussed in Section 11.2. Sometimes one simply uses the transition between an undistorted uniform transparent texture and a (chaotic) disordered texture that scatters the light strongly, as in the "dynamic-scattering" mode.

Another way of coupling the anisotropy of liquid-crystal texture to light waves is to use "dichroic" dyes. A dichroic dye can be made of a rod-like molecule, aligned along the director of a nematic host in which it is dissolved. The dye will absorb optical radiation with the electric field vibrating along the director, and remains transparent for the perpendicular orientation. Acting with an electric field (at low frequencies) one can change the orientation of the nematic director, and thus change the absorption of the liquid crystals. Color displays are based on this principle. A detailed discussion of the effect will be given in the next section.

11.1.4. Conclusion

Liquid-crystal phases are highly *anisotropic* materials. Their textures can be easily distorted because curvature elasticity is *soft* on a macroscopic scale. The damping time to restore equilibrium is correspondingly long, so that these materials are *slow*. One can easily affect the liquid-crystal texture orientation by external electric fields. The light propagation is strongly affected, directly or indirectly, by the changes of texture. The use of liquid-crystals for optical applications is based on these electrooptic effects.

The best reference concerning the physical properties of liquid crystals remains the book by de Gennes,[1] while a good references for materials is Dubois.[46] The reader wanting more information on applications is referred elsewhere.[43-45]

11.2. Optical Properties of Textures

In this section, we discuss in more detail the optical properties of twisted textures (cholesterics and C*) and the problem of light absorption by dichroic dyes. We start by explaining some methods to obtain uniform textures. They are based on "liquid epitaxial" orientation of the liquid-crystal bulk material, from surface alignment.

11.2.1. Orientation of Uniform Textures

11.2.1.1. Mechanical Effects

More than 50 years ago, it was shown that careful "rubbing" of glass plates induces an alignment of nematic material parallel to the rubbing direction, in the plane of the plates. The same technique, using abrasive powder, works very well. The alignment is understood by the difference in elastic energy between the two "planar" configurations, parallel or perpendicular to the grooves (Figure 11.1). The "homeotropic" orientation, where molecules are normal to the glass plates, was obtained by cleaning the glass surface with a strong acid.

11.2.1.2. Use of Surfactants

With the development of electrooptic devices, systematic studies have encouraged an interest in surface coating with tensioactive materials in

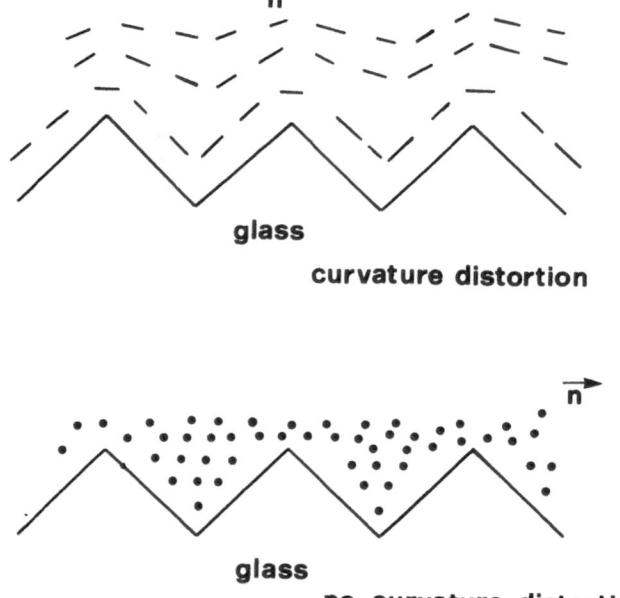

Figure 11.1. Two planar configurations with different elastic energies.

solution, like CTAB (cetyltrimethyl ammonium bromide)

$$n\,C_{16}H_{33}N^+Br^- \begin{matrix} CH_3 \\ CH_3 \\ CH_3 \end{matrix}$$

or with polymers, like HMDS (hexamethyl disiloxane) polymerized on a glass surface in the presence of a plasma discharge, such as

$$\begin{matrix} CH_3 & & & & CH_3 \\ CH_3 \!-\! Si\!-\!O\!-\!Si\!-\! CH_3 \\ CH_3 & & & & CH_3 \end{matrix}$$

These surface treatments alter the surface tension. In order to establish a more quantitative criterion, let us consider the equilibrium of a drop of liquid on a solid glass plate (Figure 11.2). If γ is the surface tension, we can write $\gamma_{LV}\cos\theta = \gamma_{SV} - \gamma_{SL}$ where the subscripts characterize the interface; γ_{LV} is characteristic of the liquid crystal. The latter quantity can be determined independently by measuring the capillary force exerted by the liquid crystal on a platinum plate in well-defined conditions.

For a typical nematic like MBBA (methoxybenzilidene butylaniline) γ_{LV} is approximately 34 cgs and increases with the dipolar character of the material ($H_2O \rightarrow 75$ cgs). The quantity $\gamma_C = \gamma_{SV} - \gamma_{SL}$ is assumed to be characteristic of the surface and can be determined using a homologous series of alkanes of various γ_{LV} and plotting $\cos\theta$ vs γ_{LV}. A critical value γ_C is obtained when the liquid wets the surface.

Interesting empirical results are the following:

1. When the liquid crystal *wets* the surface, i.e., $\gamma_{LV} < \gamma_C$, one generally obtains a "planar" orientation.
2. When the liquid crystal forms a *drop*, i.e., $\gamma_{LV} > \gamma_C$, one generally obtains a "homeotropic" orientation.

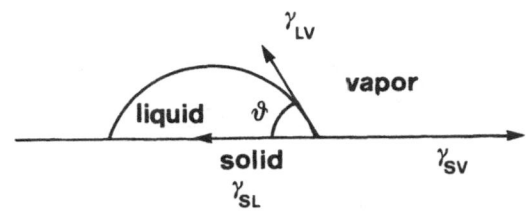

Figure 11.2. Equilibrium shape of a drop of liquid on a glass plate.

11.2.1.3. "Modern" Techniques

Unlike abrasive rubbing, it was demonstrated that the oblique evaporation of an SiO film leads to a planar orientation. This is understood by observing the shape of the SiO deposit, which creates grooves by a shadow effect (Figure 11.3).

For an almost-parallel direction of the SiO with respect to the plate (Figure 11.4) one obtains an alignment intermediate between homeotropic ($\theta = 0$) and planar ($\theta = 90$) orientation. One can combine the effect of an SiO oblique coating with the antagonistic effect of a polymer coating like PTFE (polytetrafluoroethylene), which with $\gamma_c \sim 22$ dyn/cm is one of the most effective homeotropic aligning agents. With a carefully determined polymer thickness of approximately 25 Å, one obtains finally a homeotropic orientation with a pretilted orientation θ equal to about 3°. These small pretilted orientations are very useful in electrooptic effects to suppress, for instance, an azimuthal degeneracy that may lead to different tilt domains and optical defects. Much of the skill of display makers has been invested in finding alignment techniques giving strong, reproducible, uniform, and cheap alignments.

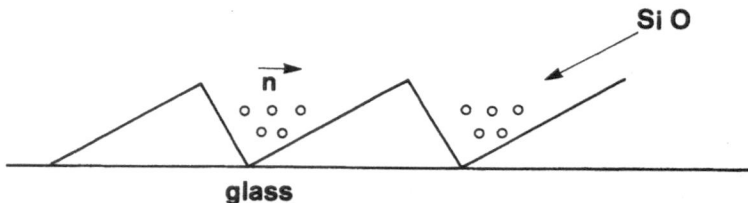

Figure 11.3. Alignment obtained by oblique evaporation of SiO film.

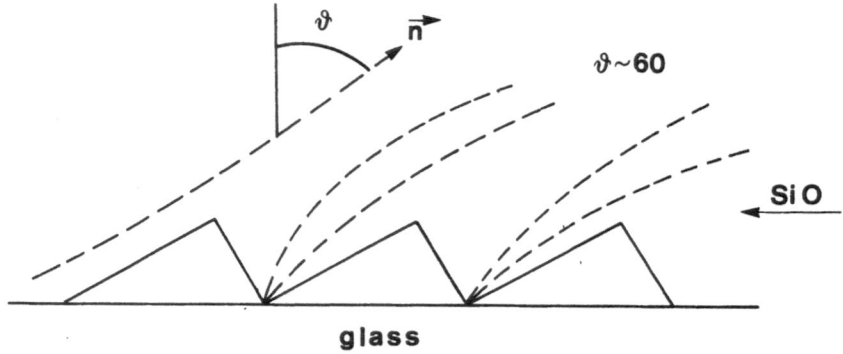

Figure 11.4. Alignment for almost-parallel direction of SiO.

11.2.1.4. Problems

Schiff bases, such as MBBA, align very well with the previously described techniques. More modern materials, like esters (benzoates) or the derivative of cyclohexane (in place of benzene rings), are more difficult to orient. Newer materials like discotics can be aligned homeotropically using surface coating with disk-shaped molecules.

Smectic phases obtained by cooling nematic phases can maintain the orientation of the higher-temperature phase, resulting at will in homeotropic or planar alignment. When no nematic phase exists, alignment is very difficult. We note that "field-aligned" samples, in a magnetic field, for instance, can never be uniformly aligned close to walls, which would induce an antagonistic alignment. From an optical point of view, these samples appear as nonaligned. Chiral C^* is in general difficult to align.

In conclusion, one can reasonably well align liquid-crystal molecules close to plates or walls, which define the edge of the liquid-crystal cell. In nematic materials, the surface orientation results in a bulk orientation giving quite easily the two interesting orientations of liquid-crystal texture, i.e., homeotropic or planar (Figure 11.5). In smectic materials, one must use both bulk and surface orientation to grow "single crystals," i.e., uniform texture. Bulk alignment can be obtained by previous alignment of an eventually higher-temperature nematic phase with a magnetic field, at least with some luck! The lack of mastery in smectic alignment is a strong limitation of the practical use of these phases in electrooptic devices.

11.2.2. Optical Properties of Twisted Textures

A twisted texture spontaneously replaces nematic uniform alignment whenever one deals with a chiral compound, where molecules are not superposable on their mirror image. The observed helical texture can be

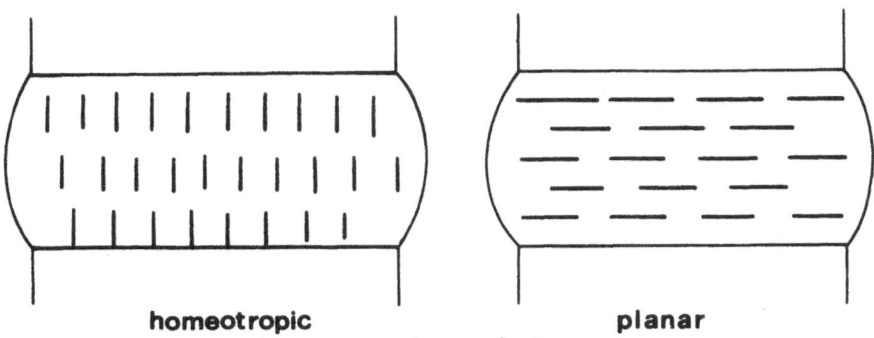

homeotropic   **planar**

Figure 11.5. Homeotropic and planar orientations of liquid-crystal texture.

explained briefly as follows: because the twisted ("cholesteric") phase is not ferroelectric, $\mathbf{n} \equiv -\mathbf{n}$ and the optical periodicity along the helical axis is $p/2$, half the pitch of the helix. We shall discuss later the case of chiral C^*. Another way to obtain a twisted texture is to place a nematic between two rubbed plates (Figure 11.6) twisted through $\pi/2$ with respect to each other. This geometry is used in twisted nematic displays.

11.2.2.1. Materials

Historically, the first materials showing a helical twisted phase were esters of cholesterol, from which the phase has derived its name "cholesterics." In practice, cholesteric phases over a wide temperature range are obtained by making eutectic mixtures of cholesterics. These materials are not very convenient; their order parameter S is weak (0.4), as well as their optical or dielectric anisotropy. New compounds, which are in fact chiral nematics, have been synthesized and give an intrinsic pitch in the 1000 Å range and strong dielectric and optical anisotropy. By mixing these

$$CH_z$$
$$\Large\text{/\!\!\backslash\!\!/} \; C^* \; \text{\backslash\!\!/}\!\!\text{-}\bigcirc\!\!-\!\!\bigcirc\!\!- C\!\equiv\!N$$
$$H$$

compounds with low-viscosity cyclohexane nematics, one obtains very appealing cholesteric materials.

One need not start from a cholesteric compound in order to obtain a cholesteric phase, but just dilute any chiral molecule inside a nematic

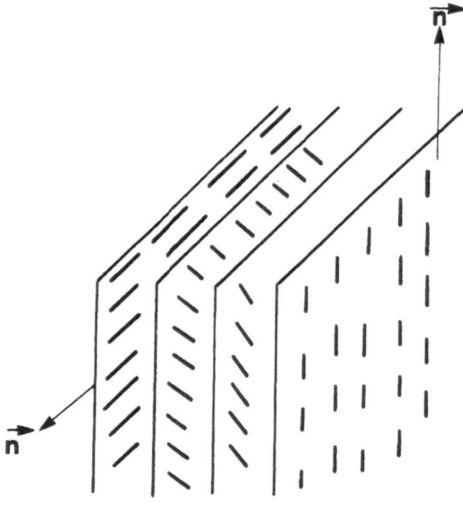

Figure 11.6. Twisted textures of nematic liquid crystals.

solvent. In the low-dilution regime, the pitch obeys a law like $pc = Cte$, where c is the concentration of the chiral compound. A chiral compound giving an intrinsic cholesteric pitch of 1 μm would then give, for instance, a 100-μ pitch at 1% concentration.

11.2.2.2. Optical Properties

We first discuss the normal-incidence regime, where light waves propagate along the helical axis z. The medium is regarded as a twisted uniaxial optical system, which is a valid approximation for cholesterics; for optical waves, the two values of $\varepsilon_\perp$ are almost equal. The Maxwell equations result in the wave equation

$$\Delta \mathbf{E} = \frac{1}{c^2} \frac{\partial^2}{\partial t^2} \mathbf{D}$$

with

$$\mathbf{D} = \varepsilon_\perp \mathbf{E} + \varepsilon_a \mathbf{n}(\mathbf{n} \cdot \mathbf{E}) = \bar{\bar{\varepsilon}} \mathbf{E}$$

$\bar{\bar{\varepsilon}}$ rotates in space resulting in a complicated situation. In a nontwisted medium, the waves traveling along z, or backward, with wave vectors $\mathbf{k}$ or $-\mathbf{k}$ (each with its two states of polarization), are degenerate. Here the degeneracy does not exist and for one given k (or frequency) there are four distinct modes of propagation. The dispersion equation $\omega(k)$ can be solved. The four modes can be shown to be described by elliptically polarized waves in the rotating frame associated with the helix. We just note two important situations.

a. *The Wave-Guide Regime.* When the pitch p and the birefringence $\Delta n = n_e - n_o$ are very large, with ordinary or extraordinary polarizations, the induced polarization remains aligned along the director. This happens when $p \cdot \Delta n \gg \lambda$, where λ is the wavelength of light in the material. In that situation, the eigenmodes of the problem are two linearly polarized vibrations, as for a birefringent untwisted plate. Along the helical axis, these polarizations follow adiabatically the slow twist of the director. This regime is important for applications, especially in the twisted nematic display.

b. *Selective-Reflection Regime.* Bragg reflection conditions arise when the wavelengths in the material are equal to the pitch, i.e., when $\lambda_e = n_e p$ and $\lambda_o = n_o p$. To understand why, let us decompose the optical wave into two circularly polarized components. The one that rotates physically in the same way as the cholesteric texture does not see any index change. Its propagation is unaffected by the texture. The other one is strongly reflected by the periodic change in refractive index. Let us decompose this wave into two rectangular components 1 and 2 (see Figure 11.7). When propagating inside the twisted medium, component 1 sees an increase in the refractive index, for instance, although component 2 sees a decrease, and *vice versa*. The reflected wave due to these index changes is then a circular wave, which rotates physically in the *opposite* sense. The change in angular momentum of light is compensated by a torque density twisting the cholesteric texture. In the tricky language of optics, a circularly *left*-polarized wave will be reflected by a cholesteric as circularly *left* wave. This property allows one to define such a cholesteric texture as a *left* one. The situation is just opposite to that of a mirror reflection, where the physical sense of the electric-field (**E**) rotation is conserved (as well as the angular momentum of the waves), although for opticians a *left*-handed polarized wave becomes after reflection a *right*-handed wave.

The optical Bragg reflection occurs in fact for all wavelengths between $n_o p$ and $n_e p$, where a forbidden band exists for wave propagation of the matched circular polarization.

11.2.2.3. Oblique Case

This problem cannot be solved analytically. Computer analysis is necessary to obtain quantitative predictions. One can make a few simple remarks.

First, when propagating at some angle to **z**, one does not see a pure harmonic index variation at the wave vector q_o. This implies higher-order Bragg reflections, which did not exist for normal propagation. The second remark concerns chiral C*. Although in normal incidence the optical periodicity of the refractive index is $p/2$, for an oblique propagation, because of the molecular tilt in the smectic planes, the fundamental period is p, which implies a new lower-frequency Bragg reflection.

Figure 11.7. Decomposition of circularly polarized wave into rectangular components.

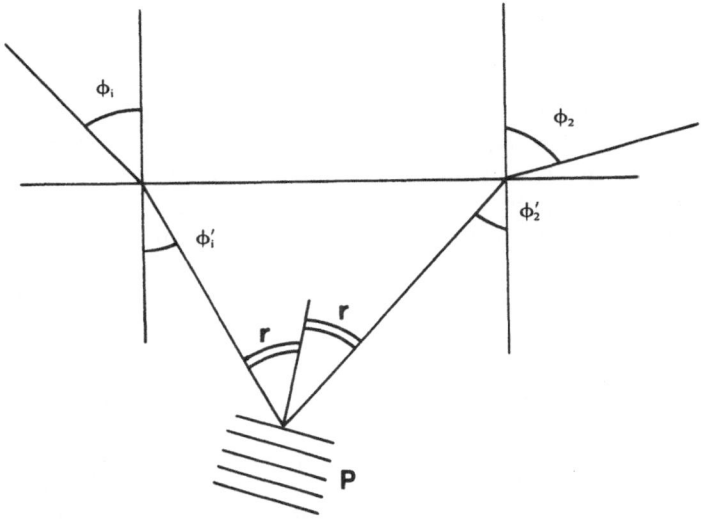

Figure 11.8. Powder-diagram geometry leading to equation (11.2).

In the limiting case of a weak birefringence cholesteric, one can easily relate the reflected wavelength λ to the "powder-diagram" geometry (Figure 11.8). Let us call n the mean refractive index. The Bragg condition yields $p \cos r = m \cdot (\lambda/n)$, where m is some integer. As $2r = \phi_i' + \phi_r'$, we derive

$$\lambda = \frac{p \cdot n}{m} \cos\left[\tfrac{1}{2}\sin^{-1}\left(\frac{\sin \phi_i}{n}\right) + \tfrac{1}{2}\sin^{-1}\left(\frac{\sin \phi_r}{n}\right) \right] \qquad (11.2)$$

which is often useful for cholesteric applications to thermography.

11.2.2.4. Applications of Cholesterics

The Bragg reflection described previously is well known. For instance, it is responsible for the bright colored (and circularly polarized!) reflection from some beetles, the chitinous skin of which presents a *solid* cholesteric ordering. The idea of using selective reflections from cholesterics is related to the extreme sensitivity of the pitch of this *liquid* phase to external agents; p varies significantly when dissolving some external molecules in the cholesteric. A change of pitch, observed through a change in reflected color, would indicate, say, the presence of some contaminating vapor. We shall discuss here only the more important applications related to the temperature sensitivity of the cholesteric pitch.

Most cholesteric materials have a temperature-sensitive pitch. The relative variation is very often large, of the order of $dp/pdT \sim 100/°C$ with $dp/dT < 0$. A sensitivity of 10^{-3} °K is easily achieved, since a 10% change in p usually exceeds the bandwidth of the forbidden band $\Delta\lambda = \Delta n \cdot p$. The generally accepted explanation of this large temperature effect is that the spontaneous twist results from the balance between a chiral-inducing source and the nematic tendency to align uniformly. This is seen in the cholesteric free energy as follows: in each term, $\mathbf{n}$ must appear with even powers, since $\mathbf{n} \equiv -\mathbf{n}$ (no ferroelectricity), while if the molecules are chiral there can be a linear term in ∇. The free energy of a cholesteric is then given by

$$\tfrac{1}{2}K_2(\mathbf{n}.\text{curl }\mathbf{n})^2 + \alpha\,\mathbf{n}.\text{curl }\mathbf{n} \equiv \tfrac{1}{2}K_2(\mathbf{n}.\text{curl }\mathbf{n} + q_0)^2 \qquad (11.3)$$

with $q_0 = \alpha/K_2$. The chiral power α is proportional to the chiral concentration. The temperature dependence of q_0 originates from the K_2 term. When decreasing T, one very often obtains a smectic phase. If the compound crystalizes first, there may exist a "virtual" smectic phase, which can be shown by analyzing the phase diagram of mixtures. The smectic order fluctuates close to a smectic phase (so-called "cybotactic" groups), often visible with X-rays. Twist is forbidden in the presence of a smectic order, because one cannot twist the molecular director without changing the smectic-layer thickness. As a result, K_2 diverges when cooling the cholesteric, and q_0 tends to vanish.

This effect has been used in thermography to determine possible spatial changes in the surface temperature. The surface is coated with a thin cholesteric film to minimize thermal inertia and thermal short circuit, but thick enough to allow more than 10 or 20 pitches of the helix, to give a reasonably sharp Bragg reflection. The material is sometimes encapsulated to prevent contamination. With a fixed optical geometry, the change in reflected color from white-light illumination follows the temperature change of the pitch. This technique has been used for nondestructive mechanical tests of materials under stresses, for visualization of RF waves in wave guides, and for medical application (trying to relate skin temperature and possible development of tumors). Historically, the first goal was infrared detection for military purposes. This application does not seem too appealing because of the large noise (the cholesteric cannot be cooled), the slow response time related to the kHz helix-stability frequency Kq_0^2/η, and hysteresis due to defects. It is interesting to note that most developed countries have undertaken similar studies on that topic, starting from US–German collaboration in Fort Belvoir. The appearance of low viscosity materials could revive the field. To complete this review we cite the sensitivity of the cholesteric pitch to an external uniaxial pressure. A medical application to podology has been proposed, to measure the transient pressure transmitted by a walking foot on the ground.

In conclusion, let us mention an aesthetic point. As the pitch of cholesterics generally increases with decreasing temperature, the hot regions in thermography appear blue or green compared to the cold, red regions. With the use of "reentrant" cholesterics, one can invert the color sequence and give a more pleasant appearance of hot becoming red and cold remaining blue. This idea was recently patented.

11.2.3. Use of Dichroic Dyes in Solution

The anisotropic properties of liquid-crystal phases can lead not only to special refraction or propagation effects on light waves, but also to absorption, through the use of oriented dichroic dyes. The principle is elementary. A dichroic molecule will absorb an optical radiation with an electric field parallel to its long axis, for instance. The dye is aligned in a nematic uniform texture. By using suitable electric fields, one can tilt the texture and pass from an absorbing to a nonabsorbing state. In white light this results in a strong color change, e.g., from red to green.

An important parameter is the angular order parameter given by

$$S = \tfrac{1}{2}\langle 3\cos^2\theta - 1\rangle$$

defined previously for the nematic and smectic phases. The order parameter of the anisotropic dye molecule S_D will be related to S. If the dye is a rigid rod, one expects, for instance, $S_D > S$ if the dye length is larger than the liquid-crystal molecule length and $S_D < S$ in the opposite case.

Let us show how the contrast of a dichroic-dye display depends on S_D. We assume that the dyes are all perfectly aligned in a sample of thickness d. If $\alpha(\theta)$ (see Figure 11.9) is the specific absorption per unit thickness, θ being the angle between the polarization $\mathbf{E}$ of the light wave and the induced dipole $\mathbf{p}$ along the dye molecule, the sample transmission can be written as

$$T = \exp[-\alpha(\theta)\cdot d]$$

For dipolar electric transition, the absorption will be proportional to $\mathbf{E}\cdot\mathbf{p}$, i.e., to $\cos^2\theta$, where $\alpha(\theta) = \alpha_M\cos^2\theta$. One can easily measure α_M by dissolving the dye in the isotropic phase of the liquid crystal. The isotropic absorption α_0 is given by

$$\alpha_0 = \alpha_M/3$$

since $\langle\cos^2\theta\rangle = \tfrac{1}{3}$.

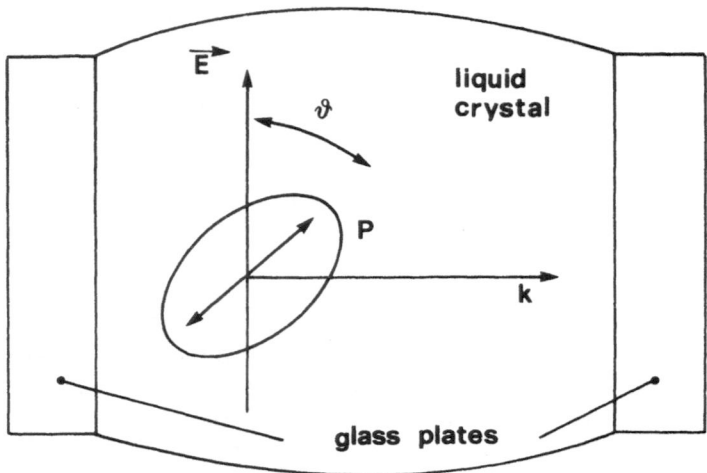

Figure 11.9. Transition dipole along the dye molecule in relation to polarization of the light wave.

We can now estimate the contrast of the display, limited by the imperfect dye alignment ($S_D < 1$), by comparing the transmissions of the absorbing and transmitting geometries, $\alpha_\parallel$ and $\alpha_\perp$, respectively (Figure 11.10):

$$\alpha_\parallel = 3\alpha_0 \langle \cos^2 \theta \rangle = \alpha_0 (2S_D + 1)$$

$$\alpha_\perp = 3\alpha_0 \langle \cos^2 \psi \rangle = \frac{3\alpha_0}{2} \langle \sin^2 \theta \rangle = \alpha_0 (1 - S_D)$$

The contrast k can be expressed simply as

$$k = \exp[-\alpha_0 (1 - S_D)d]/\exp[-\alpha_0 (2S_D + 1)]$$
$$= \exp 3\alpha_0 S_D d \qquad (11.4)$$

There is an obvious additional constraint, namely the transmitting state must be reasonably clear. Suppose we set, for instance, $T_\perp \sim 0.5$. This limits the sample thickness to

$$d \sim \frac{0.7}{(1 - S_D)\alpha_0} \qquad (11.5)$$

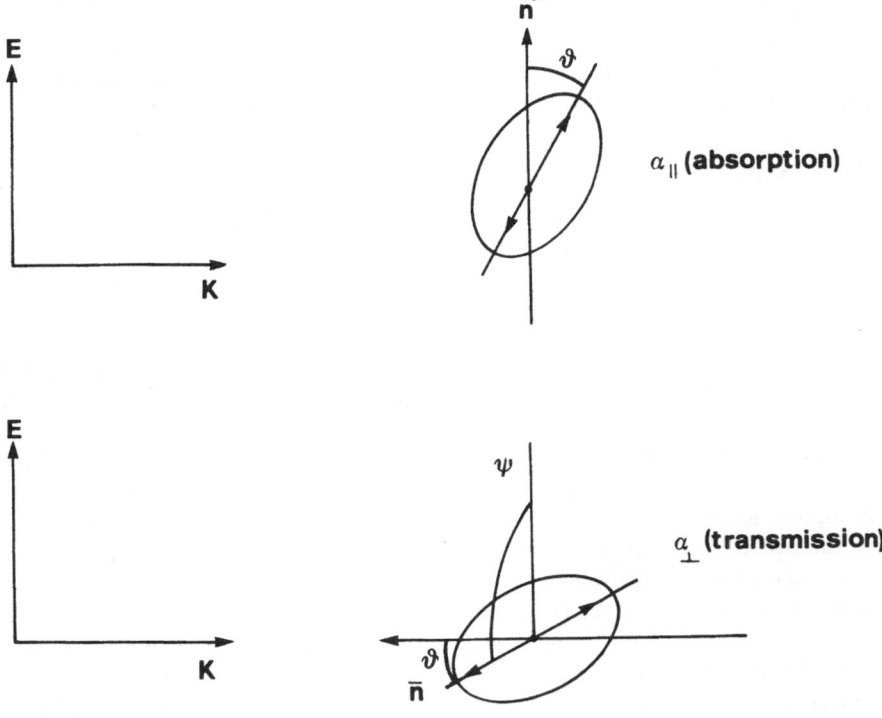

Figure 11.10. Absorbing and transmitting geometries.

Table 11.1. Relation Between Contrast k and
Order Parameter S_D of the Dye

S_D	k
0.1	1.26
0.5	8.0
0.7	128
0.8	4440

The maximum contrast is now

$$k = \exp\left(\frac{2.1 S_D}{1 - S_D}\right) \tag{11.6}$$

Table 11.1 shows that k is a very rapidly growing function of S_D.

The practical problem for dichroic-dye displays is to find a sufficiently stable dye molecule, which aligns with a good S_D. As expected, the best stable compounds have given a poor S_D, and *vice versa*. A typical case is the sudan black molecule with $S_D = 0.5$ in a stilbene liquid-crystal matrix.

In practice the first, previously discussed, proposed geometry is not the more convenient, since it implies the use of a polarizer. A simple idea was to use a cholesteric orientation to start with (Figure 11.11). There is no longer any need for a polarizer, resulting in a cheaper display. The drawback is the necessity of a higher voltage. It is noteworthy that for complete absorption, the cholesteric texture must have a pitch comparable to the optical wavelength. A twisted nematic on a much larger scale would not work because of the wave-guide regime for the incoming light. A last aesthetic problem derives from the fact that white appears in the presence of the electric field **E**, giving a white-on-color display. Various tricks are used to produce the reverse effects, by the use of dielectric negative materials, smectic matrices, complementary electrodes, and so on. With the advances made in chemistry, one can expect dichroic displays to take over most of the market supplying simple, slow, nonmultiplexed, cheap displays. For more comprehensive information on dichroic dyes the reader is referred elsewhere.[47]

11.3. Texture Distortions Under the Action of an Electric Field

In this section, we discuss texture changes due to the application of DC or AC electric fields on nematic and cholesteric materials. We present

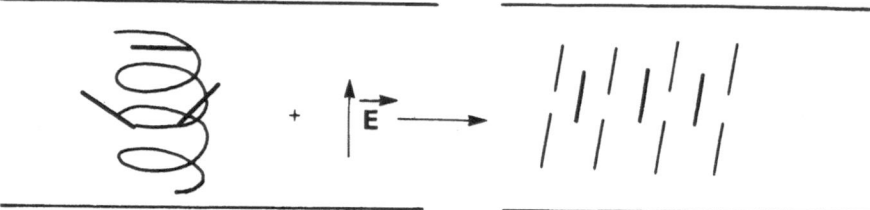

Figure 11.11. Cholesteric orientation by an electric field.

successively electrically controlled birefringence, the twisted nematic valve, the dynamic-scattering mode, and the unwinding of cholesterics.

11.3.1. Electrically Controlled Birefringence

We start from a homeotropically aligned nematic liquid cyrstal with negative dielectric anisotropy (Figure 11.12) ($\varepsilon_a > 0$). We use (as in all electrooptic effects) transparent SnO_2 electrodes connected to a DC or AC generator of voltage V. When V increases above a threshold value V_s, the homeotropic sample distorts itself to align the director more nearly perpendicular to the applied field $\mathbf{E}$. The sample becomes birefringent for light waves crossing the electrodes. As the distortion angle α increases (Figure 11.13) with $\mathbf{E}$, one can control the birefringence of the nematic slab and produce electrooptic effects.

The threshold value V_s can be easily calculated. At the threshold, the distortion is a pure bend of wave vector $q = \pi/d$, where d is the sample thickness. In the free energy F, the dielectric term dominates the elastic contribution, so one obtains

$$F = \tfrac{1}{2}\left[K_3 q^2 \alpha^2 - \frac{|\varepsilon_a|}{4\pi} E^2 \alpha^2 \right] \leq 0 \tag{11.7}$$

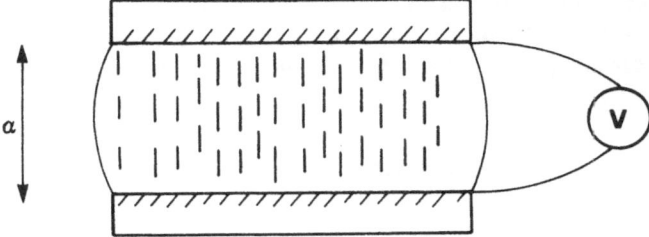

Figure 11.12. Homeotropically aligned nematic liquid crystal.

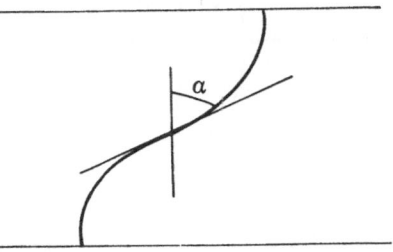

Figure 11.13. Distortion angle α induced by electric field.

i.e.,

$$V_s^2 = K_3 \pi^2 \frac{4\pi}{|\varepsilon_a|}$$

in the range of a few volts, for ε_a approximately equal to 0.5 (as in MBBA). To estimate the equilibrium value of $\alpha(V)$ above the threshold, one needs to expand F up to the fourth (and higher) powers of α. A complication arises from the fact that, when the molecules are tilted, one obviously cannot write $E = V/d$ but one must use a self-consistent treatment of the dielectric term. However, neglecting this difficulty for a rough estimate, we find two terms in α^4, one originating from the splay distortion $\frac{1}{2}K_1 q^2 \alpha^4$ and the other from the saturation of the dielectric energy $-(|\varepsilon_a|/4\pi)(\mathbf{n}.\mathbf{E})^2$. Manipulation involving a little approximative algebra allows F to be expressed as

$$F \sim \alpha^2(E_s^2 - E^2) + \alpha^4(E_s^2 + E^2) \tag{11.8}$$

The important point is that the sign of the fourth-order term is positive, which leads to a continuous second-order transition. Above V_s, the tilt is given by

$$\alpha^2 \sim \frac{V^2 - V_s^2}{V^2 + V_s^2} \tag{11.9}$$

α saturates to $\pi/2$ for a very large field. Electrically controlled birefringence can be introduced by using a nematic slab as a variable optical retardation plate (Figure 11.14). One uses, for instance, interference between ordinary and extraordinary light. To estimate the optical path

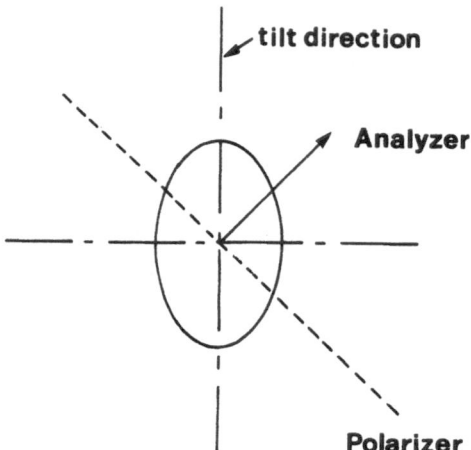

Figure 11.14. Use of a nematic slab as a variable optical retardation plate.

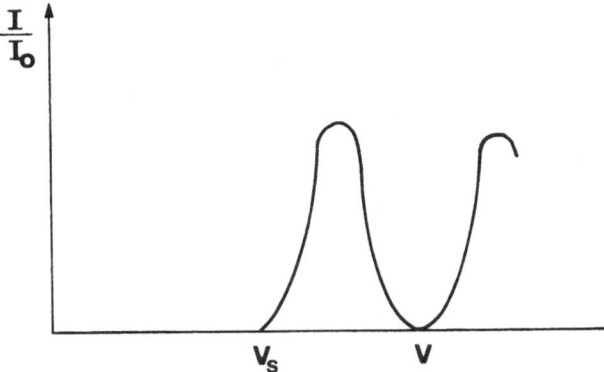

Figure 11.15. Transmitted intensity vs voltage.

difference, we take α as a mean uniform tilt value. For the extraordinary wave, the index is given by

$$\frac{1}{n^2} = \frac{\cos^2 \alpha}{n_o^2} + \frac{\sin^2 \alpha}{n_e^2}$$

(11.10)

which is replaced by the approximate, simpler expression

$$\Delta n = n_o \simeq (n_e - n_o)\sin^2 \alpha$$

(11.11)

In the absence of a field but employing an orthogonal polarizer and analyzer, no light can cross the homeotropic slab. Above the threshold, there appears a dephasing

$$\phi = \frac{d \cdot \Delta n}{\lambda} \cdot 2\pi$$

The transmitted light intensity is

$$I = I_0 \sin^2 \frac{\phi}{2}$$

(11.12)

For small angles, one gets

$$I = I_0 \left(\frac{2\pi d}{2\lambda}\right)^2 \Delta n^2$$

$$= I_0 \left(\frac{2\pi d}{2\lambda}\right)^2 (n_e - n_o)^2 \alpha^4$$

(11.13)

Close to the threshold, with $\Delta V = V - V_s$, the transmitted intensity becomes $(\Delta V/V_s)^2$ and increases sharply (Figure 11.15). With increasing V,

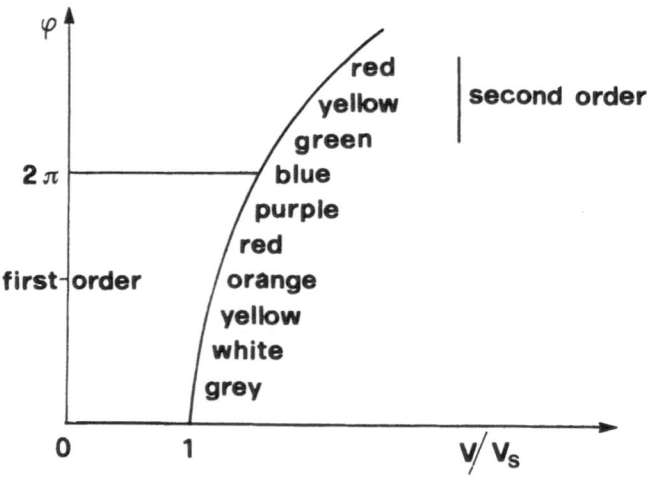

Figure 11.16. Dephasing φ vs voltage for white-light illumination.

one observes with monochromatic light a sequence of interference fringes. The first fringe maximum is obtained when the nematic slab is a half-wave plate:

$$d\Delta n = \lambda/2 \qquad (11.14)$$

$$d^2 = \frac{\lambda}{2dn_o} \simeq \frac{\Delta V}{V_s} \sim 5 \cdot 10^{-2} \qquad \text{for } d \sim 20 \ \mu\text{m} \qquad (11.15)$$

The very steep increase in I above the threshold gives the electrically controlled birefringence a large multiplexing capability (see Section 11.4).

With white-light illumination, one observes above the threshold first a scale of gray followed by the sequence of the Newton color scale (Figure 11.16). In practice, the azimuthal degeneracy allows a tilt in any direction and leads to domain formation. One minimizes the spurious domain effects by using a small pretilted homeotropic plate coating, or by using circular instead of linear polarization. In that case some point defects remain.

One problem with electrically controlled birefringence is the good thickness control necessary to obtain uniform interference effects; this may lead to an expensive system for large-area display. Another problem is the lack of large negative ε_a materials with good alignment capability. Finally, it should be noted that the applied-field polarity is irrelevant, since coupling with $\mathbf{E}$ is quadratic in the electric field. This explains why AC excitation is possible.

11.3.2. Twisted Nematic Valve

Let us consider a planar nematic (Figure 11.17), where the two plates have been twisted by $\pi/2$. If the thickness d is much larger than the optical wavelength, one easily achieves the condition for the wave-guide regime, $4d\Delta n > \lambda$ (e.g., with $d \sim 10$ μm and $\Delta n = 0.2$). An optical beam propagating upward through the twisted slab, conveniently polarized (P), will see its polarization turned by $\pi/2$ and be transmitted by analyzer A. When we apply a voltage V to the sample (with the same, previously described, set of transparent electrodes), we obtain a homeotropic alignment for a dielectrically *positive* ($\varepsilon_a > 0$) material. The wave-guide effect disappears, and the incoming light beam is blocked by the analyzer A. This is the principle of the twisted nematic light valve.

The voltage threshold (independent of sample thickness) for twisted planar to homeotropic orientation can be estimated by an approach analogous to that of the previous section. The exact result is

$$V_s^2 = \frac{1}{4\pi |\varepsilon_a|} \left(\frac{\pi}{2}\right)^2 (4K_1 - 2K_2 + K_3) \qquad (11.16)$$

The minus sign before $2K_2$ indicates that the twist energy stored initially in the twisted nematic valve decreases the threshold value by converting into bend and splay. With a typical value of ε_a approximately equal to 10, V_s can be lower than 1 V.

If the tilt of the director relative to the boundary glass plates is θ, it can be calculated as in the case of electrically controlled birefringence. A

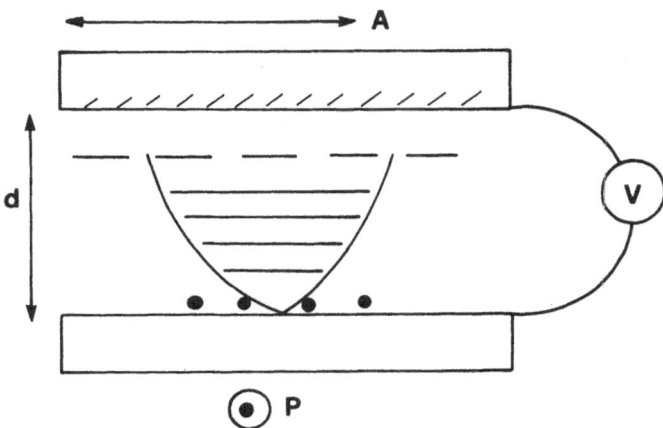

Figure 11.17. Twisted nematic valve.

typical plot of $\theta(V)$ is shown in Figure 11.18. It is important to estimate the *optical* threshold V_0 of the twisted nematic valve. The wave-guide condition is

$$\Delta n(\theta) \cdot 4d > \lambda \qquad (\text{since } p = 4d)$$

Let us assume that the optical threshold appears for $\Delta n(\theta) \cdot d \sim \lambda/2$. As $\theta(V)$ follows the same law as $\alpha(V)$ from the electrically controlled birefringence effect, we see that the optical threshold V_0 of the twisted nematic valve corresponds to the maximum of light transmission in electrically controlled birefringence, definitely above V_s. As the angular distortion θ saturates when increasing V, the transmitted light decreases slowly above V_0. The relatively wide, corresponding voltage range (Figure 11.19) is a serious limitation to the intrinsic multiplexing capability of the twisted nematic valve (see Section 11.4).

In practice, one can use parallel or orthogonal polarizers to make a high- or low-voltage transmission. Domains could appear, of opposite twist, with disclination lines in between. One prevents these defects by dissolving a small quantity of a chiral compound, to define a spontaneous sense of twist.

An important static property of the twisted nematic valve is its angle of view (Figure 11.20). The strong asymmetry is easy to understand. For light beam 1 (see Figure 11.21), the texture remains a helix. The optical threshold is very much larger than V_s. For light beam 2, V_0 and V_s are comparable.

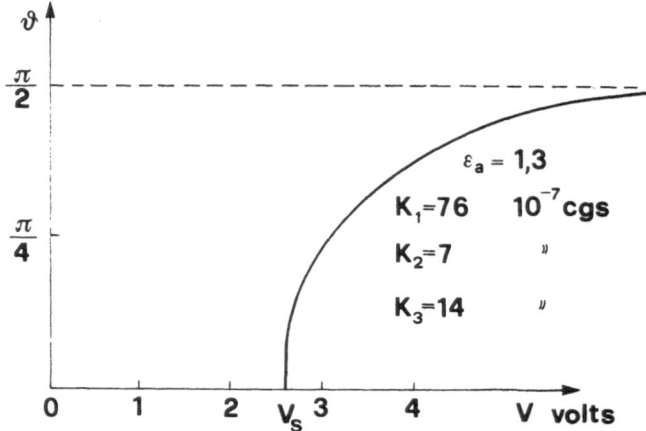

Figure 11.18. Tilt of director vs voltage.

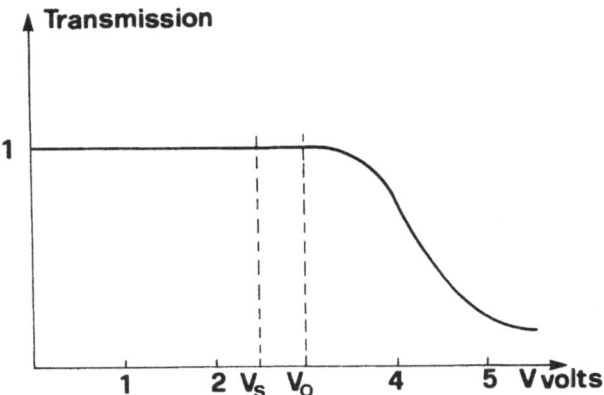

Figure 11.19. Transmission vs voltage.

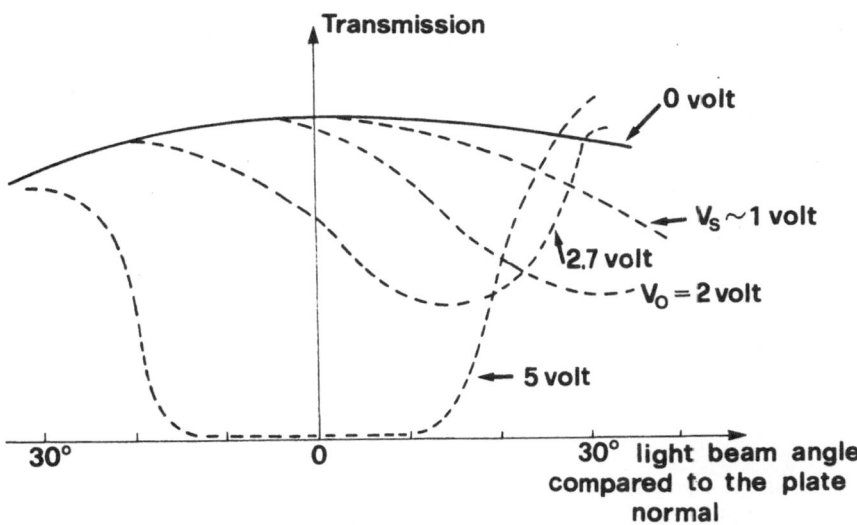

Figure 11.20. Transmission as a function of light-beam angle.

The dynamic properties of the twisted nematic valve have been studied extensively. An important result is the bump observed in the decay of transmission, e.g., in absence of field **E** (Figure 11.22). In normal incidence, the observed decay is controlled by the curvature elasticity torques. For a 10-μm thickness, the decay time is ~0.1 s. Not too far above the threshold, the peak in transmission corresponds exactly to the aniso-

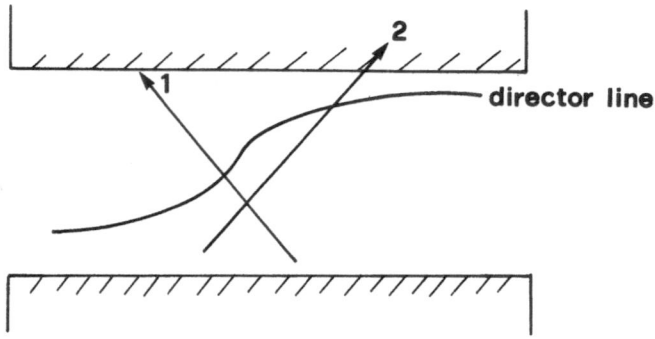

Figure 11.21. Two light beams shown together with director line.

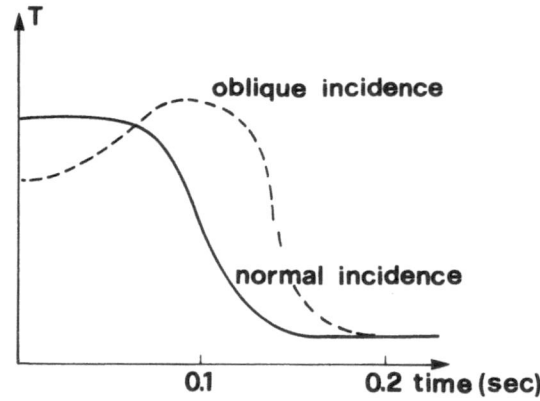

Figure 11.22. Transmission vs time for two incidences.

tropy of the transmitted light for different voltages, i.e., it is a purely static effect. Higher above the threshold, the "backflow" induced by the nonuniform rotation of molecules rebuilds a transient helix, which increases the observed bump. In the steady state, as for the electrically controlled birefringence mode, AC and DC excitations are possible.

11.3.3. Dynamic-Scattering Mode

The two previous effects were pure field phenomena. Dynamic-scattering mode implies the presence of space charges in the liquid crystal, i.e., a certain ionic conductivity. The geometry used is generally that of a planar orientation (Figure 11.23). One uses a "negative" ($\varepsilon_a < 0$) nematic material. Purely static effects can just stabilize the planar texture. The instability can be described as follows: Space charges accumulate in the presence of a small bend due to the anisotropy of the (ionic) conductivity; these space charges are dragged by the field and induce a convective flow on the scale of the thickness d. If the applied voltage is high enough, this flow can increase the initial bend fluctuation and generate a steady-state instability. The effect appears both in DC and AC excitations, provided the sign of the space charges can change when reversing the field. This limits the frequency of the applied field to the space-charge relaxation frequency $4\pi\sigma/\varepsilon$, where σ is the conductivity of the nematic liquid crystal (in practice, from 10 Hz for pure samples to 1 kHz for high-conductivity impure samples). Other instabilities appear at higher frequencies.

To obtain the dynamic-scattering mode, one must increase the voltage well above the threshold voltage for the appearance of steady-state vortices and of a periodically bent texture, i.e., well above a few volts. Typically, for a 20–30 V excitation, the excited vortices break randomly and the whole texture scatters strongly an incoming light beam. The dynamic-scattering mode does not require the use of polarizers; its angle of view is intrinsically wider than the twisted nematic valve with the drawback of a larger energy consumption, related to the larger voltage and the residual conductivity. All studies of the dynamic-scattering mode are empirical; no model is known that explains the transition from the well-ordered vortices to the chaotic dynamic-scattering mode. In practice, for instance, the relative voltage range of the optical effect above the threshold of the dynamic-scattering mode compares with (or is a little smaller than) that of the twisted nematic valve. The dynamic-scattering mode does not have a very large multiplexing capability.

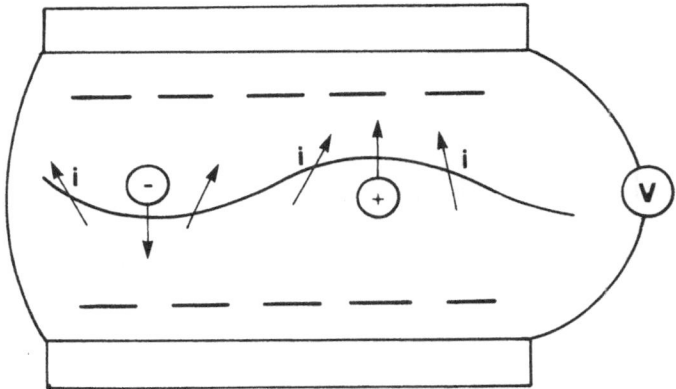

Figure 11.23. Geometry of the dynamic-scattering mode.

An important remark about the mechanism described to understand the onset of flow in a perfectly dielectrically stable texture is that this mechanism can appear in cholesteric or smectic textures and can sometimes provoke spurious destabilizations. To prevent such a situation, one uses simply an AC field of sufficiently high frequency (larger than the charge relaxation frequency).

Recent investigations in "discotic" materials (in the fluid nematic phase) have shown the existence of electrohydrodynamic instabilities analogous to that of "calamitic" materials.

11.3.4. Texture Changes in Cholesterics

In the absence of an electric field, a cholesteric phase (see Figure 11.24) can exist in a "planar" texture (a) with helical axis perpendicular to

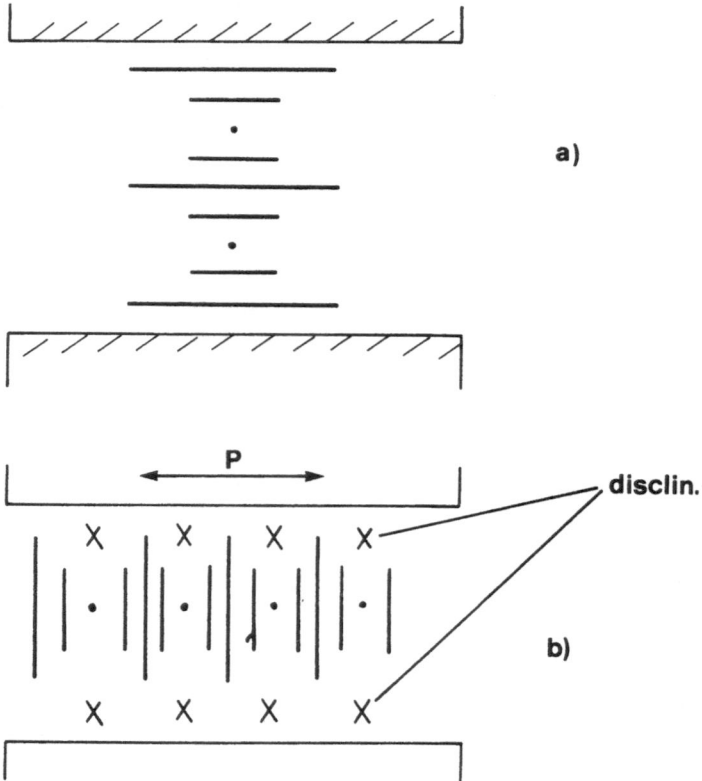

Figure 11.24. Textures of cholesteric phase.

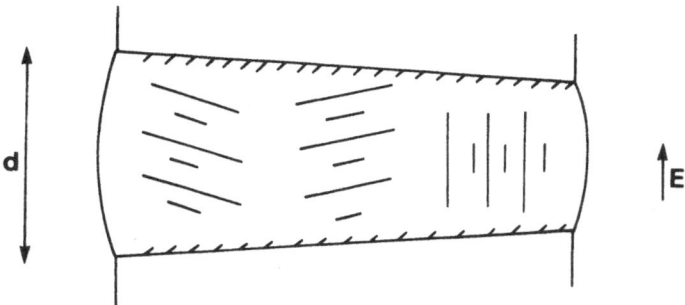

Figure 11.25. Focal-conic disordered texture.

the plates, or in a "homeotropic" texture (b). The latter cannot exist without disclinations. We shall assume a homeotropic orientation on the plates. This orientation matches the regions where the director is vertical, but cannot match continuously in between.

We shall now assume a homeotropic orientation of a cholesteric texture with positive dielectric anisotropy. The texture can be unwound under the action of a DC or AC transverse electric field. The threshold can be estimated by balancing the increase in elastic free energy $\frac{1}{2}K_2 q_0^2$ by the decrease in the dielectric term $-\frac{1}{2}(\varepsilon_a/4\pi)E^2$. One obtains a field threshold

$$E = \frac{\pi}{2}\left(\frac{K_2 4\pi}{\varepsilon_a}\right)^{1/2} q_0 \qquad \text{with } q_0 = \frac{2\pi}{p}$$

As the sample thickness is usually a few times the pitch p, the applied voltage across the sample remains within the range of a few volts. The cholesteric unwinding implies the suppression of the boundary disclination lines. This effect is negligible for thick samples, where bulk energy dominates. For thin samples ($d \sim p$), boundary effects lead to an important bistable behavior. We start from a "focal-conic" disordered texture, which can be considered as a polydomain cholesteric texture (Figure 11.25). The glass plates have been treated to induce homeotropic orientation. Due to the intense scattering from the distorted domains, the liquid-crystal cell does not transmit much light. On increasing the applied field, one induces the striped texture (Figure 11.26) of the homeotropic texture alignment and above $E_\uparrow$, one aligns the liquid crystal in a nematic homeotropic texture by unwinding the helix. On decreasing the field, the texture remains nematic down to $E_\downarrow$, where the texture becomes conical (Figure 11.27) (i.e., a "planar" cholesteric texture with continuous bend to match the homeotropic boundaries). On again decreasing E, the conical texture breaks again spontaneously to a disordered focal-conic texture. The remarkable feature

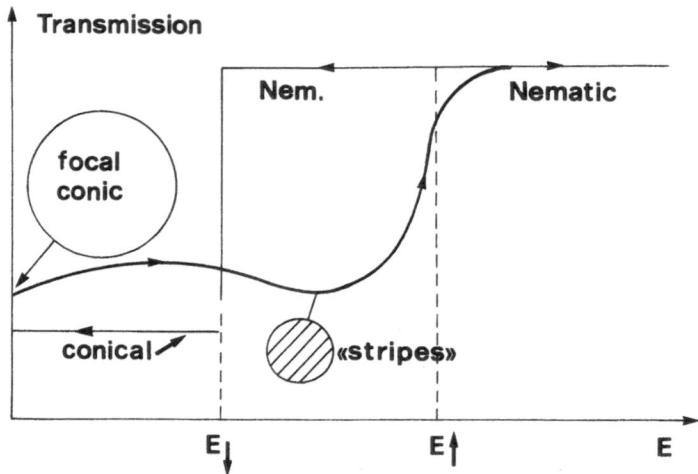

Figure 11.26. Transmission vs electric field.

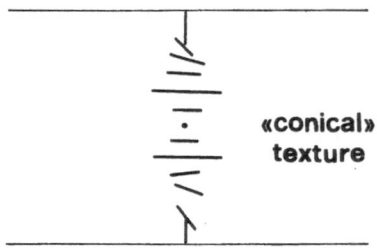

Figure 11.27. Conical texture.

of this field effect is the *bistable* behavior of the texture between $E_\uparrow$ and $E_\downarrow$. Depending on the previous history of the sample, one can get transparent nematic or scattering cholesteric textures. This hysteresis can be explained by the difficulty in nucleating surface disclinations when the elastic free energy stored in the bulk is too small, i.e., when d is comparable to p. Of course, scratches on the surface allow easy nucleation of defects and the bistable range is reduced or suppressed. This bistable range is characterized by the ratio

$$\frac{E_\uparrow}{E_\downarrow} = \pi \left[\frac{K_2 K_3}{4K_2^2 - K_3^2 (p/d)^2} \right]^{1/2} \sim \frac{\pi}{2} \frac{1}{(1 - p/2a)^{1/2}} \qquad (11.17)$$

The texture always remains nematic for thinner samples, $d \sim p/2$.

A more efficient way to couple light waves for making a display with the bistable cholesteric texture change is to dissolve in it a dichroic dye, rather than to use the simple scattering of the focal-conic texture. To obtain a colored appearance when the field is on, one uses a negative material ($\varepsilon_a < 0$). A small thickness is chosen to induce a spontaneous homeotropic texture in the absence of a field. The texture becomes cholesteric under the application of the electric field

$$E_\downarrow = \pi \frac{K_3}{\varepsilon_a} \left(\frac{2}{p} \frac{K_2}{K_3} - \frac{1}{d^2} \right)^{1/2} \tag{11.18}$$

giving the absorbing colored state. One difficulty is that, in nematic materials, the dye order parameter is smaller, resulting in a lower contrast. The advantages of such a system compared to the twisted nematic valve are a better multiplexing capability (up to 100 lines, with a few seconds of memory time and a response time of approximately 10 ms), a wider angle of view, the absence of polarizers (brighter display, robust, and low cost), and the possible use of active substrates.

More on the topics treated here can be found elsewhere: on electrically controlled birefringence,[48] on the twisted nematic valve,[49,50] on the dynamic-scattering mode,[51] and on cholesterics.[52]

11.4. Multiplexing of Liquid-Crystal Displays

In this section, we discuss some techniques used in liquid-crystal matrix imaging, in connection with some material physical properties. We first describe briefly the dynamic response of texture instabilities under the action of an electric field. We then explain the principle of multiplexing techniques with liquid crystals and of parallel image addressing.

11.4.1. Response Time of Liquid-Crystal Texture Instabilities

We shall content ourselves here with a sketch of somewhat naive arguments to predict the dynamic behavior of textures.

11.4.1.1. Decay Time

Suppose we have a distorted texture under the action of an electric field **E**. We turn off **E** instantaneously at a certain time t_0. How will the optical effect induced by the field change with time? The decay of the optical effect is obviously related to the return to equilibrium of the distorted texture under a curvature elasticity torque. If θ denotes the director distortion, the equilibrium equation of torques in the absence of inertia has already been written in the form

$$Kq^2\theta = \eta\dot{\theta} \tag{11.19}$$

where η is some viscosity and q is the distortion wave number approximately equal to π/d (d being the sample thickness). The damping time τ_D, defined by $1/\tau_D = Kq^2/\eta$, is of the order of 100 ms for $d \sim 10$ μm. It should be noted that this time is comparable with the persistence time of the eye, also related obviously to the transmission time of the television frame.

11.4.1.2. Rise Time

In the presence of a destabilizing field **E**, the dynamic equation becomes

$$\left(-\frac{\varepsilon_a}{4\pi}E^2 + Kq^2\right)\theta = \eta\dot{\theta}$$

One can define a field-dependent response time by

$$\frac{1}{\tau_R} = \frac{1}{\tau_D} - \frac{1}{\tau_E} \qquad \text{with} \quad \frac{1}{\tau_E} = \frac{\varepsilon_a E^2}{4\pi\eta} \tag{11.20}$$

The response time is infinite at the threshold; this is the mechanical equivalent of thermodynamic slowing-down close to a second-order phase transition. Above the threshold V_S, the *rise* time can be written as

$$\frac{1}{\tau_R} \simeq \frac{1}{\tau_D}\frac{V^2 - V_S^2}{V_S^2}$$

and the texture distortion is expected to grow as $\theta = \theta_0 \exp(t/\tau_R)$, where θ_0 is some initial distortion due to a small pretilt, or simply to thermally excited angular fluctuations. To achieve the full, expected optical effect, such as a given birefringence $\Delta n \sim \lambda/2d \sim \theta^2$, one needs a response time

$$t_R \sim 2\tau_R \log\left(\frac{\lambda}{2d\Delta n_0\theta_0^2}\right) \sim 10\tau_R$$

for a typical pretilt of 1°. Increasing the voltage, one can always decrease $\tau_R \sim \tau_D(V_S^2/V^2)$. Suppose, for instance, τ_D is approximately 40 ms (a television-frame time); then τ_R can be reduced to a "line" time of about 64 μs by taking $V \sim 80V_S \sim 250$ V. In principle, the short-pulse high-voltage excitation of a liquid-crystal cell would allow a large multiplexing capability (see below) of the order of $N \sim \tau_D/t_R$. Unfortunately high voltages are not convenient for liquid-crystal displays, so they are not widely used.

Another important response time is that of the dielectric-relaxation liquid-crystal cell. Most instabilities are field-induced; however, one certainly requires that the applied field $\mathbf{E}$ is forgotten after a dielectric-relaxation time $\tau = (4\pi\sigma/\varepsilon)^{-1} = RC \sim \tau_D$. Taking $\tau_D \sim 40$ ms and $\varepsilon \sim 10$, this implies for a 10 μm $\times$ 1 cm^2 cell a finite resistance $R \sim 4 \cdot 10^6$ Ω; the lowest electric consumption for V equal to a few volts is in the μW range. This figure must be compared to the 500 mW necessary to activate a seven-segment LED. This comparison demonstrates the power-consumption superiority of a *passive* rather than an *active* display.

The τ_D adjustment to the frame time ($T \sim 40$ ms) is achieved by a thickness adjustment ($\tau_D \sim d^2$) and the choice of a liquid crystal with sufficiently low viscosity. The recent use of cyclohexane compounds (instead of the two classical phenyl rings) has done much to advance that aim. Another approach was to use the change in dielectric anisotropy, which is demonstrated by certain compounds in the low-frequency range around 10 kHz. At low frequency, these compounds are positive ($\varepsilon_a > 0$), and negative ($\varepsilon_a < 0$) at high frequency. If a DC field is destablilizing, a high-frequency ($f > f_r$) field is stabilizing and can drastically reduce τ_D. Devices using this technique have been demonstrated but appear now to have been abandoned, due to their large temperature sensitivity (f_r varies rapidly, approximtely 1 kHz $\cdot$ deg C) and their intrinsic larger consumption.

11.4.2. Multiplexing Liquid Crystals

Simple displays, like the seven segments used in liquid-crystal watches, can be realized by simply using one separate electrode and one connection for each segment (and a common background electrode). This technique is no longer feasible for more complex images. One uses matrix addressing with N lines and M columns, which reduces the NM connections to $M + N$, i.e., by three orders of magnitude for a television frame. The principle is simple. The points to be displayed are represented schematically by the intersections of the N rows and M columns (Figure 11.28). The rows are excited sequentially by voltage V_1 (zero when there is no excitation). The M columns are excited in parallel by voltages $\pm V_2$. Rows and columns represent the upper and lower electrodes of the resulting MN liquid-crystal cells. A given cell of the excited row will be turned "on" if the voltage difference $V_1 - (-V_2) = V_1 + V_2$ is larger than the threshold voltage V_s; for the other points not to remain in the "off" state, one requires simply $V_1 - V_2 < V_s$. In "static" operation, the only constraint to

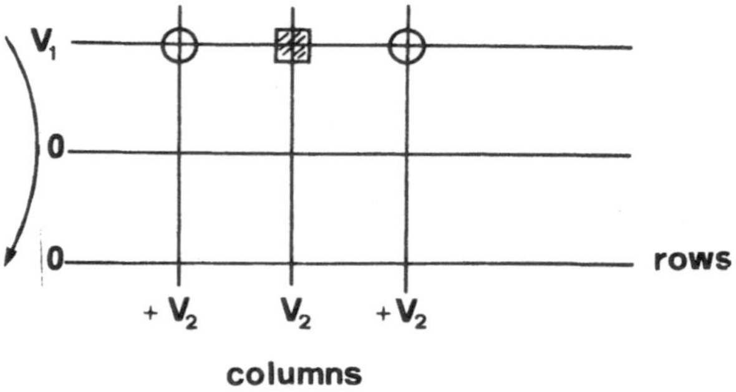
V1
0
0
rows
+V2
V2
+V2
columns

Figure 11.28. Matrix addressing.

achieve matrix addressing is the existence of a voltage threshold for the electrooptic effect that is used.

Static excitation has no interest, since the N rows are sequentially excited. To realize multiplexing, one needs to fulfill additional conditions. The excited material must retain the information until all other elements are excited, but it must be turned "on" and "off" within the period of one frame in order to prevent smearing. If something more subtle than an "on-and-off" working cycle is required, such as a gray scale in television imaging, one needs a continuously variable contrast. Multiplexing techniques have been invented to fulfill these requirements in $X-Y$ matrix addressing.

The *time-dependent mode* was first used to multiplex liquid-crystal displays. The signal seen by one given liquid-crystal cell is made up of $(N-1)$ apparently random pulses $\pm V_2$, and of a larger pulse $V_1 \pm V_2$ when its row is excited, $\pm$ meaning that the cell must be "on" or "off." If $V_1 + V_2$ is large enough so that the electrooptic effect is recorded during the time T/N of a line excitation, and so that the decay time in the presence of V_2 is adjusted to satisfy $\tau_D(V_2) \sim T$ where T is the frame time, then the multiplexing condition for a simple "on" or "off" cycle is fulfilled. As estimated earlier, the large necessary voltages are not very convenient if one is willing to use integrated circuits for the $X-Y$ video signals. Liquid-crystal multiplexing is generally achieved in the *voltage-selection mode*, described here for the "on" and "off" regimes. The basic idea is that, because of the relatively low voltage used, the reponse time of the liquid-crystal cell is larger than, or comparable with, the frame time T. The liquid-crystal cell is sensitive to the mean-square value of the applied voltage. In the case of one frame for which a given cell must be "on," this mean value is given by

$$V_{\text{on}}^2 = \frac{1}{N}[(V_1 + V_2)^2 + (N-1)V_2^2] \tag{11.21}$$

while the mean value is given by

$$V_{\text{off}}^2 = \frac{1}{N} [(V_1 - V_2)^2 + (N - 1)V_2^2]$$ (11.22)

when the cell must be "off." The value of V_{on} must be well above the threshold value, around the voltage $V_s + \Delta V$ (Figure 11.29), where the optical effect saturates. The value of V_{off} cannot be larger than V_s. For a given multiplexing power N, $V_{\text{on}}/V_{\text{off}}$ is maximum for $V_1/V_2 = N^{1/2}$. Writing $V_{\text{off}} = V_s$ results in

$$V_2 = \frac{V_s}{2^{1/2}(1 - N^{-1/2})^{1/2}} \sim \frac{V_s}{2^{1/2}}$$ (11.23)

for large N, with $V_1 = V_s/2^{1/2}N^{1/2}$. With increasing N, the relative voltage spread $r = \Delta V/V_s$ of the optical effect used is given by

$$r = \left(\frac{N^{1/2} - 1}{N^{1/2} + 1}\right)^{1/2} - 1$$ (11.24)

Table 11.2 shows that if one is willing to *maintain the full contrast* of the

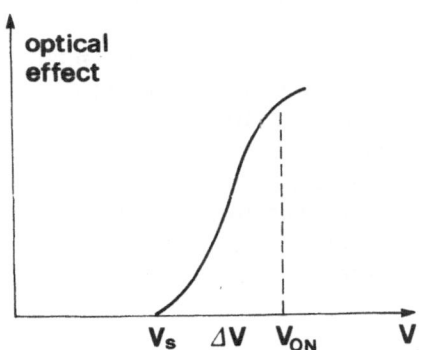

Figure 11.29. Illustration of the voltage-selection mode.

Table 11.2. Variation of Multiplexing
Powe N with Relative Voltage Spread
$r = \Delta V/V_s$ of the Optical Effect

N	r
2	1.2
4	0.7
8	0.45
128	0.093
256	0.064
512	0.045

optical effect used for display, one must limit the multiplexing power N in relation to the natural relative voltage spread of the optical effect. In the case of the twisted nematic valve, for instance, r is approximately 1 or 2 and N should not be larger than 3. It is the number chosen by Hewlett–Packard for its alphanumeric twisted-nematic-valve display of the portable calculator HP 41 C. The television screeen shown by Hitachi (78) was multiplexed by $N = 60$ with a large loss of contrast. The dynamic-scattering model is a little better and allows an intrinsic multiplexing power of about 10. The best electrooptic effect with respect to multiplexing capability is electrically controlled birefringence, as demonstrated by the LETI (Grenoble, France, 1979) with $N = 256$ and 512.

To improve multiplexing capability, one must realize electrooptic effects with a smaller threshold spread $\Delta V/V_s$. This can be achieved by the use of variable dielectric anisotropy compounds. A high-frequency field stabilizes the texture and increases V_s. A two-frequency multiplexing display has been demonstrated but is not very convenient, as explained previously. A better idea is to increase V_s by introducing a nonlinear element in series with the liquid crystal. Diodes were first used, but have now been abandoned and replaced by varistors (General Electric). The I/V characteristic of a varistor (Figure 11.30) shifts the optical-effect threshold from V_s to $V_s + V_B$. With V_B of the order of 50 V, one gains almost two orders of magnitude in the multiplexing power N. The varistor ceramic is also used to make a storage capacitor C parallel to the liquid crystal (LC) so that the RC response time can be adjusted (Figure 11.31). The advantage of this device is that it remains a two-wire dipole. One difficulty is the impossibility of using polarizers above the background opaque ceramic varistor, which forces one to use optical effects such as the dynamic-scattering mode or dichroic dye absorption. Work is under progress to achieve a larger matrix screen employing this technique.

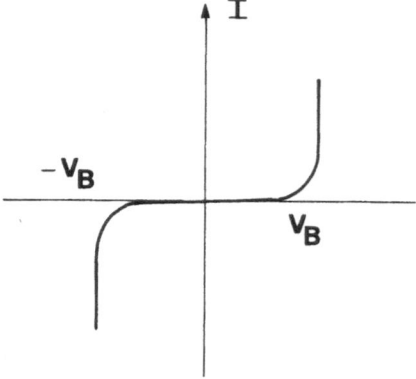

Figure 11.30. I/V characteristic of a varistor.

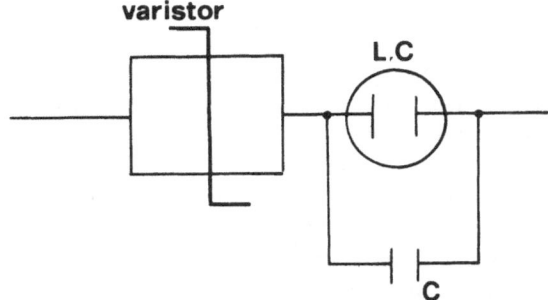

Figure 11.31. Use of a varistor in combination with a liquid crystal.

The best solution to achieve multiplexed $X-Y$ addressing is to separate the display and multiplexing functions. The display is always produced by a texture instability in a liquid-crystal cell. The memory time of the display is controlled by a capacitor in parallel with the liquid-crystal cell. The addressing is made by one transistor per cell (Figure 11.32). This has been realized by Hughes, using a MOS technique, on a single crystal of silicon. One now needs three wires (grid, drain, source of MOS transistor). As is usual with active substrates, one must work in reflection and without polarizers, i.e., with the dynamic-scattering mode or dyes. A screen of four elements, each 1 in. ×1 in. and $N = 100$, has been demonstrated by Hughes using the dynamic-scattering mode. Japanese prototypes of commercial television sets using this technique exist. The major difficulty consists in making large defect-free single crystals, which results in a high-cost screen.

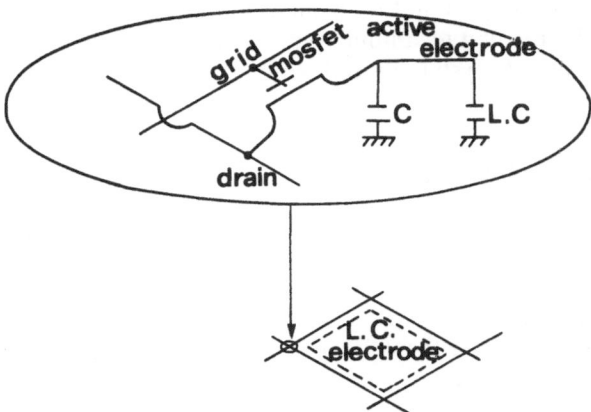

Figure 11.32. Addressing for display.

An elegant way to realize large screens is to use the technique of thin-film transistors (TFT) made, e.g., on a transparent glass plate. This method has been promoted by Westinghouse. The principle of the display remains the same, with one transistor and one storage capacitor per elementary liquid-crystal cell. The semiconductor used is typically CdSe (or PbS) with an Al_2O_3 or Ta_2O_5 insulator, and molybdenium electrodes. Although the principle has been demonstrated to work, there is much difficulty in preventing failures in the TFT. We note that a short circuit on one TFT will provoke a full blind row or column. Research is currently under way to improve the quality of TFT.

Let us say a few words about the method used to realize variable contrast in multiplexed liquid-crystal display. The idea is to use the variation in the optical contrast above the threshold, characteristic of the electrooptic effect used. The simplest approach would be to vary the voltage V_1 to cover, with V_{on}, the range ΔV above the threshold value V_S. In practice, one does not use DC but AC voltages. This improves the lifetime of the liquid-crystal cell, where the material may be dissociated electrochemically at the electrodes. As regards the video circuits, it is convenient to maintain constant the amplitudes V_1 and V_2 of the applied signals. A simple way to produce a variable contsrast is to adjust the relative phase of the voltages of the two rows and columns. In-phase and out-of-phase correspond to the plus and minus signs ($V_1 \pm V_2$), with all possible intermediate values. When the frame frequency is large enough (as for television), one can keep "DC" voltages by changing the total sign of the signal from one frame to the other, to create an AC excitation in the simple "on" and "off" modes.

Table 11.3 summarizes some results obtained in multiplexed liquid-crystal addressing.

In conclusion, we note that using a thermoelectric effect on a smectic liquid crystal is the first technique to present full television speed capability with simple liquid-crystal multiplexing. It is described in Section 11.5 below.

11.4.3. Parallel Addressing

A few words ought to be said about the use of liquid crystals for optical recording of images. The principle of these systems is simple. A liquid-crystal cell is sandwiched between photoconductive material (Figure 11.33). In the dark, the resistivity of this material is higher than that of the liquid-crystal cell, where the voltage drop is weak, lower than the threshold voltage V_S. In the presence of light possessing sufficient intensity, the material conductivity increases and the liquid-crystal cell is locally excited. One generally uses an image-forming system, like a lens or a bunch of

Table 11.3. Summary of Some Results Obtained in Multiplexed Liquid-Crystal Addressing

Effect	Laboratory	System	N	Active area	Frame frequency	Gray scale
Dynamic-scattering mode	Hitachi (74)	Liquid-crystal multiplex	400	40×50 cm^2	1 Hz	0
	Hughes (76)	I.C. M.O.S	100	$1'' \times 1''$ $2'' \times 2''$	25 Hz	16
Electrically controlled birefringence	LETI (77) (79)	Liquid-crystal multiplex	128 256	6.4×6.4 mm^2 23×23 mm^2	5 Hz 15 Hz	8 32
Twisted nematic valve	Westinghouse (77)	Thin-film transistors	120	102×102 mm^2	20 Hz	8
	Hitachi (77)	Liquid-crystal multiplex	60	90×120 mm^2	10 Hz	16
Thermolectric effect on smectics	Thomson (78) (79)	Liquid-crystal multiplex	100 256	4×5 mm^2	>25 Hz	8

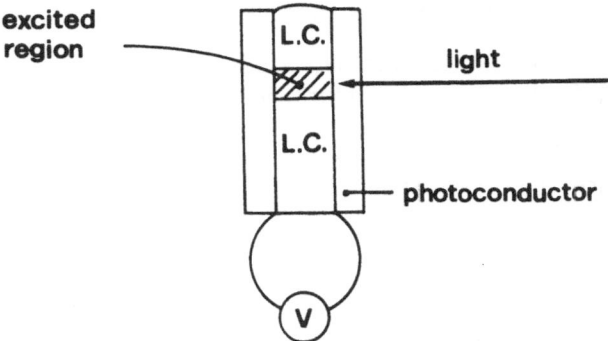

Figure 11.33. Use of liquid crystals for optical recording.

optical fibers, to guide the light to the sandwich cell. The original image is often given by a cathodic tube and the system is used as a brightness amplifier, e.g., for large-screen projection. The first optical effect used was the dynamic-scattering mode with some image smearing. A more recent realization, using more rapid cholesteric unwinding, has been demonstrated to present the television speed capability. An obvious problem in all these systems originates from the necessity to separate the image-forming light and the projection beam. This is generally achieved by using wavelength-matched interference mirrors, a reflection mode for the projection beam, and a writing light on the photoconductor of the more different possible wavelengths, such as in the ultraviolet.

As to references on this section, a general survey is given by Robert,[53] who together with Dargent examines multiplexing,[54,55] while Ching An Chao et al.[56] discuss television with cholesterics.

11.5. Applications of Smectics

The development of the physics of smectic liquid crystals has lagged behind that of nematics and cholesterics. Moreover, difficulty in obtaining good orientation and mechanical hysteresis related to the semisolid character of the phase have delayed the use of smectic liquid crystals. Recent work has demonstrated interest in smectics from the standpoint of applications. We first discuss some important properties of smectic textures. We shall describe the electric-field-induced effects and the thermal addressing of images. We end with the use of ferroelectric C* in bistable optical switching.

11.5.1. Mechanical Properties of Smectic Textures

Smectics are characterized by their mixed elasticity, where curvature elasticity and solid-like layer compression can balance. A typical example is given below because of its practical consequences. Let us consider a smectic A texture made of a positive material ($\varepsilon_a > 0$), under the destabilizing influence of an electric field in the layer plane (Figure 11.34). The

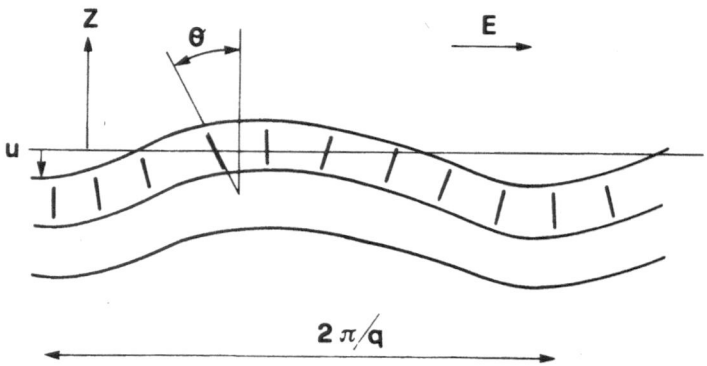

Figure 11.34. Smectic A texture in an electric field.

critical field for an undulation instability is estimated by balancing the dielectric gain in free energy $(\varepsilon_a/2 \cdot 4\pi)E^2\theta^2$ with the elastic free energy $\frac{1}{2}B(\partial U/\partial z)^2 + \frac{1}{2}Kq^2\theta^2$. Taking $q = \pi/d$ (d is the sample thickness) and $B = K/\lambda^2$ (λ is a molecular length), the elastic free energy becomes $\frac{1}{2}K[(\pi/d)^2 u^2\lambda^{-2} + q^2\theta^2]$. Because the molecules remain normal to the layers, $\theta = iqu$, i.e., the elastic term expressed in terms of θ^2 is minimum for $q^2 = \pi/\lambda d$, which is the wave vector of the expected instability. In practice, such an instability is a "ghost" instability, i.e., strong nonlinear effects limit the undulation amplitude to a very small value. Nevertheless, the point is that, because two lengths are implied in the problem (d and λ), the period of the spatial instability is not defined only by the sample thickness; the quantity $(\lambda d)^{1/2}$, much smaller than d, is the minimum extension of the texture distortion along the layers, i.e., the *resolving power* of smectic displays is much smaller than the corresponding one (d) for nematic materials. As previously mentioned, another consequence of the stiffer elasticity of smectic layers is the expected shorter response time of texture distortion. In fact, this property is not often observed. The dynamics of smectics appears slow, resulting in an extremely long *storage mode*, just because of mechanical hysteresis of polycrystalline broken textures.

11.5.2. Electric-Field Effects on Smectic A

The smectic A phase can be prepared in homeotropic or planar textures. A convenient material is 8 CB

$$C_8H_{17}\langle\bigcirc\rangle-\langle\bigcirc\rangle-C\equiv N$$

with an S_A phase from 21 to 32 °C and a nematic phase above, which allows easy alignment. Two mechanisms can be involved in destabilizing the texture.

1. Electrohydrodynamic Instability. The compound used is strongly positive ($\varepsilon_a \sim 10$) so that a high-frequency field alone would stabilize a homeotropic texture. We have explained above that low-frequency excitation can destabilize the texture by creating curvature space charges. A few hundred volts in 10 ms can *break* the texture and give a scattering "focal-conic" state (instead of the "ghost" transition). The scattering texture remains when the field is off, resulting in a storage mode (Figure 11.35). To erase, one uses the high-frequency stabilizing effect.

2. Dielectric Effects. With a positive material, one can electrically induce a planar to homeotropic transition. With a negative material, one can induce a homeotropic to planar transition. Both effects are irreversible electrically. However, by using a smectic material of sufficiently low

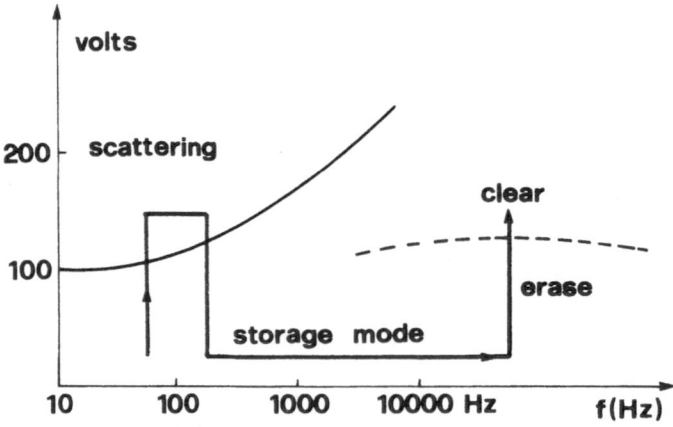

Figure 11.35. Storage mode.

relaxation frequency, one can make a reversible device that can be switched on and off. Such a matrix can be used to dissolve a dichroic dye, or to realize a light-absorbing color display.

It is of interest to understand the mechanism of texture break, which initiates the focal-conic scattering texture. Optical observations show that the texture change is produced by the nucleation of periodic surface defects (Figure 11.36). If this surface energy is $\Delta F/cm^2$, we can write at the threshold

$$\Delta F \sim \frac{1}{2} \frac{\varepsilon_a}{4\pi} E^2 \vartheta^2 d$$

where d is the sample thickness. The field threshold is indeed proportional to $d^{-1/2}$.

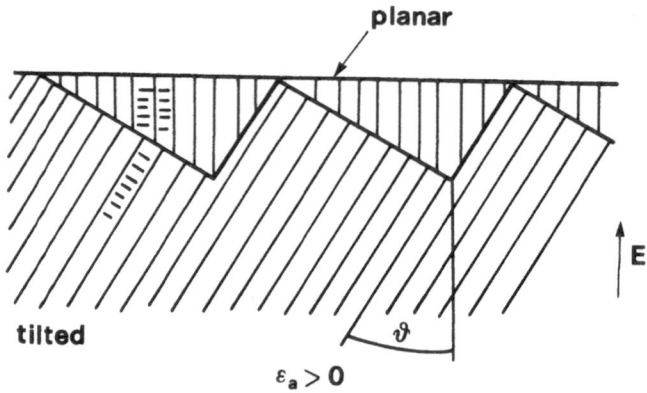

Figure 11.36. Mechanism for texture break.

Most of the texture changes in smectics can be earased by heating, close to or above the smectic → nematic transition. This difficulty has led to the idea of thermal writing on smectics, in fact applying an idea already developed with cholesteric materials.

11.5.3. The Thermally Excited Optical Smectic Valve

Such a device consists of a thin homeotropic slab of smectic A, locally heated by the absorption of an IR YAG Laser ($\lambda = 1.06$ μm) of a few mW (Figure 11.37). In the absence of an applied voltage across the cell, the heating first creates a nematic or isotropic phase, which by cooling down gives a strongly scattering focal-conic texture instead of the transparent initial texture. By moving the writing laser beam, one can record typically $2 \cdot 10^6$ points in 3 s. The size of each point can vary from 20 to 200 μm, with a definition of 4000 lines. Once written, a cell can retain the written image for longer than six months. To erase globally, one can use a large electric field on a positive material ($V \sim 100$ V on $d = 10$ μm, $\varepsilon_a \sim 10$). To erase selectively, one just replaces the writing laser beam on the point to be erased in the presence of the AC electric field, which aligns the nematic phase, and consequently the smectic phase, back to homeotropic texture. A scale of gray can be achieved by writing in the presence of an applied weak AC field, which allows one to suppress the modulator on the writing beam. The projection optics is standard, although it works in reflection.

Industrial realizations have been made by IBM, Thomson–CSF, and Western Electric. The X, Y deflection system is made with mechanically controlled mirrors. Color images are possible, with three cells but just one optical system. Applications included the high-resolution large-screen display and, more specifically, the realization of integrated circuit masks.

11.5.4. The Thermal Electric $X-Y$ Addressed Smectic

A homeotropic material is placed in an $X-Y$ matrix. The background electrode is made of Al striped wires, insulated from a bulk silicon plate by an SiO_2 layer. These row wires are heated sequentially by an electric-current pulse. The heat from this pulse causes a thin smectic layer around the wire to go up to the isotropic phase. By choosing the heat power, one can decrease this "erasing" time down to $\tau = 64$ μs, the access time of a line in a 625–line television-imaging system. The temperature decay down to the smectic phase can also be adjusted to τ (Figure 11.38), by suitable choice of the insulating SiO_2 layer from the bulk Si thermal leak. In the absence of an electric field, after time 2τ, all the points of the excited line would be in a disordered focal-conic texture, strongly scattering an incoming light beam. The various column electrodes allow the application

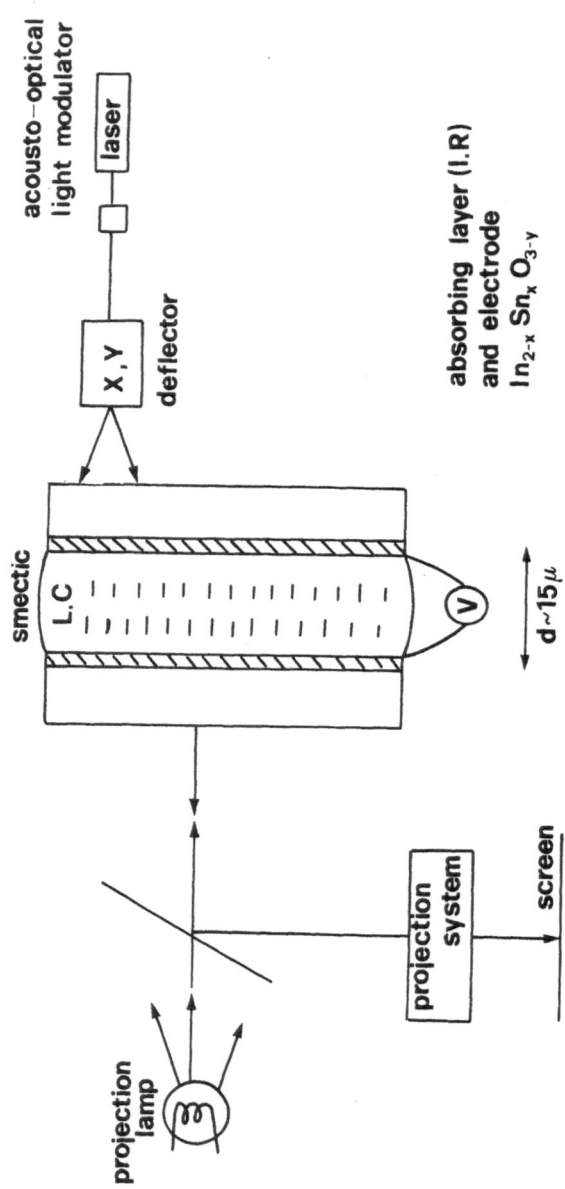

Figure 11.37. Thermally excited optical smectic valve.

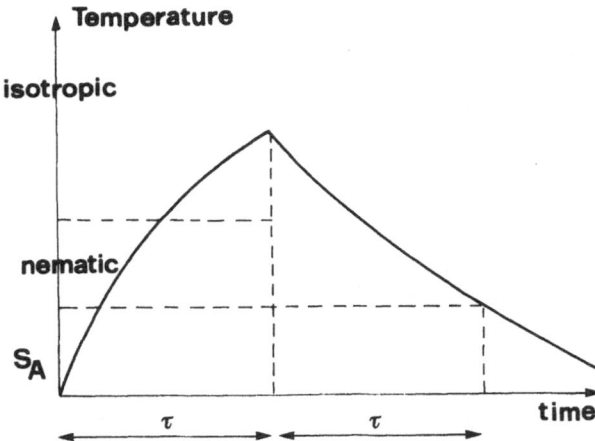

Figure 11.38. Temperature vs time in a smectic.

of an electric field. Cooling in the presence of the AC electric field restores the initial homeotropic orientation for a well-chosen dielectrically positive ($\varepsilon_a > 0$) material. To allow multiplexing, it is sufficient to send in parallel on the columns electric information related to the desired state of the excited row, during the cooling time interval τ. The present contrast is around 12, with a scale of gray, and full television speed capability. A 256×256 matrix has been demonstrated. The observation is made in reflection because of the opaque Si substrate (which could be used for integrated excitation electronics). When the video stops, there is an image storage mode, which lasts for days. A nice trick of the thermal effect is the double "erase" plus "writing" effect which, with the almost infinite storage time, does not limit the multiplex capability. The only problems are the small size (necessary to achieve short thermal-response time) and the important power (approximately 10 W) used in the heating process. A reduction in speed requirement would yield a proportional reduction in power consumption, but because of its principle the thermal electric X–Y display can hardly be portable.

11.5.5. Ferroelectric C*. The Bistable Optical Switch

The chiral C* phase is the only true liquid crystal to be potentially ferroelectric. The only symmetry element of the C* phase is the C_2-axis, which allows molecules to "flip-flop" head to tail. If there is any transverse electric molecular dipole moment (as on almost all liquid-crystal molecules), its component on the C_2-axis cannot be averaged to zero. A bulk polarization **P** grows. A typical maximum value for **P**, with molecular dipole $\mathbf{P_M}$ of the order of a few debyes, is $\mathbf{P} \sim 10^{-2} \, \mathbf{P_M}$. The bulk polarization **P** is proportional to the tilt angle θ, since it is zero in the S_A phase and is an odd function of θ. Because of the chirality, the equilibrium state of a C^* is a conical spiral: the molecular projection on the layer plane rotates around the layer normal **z**, resulting in a spontaneous twist (and bend) as in a cholesteric material. The pitch of this helical texture is in the μm range. Optically, starting from a smectic A in a planar alignment, one can see a striped texture when cooling down into the C^* phase.

As in cholesterics, one can unwind the C^* helix with an electric field. The coupling term is $-\mathbf{P} \cdot \mathbf{E}$, which indicates that the transverse dipole likes to align along **E**. This effect is linear in E, and is expected to dominate at low fields over the dielectric effects previously discussed, which behave like E^2. This unwinding has indeed been observed. It is used to measure P, by expressing the balance between the elastic twist term $K_c q_0^2$ and the electric $-PE$ term. The twist elastic constant K_e is expected to behave as θ^2. If we set $P = P_0\theta$, this results in

$$P_0 \sim \frac{K_2 q_0^2 \theta}{E_s}$$

K_2 is a typical twist curvature elastic constant equal to approximately 10^{-6} cgs; q_0 is the helix wave vector and E_s the threshold field that unwinds the helical texture. In practice, the change from the striped to the nonstriped texture does not create much contrast and is not very appealing as regards applications, although E_s may correspond to less than 1 V across a few tenths of 1 μm.

A simple bistable optical device using a C^* has been demonstrated recently. A sample of C^* is aligned in a planar texture (Figure 11.39), with

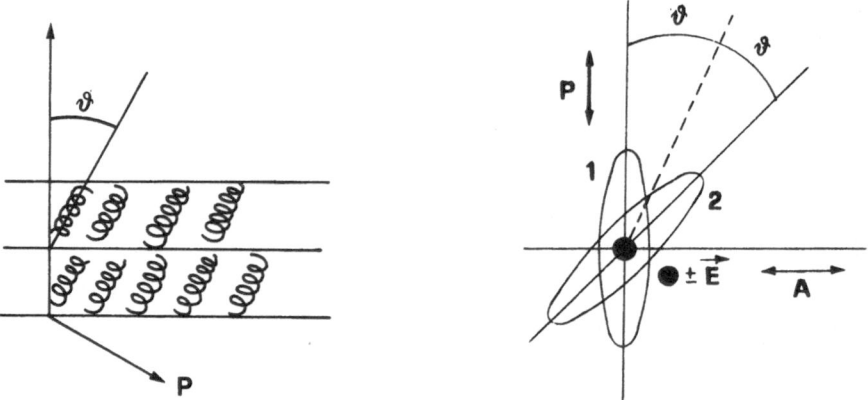

Figure 11.39. Alignment of C^* in a planar texture.

the bulk polarization **P** normal to the plates in an upward direction. The helix has been suppressed, as in a cholesteric, by choosing a small thickness ($d <$ pitch). Seen from above, the smectic slab appears as a birefringent crystal with two eigenpolarizations **P** and **A**. A light beam normal to the plates, polarized by Polarizer P, is blocked by analyzer A. One applies an electric-field impulse opposed to the uniform bulk polarization **P**. If E is large enough, the molecules rotate on their cone and realign parallel to the plates with an opposite bulk polarization $-$ **P**. Seen from above, the eigendirections have rotated through an angle 2ϑ. One can adjust d to realize a half-wave plate. The temperature can be chosen such that ϑ is about 22°, i.e., 2ϑ is approximately 45°, to obtain the largest effect. With a birefringence Δn_0 equal to about 0.3, a thickness of about 1 μm fulfills the half-wave condition for λ approximately 0.6 μm and is smaller than the typical 5-μm pitch. For that geometry, all the incoming-light intensity passes through the crossed polarizer and analyzer. On reversing the applied field, the molecules resume their initial orientation. No more light is transmitted through the cell. The potential barrier between the two, up and down states is of the order of the C* twist energy when the thickness d is comparable to the pitch t. For smaller values, one can guess that it varies as $(d/t)^2$. Using a coupling energy EP close to the potential barrier, one expects a rise time of about the damping time of a helix of pitch d, in the ms range. A larger E should result in a rise time varying as E^{-1}. In practice, one observes for $V = 0.2$ V across 1.2 μm a 4 ms rise time, and for $A = 20$ V a 1 μs rise time. Larger fields cannot be used, because the coupling to the dielectric anisotropy would become dominant, varying as E^2. Anyway, a 1 μs optical switch with a reasonably low voltage (20 V) is an interesting device.

Finally, we note that a good review of smectic materials is given by Raynes[57] and the paper of Hareng and Le Berre[58] may also be usefully consulted. Valuable sources on C* are the works by Durand and Martinot-Lagarde[59] and Clark and Lagerwall.[60]

References for Part II

1. P. G. de Gennes, *The Physics of Liquid Crystals*, Clarendon Press, Oxford (1974).
2. S. Chandrasekhar, *Liquid Crystals*, University Press, Cambridge (1977).
3. S. Chandrasekhar, *Reports on Progress in Physics* **39**, 613 (1976).
4. M. J. Stephen and J. P. Straley, *Rev. Mod. Phys.* **46**, 617 (1976).
5. E. B. Priestley, P. J. Wojtowicz, and Ping Sheng, eds. *Introduction to Liquid Crystals*, Plenum, New York (1975).
6. G. W. Gray and P. A. Winsor, eds. *Liquid Crystals and Plastic Crystals, Vols.* 1 *and* 2, Ellis Horwood, Chichester (1974).
7. G. W. Gray, 8*th Int. Liquid Crystal Conf.*, Kyoto, 1980, *Mol. Cryst. Liq. Cryst.* **63**, 3 (1981), and references contained therein.
8. S. Chandrasekhar and R. Shashidhar, in: *Advances in Liquid Crystals*, Vol. 4, G. H. Brown, ed. Academic Press, New York, p. 73 (1979).
9. P. E. Cladis, R. K. Bogardus, W. B. Daniels, and G. N. Taylor, *Phys. Rev. Lett.* **39**, 720 (1977).
10. S. Chandrasekhar, 8*th. Int. Liquid Crystal Conf.*, Kyoto, 1980, *Mol. Cryst. Liq. Cryst.* **63**, 171 (1981) and references contained therein.
11. P. G. de Gennes, *Mol. Cryst. Liq. Cryst.* **12**, 193 (1971).
12. F. C. Frank, *Disc. Far. Soc.* **25**, 19 (1958).
13. J. Nehring and A. Saupe, *J. Chem. Soc. Far. Trans.* 2, **68**, 1 (1972).
14. R. B. Meyer, *Mol. Cryst. Liq. Cryst.* **16**, 355 (1972).
15. G. S. Ranganath, *Mol. Cryst. Liq. Cryst.* **34**, 71 (1976).
16. G. S. Ranganath, Liquid Crystals, *Proc. Int. Liquid Crystals Conf.*, Bangalore, 1979, ed., S. Chandrasekhar, Heyden, London, p. 213 (1980).
17. J. Rault, *Phil. Mag.* **28**, 11 (1973).
18. F. M. Leslie, *Advances in Liquid Crystals*, Vol. 4 (1979).
19. M. Miesowicz, *Nature* **158**, 27 (1946).
20. Ch. Gahwiller, *Mol. Cryst. Liq. Cryst.* **20**, 301 (1973).
21. V. Tsvetkov, *Acta Physicochim.* (*USSR*) **10**, 557 (1939).
22. H. C. Tseng, D. L. Silver, and B. A. Finlayson, *Phys. Fluids* **15**, 1213 (1972).
23. S. Jen, N. A. Clark, P. S. Pershan, and E. B. Priestley, *Phys. Rev. Lett.* **31**, 1552 (1973).
24. M. A. Cotter, in: *The Molecular Physics of Liquid Crystals*, eds., G. R. Luckhurst and G. W. Gray, Academic Press, London (1979), p. 181.
25. T. W. Stinson, J. D. Litster, and N. A. Clark, *J. de Physique* **33**, C1–69 (1972); E. Gulari and B. Chu, *J. Chem. Phys.* **62**, 798 (1975).

26. T. W. Stinson and J. D. Litster, *Phys. Rev. Lett.* **30**, 688 (1973).
27. R. Nityananda, U. D. Kini, S. Chandrasekhar, and K. A. Suresh, *Proc. Int. Liquid Crystals Conf.*, Bangalore, December 1973, ed., S. Chandrasekhar, *Pramana Suppl.* **1**, p. 325.
28. W. Helfrich, *Phys. Rev. Lett.* **23**, 372 (1969).
29. U. D. Kini, Seventh Int. Liquid Crystal Conf., Bordeaux, 1978, *J. de Physique* **40**, C3-62 (1979).
30. S. Bhattacharya, C. E. Hong, and S. V. Letcher, *Phys. Rev. Lett.* **41**, 1736 (1978).
31. W. L. McMillan, *Phys. Rev.* **A4**, 1237 (1971) and **A6**, 936 (1972).
32. P. P. Karat and N. V. Madhusudana, *Mol. Cryst. Liq. Cryst.* **55**, 119 (1979).
33. J. D. Litster, J. Als-Nielsen, R. J. Birgeneau, S. S. Dana, D. Davidov, F. Garcia-Golding, M. Kaplan, C. R. Safinya, and R. Schaetzing, *J. de Physique*, **40**, C3-339 (1979).
34. S. Chandrasekhar, R. Shashidhar, and K. V. Rao, *Third Liquid Crystal Conf. of Socialist Countries*, Budapest, Hungary, August (1979), L. Bata, ed., Pergamon Press, Oxford, p. 173 (1981).
35. S. Chandrasekhar, K. A. Suresh, and K. V. Rao, *Proc. Int. Liquid Crystals Conf.*, Bangalore, December 1979, ed., S. Chandrasekhar, Heyden, London, p. 131 (1980).
36. A. M. Levelut, F. Hardouin and G. Sigaud, *Proc. International Liquid Crystals Conf.*, Bangalore, 1979, ed., S. Chandrasekhar, Heyden, London, p. 143 (1980).
37. P. G. de Gennes, Solid State Commun. **10**, 753 (1972); *Mol. Cryst. Liq. Cryst.* **21**, 49 (1973).
38. K. C. Chu and W. L. McMillan, *Phys. Rev.* **A15**, 1181 (1977).
39. J. Prost and P. S. Pershan, *J. Appl. Phys.* **47**, 2298 (1976).
40. R. B. Meyer, *Mol. Cryst. Liq. Cryst.* **40**, 33 (1977).
41. B. I. Ostrovskii, A. Z. Rabinovich, A. S. Sonin, and B. A. Strukov, *Sov. Phys. JETP*, **47**, 911 (1978).
42. L. M. Blinov, L. A. Beresnev, N. M. Shtukov, and Z. M. Elashvili, *J. de Physique* **40**, C3-269 (1979).
43. T. Kallard, *Liquid Crystals and their Applications* Optosonic Press (1970).
44. Meier, Sackmann, and Grabmeier, *Applications of Liquid Crystals*, Springer-Verlag, Berlin (1975).
45. Kelter and Hatz, *Handbook of Liquid Crystals*, (1980).
46. Dubois, *Ann. Phys.* **3**, 131 (1978).
47. D. White and G. Taylor, *J. Appl. Phys.* **45**, 4718 (1974).
48. M. Hareng, G. Assouline, and E. Lieba, *Appl. Opt.* **11**, 2920 (1972).
49. M. Schadt and W. Helfrich, *Appl. Phys. Lett.* **18**, 127 (1971).
50. F. Charadjedaghi and J. Robert, *Rev. Phys. Appl.* **11**, 467 (1976).
51. G. Durand, *Les Houches Lectures*, p. 403, Gordon and Breach, London (1976).
52. W. Greubel, *Appl. Phys. Lett.* **25**, 5 (1974).
53. J. Robert, *Sci. Avenir* **29**, 80 (1980).
54. J. Robert and B. Dargent, *IEEE Trans. Electron Devices* **24**, 694 (1977).
55. J. Robert, *IEEE Trans. Electron Devices* **26**, 1128 (1979).
56. Ching An Chao, Shou Sheng Tung, and Liang Luan, Science Report of Tsing Hua University, Peking (September 1978).
57. E. Raynes, in: *Proc. European Conf. on Smectics and Low Dimensional Systems* (H. Helfrich, ed.), Springer-Verlag, Berlin (1980).
58. M. Hareng and S. Le Berre, *Ann. Phys.* **3**, 317 (1978).
59. G. Durand and P. Martinot-Lagarde, *Ferroelectrics* **24**, 89 (1980).
60. N. Clark and S. Lagerwall, in: *Proc. European Conf. on Smectics and Low Dimensional Systems* (H. Helfrich, ed.), Springer-Verlag, Berlin (1980).

III

Low-Dimensional Solids

Chemical Bonding

N. H. March

12.1. Outline

In this chapter, we present some features of the theory of chemical bonding that are particularly relevant to the molecular aspects of the electronic structure of low-dimensional solids.

Attention is first given to the linear polyenes. These are treated both by Hückel theory and by the free-electron model, the experimental contact here being the optical absorption as a function of chain length. The need for alternating bond lengths, and its theoretical foundation, will then be discussed. The relation of this alternation to many-electron theory will be considered briefly and, in particular, the work of Lieb and Wu will be summarized. As a quite different example of a one-dimensional solid, the platinum chain compounds are considered next. Here, the focal point is the influence of chemical bonding through the ligand-field treatment of Pt coordinated to CN groups. It will be stressed that this determines the degree of hybridization on the Pt atoms, which in turn has important consequences for the electronic structure of the Pt chain.

A further, important use of ligand-field theory is then discussed in connection with the disulfides and diselenides of some transition metals. These materials have layer structures, and trigonal-prismatic coordination of the metal atoms occurs. Therefore, following the work of Huisman et al.,[1] the ligand field splitting of d-levels in trigonal-prismatic coordination is treated and the splitting compared with that in octahedral coordination.

N. H. March · Theoretical Chemistry Department, University of Oxford, 1 South Parks Rd., Oxford OX1 3TG.

It is stressed that d-covalency provides a stabilizing factor for trigonal-prismatic coordination for atoms with d^0-, d^1-, and d^2-configurations. As a second class of layer structures, graphite and boron nitride are then treated. It is stressed first that graphite is to boron nitride as benzene is to borazole. Therefore, these two molecules are treated in linear combination of atomic orbitals (LCAO) theory, and the implications of these treatments for the electronic band structures of graphite and boron nitride are then considered.

Next, we discuss in some detail the way molecular calculations can be used, first in $(SN)_x$, where the work of Salahub and Messmer[2] is the focal point, and then in TTF-TCNQ, to interpret solid-state features. Finally, the foundation of such treatments, based on potential theory, is referred to in the course of a brief introduction to electron-density theory of molecules and solids.

12.2. Linear Polyenes

Throughout this chapter, we shall not hesitate to present relatively elementary treatments of electronic structure, e.g., either LCAO or free-electron methods, when they expose essential features of the way the electrons behave. We shall show below that, used in the linear polyenes $C_{2N}H_{2N+2}$, both these methods expose a conflict with experiment when they are applied in conjunction with the assumption that all bond lengths are the same. With that common assumption, we shall see that both methods predict that the gap in the excitation spectrum of the polyenes tends to zero as the number of carbon atoms in the chain tends to infinity. This disagrees with observations on the absorption spectra and draws attention to the need for alternation on bond lengths and/or many-electron effects. Both of these topics are treated, the first following the lead of Longuet-Higgins and Salem[3] and the second via the aforementioned many-electron solution of Lieb and Wu.[4]

Turning to the linear polyenes $C_{2N}H_{2N+2}$, we note first that Lennard-Jones,[5] in a pioneering paper, recognized the interest in ascertaining their properties in the limit as the number of C atoms tended to infinity. This is evidently then a prototype of a one-dimensional solid. In his work, Lennard-Jones employed molecular orbital (MO) theory in the LCAO Hückel approximation. While we shall discuss this method below, it will be useful to obtain a first orientation on the problem of $2N$ C atoms, with one π-electron per atom, by starting from free-electron theory.

12.2.1. Free-Electron Model: Absorption Spectra as Function of Chain Length

For the π-electrons enclosed in a box of length l, the free-electron model immediately gives the discrete energy levels, ε_r say, as

$$\varepsilon_r = \frac{r^2 h^2}{8ml^2} \qquad r = 1, 2, 3, \ldots \tag{12.1}$$

where m is the electron mass. There is a certain degree of flexibility in choosing l, but if we have $2N$ conjugated C atoms with bond length d and we put infinite barriers at distance d from the end C atoms, then $l = (2N+1)d$.

We shall now consider the absorption spectra of the linear polyenes. Evidently, in this free-electron model, the N lowest levels given by equation (12.1) are doubly occupied by π-electrons in the ground state. The first excited state is then constructed by exciting an electron from the level $r = N$, which is the highest occupied level in the above-ground state, to the lowest unoccupied level with $r = N+1$. Evidently the excitation energy, $\Delta\varepsilon_N$, say, is given from equation (12.1) by

$$\Delta\varepsilon_N = \varepsilon_{N+1} - \varepsilon_N$$
$$= \frac{h^2}{8m(2N+1)d^2} \simeq \frac{20}{(2N+1)} \text{ eV} \tag{12.2}$$

where the numerical estimate in equation (12.2) has been obtained[6] by taking $d \simeq 1.4$ Å, which is a reasonable average bond length for conjugated polyenes.

The important point to make here is that, as $N \to \infty$ in equation (12.2), $\Delta\varepsilon_N \to 0$ as $1/N$, and the absorption wavelength λ increases proportionally to N.

12.2.2. Linear Combination of Atomic Orbitals (LCAO) Hückel Approximation

That this same conclusion follows from the LCAO Hückel approximation, with equal bond lengths d, is clear, for example, from the book by Salem,[7] where the energy levels are derived as

$$\varepsilon_r = \alpha + \beta \cos\left(\frac{r\pi}{2N+1}\right) \tag{12.3}$$

Here α is a property of one carbon atom, while β is the resonance integral determined by the properties of two adjacent atoms. Evidently, the excitation energy is again $\varepsilon_{N+1} - \varepsilon_N$, and evaluating this difference using equation (12.3) yields

$$\Delta\varepsilon_N = -4\beta \sin\left[\frac{\pi}{2(2N+1)}\right] \qquad (12.4)$$

which, for large N, gives $\Delta\varepsilon_N = -2\beta\pi/(2N+1) \propto N^{-1}$.

We note that if, in the LCAO Hückel formula (12.3), we employ the approximation $\cos x \simeq 1 - x^2/2$, we regain the free-electron result (12.2) by appropriate choice of the resonance integral β. This would lead to $\beta \simeq -2$ eV in the present case, while if we estimate β empirically (actually to fit dimethyl butadiene[6]), we find -4.4 eV. So tight-binding (LCAO) theory and free-electron theory correspond at least qualitatively.

However, when we turn to experiment, we find that plotting the absorption wavelength λ against N leads to the curve[6] shown in Figure 12.1. It is quite clear from this figure that $\lambda \propto N$ is a gross overestimate and the experiments can only be understood if λ tends to a finite value as $N \to \infty$. This corresponds to the existence of an energy gap in the excitation spectrum of the linear chain of C atoms, which evidently conflicts with the theoretical results discussed above.

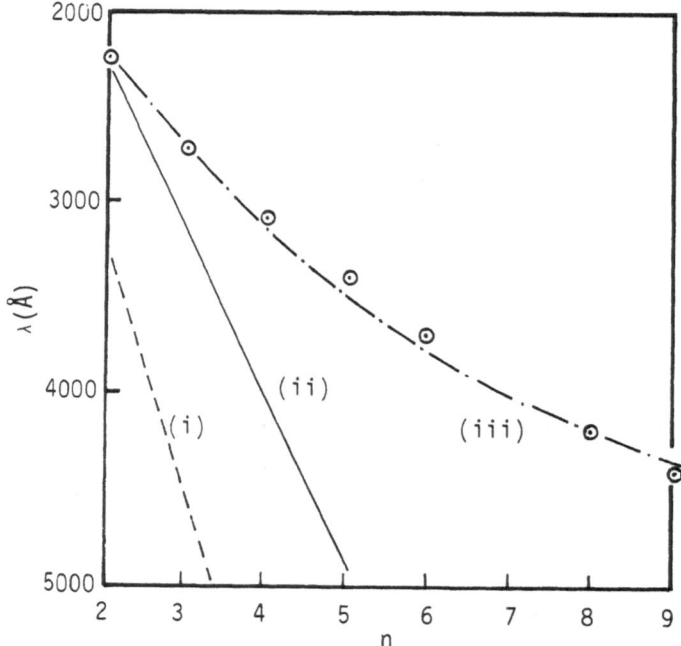

Figure 12.1. Absorption wavelength of polyenes as a function of chain length. The various curves show approximate theories: (i) free electron, (ii) LCAO Hückel, (iii) with introduction of energy gap (redrawn from Murrell[6]).

12.2.3. Alternating Bond Lengths

Once the above disagreement between theory and experiment became evident, it was quickly recognized by various workers (for example, Kuhn[8] who used the free-electron model, or Dewar[9] who employed Hückel theory) that the conflict could be resolved if one gave up the constraint of equal bond lengths d and allowed the bond lengths to alternate. When viewed as a problem in solid-state theory, this evidently introduces a new periodicity and hence an energy gap (cf Chapter 13). Then the absorption wavelength λ is constant, determined by the energy gap.

In Appendix A12.1 we summarize the way in which the Lennard-Jones treatment,[5] based as it was on equal bond lengths d, was extended by Longuet-Higgins and Salem,[3] and independently by others,[10] to include alternating bond lengths, x_1 and x_2 say. Evidently, the energy gap which exists for $x_1 \neq x_2$ must tend to zero in the Lennard-Jones limit $x_1 = x_2 = d$. As discussed in Appendix A12.1, the energy $E(x_1, x_2)$ can be split into a part due to the σ-electrons and a part due to the π-electrons:

$$E(x_1, x_2) = E_\sigma(x_1, x_2) + E_\pi(x_1, x_2) \tag{12.5}$$

While the term from the σ-electrons can be written in the form $\sum_{i=1,2} f(x_i)$, it is not possible to express E_π in this additive form. However, in Hückel theory, equation (12.5) can be written as

$$E(x_1, x_2) = \sum f(x_i) + E_\pi(\beta_1, \beta_2) \tag{12.6}$$

where we have introduced resonance integrals depending on the bond lengths, i.e., $\beta_i = \beta(x_i)$. It follows from the Lennard-Jones treatment that if we compare the LCAO results and the free-electron method, we find $\beta(x_i) \propto x_i^{-2}$, but Longuet-Higgins and Salem[3] adopted a more realistic exponential decrease. Obtaining E_π in this way, one can carry out the minimization of $E(x_1, x_2)$ with respect to x_1 and x_2. The Lennard-Jones choice $x_1 = x_2$ then turns out to correspond to unstable equilibrium. Within the framework of Hückel theory the minimum occurs for $x_1 - x_2 \cong 0.03$ Å.

In summary, the Hückel theory shows that if one allows alternating bond lengths, then the minimum energy occurs when the bond lengths differ by about $\frac{3}{100}$ Å. It has not, of course, been excluded that a more general relaxation of the positions of the C nuclei from the equal bond-length configuration could not lead to an even lower energy. Naturally, the arguments presented above and in Appendix A12.1 have all the defects of the Hückel theory, including the serious approximation that the total energy is the sum of orbital energies (cf, however, Section 12.8.3

below). From a solid-state viewpoint, what we have been discussing in terms of alternating bond lengths can be thought of as a manifestation of Peierls' theorem that a one-dimensional metal cannot exist (cf the discussion in Chapter 13).

12.2.4. Effect of Electron–Electron Interactions

It would be of obvious interest if electron–electron interactions could be introduced analytically into the alternating bond arguments presented above within a molecular-orbital framework. Though this has not yet been done, we conclude this section on linear polyenes by reporting briefly the work of Lieb and Wu[4,11,12] in which a model in one dimension, namely, a single tight-binding band with only intra-atomic (i.e., short-range) interactions between the electrons, has been solved exactly for the case of equal bond lengths. For the details we must refer the reader to the original paper. However, the allowed energies E take the form

$$E \propto \sum_j \cos k_j = \int_{-\pi}^{\pi} \rho(k, U) \cos k \, dk \qquad (12.7)$$

where the quasi-momenta k_j are determined by a rather complicated set of equations. As the length of the chain tends to infinity, and for the case of a half-filled band, the ground-state energy takes the form

$$E = -4N \int_0^{\infty} \frac{J_0(\omega) J_1(\omega) d\omega}{[1 + \exp(\frac{1}{2}\omega U)]\omega} \qquad (12.8)$$

where U measures the strength of the interaction while J denotes a Bessel function of the first kind.

For small interaction strength U, one regains the Hartree–Fock result

$$\frac{E}{N} = -\frac{4}{\pi} + \frac{1}{4}U + \cdots \qquad (12.9)$$

The full expression (12.8) for E has been evaluated numerically as a function of the Coulomb interaction strength U by Johansson and Berggren,[13] leading to exact knowledge of the correlation energy in this, admittedly, very simple model. In addition to the ground state, the electronic excitation spectrum is also known for this model and appears relevant to that of a one-dimensional chain with conjugated bonds. We therefore summarize the numerical consequences of the model for the density of states $\rho(k, U)$ in equation (12.7), this being plotted[12] in Figure 12.2 as a function of the interaction strength U. The effect of the

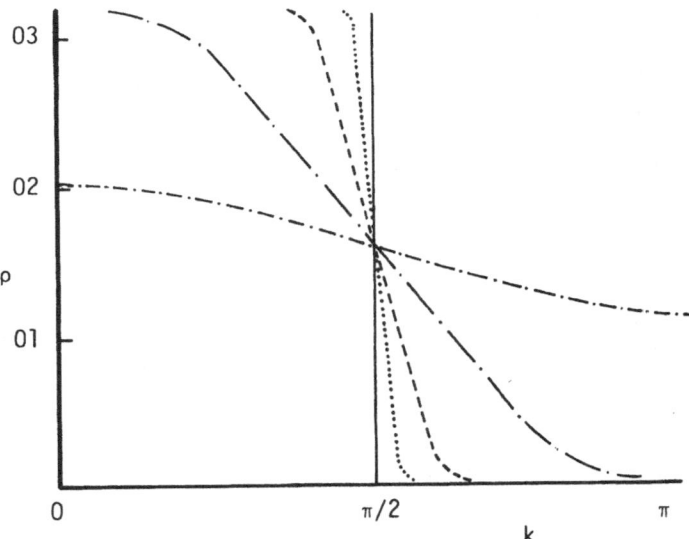

Figure 12.2. Density of states $\rho(k)$ in equation (12.7) as a function of the interaction U. This interaction is measured in units of the bandwidth. Only for $U = 0$ is there discontinuous behavior corresponding to a Fermi surface: —— $U = 0$, $\cdots\cdots$ $U = 0.01$, ---- $U = 0.1$, —·—· $U = 1.0$, –·–· $U = 10.0$.

electron–electron interactions is profound: only at the singular point $U = 0$ is there discontinuous behavior corresponding to a Fermi surface. Eventually, as $U \to \infty$, $\rho(k)$ tends to the constant value $\pi/2$ in the Brillouin zone but, as Figure 12.2 shows, $\rho(k)$ remains small near the zone boundary $k = \pi$ until the electron–electron interaction energy considerably exceeds the bandwidth.

Lieb and Wu also studied further the insulating character of the system as induced here by the Coulomb interaction away from $U = 0$, by considering the difference between the energy μ_+, say, required to add an electron, and that required to remove one, μ_-. The difference $\mu_+ - \mu_-$, which expresses the insulating character, is plotted[12] in Figure 12.3. One interesting point is that, although there is a correlation gap in the energy when $U \neq 0$, it has been shown that, to $O(U)$, $\mu_+ = \mu_- = \frac{1}{2}U$ and hence the gap is $O(U^2)$ at small U. Even when the electron–electron interaction energy is equal to the bandwidth, however, Figure 12.3 shows that $\mu_+ - \mu_-$ remains quite small.

It is clear that, in the polyenes, it is of considerable interest that a gap in the excitation spectrum, required by experiment, can arise from both a correlation gap, the sole effect considered in the Lieb–Wu treatment, as

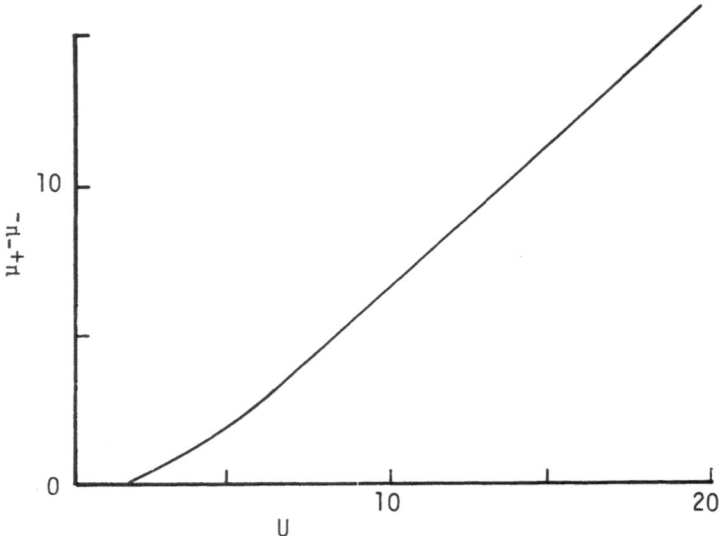

Figure 12.3. Insulating character as a function of the strength of interaction as measured by $\mu_+ - \mu_-$. Note the flatness of the curve for small U showing the absence of a term of $O(U)$ in $\mu_+ - \mu_-$.

well as from a new periodicity, due to bond alternation. Clearly, in a final theory, both effects must be incorporated, and relevant work is summarized elsewhere.[14-17]

12.3. Platinum Chain Compounds

There has been a good deal of interest, both experimental and theoretical, in the properties of a group of mixed-valence Pt salts, of which $K_2Pt(CN)_4Br_{0.3} \cdot 3H_2O$, known as KCP, has been the most extensively studied. A large number of properties of KCP show the behavior expected on the basis of a one-dimensional physical model.[18,19] For example, the so-called Peierls' distortion referred to briefly in the earlier discussion of the linear polyenes, expected to occur only in one-dimensional systems, has been observed experimentally in KCP by Renker et al.[20] and by Comes et al.[21]

It is clear that the relevant features of the electronic structure of KCP associated with the pseudo-one-dimensional properties arise primarily from the chains of Pt atoms and that therefore a one-dimensional band-structure calculation is appropriate.

12.3.1. Tight-Binding (LCAO) Results

The usual first-order approximation to the band structure of transition metals, which involve rather localized d-electrons, is the LCAO (tight-binding) model.[22] Surprisingly, however, there is a substantial body of experimental evidence to indicate that the band structure of KCP is free-electron in character, i.e., there is a parabolic band with an associated effective mass equal to the free-electron mass.

In discussing this, at first sight, surprising result, we shall focus below on the work of Messmer and Salahub.[23] In previous work, s-, p-, and d-electrons were considered and calculated bands were derived from them (three-band model). Such a calculation led to the prediction of interband transitions, but the difficulty was that no such transitions in the relevant frequency range were observed experimentally. Therefore, the above band picture cannot be correct, in spite of the fact that a variety of techniques all give the same answer with the same starting assumptions.

12.3.2. Role of Ligands

Messmer and Salahub pointed out that the error arises, in an important part at least, from the neglect of ligands. Each Pt atom has associated ligands to four CN groups in $K_2Pt(CN)_4Br_{0.3} \cdot 3H_2O$. One needs to treat $Pt(CN)_4$ carefully, in which there is a square planar arrangement of four ligands around a central Pt ion.

Relevant calculations on this ligand field problem are described in Interrante and Messmer[24] and have led to the results obtained by Messmer and Salahub. The important point we wish to stress here is that, unlike the LCAO approach based on s-, p-, and d-functions, the ligands fix, from the outset, the $s-d$ hybridization at around d 84%, s 16%, independent of wave number k. This naturally leads to a two-band model, rather than to the (inappropriate) three-band model referred to above, and the calculations of Messmer and Salahub[23] show clearly that the resulting bands then have the nearly free-electron character required by the observations.

12.4. Two-Dimensional Layer Compounds I

Having established the importance of the ligands for the description of KCP, we turn immediately to a related topic: the relevance of ligand field theory to the properties of transition metal–dichalcogenide layer compounds. These have the formula MX_2, where M is a transition metal atom from the group IVB, VB, or VIB columns of the Periodic Table, and X is one of the chalcogens: sulfur, selenium, or tellurium.

12.4.1. Structure of Transition Metal–Dichalcogenide Layer Compounds

These compounds can be viewed as strongly bonded two-dimensional $X—M—X$ layers or sandwiches, loosely coupled by weak van der Waals forces. Within a single $X—M—X$ sandwich, the M and X atoms form two-dimensional hexagonal arrays.

Depending on the relative alignment of the two X-atom sheets within a single $X—M—X$ sandwich, two distinct two-dimensional crystal structures are obtained. In one of these, the metal atoms are octahedrally coordinated by six neighboring X atoms. In the other, the metal atoms have coordination which is trigonal prismatic.

Variation in the stacking sequence and registry of successive $X—M—X$ sandwiches along the hexagonal c-axis lead to a large number of crystal structures or polytypes in three dimensions. These are referred to as the 1T, 2H, 3R, 4Ha, 4Hb, and 6R phases. In this notation, the integer indicates the number of $X—M—X$ sandwiches per unit cell along the hexagonal c-axis, and T, H, and R denote trigonal, hexagonal, and rhombohedral symmetries, respectively.

The 1T phase contains $X—M—X$ sandwiches in which the metal atoms are octahedrally coordinated. The coordination is trigonal prismatic in the 2H, 3R, and 4Ha phases. In the 4Hb and 6R polytypes, the coordination within successive sandwiches alternates between octahedral and trigonal prismatic.

12.4.2. Trigonal-Prismatic Coordination

Trigonal-prismatic coordination has received less attention than, for instance, octahedral and tetrahedral coordination. This is probably due to the fact that its occurrence in isolated complexes is unusual, though complexes $Mo(S_2C_2H_3)_3$ and some other related complexes have such coordination.

However, as mentioned above, it is far more common for such coordination to occur in solid compounds of transition metals. Some of the essential facts for the discussion to follow may be summarized[25] as follows:

1. Disulfides and diselenides of Nb, Ta, Mo, and W have layer structures, with trigonal-prismatic coordination of the metal atoms.
2. Examination of relevant data shows that trigonal-prismatic coordination is more common for $4d$ and $5d$ transition metal atoms than for $3d$ metal atoms.
3. In nearly all cases, metal atoms with trigonal-prismatic coordination have a formal configuration d^1 or d^2.
4. In some compounds, trigonal-prismatic coordination is found at low temperatures, while octahedral coordination is found at high temperatures.

Questions which naturally arise from the above survey are:

1. Why is the trigonal-prismatic coordination stable with respect to the more common octahedral coordination?
2. In particular, why is this apparently so only for atoms with configurations d^0, d^1, and d^2?

The discussion given below follows that of Huisman et al.[1] This work was preceded by the important early studies on the same problem by Goodenough.[26]

We shall begin the treatment of this problem by considering the crystal-field splitting of d-levels.

12.4.3. Crystal-Field Splitting of d-Levels

The simplest starting point for discussing the splitting of d-levels is by a crystal-field calculation. In such an approach the ligand atoms surrounding the central atom are replaced by point charges Ze. The electrostatic potential V produced by these charges can be expanded in a series of spherical harmonics $Y_l^m(\theta, \phi)$. For a trigonal prism one can show that

$$V = 7 \left(\frac{4\pi}{5}\right)^{1/2} Ar^2 Y_2^0(\theta, \phi) + 7(4\pi)^{1/2} Br^4 Y_4^0(\theta, \phi) \qquad (12.10)$$

with

$$A = \frac{3}{7}(3\cos^2\theta_0 - 1)\frac{e^2 Z}{R^3} \qquad (12.11)$$

and

$$B = \frac{1}{28}(35\cos^4\theta_0 - 30\cos^2\theta_0 + 3)\frac{e^2 Z}{R^5} \qquad (12.12)$$

Here R is the metal–ligand distance and θ_0 is the angle between the trigonal axis and the line connecting central and ligand atoms.

The d-electron wave functions of the central atom can be written as $R(r)Y_l^m(\theta, \phi)$. The energy levels ε_m are the diagonal matrix elements of eV and are given by

$$\varepsilon_0 = 2A\overline{r^2} + 6B\overline{r^4} \qquad (12.13)$$

$$\varepsilon_{\pm 1} = A\overline{r^2} - 4B\overline{r^4} \qquad (12.14)$$

$$\varepsilon_{\pm 2} = -2A\overline{r^2} + B\overline{r^4} \qquad (12.15)$$

with

$$\overline{r^k} = \int_0^\infty |R(r)|^2 r^{k+2}\, dr \qquad (12.16)$$

For all reasonable values of θ_0 and $\overline{r^2}/\overline{r^4}$, the $m = \pm 1$ level has the highest energy.

However, whether the $m = 0$ or the $m = \pm 2$ level has the lowest energy depends in a rather sensitive way on the values of θ_0 and $\overline{r^2}/\overline{r^4}$.

The values of θ_0 observed in compounds of the MoS_2 type are close to the value $\theta_0 = 49° \, 6'$ $(\cos^2 \theta = \frac{3}{7})$ for an ideal prism (with all edges equal). The values of $\overline{r^2}/\overline{r^4}$ can be calculated if the radial part of the wave function is known. Using simple Slater wave functions, one finds that the $m = 0$ and $m = \pm 2$ levels lie close to each other for compounds of MoS_2 type. However, the use of analytical approximations to numerical Hartree–Fock wave functions leads to an appreciable splitting with the $m = \pm 2$ level (e') as the lowest energy level.

Unfortunately, as we shall see below, it is necessary to transcend such crystal-field calculations to obtain a wholly reliable description of the ligand-field splitting. We shall see that a molecular-orbital calculation indicates that the $m = 0$ (a_1') level has the lowest energy.

12.4.4. Molecular-Orbital Calculations

A simple molecular-orbital calculation of the Hückel type can be carried out for trigonal-prismatic and for octahedral coordination.[1] In the first instance, only metal d-orbitals and ligand p-orbitals are considered. The metal d-orbitals are expressed in a coordinate system centered on the metal atom, with z parallel to the trigonal axis (see Fig. 1 of ref. 1). For

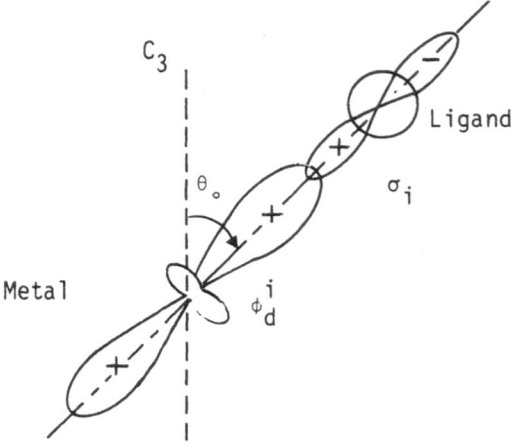

Figure 12.4. Coordinate system in which d-orbitals are expressed.

each ligand i, a coordinate system is chosen with z_i pointing in the direction of the metal atom and x_i in the $z - z_i$ plane. The ligand p-orbitals are p_i (p-orbital along z_i axis), π_{vi} (along x_i), and π_{hi} (along y_i). The σ_i-orbitals are responsible for σ, the π_{vi}- and π_{hi}-orbitals for π-bonding. The orbitals can be classified according to the irreducible representations of the symmetry group D_{3h} of the trigonal prism. The result for d- and σ-orbitals is shown in Table 12.1.

First, all overlap integrals are neglected and only σ-bonding is taken into account. The energy levels then depend only on the angle θ_0, on the energy difference between d- and σ-orbitals, and on the overlap energy V_σ defined by

$$V_\sigma = \int \phi^i_d H \sigma_i d\tau \tag{12.17}$$

where ϕ^i_d is a metal d-function constructed in such a way that it has maximum overlap with the σ_i-orbital of ligand i (see Figure 12.4). The energy levels thereby obtained are recorded in Table 12.2.

By rotating one triangle of ligand atoms through an angle of 60°, the trigonal prism is transformed into a trigonally distorted octahedron (symmetry D_{3d}). The energy levels for this coordination can be expressed in the same manner in terms of the overlap energy V_σ, the results being recorded in Table 12.3.

For an ideal octahedron ($\cos^2 \theta_0 = \frac{1}{3}$; symmetry O_h) the a_{1g} and e_g levels coalesce into the t_{2g} level. In Figure 12.5 the energy levels for an ideal trigonal prism ($\cos^2 \theta_0 = \frac{3}{7}$) and an ideal octahedron are compared.

Table 12.1. d- and σ-Orbitals in a Trigonal Prism
($\varepsilon = \exp(2\pi i/3)$; $\varepsilon^* = \exp(-2\pi i/3)$)[a]

	Metal, ϕ^n_d	Ligand functions	V^m_σ
A'_1	ϕ^0_d	$6^{-1/2}(\sigma_1 + \sigma_2 + \sigma_3 + \sigma_4 + \sigma_5 + \sigma_6)$	$(\frac{3}{2})^{1/2}(3\cos^2\theta_0 - 1)V_\sigma$
A''_2	—	$6^{-1/2}(\sigma_1 + \sigma_2 + \sigma_3 - \sigma_4 - \sigma_5 - \sigma_6)$	Nonbonding
E'	$\begin{cases} \phi^2_d \\ \phi^{-2}_d \end{cases}$	$6^{-1/2}(\sigma_1 + \varepsilon\sigma_2 + \varepsilon^*\sigma_3 + \sigma_4 + \varepsilon\sigma_5 + \varepsilon^*\sigma_6)$ $6^{-1/2}(\sigma_1 + \varepsilon^*\sigma_2 + \varepsilon\sigma_3 + \sigma_4 + \varepsilon^*\sigma_5 + \varepsilon\sigma_6)$	$\frac{3}{2}\sin^2\theta_0 V_\sigma$
E''	$\begin{cases} \phi^1_d \\ \phi^{-1}_d \end{cases}$	$6^{-1/2}(\sigma_1 + \varepsilon^*\sigma_2 + \varepsilon\sigma_3 - \sigma_4 - \varepsilon^*\sigma_5 - \varepsilon\sigma_6)$ $6^{-1/2}(\sigma_1 + \varepsilon\sigma_2 + \varepsilon^*\sigma_3 - \sigma_4 - \varepsilon\sigma_5 - \varepsilon^*\sigma_6)$	$3\sin\theta_0\cos\theta_0 V_\sigma$

[a] All overlap integrals are neglected and only σ-bonding is taken into account. The energy levels then depend only on the angle θ_0, on the energy difference between d- and σ-orbitals, and on the overlap energy V_σ, defined as $V_\sigma = \int \phi^i_d H \sigma_i d\tau$ where ϕ^i_d is a metal d-function constructed in such a way that it has maximum overlap with the σ_i-orbital of ligand i. V^m_σ is the overlap energy[1] of d-orbital ϕ^m_d with a ligand function $\sum_i a_i\sigma_i$.

Table 12.2. Energy Levels for a Trigonal Prism (σ-bonding only, after Huisman et al.[1])[a]

$$E(e''^*) = \tfrac{1}{2}\Delta[-1+(1+\alpha_3)^{1/2}]$$
$$E(e'^*) = \tfrac{1}{2}\Delta[-1+(1+\alpha_2)^{1/2}]$$
$$E(a_1'^*) = \tfrac{1}{2}\Delta[-1+(1+\alpha_1)^{1/2}]$$
$$E(a_2'')_{nb} = -\Delta$$
$$E(a_1') = \tfrac{1}{2}\Delta[-1-(1+\alpha_1)^{1/2}]$$
$$E(e') = \tfrac{1}{2}\Delta[-1-(1+\alpha_2)^{1/2}]$$
$$E(e'') = \tfrac{1}{2}\Delta[-1-(1+\alpha_3)^{1/2}]$$

with

$$\alpha_1 = 6(V_\sigma/\Delta)^2(3\cos^2\theta_0 - 1)^2$$
$$\alpha_2 = 9(V_\sigma/\Delta)^2\sin^4\theta_0$$
$$\alpha_3 = 36(V_\sigma/\Delta)^2\sin^2\theta_0\cos^2\theta_0$$

[a] Antibonding levels are denoted by *, non-bonding levels are labeled nb; Δ is the energy difference between d- and σ-orbitals while V_σ is the overlap energy defined by equation (12.17).

Table 12.3. Energy Levels for a Trigonally Distorted Octahedron (σ-bonding only, after Huisman et al.[1])[a]

$$E(e_g^*) = \tfrac{1}{2}\Delta[-1+(1+\alpha_2+\alpha_3)^{1/2}]$$
$$E(a_{1g}^*) = \tfrac{1}{2}\Delta[-1+(1+\alpha_1)^{1/2}]$$
$$E(e_g)_{nb} = 0$$
$$E(a_{2u})_{nb} = -\Delta$$
$$E(e_u)_{nb} = -\Delta$$
$$E(a_{1g}) = \tfrac{1}{2}\Delta[-1-(1+\alpha_1)^{1/2}]$$
$$E(e_g) = \tfrac{1}{2}\Delta[-1-(1+\alpha_2+\alpha_3)^{1/2}]$$

[a] Notation is as in Table 12.2.

12.4.5. Effect of π-Bonding

The effect of π-bonding was also investigated by Huisman et al.[1] The contribution of bonding can be expressed in terms of an overlap energy

$$V_\pi = \int d^{i,\pm 1} H\pi^{i,\pm 1}\,d\tau \tag{12.18}$$

where $d^{i,\pm 1}$ and $\pi^{i,\pm 1}$ are d-orbitals of the central atom and p-orbitals of ligand i, respectively, with magnetic quantum numbers $m = \pm 1$ with respect to the metal–ligand i-direction. The results of the calculations are shown in Figure 12.6. For σ-bonding only, the $a_1'^*$ level is the antibonding level with the lowest energy. The inclusion of π-bonding, however, decreases the energy difference between the $a_1'^*$ and e'^* levels.

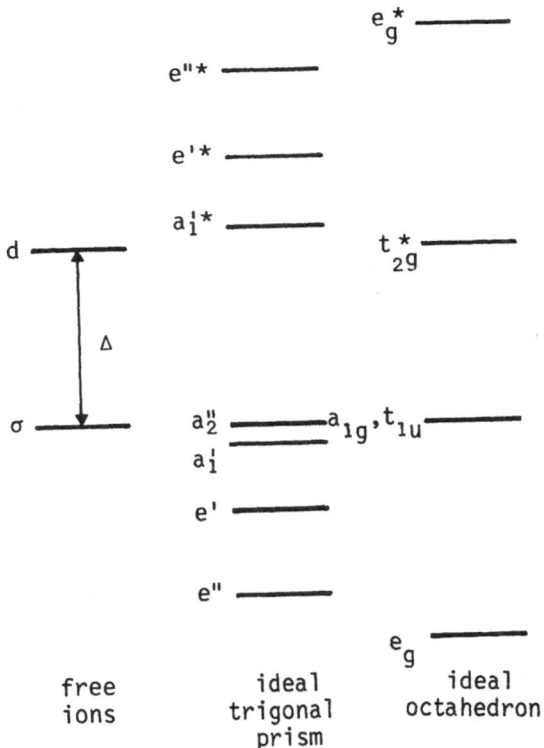

Figure 12.5. Comparison of the energy levels of an ideal trigonal prism and an ideal octahedron, for $V_\sigma/\Delta = 1$.

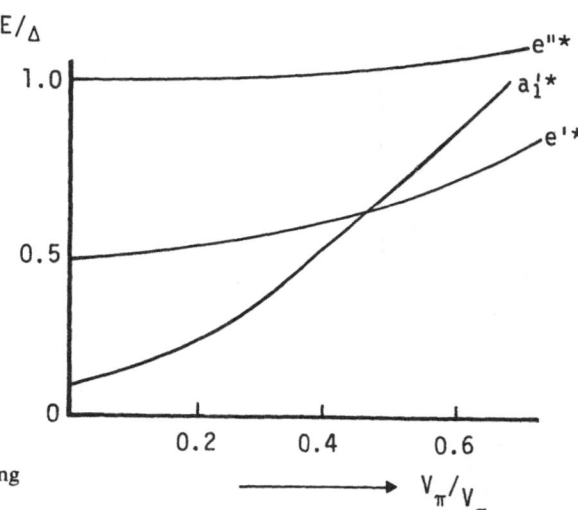

Figure 12.6. Effect of π-bonding on energy levels.

12.4.6. Role of d-Covalency in Stabilizing Trigonal-Prismatic Coordination

Huisman *et al.*[1] proposed an explanation for the stability of the trigonal-prismatic coordination based on the molecular-orbital calculations discussed above.

However, a cautionary remark is called for here. It is known that one cannot obtain a fully quantitative estimate of the total energy by summing the orbital energies of occupied one-electron states. This point is discussed in detail in Section 12.8.3. But this procedure of using the sum of one-electron energy levels may well be better for energy differences, because a comparison of a trigonal prism and an octahedron leads one to expect that most of the electron–electron Coulomb repulsion energies will be quite similar in the two systems. The most important difference will be due to a different anion–anion repulsion. Huisman *et al.*[1] point out that this energy favors octahedral coordination.

For atoms with a d^0-configuration, only the bonding and nonbonding levels are occupied. The contribution of d-covalency to the energy difference $(E_p - E_o)_{cov}$ between an ideal trigonal prism and an ideal octahedron for a d^0 central atom can be calculated from Tables 12.2 and 12.3. In Figure 12.7, $(E_p - E_o)_{cov}$ is plotted as a function of the parameter

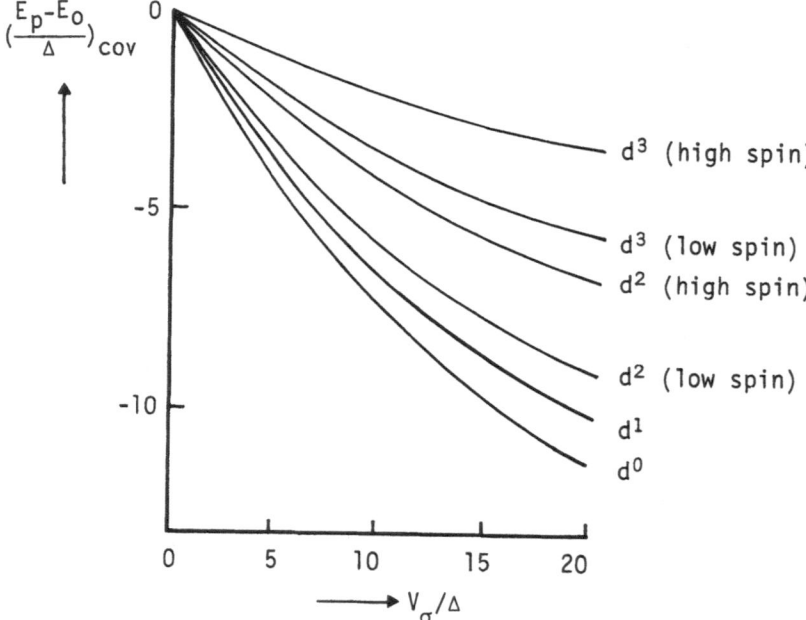

Figure 12.7. $(E_p - E_o)_{cov}$ plotted as a function of the parameter which measures d-covalency. Note that trigonal prismatic coordination is favored in d^0-, d^1-, and low-spin d^2-configurations.

(V_σ/Δ) which is a measure of d-covalency. Since $(E_p - E_o)_{cov}$ is negative, the d-covalency stabilizes trigonal-prismatic coordination with respect to octahedral coordination.

As stressed in Huisman et al.,[1] it turns out that the contribution of electrostatic interactions between the ions to the energy difference $E_p - E_o$ is positive, i.e., the Madelung energy is in favor of octahedral coordination. In ionic compounds, the contribution of the Madelung energy will dominate and octahedral coordination is to be expected. In strongly covalent compounds, however, the contribution of d-covalency may stabilize the trigonal-prismatic coordination.

The energy difference $(E_p - E_o)_{cov}$ for ions with configurations d^n ($n = 1, 2, 3$) can be calculated from the data in Tables 12.2 and 12.3. For d^2- and d^3-ions in a trigonal prism, both low-spin and high-spin configurations are compared with the high-spin configuration in an octahedron.

Taking into account that the Madelung energy favors octahedral coordination, it can be seen from Figure 12.7 that trigonal-prismatic coordination is most likely in predominantly covalent compounds of ions with configurations d^0, d^1, and low-spin d^2. It is less probable for ions with high-spin d^2- and low-spin d^3-configurations and it is not expected for high-spin d^3-ions and ions with more than three d-electrons.

We shall conclude this discussion of the transition metal dichalcogenides by referring to the connection of the work of Huisman et al.[1] with the band structure calculations of Matthiess[27] for 1T-TaS₂ (octahedral coordination) and 2H-TaS₂ (trigonal-prismatic coordination). To do so, we have plotted in Figure 12.8 the ligand-field d-energy levels of an ideal trigonal prism, and an ideal octahedron. Matthiess compares these centers of gravity of the various d subbands calculated by the augmented plane-wave LCAO method and finds they are in excellent agreement with Huisman et al.[1]

12.5. Two-Dimensional Layer Compounds II

We next discuss, though much more briefly than the treatment of the previous section, chemical bonding in relation to the electronic structure of the layer compounds graphite and boron nitride. Since the bonding within a hexagonal layer is strong, and the van der Waals interaction between layers is weak, it is again useful to discuss the electronic structure of a two-dimensional layer. When necessary, the effect of interlayer interactions can be added later.

As already pointed out, the relevant molecular structures are benzene and borazole, respectively. It is natural therefore to appeal to a simple treatment of bonding in these two molecules and then to extend the

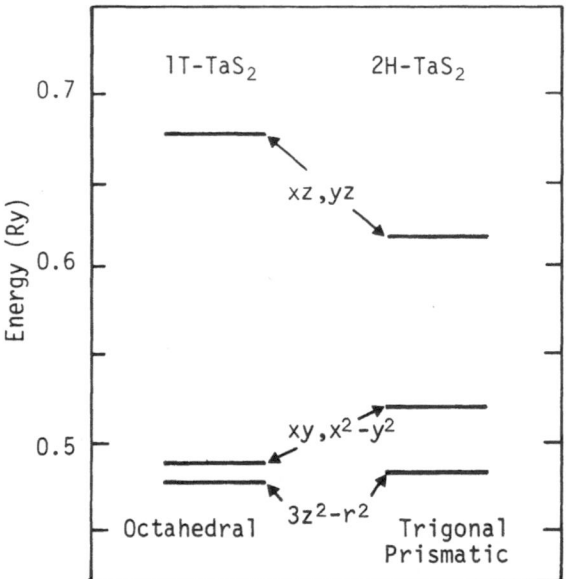

Figure 12.8. Ligand-field d-energy levels of an ideal trigonal prism, and an ideal octahedron derived from the centers of gravity of d-subbands calculated by Matthiess.[27]

discussion to apply to graphite and boron nitride. It is worthwhile here to treat benzene and borazole by a common approach to their π-electron levels. The starting point used is the work of Roothaan and Mulliken,[28] to whose paper the interested reader is referred for details of the results quoted below.

12.5.1. Electronic Structure of Borazole

These workers treat the electronic structure of borazole in a manner which proved very favorable in the later work of Coulson and Taylor[29] to effect the generalization of their molecular results to the electronic band structure of boron nitride.

Therefore, we shall summarize their results for borazole below (referring to benzene as a limiting case in the discussion). We note that one purpose of the study of Roothaan and Mulliken[28] was to treat the ultraviolet spectra of benzene and borazole by the semiempirical molecular orbital method.

Neglecting interactions of nonneighboring ring atoms, one can write the secular equations for the six approximate molecular orbitals, derivable as linear combinations of $2p_\pi$ atomic orbitals of the ring atoms, for benzene and borazole, respectively, in the forms

$$\begin{vmatrix} A & 1 & 0 & 0 & 0 & 1 \\ 1 & A & 1 & 0 & 0 & 0 \\ 0 & 1 & A & 1 & 0 & 0 \\ 0 & 0 & 1 & A & 1 & 0 \\ 0 & 0 & 0 & 1 & A & 1 \\ 1 & 0 & 0 & 0 & 1 & A \end{vmatrix} = 0 \quad \text{(benzene)} \qquad (12.19)$$

and

$$\begin{vmatrix} A' & 1 & 0 & 0 & 0 & 1 \\ 1 & A'' & 1 & 0 & 0 & 1 \\ 0 & 1 & A' & 1 & 0 & 0 \\ 0 & 0 & 1 & A'' & 1 & 0 \\ 0 & 0 & 0 & 1 & A' & 1 \\ 1 & 0 & 0 & 0 & 1 & A'' \end{vmatrix} = 0 \quad \text{(borazole)} \qquad (12.20)$$

In the case of benzene, for example, A involves not only the desired energy E but also Coulomb and resonance integrals, and the overlap S between adjacent C atoms, with appropriate generalization for borazole involving evidently both B and N atomic orbitals.

These secular equations can be reduced by suitable similarity transformations to the forms

$$\begin{vmatrix} A' & 2 & 0 & 0 & 0 & 0 \\ 2 & A'' & 0 & 0 & 0 & 0 \\ 0 & 0 & A' & 1 & 0 & 0 \\ 0 & 0 & 1 & A'' & 0 & 0 \\ 0 & 0 & 0 & 0 & A' & 1 \\ 0 & 0 & 0 & 0 & 1 & A'' \end{vmatrix} = 0 \qquad (12.21)$$

where, to get from the borazole result (12.21) to benzene, one merely has to set $A' = A'' = A$. (It is, in fact, possible to reduce the secular equation for benzene completely by a different transformation based on its higher symmetry. However, in the form presented above the relation between benzene and borazole, to be exploited in a two-dimensional layer, comes out very clearly.)

The molecular orbitals corresponding to the roots $A = -2$ and $A = -1$ (twice) in benzene, or the corresponding orbitals in borazole, are each occupied in the ground state by two electrons.

Without more ado, let us write the above in a form such that we can describe the effect on the π-levels of bringing molecules together into a layer compound, the molecular levels being evidently broadened into bands. To see this broadening, we follow Coulson and Taylor and rewrite the above results in an explicit form for borazole as

$$(\alpha_B - E)(\alpha_N - E) - g^2(\beta - ES)^2 = 0 \tag{12.22}$$

where we have now made explicit the Coulomb integrals α_B and α_N on boron and nitrogen atoms, respectively, and β denotes the resonance integral.

As an example, to go to benzene we merely let $\alpha_B = \alpha_N = \alpha_C$ and we find discrete allowed values of g^2 (corresponding to discrete values of A above) as $g^2 = 1$ (twice) and $g = 4$. We stress now that the electronegativity difference between B and N will be reflected in the difference between α_B and α_N, and the fact that boron is more electronegative than nitrogen is reflected in the inequality $\alpha_B > \alpha_C > \alpha_N$. We shall see below that this is a crucial point in establishing an essential difference between the electronic structure of a graphite layer and of a layer of boron nitride, the focal point of this discussion.

12.5.2. Broadening of π-Levels into Bands

To deal with the detailed tight-binding calculation of graphite, the reader is referred to the paper by Coulson and Taylor. However, the essential point which should be clear from the above discussion is that the discrete levels of the molecule, corresponding to discrete values of g^2 given above, will be broadened into bands, the detailed nature of the structure of the graphite layer merely giving the spread of values of g^2 that is found to run through the range 0 to 9 after detailed calculation. We next define sums and differences of Coulomb integrals as

$$E_0 = \tfrac{1}{2}(\alpha_B + \alpha_N) \tag{12.23}$$

and

$$\delta = \tfrac{1}{2}(\alpha_B - \alpha_N) \tag{12.24}$$

If we finally introduce the abbreviations $Z = E - E_0$, merely shifting the zero of energy, and $\gamma = \beta - E_0 S$, then it is elementary to obtain from equation (12.22) that

$$Z = \frac{-2\gamma g^2 S \pm [4\gamma^2 g^4 S^2 + 4(\delta^2 + g^2\gamma^2)(1 - g^2 S^2)]^{1/2}}{2(1 - g^2 S^2)} \tag{12.25}$$

If we denote the result of taking the plus sign in equation (12.25) by $Z_+(g)$ and the result of taking the minus sign by $Z_-(g)$, then allowing g^2 to traverse its range of values discussed above it is straightforward to show that $Z_+(g)$ has its lowest value equal to δ, having a range which is continuous up to a maximum value, while $Z_-(g)$ decreases to its minimum from the value $-\delta$. Thus, for boron nitride, the levels are separated into two subbands, with an energy gap 2δ which can be seen from the above discussion to be directly connected to the electronegativity difference between boron and nitrogen. On the other hand, when one passes to the limit of graphite, then $\alpha_B = \alpha_N = \alpha_C$, and δ tends to zero. Thus, while in boron nitride there is an energy gap, in graphite the gap 2δ tends to zero and the π-bands touch.

Though, of course, the above treatment is oversimplified in a number of respects, the important conclusion is that the electronegativity difference between boron and nitrogen is responsible for the known insulating character of boron nitride. Even the order of magnitude of the energy gap can be estimated successfully, using the parameters extracted for the borazole molecule by Roothaan and Mulliken from their study of the ultraviolet spectrum (the gap comes out to be ~ 4.5 eV with reasonable parameters). More accurate, subsequent band-theory work[30] is in general accord with this conclusion. For graphite the bands touch, and we expect a semimetallic character. The interaction between layers will be more significant on the electronic density of states in graphite than in boron nitride, but we shall not enter into further details here. The subsequent work of Corbato,[31] however, should be referred to, as well as the use of related types of treatment to understand the optical properties of graphite.[32]

From the chemically oriented discussion of one- and two-dimensional solids, we turn to some further interesting examples which are better treated via scattered wave theory.

12.6. Molecular Scattered Wave Calculations I: Treatment of TTF-TCNQ

Batra et al.[33] have approximated the energy-level structure of solid tetrathiofulvalene-tetracyanoquinodimethane (TTF-TCNQ) by superimposing the energy levels of the radical ions TTF$^+$ and TCNQ$^-$.

The energy levels of these ions were obtained by carrying out self-consistent statistical-exchange multiple scattering calculations using overlapping-atomic-sphere molecular models. They are thereby able to account for most of the prominent features of the experimental photoemission spectrum of solid TTF-TCNQ by shifting the free-cation TTF$^+$ and the free-anion TCNQ$^-$ energy levels upward and downward by 3.8 eV, respectively, the reasons for this shift being discussed below.

12.6.1. Molecular Structure of Cation and Anion

The geometrical models used for the molecular ions TTF$^+$ and TCNQ$^-$ are shown in Figures 12.9 and 12.10, respectively. As discussed in Appendix A12.2, for each molecular ion the set of atomic spheres is surrounded by a circumscribing outer sphere (labeled OUT 1 in Figures 12.9 and 12.10). The outer sphere is taken to be externally tangential to the outermost atomic spheres. Charge distributions and potential fields are spherically averaged within each atomic sphere and outside the outer sphere, and volume averaged in the intersphere region. The theoretical basis for the overlapping-atomic-sphere version of the method in Appendix A12.2 has been discussed by Herman *et al.*[34]

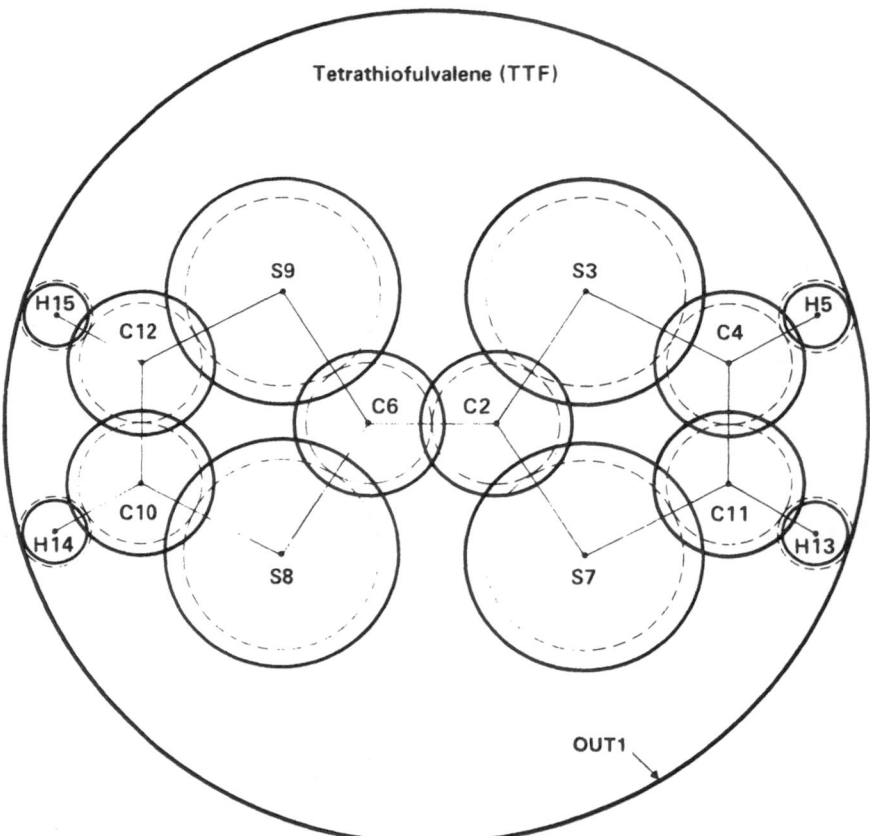

Figure 12.9. Nonoverlapping atomic sphere model (dashed circles) and overlapping-atomic-sphere model (solid circles) used by Batra *et al.*[33] for TTF ($\equiv C_6S_4H_4$) and TTF$^+$ molecules. The outer sphere is labeled OUT 1 and the atomic spheres are numbered 2 to 15.

The molecular ion calculations of Batra *et al.* were carried out self-consistently using the statistical-exchange parameter $\alpha = 0.75$ (cf Appendix A12.2). Spherical harmonics up to $l_{max} = 4$ and 2 were employed outside the bounding sphere and inside the atomic spheres, respectively. The atomic dimensions of TTF$^+$ and TCNQ$^-$ are known to be slightly different from those of TTF^0 and TCNQ^0, but Batra *et al.* ignore these small differences and take the dimensions from X-ray data on crystalline TTF[35] and crystalline TCNQ.[36] In each case the molecular ion was assumed to be planar and to have D_{2h} symmetry.

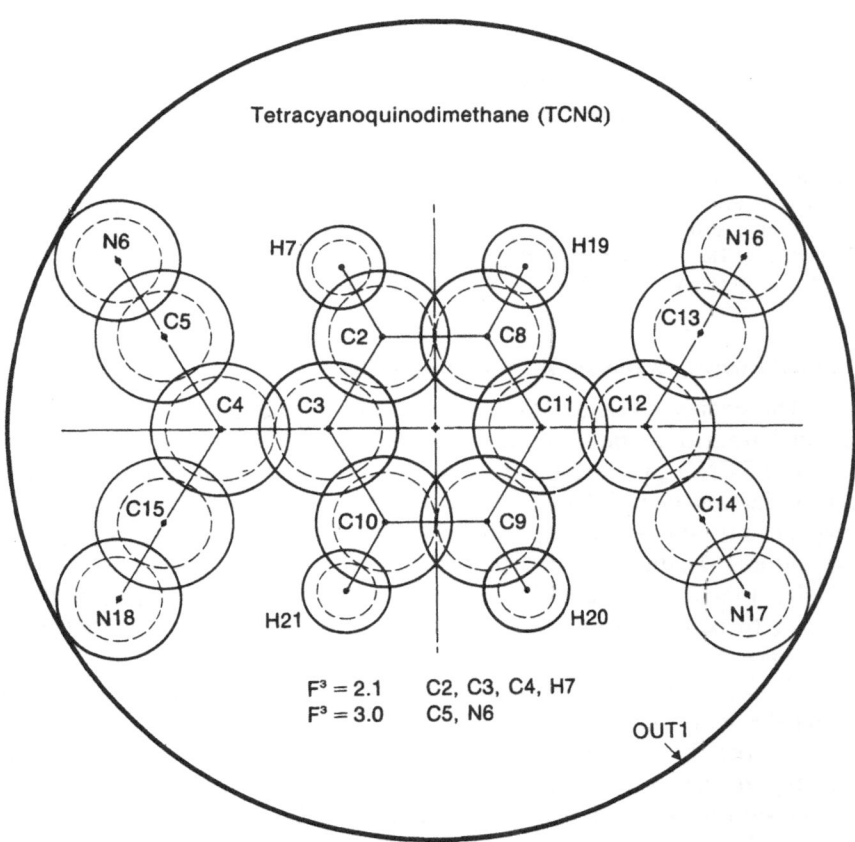

Tetracyanoquinodimethane (TCNQ)

$F^3 = 2.1$ C2, C3, C4, H7
$F^3 = 3.0$ C5, N6

OUT1

Figure 12.10. As in Figure 12.11 but for TCNQ ($\equiv C_{12}N_4H_4$) and TCNQ$^-$ molecules. Atomic spheres are numbered 2 to 21. The ratio of the volumes of the overlapping and corresponding nonoverlapping spheres is denoted by F^3 (after Batra *et al.*[33]).

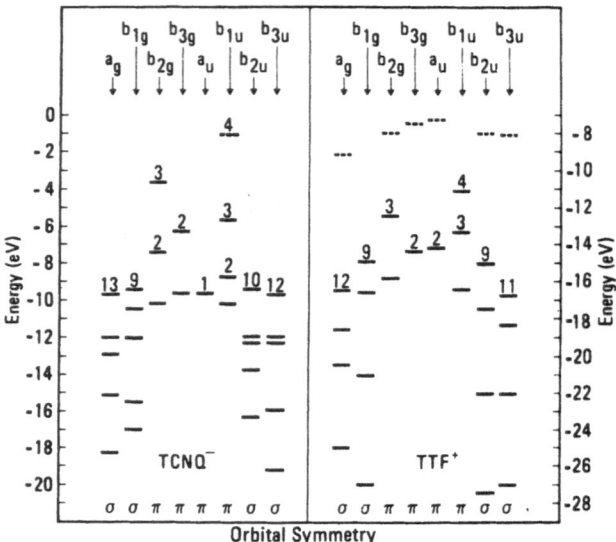

Figure 12.11. Energy levels of TTF$^+$ and TCNQ$^-$ as calculated by Batra *et al.*[33] For each irreducible representation of the D_{2h} point group, the levels are labeled by serial numbers in order of increasing energy. Occupied levels are denoted by solid lines, unoccupied levels by dashed lines. The ionization level ($4b_{1u}$) in TTF$^+$ is singly occupied and the affinity level ($3b_{2g}$) in TCNQ$^-$ is singly occupied.

12.6.2. Energy Levels of TCNQ$^-$ and TTF$^+$

The energy levels of TCNQ$^-$ and TTF$^+$ calculated using Johnson and Slater's transition state method[37,38] are shown in Figure 12.11.

For TCNQ$^-$, the orbital $3b_{2g}$ is singly occupied and $4b_{1u}$ is unoccupied. All ionization energies in TCNQ$^-$ are found to be shifted upward with respect to the corresponding values for TCNQ^0 in Herman and Batra,[39] by an amount ranging from 3.5 to 4.2 eV. The electron affinity of TCNQ^0, which is the same as the ionization potential for TCNQ$^-$, is calculated to be 3.6 eV, to be compared with the experimental value[40,41] of 2.7 ± 0.1 eV. The two lowest optical transitions in TCNQ$^-$ are $3b_{1u} \rightarrow 3b_{2g}$ and $3b_{2g} \rightarrow 4b_{1u}$ which occur, respectively, at 2.0 and 2.5 eV and are polarized along the long molecular axis.

All the TTF$^+$ energy levels calculated by Batra *et al.* are shifted downward with respect to the neutral spectrum by an amount ranging from 4.1 to 4.6 eV. The orbital $4b_{1u}$ is singly occupied and has an ionization energy of 11.2 eV. Transition state calculations for the three lowest-lying symmetry-allowed excitations yield $4b_{1u} \rightarrow 13a_g$, with energy 2.0 eV and polarized along the axis perpendicular to the molecular plane; $4b_{1u} \rightarrow 4b_{2g}$, 3.2 eV polarized along the long molecular axis; and $4b_{1u} \rightarrow 3b_{3g}$, 3.6 eV polarized along the short molecular axis.

12.6.3. Approximation to Energy Levels of a Crystal

Batra *et al.* then consider a simplified model for solid TTF-TCNQ composed of TTF$^+$ and TCNQ$^-$ ions. They neglect polarization effects and assume simply that when the crystal is formed, the energy levels for each constituent ion are moved upward or downward by the Madelung potential produced by the ionic lattice. As an idealization, they assume that the Madelung potential affects all orbitals on a given molecule equally, so that all the TTF$^+$ or TCNQ$^-$ levels are shifted rigidly by the same amount. They also assume that the energy bands of interest have negligible width and that the crystal is a semimetal with the Fermi level passing through the coincident narrow bands arising from the TTF$^+$ $4b_{1u}$ and TCNQ$^-$ $3b_{2g}$ levels. In the free ions, these levels have energies -11.2 and -3.6 eV, respectively, according to the calculations of Batra *et al.* In order to bring these levels into coincidence, they therefore move the TTF$^+$ levels upward in energy by the amount 3.8 eV already mentioned and the TCNQ$^-$ levels downward in energy by the same amount. The superimposed and shifted energy-level structures for this model of TTF-TCNQ are shown in Figure 12.12(c).

12.6.4. Photoemission Spectrum

The above model has been used by Batra *et al.* to interpret the photoemission spectrum of solid TTF-TNCQ. Or rather, it is assumed that a model intermediate between those for TTF^{0}TCNQ^0 and TTF$^+$TCNQ$^-$, the energy levels of both being shown in Figure 12.12, will be appropriate.

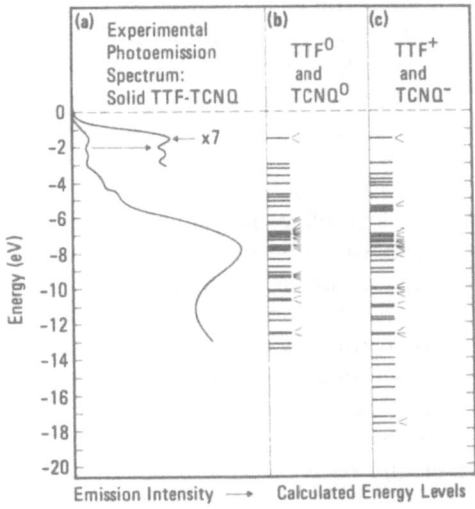

Figure 12. 12. Comparison of theoretical and experimental spectra, after Batra *et al.*[33]: (a) experimental photoemission spectrum, (b) superposition of TTF^0 and TCNQ^0 levels, (c) superposition of TTF$^+$ and TCNQ$^-$ levels.

The photoemission spectrum as measured[42] is shown alongside the theoretically calculated energy levels in Figure 12.12.

In that plot, the highest experimental peak has been adjusted to the highest occupied level (TTF $4b_{1u}$ and TCNQ $3b_{2g}$) on the assumption that this peak is produced by photoemission from the highest occupied bands in the solid. This, as can be seen from Figure 12.12, leads to a reasonable correspondence between the remainder of the experimental spectrum, with the exception of the next-to-highest peak, and the theoretical level structure. For instance, there are a large number of theoretically predicted energy levels in the vicinity of the main experimental peak.

It is then argued that the next-to-highest peak, which does not correspond to any of the theoretically predicted levels, is associated with the TTF-TCNQ surface. What they propose is that this peak is due to electronic transitions from neutral TCNQ or neutral TTF molecules lying on the surface immediately above TTF$^+$ or TCNQ$^-$ ions, respectively. We do not know, at the time of writing, of any independent evidence, theory, or experiment bearing on this point and we shall therefore not go into further details here.

12.7. Molecular Scattered Wave Calculations II: Treatment of Polymer $(SN)_x$

Chemical bonding arguments are again of considerable value in treating the electronic structure of the polymer $(SN)_x$, as is clear from the work of Salahub and Messmer.[2] The discussion below is largely based on the work of these authors. Although this inorganic polymer sulfur nitride has been known for more than 60 years, it is only over the last 20 years that detailed studies of its rather fascinating properties have been carried out.

It is known that $(SN)_x$ can be formed via solid-state polymerization from S_2N_2. In the reaction, the originally colorless S_2N_2 first turns a dark blue color and subsequently bronze as $(SN)_x$ forms; we shall return to these facts below.

As to structure, the polymer forms as bundles of fibers and this has created problems in obtaining an accurate crystal structure. While in early work it was supposed that $(SN)_x$ consisted of zigzag chains of alternating sulfur and nitrogen atoms with a large alternation in S—N bond lengths, both electron diffraction[43,44] and X-ray studies[45–47] yield a chain structure of the form shown in Figure 12.13. However, important multiple-scattering effects render the interpretation of the electron-diffraction data difficult, and therefore the results of X-ray studies seem preferable. The X-ray analysis indicates that the $(SN)_x$ chain is nearly planar and has almost uniform S—N bond lengths.

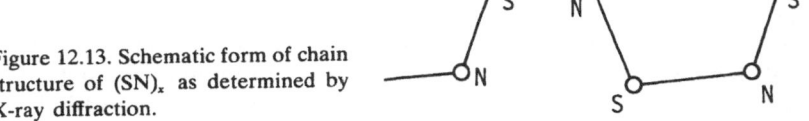

Figure 12.13. Schematic form of chain structure of (SN)$_x$ as determined by X-ray diffraction.

Measurements of electrical and optical properties of (SN)$_x$ have revealed a highly anisotropic character, and these have prompted the suggestion that (SN)$_x$ might belong to the class of pseudo-one-dimensional compounds, such as KCP and TTF-TCNQ, discussed in some detail above. However, unlike these latter one-dimensional compounds, (SN)$_x$ does not undergo a metal–insulator transition as the temperature is lowered, but in fact remains metallic down to 0.33 K where it undergoes a transition to the superconducting state.[48] This tells us that (SN)$_x$ is not one dimensional in the same sense as TTF-TCNQ and KCP and therefore that two- or three-dimensional effects are of importance.

Salahub and Messmer[49] used the Xα scattered-wave method summarized in Appendix A12.2 to perform self-consistent field calculations on the molecules SN, S$_2$N$_2$, S$_4$N$_4$, NO, and N$_2$O$_2$, which led to a greater understanding of the formation of an (SN)$_x$ chain and also indicated the importance of the polar SN bonds in these molecules. Subsequently, these workers have reported the results of molecular-orbital cluster calculations and band-structure calculations,[2] which allow one to obtain a better understanding of the electronic structure of (SN)$_x$. Below, we summarize their main results for (1) the S$_2$N$_2$ species, which is an intermediate in the polymerization, (2) an S$_4$N$_4$ chain, which is important not only in its own right but also as a reasonable cluster model for the infinite chain, (3) the interaction of two S$_4$N$_4$ chains, and (4) one-, two-, and three-dimensional band structures in (SN)$_x$.

12.7.1. Square and Open S$_2$N$_2$

A dark-blue colored paramagnetic intermediate is observed in the polymerization reaction of solid S$_2$N$_2$ to (SN)$_x$.[45,46] The reaction may be represented schematically as in Figure 12.14, where the originally (nearly) square planar S$_2$N$_2$ molecule opens up by stretching one S—N bond yielding a high-spin species, which then reacts with another similar species to propagate the chain. In order to characterize the intermediate, Salahub and Messmer have performed self-consistent field, Xα scattered-wave calculations on S$_2$N$_2$ in the geometry inferred from the crystal structure of (SN)$_x$.

Figure 12.15 is taken from their work, and the calculated orbital energies of square planar S$_2$N$_2$ and the open S$_2$N$_2$ structure are shown in the first two columns. The most striking difference in the two sets of orbital energies is the occurrence of a new σ-level in the open structure at -0.45 Ry. This level has no bound counterpart for the square-planar geometry. It becomes the highest occupied molecular orbital in the open geometry and leaves an empty π-orbital at slightly higher energy. This π-orbital was occupied in the square geometry.

As to the origin of this level, Salahub and Messmer made a contour plot of the corresponding molecular orbital and hence demonstrated that it is predominantly an antibonding combination of p-functions on sulfur and nitrogen interacting along the stretched S—N bond. If one forces these two atoms together again, to revert to the square-planar geometry, it is found that the node forces the energy of the level to rise so much that it becomes unbound in the square-planar configuration. Or, looked at from the starting point of square S_2N_2 being opened up, this level descends in energy and, at some point, becomes degenerate with the highest π-level. As this degeneracy occurs, one will have partially filled orbitals and a high-spin (paramagnetic) state can result.

We have already stated that square S_2N_2 is colorless while the open form observed in the polymerization is blue. By inspecting Figure 12.15 it can be seen that for the square geometry, the lowest absorption bands are likely to arise from excitations from the two π-levels near the Fermi level (E_F) to the unoccupied π-level at -0.20 Ry. For the open configuration the lowest transitions involve excitations from the two σ-levels near E_F. It is possible to excite an electron to the empty π-level just 0.02 Ry above E_F; however, as discussed above in the actual open-shell species responsible for the blue color, this level should become degenerate with the higher σ-level. The transitions of interest therefore involve excitation to the vacant π-level at -0.29 Ry.

Salahub and Messmer[2] have calculated the relevant transition energies using Slater's transition-state procedure, which yields an average of singlet and triplet excitation energies. These values should therefore be underestimates of the singlet–singlet transition energies. The singlet–triplet splittings should be larger for the square geometry, where the excitations involved are of the $\pi^* \rightarrow \pi$ type and the relevant exchange interaction should be appreciable. For the open configuration, the excitations are of the $\pi^* - \sigma$ type and one would expect quite small differences in the energies of the singlet and triplet states.

For square S_2N_2, Salahub and Messmer calculate transitions at 3.66 and 3.69 eV which lie in the ultraviolet, consistent with the lack of color of S_2N_2. For the open geometry they calculate transitions at 2.69 and 2.75 eV (~ 4600 Å) in the visible region of the spectrum. Their calculations on square and open geometries of S_2N_2 therefore lead them to a consistent explanation of the change from a colorless diamagnetic species to a colored paramagnetic form.

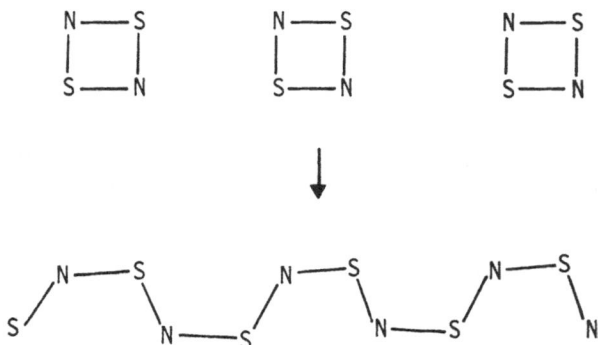

Figure 12.14. Schematic representation of the formation of an $(SN)_x$ chain from S_2N_2 molecules.

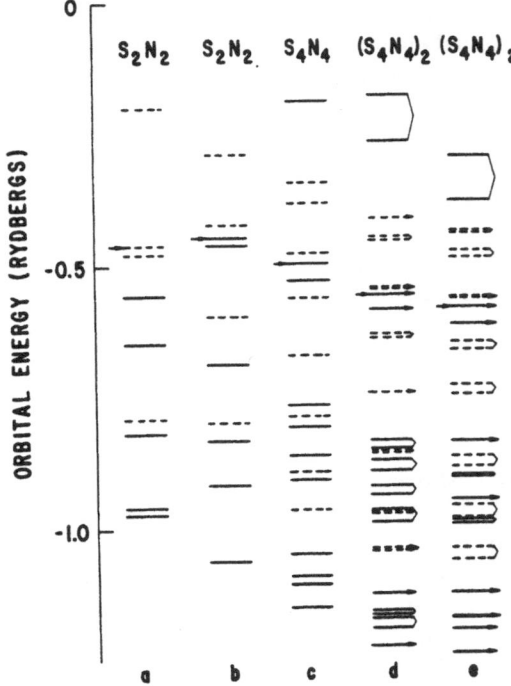

Figure 12.15. Calculated orbital energies (Ry) for molecular orbitals of (a) square S_2N_2, (b) open S_2N_2, (c) an S_4N_4 chain, (d) $(S_4N_4)_2$, ($\bar{1}$ 0 2) plane, (e) $(S_4N_4)_2$, (1 0 0) plane (after Salahub and Messmer[2]). Dashed lines denote π-levels. The Fermi level is denoted by an arrow to the left of each column. Arrows on energy levels in columns d and e indicate the presence of two close-lying levels. Several low-lying σ-levels are not shown.

12.7.2. S$_4$N$_4$ Chain

As Salahub and Messmer emphasize, their calculations on an S$_4$N$_4$ chain serve two purposes. First, there is mass spectroscopic evidence that in the sublimation of (SN)$_x$ it is an S$_4$N$_4$ unit which is volatilized. This S$_4$N$_4$ unit has been characterized as a biradical. Under the assumption that there is no change in the geometry of an S$_4$N$_4$ chain fragment upon sublimation, their calculations are appropriate to this species. Furthermore, as mentioned above, the S$_4$N$_4$ chain also serves as a simple model for the infinite chain.

Their orbital energies are shown in the third column of Figure 12.15. By contour plots of the two highest orbitals which are of σ-symmetry, Salahub and Messmer show that they are "dangling bond" orbitals and thus have no counterpart in the infinite chain. They prove to be highly localized on the end sulfur atom (64%) or on the end nitrogen atom (70%), respectively. An unoccupied π-level occurs very near to these two levels. Since the exact geometry of the observed species is not known and may differ somewhat from the geometry appropriate to (SN)$_x$ that is used in the calculations, the actual eigenvalues for the S$_4$N$_4$ fragment may be altered to some extent from the values plotted in Figure 12.15. Hence it is not possible to be sure whether this π-level should be occupied or vacant in the actual S$_4$N$_4$ fragment. If it is occupied, then the σ-orbitals referred to above, which are also nearly degenerate, could hold one electron each and then one would have a situation very like the normal picture of a biradical, that is, one electron in each of two highly localized orbitals at either end of the molecule. One can then understand how, in an encounter between two of these species with the sulfur end of one approaching the nitrogen end of the other, a bond could be formed leaving again an unpaired electron on either end of an S$_8$N$_8$ chain, and so on.

12.7.3. X-Ray Photoelectron Spectrum

Figure 12.16(a) shows the measured photoelectron spectrum[50,51] of (SN)$_x$. In Figure 12.16(b), a "density-of-states" curve, due to Salahub and Messmer,[2] is reproduced. They obtained this by replacing each of the calculated discrete eigenvalues with a Gaussian of width chosen to be 0.06 Ry. The dangling bond states have not been included, since these will not survive in the infinite chain limit. It is immediately seen that the agreement with the measured X-ray photoelectron spectrum is striking. All the main features, and even the relative intensities, are accounted for in a reasonably satisfactory way. One concludes from the agreement that the intensity matrix elements do not vary greatly over the region of the spectrum being considered. In the lower part of Figure 12.16, the weak shoulder near 0

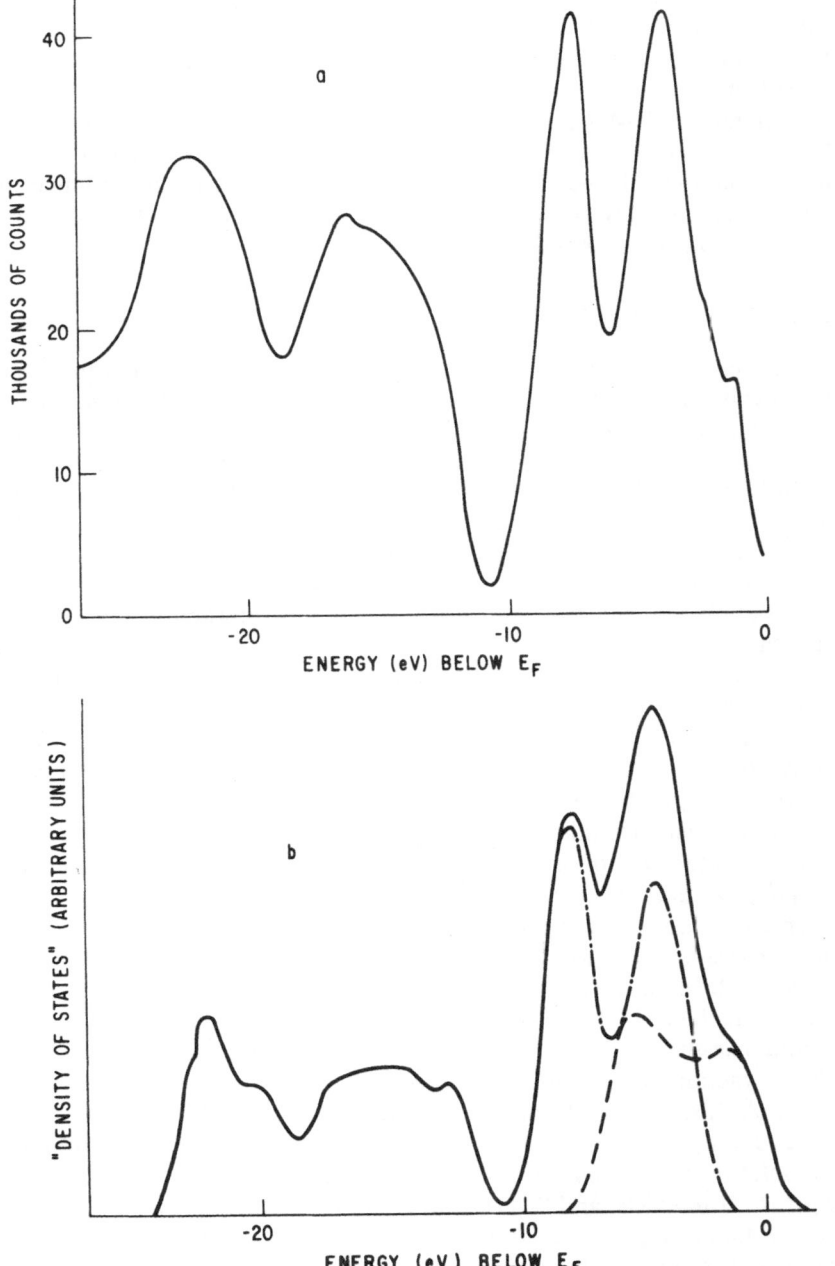

Figure 12.16. (a) X-ray photoelectron spectrum[49] of $(SN)_x$, (b) "density-of-states" for an S_4N_4 fragment (after Salahub and Messmer[2]).

is due to the π-conduction band of $(SN)_x$. The next peak results from overlapping π- and σ-bands while the other three peaks are due to σ-electrons.

Salahub and Messmer emphasize that the agreement demonstrated above between the experimental X-ray photoemission spectrum of $(SN)_x$ and the theoretical calculations for an S_4N_4 cluster is good evidence that many of the essentials of the electronic structure of $(SN)_x$ are already contained in such a simple cluster model.

12.7.4. Interaction of Two S_4N_4 Chains

Because of the above conclusion, it may be anticipated that the results obtained by allowing for interaction between S_4N_4 units will prove relevant to the interaction of infinite $(SN)_x$ chains. The density-of-states curves corresponding to the $(S_4N_4)_2$ units discussed below have been generated by Salahub and Messmer and they find only insignificant changes in the lower part of Figure 12.16. This agrees with Mengel et al.,[50] who concluded that none of the X-ray photoemission structure appears to be directly related to interchain coupling effects.

Salahub and Messmer have calculated the interaction between two S_4N_4 chains for two cases:

1. The model for the interchain interaction in a $(\bar{1}\ 0\ 2)$ plane (the mean plane of the $(SN)_x$ chains). The changes in the one-electron levels of S_4N_4 brought about by interchain coupling are shown for the $(\bar{1}\ 0\ 2)$ dimer in the fourth column of Figure 12.15. Each level of S_4N_4 is split into two levels in $(S_4N_4)_2$ corresponding to bonding and antibonding combinations of the S_4N_4 molecular orbitals. The energy difference between these levels depends on the type of orbital involved. It is found to be very small for the low-lying σ- and π-levels. The largest interaction occurs for the highest occupied σ molecular orbitals. Here the splitting is around 0.3 eV and rough extrapolation to the infinite case would probably increase this "bandwidth" by a factor of about 2. There is also significant splitting of some of the higher-lying π-levels, of about 0.1 eV, which one would expect to influence the conduction bands of $(SN)_x$. However, as discussed below, this interaction is not the most important for the π-bands.

2. A S_4N_4 dimer relevant to the interaction in a (100) crystallographic plane. Here, one would expect intuitively that the largest interchain interactions would occur for the π-levels since a favorable end-on overlap of p-functions exists, and this is what Salahub and Messmer find. From the fifth column of Figure 12.15 it can be seen that the splitting of the levels is very small for this configuration, while there are appreciable splittings for the π-levels (up to 0.4 eV). Below, it will be shown how this splitting affects the conduction bands of $(SN)_x$ and is in fact responsible for the metallic conductivity.

12.7.5. Band-Structure of $(SN)_x$

Early band-structure calculations on $(SN)_x$ are reviewed by Salahub and Messmer, to whose account the interested reader is referred. We focus below on the results of these workers. They argue that, in the absence of self-consistent field calculations, a reliable procedure is to use the extended Hückel method,[52] but to choose the parameters to give agreement with self-consistent field calculations of the related molecules.

In their work,[2] the adjustment was done using the self-consistent field results for SN and for square S_2N_2. The extended Hückel and self-consistent-field eigenvalues are plotted in Figure 12.17. The relative positions of the energy levels are quite good. The calculated (extended Hückel) charge transfer from S to N is 0.5 electrons for SN and 0.3 electrons for S_2N_2, in reasonable agreement with the self-consistent field $X\alpha$ value of 0.5 electrons for both these molecules.

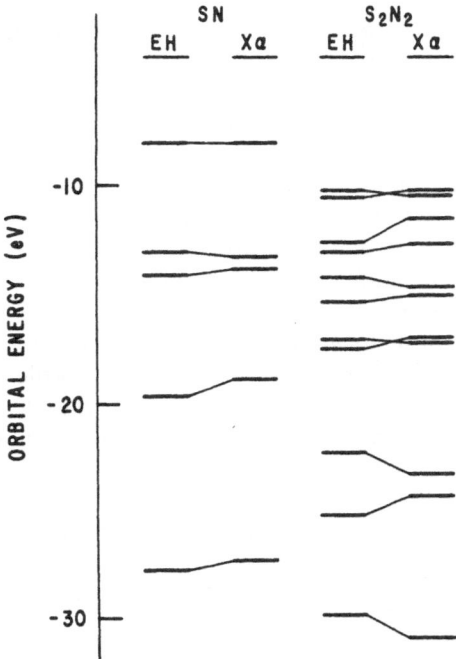

Figure 12.17. Comparison of extended Hückel and self-consistent-field $X\alpha$ scattered wave orbital eigenvalues for SN and S_2N_2 (square). The SCF-$X\alpha$ levels have been rigidly moved so that the Fermi level coincides with that of the extended Hückel calculations (after Salahub and Messmer[2]).

Salahub and Messmer then performed extended Hückel calculations of the band structure of $(SN)_x$ with the parameters thus obtained. As in the calculations on the molecular fragments discussed above, their two-dimensional band-structure calculations indicate that interchain interactions in a (1 0 0) plane are important for the conduction bands of $(SN)_x$. While such band-structure calculations predict that a single $(SN)_x$ chain should be a one-dimensional metal and hence would be subject to a Peierls distortion, which would yield a one-dimensional semiconductor, interchain coupling is shown by Salahub and Messmer to lead to band doubling and the net result is that three-dimensional $(SN)_x$ is predicted to be metallic, in agreement with experiment.

12.8. Electron Density Description of the Ground State of Molecules

In the previous section, we discussed mainly the redistribution of electronic charge density $\rho(\mathbf{r})$ when atoms are chemically bonded on the basis of molecular orbitals built up by a linear combination of atomic orbitals. Here we wish to focus on an approach which leads to a description of the ground state of molecules by means of the electron density,[53-55] rather than by wave function theory. In the course of the discussion, we shall:

1. Provide some basis for the Hückel assumption that the total energy of a molecule is the sum of the orbital energies. In fact, in the simplest density description of the type just referred to, we shall see that these quantities are proportional to one another for molecules at equilibrium and we shall be able to calculate the proportionality factor.

2. Give a justification from first principles of the one-body potential theory employed in the previous section to treat S_2N_2, TTF, and TCNQ. Of course, this is the basic local potential energy, not the additional assumption involved in partitioning this potential energy, which in the end is done to expedite the numerical calculations and is not chemically very satisfying.

This description of a molecular bond by its electron density can also be used, for example, to treat electron and X-ray diffraction from chemically bonded silicon, both in crystalline and amorphous forms,[56] though we cannot go into the details of this particular application here.

12.8.1. Simplest Theory of Inhomogeneous Electron Gas

The idea behind the theory of the inhomogeneous electron gas, such as exists in atoms, molecules, and solids, is to start from the properties of the homogeneous electron gas and, by neglecting at first density gradients ($\nabla\rho$) and electron correlation, to build an approximate theory of the inhomogeneous electron density $\rho(\mathbf{r})$. Thus, for the homogeneous gas we may write the Fermi energy E_f in the form

$$E_f = \frac{p_f^2}{2m} + V \tag{12.26}$$

where p_f is the maximum momentum and V is evidently constant, independent of $\mathbf{r}$, in this uniform system. The electron density ρ_0 is obtained from the maximum momentum p_f by using the theorem that a cell in phase space of volume h^3 can hold two electrons, leading to the relation

$$\rho_0 = \frac{8\pi}{3h^3} p_f^3 \tag{12.27}$$

Now we go over to the inhomogeneous gas, with electron density $\rho(\mathbf{r})$ in the ground state. In the approximation in which we neglect density gradients and electron correlation, we can replace equation (12.26) by

$$\mu = \frac{p_f^2(\mathbf{r})}{2m} + V(\mathbf{r}), \quad V(\mathbf{r}) = V_{\text{Coulomb}} + V_{\text{exchange}} \tag{12.28}$$

μ being the energy of the fastest electron. We have here asserted that while the right-hand side consists of the sum of two terms, each of which depends on position $\mathbf{r}$, the left-hand side is a constant throughout the entire charge cloud of the molecule. This is so because if there were any variation in space of μ, then electrons could redistribute themselves to lower the energy. We have written μ in anticipation of the fact that it is the chemical potential. Equation (12.27) becomes

$$\rho(\mathbf{r}) = \frac{8\pi}{3h^3} p_f^3(\mathbf{r}) \tag{12.29}$$

and hence, by eliminating $p_f(\mathbf{r})$ between equations (12.28) and (12.29), we find

$$\mu = \frac{1}{2m} \left(\frac{3h^3}{8\pi}\right)^{2/3} [\rho(\mathbf{r})]^{2/3} + V(\mathbf{r}) \tag{12.30}$$

12.8.2. Value of the Chemical Potential in a Neutral Atom or Molecule

We have stressed above that the constancy of μ reflects the fact that, in a molecular formation, electrons have flowed from one atom to another until the Fermi level is the same at each position $\mathbf{r}$. This idea shows that μ

is intimately related to the concept of electronegativity used earlier in this chapter, and indeed Parr and his co-workers[57] have argued convincingly for such a connection between μ and electronegativity.

In the simplest density description embodied in equation (12.30), however, we can evaluate μ in, say, a neutral atom by letting $\mathbf{r}$ tend to infinity. In this case, we expect $\rho(\mathbf{r})$ to tend to zero, $V_{Coulomb}(\mathbf{r}) \to 0$, and if we use the form of exchange potential discussed in Appendix A12.2 and utilized in the calculations reported earlier of $(SN)_x$, for example, then $V_{exchange} \propto [\rho(\mathbf{r})]^{1/3}$ and therefore again tends to zero at infinity. Since the right-hand side is zero at infinity, it follows that μ is zero and it is clear that, in this simplest theory, electronegativity is not dealt with. Nonetheless, in spite of this deficiency of the result (12.30), a valuable approximate relation emerges from it.

12.8.3. Relation between Total Energy and Sum of Orbital Energies

Hückel theory, as emphasized above, assumed that the total energy was given by the sum over occupied states of the orbital energies ε_i. However this, in general, is not correct as it stands (see, however, Pucci and March,[58] who discuss its dependence on dimensionality) because it counts the electron–electron interactions twice over. Nevertheless, we shall now see how the density description given above supports the Hückel-type procedure under some circumstances to be clarified below.

First, without as yet assuming $\mu = 0$, multiply equation (12.30) by the density $\rho(\mathbf{r})$ and integrate over the whole of space using

$$\int \rho(\mathbf{r})d\tau = N \tag{12.31}$$

N evidently being the total number of electrons in the molecule. Then we find

$$N\mu = \frac{1}{2m}\left(\frac{3h^3}{8\pi}\right)^{2/3}\int [\rho(\mathbf{r})]^{5/3}d\tau + \int \rho(\mathbf{r})V(\mathbf{r})d\tau \tag{12.32}$$

But now the first integral on the right-hand side is related directly to the total kinetic energy, T say. This is because we know that for a uniform Fermi gas the kinetic energy per electron is $\frac{3}{5}p_f^2/2m$ and hence for the inhomogeneous gas we can write, again using free-electron relations locally,

$$\text{Kinetic energy/unit volume} = \frac{3}{5}\frac{p_f^2(\mathbf{r})}{2m}\rho(\mathbf{r}) \tag{12.33}$$

If equation (12.29) is used to eliminate $p_t(\mathbf{r})$, we obtain

$$T = c_k \int [\rho(\mathbf{r})]^{5/3} d\tau, \qquad c_k = \frac{3}{10m} \left(\frac{3h^3}{8\pi}\right)^{2/3} \tag{12.34}$$

Returning to equation (12.32), we have

$$N\mu = \tfrac{5}{3}T + \int \rho(\mathbf{r}) V(\mathbf{r}) d\tau \tag{12.35}$$

For the wave function ψ_i satisfying the one-electron Schrödinger equation with potential energy $V(\mathbf{r})$, we can now write the corresponding orbital energy ε_i as

$$\varepsilon_i = \frac{\hbar^2}{2m} \int \psi_i^* \nabla^2 \psi_i d\tau + \int \psi_i^* V(\mathbf{r}) \psi_i d\tau \tag{12.36}$$

and hence the orbital energy sum, E_s say, is immediately given by

$$E_s = \sum_{\substack{\text{occupied} \\ \text{states}}} \varepsilon_i = T + \int \rho(\mathbf{r}) V(\mathbf{r}) d\tau \tag{12.37}$$

since

$$\rho(\mathbf{r}) = \sum_{\substack{\text{occupied} \\ \text{states}}} |\psi_i|^2$$

Subtracting equations (12.35) and (12.37), we derive

$$N\mu - E_s = \tfrac{2}{3}T \tag{12.38}$$

If we apply the above result to a neutral atom, where we argued earlier that in this simplest (Thomas–Fermi) density description the chemical potential μ was zero, we find, using the virial theorem in the form $E = -T$, that

$$E = \frac{3}{2} \sum_{\substack{\text{occupied} \\ \text{states}}} \varepsilon_i \tag{12.39}$$

a result first given for atoms by March and Plaskett.[59]

Therefore, though we knew that the Hückel result was an oversimplification in general when it replaced E by the orbital sum E_s, the message from that theory, that the total energy can be obtained directly from the orbital energy sum, is indeed correct in this simplest density description. Actually, for molecules, the same result follows after including the nuclear–nuclear potential energy in the total energy expression, provided one uses the virial theorem at equilibrium in the form $E = -T$. Of course, this is an exact result, and therefore it is quite permissible to use it. Unfortunately, however, it is known from Teller's theorem[60] that the Thomas–Fermi theory does not lead to molecular binding. Nevertheless, if one again employs the Euler equation (12.30), with the virial theorem as stated, plus the assumption that the chemical potential is zero as for neutral atoms in the Thomas–Fermi density description, then equation (12.39) is regained. That such a relation was useful for molecules at equilibrium was pointed out by Ruedenberg.[61] The proof from the statistical theory, outlined above, was given by March.[62]

As an illustration of the above relation, we compare in Table 12.4 the Hartree–Fock orbital energy sum and the total energy in extended chain systems; cases with hydrogen have been omitted. The alternating bond lengths discussed for the linear polyenes in Section 12.2 are reflected by various numbers in Table 12.4. The ratio is near to $3/2 = 1.50$, but there are corrections to it. We shall briefly indicate their origin below by summarizing the (formal) theory, including "density gradients" neglected in the Thomas–Fermi theory and also electron correlation.

12.8.4. Inclusion of Density Gradients and Electron Correlation

This time, we can usefully write the total energy E as a function(al) of the ground-state density $\rho(\mathbf{r})$ in the form

$$N\mu - E_s = \tfrac{2}{3}T + \int \left(\frac{\delta T}{\delta \rho} - \tfrac{5}{3}t\right) d\tau \tag{12.40}$$

where the last term in equation (12.40) is identically zero in the Thomas–Fermi limit, i.e., $t = c_k \rho^{5/3}$. But in equation (12.40), T is again the

Table 12.4. Ratio of Hartree–Fock Orbital Energy Sum to Total Energy in Extended Chain Systems (after Kertesz et al.[63])

	Ratio (theory of equation (12.39) gives 3/2)	Alternating bond distances
C atomic chain	1.464(5)	1.20 and 1.46 Å
	1.464(1)	1.19 and 1.34(5)
	1.460(8)	1.14(5) and 1.39(5)
	1.456	1.09(5) and 1.44(5)

single-particle kinetic energy, the correlation kinetic energy being assumed to be incorporated in E_{XC}.* Then the virial theorem at equilibrium reads

$$E = -T_{exact} = -T - T_{correlation} \tag{12.41}$$

where we have thereby defined the correlation kinetic energy.

For the uniform electron gas, after switching on the electron–electron interactions, $T_{correlation}$ is positive since electrons are promoted from states inside the Fermi sphere of radius p_f to states outside, leaving holes within the sphere. In general, we expect $T_{correlation}$ to be positive, while in terms of density gradients it can be shown that[55]

$$\int \left(\rho \frac{\delta T}{\delta \rho} - \tfrac{5}{3}t \right) d\tau = \frac{-\hbar^2}{108m} \int \frac{(\nabla \rho)^2}{\rho} d\tau + \text{higher-order gradient terms} \tag{12.42}$$

Finally, μ is negative for electrons in molecules, since they are evidently in bound states. Thus in the expression

$$E = \tfrac{3}{2}E_s + \tfrac{3}{2} \int \left(\rho \frac{\delta T}{\delta \rho} - \tfrac{5}{3}t \right) d\tau - \tfrac{3}{2}N\mu + \tfrac{2}{3}T_{correlation} \tag{12.43}$$

at least the "correction" terms to equation (12.39) tend to cancel. More work is required to study the accuracy of the cancellation.

In this connection, it is worth stressing that the rules formulated by Walsh[64] for determining molecular shapes work well, by comparing the energy of, say, the H_2O molecule at different angles, the energy being replaced by the sum of the orbital energies. It might appear that only at the equilibrium angle could one use the virial theorem in the form $E = -T_{exact}$. However, it turns out[65] that if at each angle the energy is minimized with respect to the bond length, then $T_{exact} = -E$ still holds.

For present purposes, the two important conclusions derived here that are applicable in the density description of atoms and molecules are:

1. One-body potential theory can be justified for calculating electron density and ground-state energy. This procedure involves the contribution to the one-body potential, $\delta E_{XC}/\delta \rho$, from exchange and correlation. The work on $(SN)_x$ and TTF-TCNQ discussed in Section 12.6 replaces this by the Slater form (A12.2.3), where α is to be adjusted, if possible, to incorporate some account of correlation.

2. There is an (approximate) relation between total energy and sum of eigenvalues given in equation (12.39). The corrections to this are exhibited in equation (12.43) and appear to be relatively small (see one example of this in Table 12.4). Therefore the procedures used in Hückel theory in Sections 12.2 and 12.4 are far more justified than we had a right to expect.

---

* The energy associated with exchange and correlation interactions: see also conclusion 1 below.

One final point is noteworthy. The chemical potential μ in the density description is certainly connected with the chemical concept of electronegativity. Thus, in equation (12.43), one reason for deviations from $E = \frac{3}{2}E_s$ is the nonzero μ. By examining deviations from this latter relation for a variety of molecules treated by fully wave-mechanical self-consistent field methods, of course, always at equilibrium, to which the above arguments are restricted, it has been shown[66] that the deviation

$$\Delta = E - \tfrac{3}{2}E_s \qquad (12.44)$$

does indeed correlate with electronegativity, Δ being largest and most negative for molecules containing the most electronegative atoms O and F.

Appendix A12.1. Alternation of Bond Lengths in Long Conjugated Chain Molecules

Ooshika[10] and independently Longuet-Higgins and Salem[3] argued that alternation of bond lengths in an infinite chain was the stable situation, and not equal bond lengths, as had been assumed by Lennard-Jones.[5] Ooshika's results, at least at first sight, seem to depend on the semiempirical evaluation of several energy integrals, so we shall present below the essence of the argument of Longuet-Higgins and Salem. This argument demonstrates that the LCAO theory, allowing for σ-bond compression, leads inevitably to bond alternation even if the π-electrons are assumed to move independently.

A12.1.1. Theoretical Assumptions

Equation (12.6) of the main text implies that one starts out from the assumption that the binding energy of a polyene can be expressed as the sum of two parts: one from the σ-bonds and the other from the π-electrons. The σ-electron energy, as in equation (12.6), is taken as a sum of independent contributions from the CC bonds (the CH bonds being assumed irrelevant to the arguments). The π-electron energy, in contrast, is to be determined by the LCAO molecular orbital theory and is a function of the resonance integrals β_i of all the CC bonds.[67] Below, we follow Longuet-Higgins and Salem[3] in assuming the resonance integral to decrease exponentially with increasing bond length.

The conditions for static equilibrium are clearly

$$\frac{\partial E}{\partial r_i} = 0, \qquad \text{all } i \qquad (A12.1.1)$$

which implies from equation (12.6) that

$$0 = \frac{\partial E_\sigma}{\partial r_i} + \frac{\partial E_\pi}{\partial r_i} = \left(\frac{df}{dr}\right)_{r_i} + \frac{\partial E_\pi}{\partial \beta_i}\left(\frac{d\beta_i}{dr}\right)_{r_i} \qquad (A12.1.2)$$

or, if we introduce p_i as the mobile order of the ith bond,[68] equation (A12.1.2) can be written as

$$f'_i + 2p_i\beta'_i = 0 \qquad (A12.1.3)$$

However, by assumption, $f(r)$ and $\beta(r)$ are unique functions of r and hence it follows from equation (A12.1.3) that there must be a fixed relation between the mobile bond order and the length of a bond in static equilibrium.

In the sp^2 hydrocarbons, ethylene, benzene, and graphite, the bond orders are determined by symmetry. The mobile orders and lengths of the bonds in these systems are given in Table A12.1.[3] Empirically, therefore, the mobile order and equilibrium length in Å of a bond between two sp^2 carbon atoms are related by the approximate formula

$$r(\text{Å}) = 1.50 - 0.15p \qquad (A12.1.4)$$

Longuet-Higgins and Salem[3] utilize this formula for determining the function $f(r)$ in equation (A12.1.3).

Evidently, equation (A12.1.3) applies to any configuration of static equilibrium, whether stable, metastable, or unstable. If the configuration considered is to correspond to stable equilibrium, then a further set of conditions must hold, namely, that $\partial^2 E/\partial q^2$ must be positive for any conceivable displacement q of the atoms from equilibrium.

A12.1.2. Linear Polyenes

In fact, Longuet-Higgins and Salem discuss the cyclic polyenes, the linear polyenes being more difficult mathematically. They point out that one would expect the bond alternation, which they demonstrate for the cyclic polyenes, to be at least as great in a linear even-membered polyene

Table A12.1. Mobile Orders and Lengths of the Bonds in the sp^2 Hydrocarbons, Ethylene, Benzene, and Graphite

	r (Å)	p (mobile order)
Ethylene	1.353 1.344	1.0
Benzene	1.397	0.667
Graphite	1.421	0.525

as in the infinite cyclic polyene. The reason is that one can (mathematically) obtain a pair of even linear polyenes by severing two "single" bonds of an even cyclic polyene. The effect is to raise and lower the mobile orders of alternate bonds as one proceeds away from the points of detachment, and this would be expected to lead to a more pronounced alternation of bond lengths near the ends of the fragments, though this effect would not penetrate very far into the conjugated chain. Longuet-Higgins and Salem point out that it is possible to obtain a limited confirmation of their above-expressed ideas by considering the energy of an infinite linear polyene in which the bond lengths are restricted to two alternate values r_1 and r_2. Following Lennard-Jones,[5] the orbital energies are of the form

$$E_{\pm j} = \pm(\beta_1^2 + \beta_2^2 + 2\beta_1\beta_2 \cos\theta_j)^{1/2} \qquad (A12.1.5)$$

where θ_j is a solution of the transcendental equation

$$\beta_2 \sin(n+1)\theta + \beta_1 \sin n\theta = 0 \qquad (A12.1.6)$$

Writing

$$\tan n\theta = \frac{-\beta_2 \sin\theta}{\beta_1 + \beta_2 \cos\theta} \qquad (A12.1.7)$$

and noting that as θ increases by π/n the left-hand side assumes all possible values, one can show that equation (A12.1.7) has one solution in each interval of length π/n between 0 and π, so that the values of θ_j are evenly distributed over this range. The total energy of the π-electrons in the limit $n \to \infty$ can then be shown to have the form

$$E_\pi = -\frac{n}{\pi} \int_{-\pi}^{\pi} (\beta_1^2 + \beta_2^2 + 2\beta_1\beta_2 \cos\phi)^{1/2} d\phi \qquad (A12.1.8)$$

For the cyclic polyenes, n is replaced by $(2n+1)$ in equation (A12.1.8). The resulting bond alternation should then be the same in the middle of the long-chain polyene as in the infinite cyclic polyene.

Appendix A12.2. Partitioning of the Space of Molecule (and One-Body Potential)

We have seen that the LCAO molecular orbital approach affords a valuable chemical method of constructing realistic wave functions and calculating energy levels. However, if we wish to transcend Hückel theory and proceed without empirical parametrization, the computations are very heavy. Therefore Johnson and Slater,[37] especially, have developed a method which, while chemically less satisfying, is very much less difficult computationally.

The idea is stated simply: it is to partition the space of the molecule into three regions:

I. Atomic regions within nonoverlapping spheres centered on the constituent atoms.
II. Interatomic regions between inner atomic spheres and an outer sphere surrounding the entire molecule.
III. The extramolecular region.

One is concerned with solving the one-electron Schrödinger equation (written below in Rydbergs)

$$[-\nabla^2 + V(\mathbf{r})]\psi(\mathbf{r}) = E\psi(\mathbf{r}) \tag{A12.2.1}$$

in each of the regions I–III, for a local potential energy function $V(\mathbf{r})$ given by

$$V(\mathbf{r}) = V_{\text{Hartree}} + V_{X\alpha}(\mathbf{r}) \tag{A12.2.2}$$

where

$$V_X = -6\alpha[(3/8\pi)\rho(\mathbf{r})]^{1/3} \tag{A12.2.3}$$

This exchange potential. it is argued, may be modified to include some account of electron correlation by appropriate choice of α, which is often selected so that the separated atoms are described in the best possible manner.

In region I, the potential energy V is assumed spherical within each sphere. The usual spherical harmonics $Y_l^m = Y_L$ then appear and we can express the molecular orbital wave functions within each spherical atomic region I, of radius b_j say, in the form

$$\Psi_1^j(\mathbf{r}) = \sum_L C_L^j R_1^j(E, r) Y_L(r), \qquad 0 < r \leqslant b_j \tag{A12.2.4}$$

Here the quantities C_L^j are to be determined while the radial wave functions satisfy

$$\left[\frac{-1}{r^2}\frac{d}{dr}\left(r^2\frac{d}{dr}\right) + \frac{l(l+1)}{r^2} + V^j(r) - E\right] R_1^j(E, r) = 0 \tag{A12.2.5}$$

for the spherical potential energy $V^j(r)$. The radial wave function must be finite at the origin $r = 0$ of each atomic sphere.

Next, in region III, one again assumes a spherical potential measured from the molecular center and the solution once more takes the form

$$\Psi_{\text{III}}(r) = \sum_L D_L^0 R_l^{\text{external}}(E, r) Y_L(r), \qquad b_0 \leqslant r < \infty \tag{A12.2.6}$$

where b_0 is the radius of the outer sphere. The functions R_l^{external} are obviously solutions of the above radial wave equation of the spherically averaged potential energy in region III.

Finally, in region II, of volume Ω_{II} say, one takes the volume average of $V(\mathbf{r})$ to obtain the constant potential V_{II} as

$$V_{II} = \frac{1}{\Omega_{II}} \int_{\Omega_{II}} V(\mathbf{r}) d\tau \qquad (A12.2.7)$$

Then the solutions are free-electron-like in region II. They may be written explicitly in multicenter partial wave form as

$$\Psi_{II}(\mathbf{r}) = \sum_L B_L^0 j_l(Kr_0) Y_L(\mathbf{r}_0)$$

$$+ \sum_j \sum_L A_L^j f_l(Kr_j) Y_L(\mathbf{r}_j), \qquad K = (E - V_{II})^{1/2} \qquad (A12.2.8)$$

where $\mathbf{r}_j = \mathbf{r} - \mathbf{R}_j$ and $\mathbf{R}_0$ is the vector defining the central atom. In the above equation,

$$f_l(Kr) = h_l^{(1)}(Kr), \qquad E < V_{II}, \quad K \text{ imaginary}$$

$$= n_l(Kr), \qquad E > V_{II}, \quad K \text{ real} \qquad (A12.2.9)$$

The functions j_l, $h_l^{(1)}$, and n_l are, respectively, the spherical Bessel function, spherical Hankel function of the first kind, and spherical Neumann function. The wave functions and their derivatives are required to be continuous across the adjacent spherical boundaries. This leads, as an example, to a relation between the above coefficients of the form

$$A_L^j = - iKb_j^2 [j_l(kb_l), R_l^j(Eb_j)] C_L^j \qquad (A12.2.10)$$

where

$$[j(x), R(x)] = j(x) \frac{dR(x)}{dx} - R(x) \frac{dj(x)}{dx} \qquad (A12.2.11)$$

One can finally write secular equations for the independent coefficients, thus yielding the orbital energies also.

This was the procedure employed to obtain the energy-level spectra for TTF-TCNQ and S_2N_2 molecules, etc., reported in the body of this chapter.

Note Added in Proof. Important advances in the theory of electron–phonon interaction in polyacetylene are reviewed in the 1983 Varenna lectures of J. R. Schrieffer. In addition, the writer and Prof. Schrieffer have focused on the importance of both Coulomb and Fermi holes around π-electrons in polyacetylene and polydiacetylene in interpreting the linear dispersion of π-plasmons (unpublished work).

Phase Transitions and Dimensionality

R. B. Stinchcombe

This chapter is intended to provide an introductory and largely self-contained review of selected topics in the area of phase transitions and dimensionality. Suitable background material may be found on the solid-state aspects in the book by Ashcroft and Mermin[69] and on statistical physics in the work of Landau and Lifshitz.[70] The writer found various other books[71-75] particularly useful in the preparation of this chapter.

13.1. Cooperative Behavior, and Phase Transitions

13.1.1. Basic Aspects of Cooperative Behavior and Phase Transitions

It is well known that systems with many constituents can exhibit cooperative behavior due to the influence of "statistics" or of interactions (or both). An example of the effect of statistics is condensation in the noninteracting Bose gas, where the Bose particles crowd into their lowest state at low temperatures. An example of a cooperative effect brought about by interactions is the calming down of a large lecture audience at the start of a lecture; another is the well-known "cocktail-party phenomenon," where the noise level can increase dramatically. In both of these cases, each person is responding to the background noise level produced by all the

R. B. Stinchcombe · Institut Laue-Langevin, Grenoble, France; and *Permanent address:* Theoretical Physics Department, University of Oxford, 1 Keble Rd., Oxford OX1 3NP, UK.

others. In other words, the individual interaction of one person with another has been added to provide a mean result in which individual aspects are lost. This "mean-field" view of cooperative behavior and phase transitions is one of the basic methods to be discussed later (Section 13.3).

Another example sometimes given is the set of people who sit, equally spaced, at a round table set for dinner with napkins equidistant between the diners. In a world without broken left–right symmetry, which napkin should they take? Each diner can polarize both his neighbors and then everybody ends up taking the same (e.g., their left) napkin. This is an example both of "order" and of "broken symmetry," which are important and related ideas in the theory of phase transitions. The example also illustrates an aspect, which will be of great importance in our discussion of *low*-dimensional systems, that of fluctuations destroying order: it only needs one of the chain of diners to respond in a less-than-ideal fashion and the order is lost.

Spin systems are the simplest examples allowing a quantitative discussion. Of these the simplest is the Ising chain of equally spaced spins, each of which may point up or down ($\sigma = \pm 1$); each interacts with its neighbors in such a way that the parallel orientation of a neighboring spin pair is of less energy (by a fixed amount $2J$, say) than the antiparallel orientation. It is clear that the lowest energy state(s) of the chain has all spins parallel, and this broken symmetry (ordered) state will occur at zero temperature. But if there are $N + 1$ spins in the chain the ordered arrangement can be upset by introducing a break in it, by reversing all the spins to the right of a selected point. This costs energy $2J$, but since the break can be made at any of N positions the entropy gain is $K \ln N$ and so the free-energy change is[76]

$$\Delta F = 2J - KT \ln N \qquad (13.1)$$

At any nonzero temperature, an infinite system ($N \to \infty$) will therefore always find it favorable to lower the free energy by destroying the order.

Peierls[77] proved by similar considerations that the corresponding *two*-dimensional Ising model (spins on a square net) *is* ordered at low, but nonzero temperatures: Peierls was able to show that if one starts from the ordered state in which all spins are aligned and reverses all the spins to the right of a border of length L, in a large system both the energy and the entropy change are proportional to L. Therefore, for temperatures less than some N-independent multiple of J/K, the energy change will win out and it will be unfavorable to remove the order.

In Section 13.2 we shall briefly review some of the many interacting systems that exhibit phase transitions. The above discusion (correctly) suggests that there is always a least dimensionality for the transitions to occur at nonzero temperatures. We should also bear in mind that in many systems we cannot separate the effects of statistics and interactions (this is, of course, the case for the λ-transition in He^4).

The most familiar phase transitions are the liquid–gas and magnetic transitions. Models will be given for these in Section 13.2, but we would like to use here their well-known properties to illustrate basic features.[78]

In the liquid–gas system the transition is from the high-temperature (or low-pressure) gas phase to the liquid phase, which has short-range order absent in the gas. Further lowering of temperature or increase of pressure normally produces a further transition, to the solid phase, which has crystalline order. A typical phase diagram is shown in Figure 13.1. Having used the variables T and P, the third variable V is dependent and could have been used to distinguish the phases. For the liquid–gas transition a better choice is the difference $\rho_L - \rho_g$ between the liquid and gas densities: this is zero in the high-temperature ("disordered") gas phase and becomes nonzero in the ("ordered") liquid phase. This is the behavior typical of an *order parameter*, which becomes nonzero in the ordered phase. An order parameter that is only nonzero in the solid phase could be constructed from an appropriate Fourier component of the spatially varying density. The gas–liquid–solid example is one in which there are a series of ordered phases, characterized by different order parameters. It also illustrates the difference between short-range and long-range order and correlations. A more quantitative discussion of correlations will be given in Section 13.1.3, and later we shall also meet the concept of quasi-long-range order.

The phase diagram for a typical ferromagnet is shown in Figure 13.2. On lowering the temperature a phase transition takes place, in the limit of

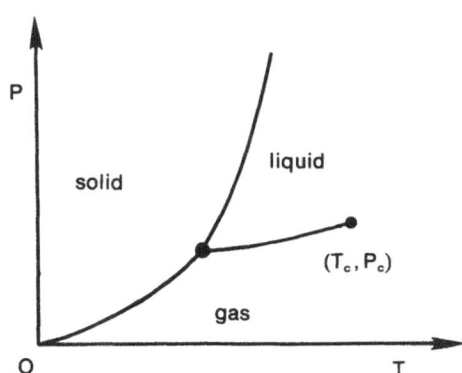

Figure 13.1. Solid–liquid–gas phase diagram.

zero external field h, at the transition temperature T_c. The transition is from a paramagnetic state, to the ferromagnetic state that has long-range order manifested by the appearance of a nonzero magnetization M, which is the order parameter (Figure 13.3). The transition is of *second order* (second-order derivatives of the free energy, such as the susceptibility and specific heat, diverge at the transition).

The antiferromagnet shows phase transitions at nonzero field. A typical phase diagram is shown in Figure 13.4. Here there are a series of ordered phases characterized by different order parameters related to components of the sublattice magnetization.

13.1.2. Criticality

At the transition, systems like those just discussed may become *critical* in the sense that strong fluctuations can occur because of the equal acceptability of two different phases. These fluctuations can cause the

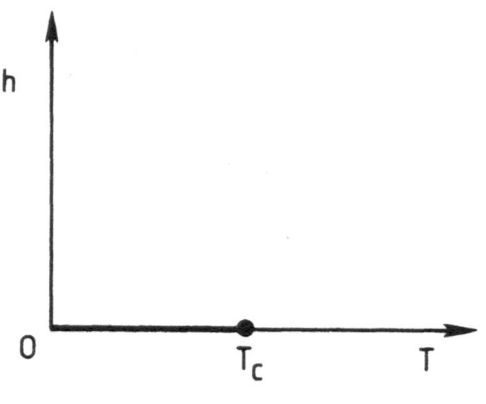

Figure 13.2. Phase diagram for a ferromagnet.

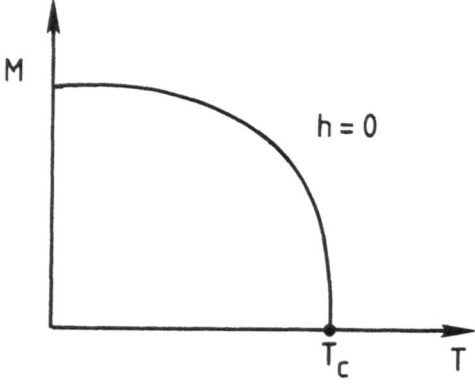

Figure 13.3. Temperature dependence of order parameter (magnetization) for a typical ferromagnet in zero field.

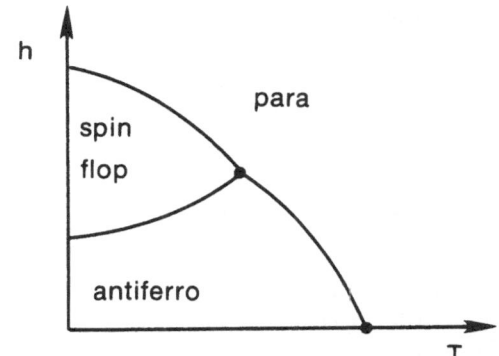

Figure 13.4. Phase diagram for a typical antiferromagnet.

divergences seen during second-order transitions, in quantities like susceptibilities or compressibilities. When the gas is as happy to be a liquid it can be easily compressed, and a small magnetic field will strongly modify the magnetization near the transition in a ferromagnet. It will be shown in Section 13.1.3 that this criticality will arise if the correlations become of infinite range. This is an important characteristic of a second-order phase transition.

The specific heat may also diverge at the transition; in magnetic systems the specific heat is a measure of short-range order and such a divergence there indicates the increasing short-range order as the transition is approached. Certain systems, like the three-dimensional Heisenberg ferromagnet, do not have diverging specific heats, though the susceptibility (a measure of long-range order) does diverge at the transition.

The bulk critical behavior can be described in terms of *critical exponents*, α, β, γ, δ. We now give their definition for the ferromagnet, for which the order parameter and critical condition are particularly simple. Similar definitions hold, with their appropriate parameters, for the liquid–gas and other second-order transitions. Near the transition, the singular parts of the specific heat C, magnetization M, and susceptibility χ are represented by[79]

$$C \propto |T - T_c|^{-\alpha}, \qquad h = 0 \tag{13.2}$$

$$M \propto (T_c - T)^{\beta}, \qquad T < T_c, \qquad h = 0 \tag{13.3}$$

$$\chi \propto |T - T_c|^{-\gamma}, \qquad h = 0 \tag{13.4}$$

$$M \propto |h|^{1/\delta}, \qquad T = T_c \tag{13.5}$$

The exponents α and γ may, in principle, take different values above and below the transition. It is not usually necessary to allow for this situation. As mentioned above, α need not be positive.

Some experimental and theoretical values of the critical exponents for various systems are given in Table 13.1, where we have also included two other exponents (η, ν) to be defined in the next subsection. The striking regularity of some of the values ("universality") and other aspects of critical behavior will be discussed in Sections 13.9 and 13.10.

13.1.3. Correlation Functions[80]

The susceptibility in a magnet is a particular kind of correlation function. It is, of course, the derivative of the average magnetization $\langle M \rangle$ (the order parameter) with respect to the external field h. Since this field is the variable conjugate to M, it occurs in the Hamiltonian through a term

$$H' = -hM \tag{13.6}$$

so the susceptibility is

$$\chi = \frac{\partial \langle M \rangle}{\partial h} = \beta [\langle MM \rangle - \langle M \rangle^2] \tag{13.7}$$

Here we have only considered the (nonvanishing) component of the magnetization along the direction of the field, and have neglected any

Table 13.1. Critical Exponents for Various Systems

Dimensionality	System or model	α	β	γ	δ	η	ν
$d = 3$	Antiferro (RbMnF$_3$)	—	0.316 ± 0.008	1.397 ± 0.034	—	0.067 ± 0.01	—
	Liquid gas CO$_2$	—	0.3447 ± 0.0007	1.20 ± 0.02	4.2	—	—
	Xe	—	0.344 ± 0.003	1.203 ± 0.002	4.4 ± 0.4	—	—
	Binary alloy (Co–Zn)	—	0.305 ± 0.005	1.25 ± 0.02	—	—	—
$d = 2$	Antiferro (Rb$_2$CoF$_4$)	(ln)	0.119 ± 0.008	1.34 ± 0.22	—	—	—
	(K$_2$CoF$_4$)	(ln)	0.123 ± 0.008	1.71 ± 0.05	—	0.2 ± 0.1	—
	Ising model (exact)	(ln)	0.125	1.75	15	0.25	1
—	Mean-field theory, Ornstein–Zernicke theory	(discontinuity)	0.5	1	3	0	0.5

possible commutation problems. The appropriate generalizations are straightforward. The square brackets in (13.7) denote the mean-square deviation of the magnetization. Since (neglecting for simplicity factors of $g\mu_B$)

$$M = \sum_i \sigma_i \qquad (13.8)$$

where σ_i is the spin at site i, the susceptibility is directly related to the correlation function defined by

$$\Gamma_{ij} = \langle \sigma_i \sigma_j \rangle - \langle \sigma_i \rangle \langle \sigma_j \rangle \qquad (13.9)$$

The relationship ("sum rule") is

$$\chi = \beta \sum_{ij} \Gamma_{ij} \qquad (13.10)$$

In simple cases, like the Ising magnet, quantities such as the internal energy and specific heat can be related to Γ_{ij} and to higher correlation functions (containing more spin operators). The corresponding discussion of, for example, the compressibility of the classical fluid involves density–density correlation functions (where σ_i is replaced by the density at point i).

Since Γ_{ij} vanishes if the spins at sites i, j are uncorrelated, it measures short-range correlations if i is near to j or long-range correlations if r_{ij} is large.

It is useful to define also time-dependent correlation functions, for example,

$$\Gamma_{ij}(t) = \langle \sigma_i(t)\sigma_j \rangle - \langle \sigma_i(t) \rangle \langle \sigma_j \rangle \qquad (13.11)$$

and the Fourier transforms of both space- and time-dependent types:

$$\chi_q \equiv \beta \int d^d r_{ij} [\exp(-i\mathbf{q}.\mathbf{r}_{ij})]\Gamma_{ij} \qquad (13.12)$$

$$\chi_q(\omega) \equiv \beta \int d^d r_{ij} dt [\exp(-i\mathbf{q}.\mathbf{r}_{ij} - i\omega t)]\Gamma_{ij}(t) \qquad (13.13)$$

(d is the dimensionality).

It is well known that the generalized susceptibility $\chi_q(\omega)$ is related to a linear response function (discussed elsewhere[80]), so that its poles give the frequencies of perturbing fields to which the system responds in a "resonant" way; thus these poles are the frequencies of the normal modes (or collective excitations) of the system.

It follows from the above equations that

$$\chi_q = \frac{1}{2\pi} \int d\omega \chi_q(\omega) \tag{13.14}$$

and that the usual susceptibility is

$$\chi = \chi_{q=0} \tag{13.15}$$

Let us suppose that the correlation function falls off at long distances like

$$\Gamma_{ij} \propto \frac{1}{r^{d-2+\eta}} \exp(-r_{ij}/\xi), \qquad r_{ij} \to \infty \tag{13.16}$$

The power-of-r prefactor is normally dominated by the exponential factor, in which ξ measures the range of correlations (ξ is the correlation length). For the special case $\eta = 0$, the Fourier transform is then

$$\chi_a \propto \frac{1}{(q^2 + 1/\xi^2)}, \qquad q \to 0 \tag{13.17}$$

This is the result of Ornstein–Zernicke theory, to be discussed briefly in Section 13.3.2. If $\xi \to \infty$, the correlations become of infinite range and we have the criticality referred to in the previous section. Then

$$\chi = \chi_{q=0} \to \infty, \qquad \xi \to \infty \tag{13.18}$$

Thus the diverging correlation length produces a divergent susceptibility, as mentioned earlier. The power-of-r prefactor becomes important when $\xi \to \infty$, since it then gives the dominant spatial dependence of Γ. The parameter η is a measure of the deviation from the simple Ornstein–Zernicke result. It is a new critical exponent. The last critical exponent we shall need, ν, describes the way in which ξ diverges at the transition:

$$\xi \propto |T - T_c|^{-\nu}, \qquad h = 0 \tag{13.19}$$

The long-range order is normally measured by $\langle M \rangle$. An equally good measure is $\langle M^2 \rangle$, since the relative fluctuation $[\langle M^2 \rangle - \langle M \rangle^2]/\langle M \rangle^2$ is of order $1/N$ except where the susceptibility χ diverges.

For systems with a vector order parameter there can be different correlation lengths for the correlation functions describing different components of the order parameter, for example, the transverse and longitudinal correlation functions Γ^{xx} and Γ^{zz} of a Heisenberg ferromagnet, where

$$\Gamma_{ij}^{\alpha\beta} = \langle S_i^\alpha S_j^\beta \rangle - \langle S_i^\alpha \rangle \langle S_j^\beta \rangle, \qquad \alpha, \beta = x, y, \text{ or } z \qquad (13.20)$$

(In this chapter we shall use interchangeably the two notations S^α or σ^α for a spin component; σ, without superscript, means σ^z.)

13.2. Systems and Models Exhibiting Transitions

13.2.1. Magnetic Systems

The phase diagram and order parameter for a typical ferromagnet were given in Figures 13.2 and 13.3. An example is EuO. We show in Figure 13.5 the inverse longitudinal correlation length for this system,[81] exhibiting clearly the divergence of ξ at the transition from the paramagnetic to the ferromagnetic phase. EuO is perhaps the best example of the Heisenberg model ferromagnet, whose Hamiltonian is

$$H = -\sum_i h S_i^z - \sum_{i \neq j} J_{ij} \mathbf{S}_i . \mathbf{S}_j \qquad (13.21)$$

The first term is the interaction with the external field, taken as usual to be in the z-direction, and the second term is the exchange interaction which (for $J > 0$) tends to align spins parallel.

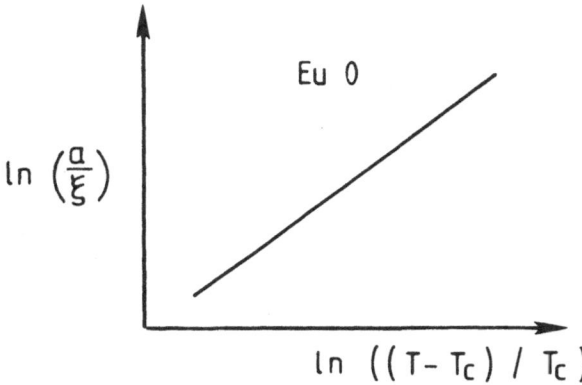

Figure 13.5. Inverse longitudinal correlation length for the ferromagnet EuO ($\nu = 0.69$).

The Ising model[82] has a simplified form of exchange term only involving the z-component of spin; thus the number n of components of the order parameter is $n = 1$, while it was $n = 3$ for the Heisenberg model. Because of the absence of transverse components of the spin, there are no commutation problems in the Ising model:

$$H = -\sum_i hS_i^z - \sum_{i \neq j} J_{ij}S_i^zS_j^z \qquad (13.22)$$

As discussed in Section 13.1, this model has a phase transition in two or more dimensions. The two-dimensional case ($d = 2$) was solved exactly by Onsager[83,84] for zero field and the resulting exponents are given in Table 13.1. There are examples of the Ising-model ferromagnet and, as for the Heisenberg model, many more examples of the antiferromagnetic case ($J < 0$). For a nearest-neighbor interaction with $J < 0$, adjacent spins tend to align antiparallel, and two sublattices of oppositely oriented spins may form. Rb_2CoF_4, K_2CoF_4 are examples of Ising antiferromagnets that can be considered as two dimensional,[85–87] since they have layers of strongly coupled spins with weak coupling between layers. Critical exponents for Rb_2CoF_4, K_2CoF_4 are given in Table 13.1 for comparison with the Onsager values. Also given are the exponents for the (three-dimensional) Heisenberg antiferromagnet $RbMnF_3$.[88]

In many respects the properties of the Ising and Heisenberg models are similar. But there are some crucial differences, which will be discussed later in Sections 13.4, 13.5, and 13.10, and they relate to the different character of the low-lying excitations, the lack of ordering in the two-dimensional Heisenberg model, and the different critical behavior of the two models.

Some magnets are intermediate cases, having Hamiltonians of the form

$$H = -\sum_i hS_i^z - \sum_{i \neq j} J_{ij}[\alpha \mathbf{S}_i . \mathbf{S}_j + (1 - \alpha)S_i^zS_j^z] \qquad (13.23)$$

with $0 < \alpha < 1$. In other cases, the same form of Hamiltonian (13.23) applies with $\alpha > 1$. If the field is zero, and α is so large that the longitudinal term in (13.23) can be neglected, we have the XY model

$$H = -\sum_{i \neq j} J_{ij}[S_i^xS_j^x + S_i^YS_j^Y] \qquad (13.24)$$

(where we redefined the exchange constant). This is a model with a two-component ($n = 2$) order parameter and its two-dimensional version has a special transition (the Kosterlitz–Thouless transition)[89-91] to a state without the usual long-range order (Section 13.8).

Many examples exist of all these systems in various dimensionalities.[92,93] We mention here only the following examples (1) of one-dimensional systems (weakly coupled chains): TMMC (Heisenberg model), $CsCl_3$ (XY model, at very low temperatures); and (2) of two-dimensional (layer) systems: K_2CoF_4, Rb_2CoF_4 (Ising models), Rb_2MnF_4 (Heisenberg model).

Diluted magnets exhibit percolation phenomena (Section 13.2.11). Another fascinating case is the spin glass. Here the competition between exchange interactions of opposite sign can lead to an ordering without the familiar long-range order.

13.2.2. Alloys

In the simplest model of a binary alloy, each site i of a regular lattice can be occupied by either of two species of atoms (A or B, say). The interaction energy for each nearest-neighbor pair of atoms is assumed to depend on whether they are the same (energy J) or different (energy $-J$). The occupation of each site i can be represented by a spin operator σ_i, where $\sigma_i = 1, -1$ correspond respectively to the site being occupied by an A or a B atom. The model is then clearly equivalent to an Ising antiferromagnet (if $J > 0$). The order parameter

$$\sum_{i \text{ on sublattice } 1} \langle \sigma_i \rangle \tag{13.25}$$

measures the tendency for a given sublattice to have, at low temperatures, more A than B atoms (or *vice versa*). The mean field approximation for this model is the Bragg–Williams theory.[94] Co—Zn is an example of a binary alloy having critical properties similar to those of the Ising model just described (see Table 13.1). Of course, such a model will be useless where band-electron properties are of interest.

13.2.3. Liquid–Gas Systems

A typical phase diagram was given earlier, and the critical properties of two typical examples, CO_2 and Xe, are given in Table 13.1. The "lattice gas" provides a somewhat oversimplified model of the liquid–gas system.[79] The model is again related to the Ising model. A lattice of sites i is used to represent the possible positions of a gas molecule. The site occupation is represented by a spin operator: $\sigma_i = +1, -1$ correspond respectively to site i being occupied or unoccupied. The attractive interaction between gas molecules favors occupation of neighboring sites or, equivalently, the parallel alignment of neighboring spins. This mapping of the system onto an Ising model again has obvious limitations but it can be helpful, especially since many liquid–gas systems are in the same "universality class" as the Ising model (see Section 13.9).

13.2.4. Melting

The solid–liquid transition (melting) is normally a first-order transition (being accompanied by latent heat), and the correlation length is not normally infinite at the transition. The transition is marked by the disappearance of the order parameters

$$\langle \rho_G(r) \rangle = \langle e^{iG.u(r)} \rangle \tag{13.26}$$

where G is any reciprocal lattice vector and $u(r)$ is the atomic displacement at point r. In the Lindemann picture of melting,[95] the transition takes place when the amplitude of the atomic vibrations becomes "comparable" to the atomic spacing (we give a quantitative discussion in Section 13.5.5).

Two-dimensional melting has been the subject of much recent interest.[96-99] The possibility of an ordered phase having only quasi-long-range order (in the sense to be discussed in Section 13.8.1) has been suggested. Furthermore, the mechanism for melting is provided by dislocations, in a manner suggested by the Kosterlitz–Thouless theory (Section 13.8).

13.2.5. λ-Transition in He⁴

The superfluid transition (λ-transition) in He⁴ is a Bose–Einstein condensation essentially modified by interactions. Pioneering theories are those of Landau[100] and of Feynman.[101] The basic Hamiltonian comprises the kinetic energy and the energy of interaction between the α-particles:

$$H = \sum \varepsilon_k a_k^+ a_k + \sum v_q a_k^+ a_{k'}^+ a_{k+q} a_{k'-q} \tag{13.27}$$

Landau allowed for the appearance of a condensate through the existence of a nonzero value of the order parameter

$$\langle a_0 \rangle = \sqrt{n_0} e^{i\phi} \tag{13.28}$$

where n_0 is the number of condensed particles and ϕ the phase of the wave function. Like the XY model the system has $n = 2$ (two components, the real and imaginary parts, for the complex order parameter). The Landau theory uses a mean-field approximation of the interaction term in the Hamiltonian, so that the particles appear to move in an effective field provided by the condensate.

The Feynman approach relates, among other things, the quasi-particle spectrum to the static correlation function describing neutron scattering, and leads to the concept of a roton (vortex-like excitation).

The phase ϕ of the order parameter is a gauge angle, and its appearance is related to the breaking of a (gauge) symmetry of the Hamiltonian. The breaking of a continuous symmetry has important consequences in the theory of phase transitions (and elsewhere)[102,103] and such aspects will be illustrated later (Section 13.4.4) for the ferromagnet.

The two-component nature of the order parameter implies that, like the XY model, the superfluid transition should have a different character in two dimensions than in three. We discuss the two-dimensional case in more detail in Section 13.8.2. Its essential characteristic is the removal of quasi-long-range order by vortex formation.[104]

13.2.6. Superconductivity

In some respects this is analogous to superfluidity. The superconducting ground state (of the Bardeen–Cooper–Schrieffer 1957 theory)[105] is a coherent superposition of bound (Cooper) pairs.[106] So the expectation value of the pair annihilation operator is a possible (complex) order parameter

$$\langle a_{-k\downarrow} a_{k\uparrow} \rangle \tag{13.29}$$

(the members of the pair normally have opposite spin and wave vector assignments). This is again a two-component order parameter (real and imaginary parts, or amplitude and phase). Since the single particle excitations have to break a pair, they have an energy gap Δ, which corresponds to the pair binding. Since Δ vanishes where the pair amplitude vanishes, it is often used as the order parameter in situations where the phase is not being considered; but once again there strictly is broken gauge symmetry.

The interaction binding the pairs is the attractive interaction between electrons at the Fermi surface caused by exchange of a phonon (Bardeen and Pines[107]); this is a long-range attraction. The Coulomb interaction does not favor pair formation but it affects the collective excitations. Because of the long-range nature of the interaction, mean field theory gives an excellent description of the system provided the dimensionality is not too low (such qualifications are discussed in Sections 13.3.6 and 13.5). The BCS theory is a mean field theory, and will be arrived at in

Section 13.3.4 through a reformulation due to Anderson.[108] For a more detailed treatment, see Chapter 16.

13.2.7. Peierls' Transition[109]

Let us consider a monatomic chain with one (delocalized) electron per atom. The conduction band is half-filled, as shown in Figure 13.6a. The total electron energy can be reduced by the chain distorting, as shown in Figure 13.6b, to double the unit cell and introduce a gap at the Fermi surface. Thus, as long as the increased lattice energy does not outweigh the decrease in electron energy, such a Peierls' distortion will occur. The order parameter is the distortion d, or the gap Δ.

The lattice (elastic) energy increases by an amount proportional to d^2 while the total electron energy decreases by an amount proportional to

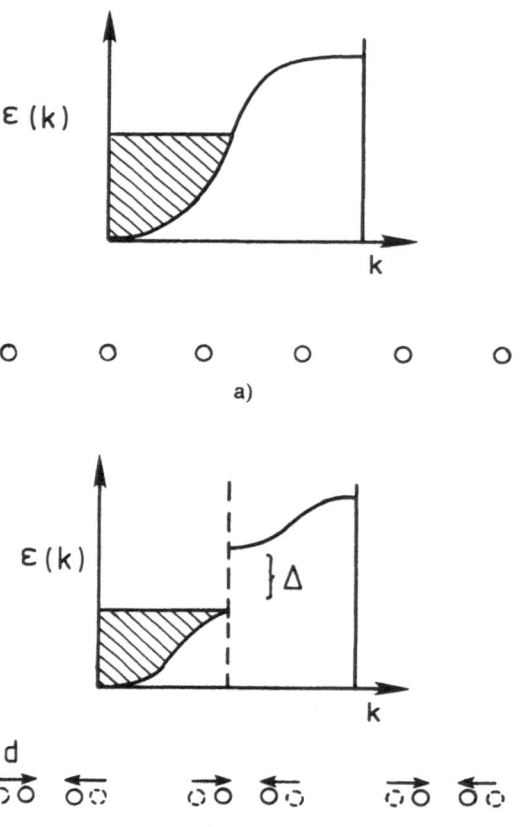

Figure 13.6. Peierls' transition: (a) undistorted and (b) distorted chain of atoms, and corresponding band structures.

$\Delta^2 \ln(1/\Delta)$. But in the simple weak-binding approximation, $\Delta \propto d$, so a distortion always occurs, to reduce the energy, in a linear chain with half-filled band.

This is a distortion of the lattice driven by its coupling to the electrons; another example is the cooperative Jahn–Teller effect described in Section 13.2.9 below. The distortion would cause the one-dimensional metal to become an insulator: thus it is one mechanism for the metal–insulator transition. Others are referred to in the next subsection.

If the band is not half-filled, but with say the Fermi distribution ending at or near $k_F = \pi/ra$, an atomic distortion repeating every rth atom will again cause a gap to develop at the Fermi surface (and elsewhere), reducing the total energy. So the distortion always occurs in the linear chain (Peierls[109]).

For higher-dimensional cases (e.g., $d = 3$) the effect may again occur.[110] The wave vector associated with the distortion has to connect points on the Fermi surface with a large density of states. This is the situation when Fermi surfaces are parallel planes (as in the one-dimensional case, or in the case of the half-filled zone for the tight-binding bcc lattice) or when Fermi surfaces nest (Figure 13.7a,b). We shall return later to a further brief discussion of the Peierls' distortion (Section 13.6).

13.2.8. Mott and Anderson Transitions

Let us consider whether it is favorable to form a conducting state from an insulating state by removing an electron from the vicinity of an atom. The electron and the hole it leaves behind will interact through the Coulomb attraction and, in narrow-band materials, will form a bound exciton so that neither electron nor hole can take part in conduction. The Coulomb attraction will, however, be screened by any other excited particles, so that if a sufficient density is excited the interaction will be too weak to form bound excitons and the material will have free carriers and behave like a conductor. This transition from insulating to conducting state

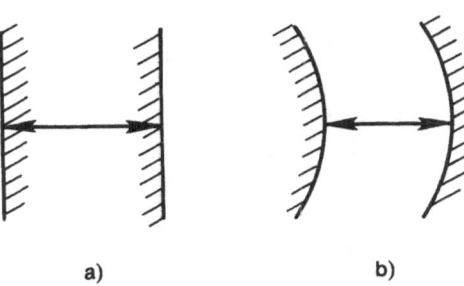

Figure 13.7. (a) Parallel Fermi surfaces, (b) nesting Fermi surfaces. The wave vector associated with the related Peierls' distortion is shown.

a)   b)

is the Mott transition.[111,112] It can be induced by pressure, which can reduce the parameter

$$\text{carrier density} \times \text{cell volume} \tag{13.30}$$

below a critical value predicted by Mott. Such an effect has been seen in doped semiconductors.

The Hubbard model[113] is a quantitative generalization of the Mott picture, but involving only the interaction between electrons on the same atom.

The Anderson picture of localization does not depend on interactions but on disorder producing randomness, for example, of single electron levels.[114,112,115] The probability of hopping of an electron from site i to a neighboring site j (in a Wannier representation) depends on the difference $\varepsilon_i - \varepsilon_j$ of the site energies and on the hopping matrix element. Whether the electron wave functions are, in a statistical sense, localized or extended depends in general on a dimensionless measure of the width of the distribution of site energies. As the composition is varied the Fermi energy can pass through a localization edge and the behavior can change between thermally-activated hopping and "metallic" conduction. It is possible to show that, however weak the disorder, all states should be localized in a one-dimensional system.[116,117] Definite conclusions have also recently been obtained concerning localization in the two-dimensional Anderson model.[118]

13.2.9. Structural Transitions; Jahn–Teller Transition

The Peierls' transition is, perhaps, the simplest transition showing a transformation from one regular crystal structure to another.

Structural transitions fall, broadly, into two types: order–disorder and displacive types.[119] The displacive types are often described by an interacting phonon picture. An example is the structural transition in $SrTiO_3$, where the order parameter is the distortion coordinate corresponding to the $q = (\frac{1}{2}, \frac{1}{2}, \frac{1}{2}) \, \pi/a$ (soft) phonon mode, which "freezes in" at the transition to cause a doubling of the unit cell. In the order–disorder types, the transition is accompanied by the ordering of some constituents among various local positions of equilibrium. A simple example is $NaNO_2$, where the equilibrium positions relate to the rotational configurations of the NO_2 ions. In such cases, where there are two equivalent positions, they can be distinguished by a pseudospin operator.[120] The $NaNO_2$ system can then be shown to be equivalent to the Ising model (13.22). This is very like the situation for the binary alloy.

The cooperative Jahn–Teller transition[120–123] is also normally discussed with a pseudospin representation. Like the Peierls' transition it is driven by the electron–lattice interaction, but in this case the electrons are localized at lattice sites. The simplest example is when only two electronic levels are in the range of interest. Which level an electron at site i occupies can then be represented by the two states $\sigma_i^z = \pm 1$ of a Pauli spin operator. A displacement Q_i of the atom can introduce a transition between the states, or affect their separation. These two possibilities are represented by an interaction term, linear in the displacements,

$$\sum_{ij} A_{ij} Q_j \sigma_i^\alpha = \sum_k A_k Q_k \sigma_k^\alpha \tag{13.31}$$

where $\alpha = x$ for the first case, or $\alpha = z$ for the second; A_{ij}, or its Fourier transform A_k, is an interaction constant. For the case where the two electron levels were degenerate, the only other term necessary in the Hamiltonian is the lattice energy

$$\sum_k \left(\frac{1}{2m} P_k^2 + \frac{1}{2} m\omega_k^2 Q_k^2 \right) \tag{13.32}$$

A canonical transformation

$$Q_k \rightarrow \tilde{Q}_k = Q_k + \frac{A_k \sigma_k^\alpha}{m\omega_k^2}, \quad P_k \rightarrow \tilde{P}_k = P_k, \quad \sigma_k^\alpha \rightarrow \tilde{\sigma}_k^\alpha = \sigma_k^\alpha \tag{13.33}$$

separates the total Hamiltonian into two terms. The first has a free phonon form (like (13.31) but with tildes on the operators). The other term is

$$-\sum_k \frac{\sigma_k^{\alpha 2} A_k^2}{2m\omega_k^2} = -\sum_{ij} J_{ij} \sigma_i^\alpha \sigma_j^\alpha \tag{13.34}$$

where J_{ij} is the inverse Fourier transform of $A_k^2/2m\omega_k^2$; this term represents the pseudospin interaction caused by phonon exchange. The last Hamiltonian is now of Ising form, and can lead to a transition temperature below which $\langle \sigma^\alpha \rangle$ is nonzero. This is the order parameter for the transition, and from (13.33) it corresponds to a net distortion of the lattice.

In the case where the two electron levels were nondegenerate, an additional term $-h \sum_i \sigma_i^z$ occurs in the original Hamiltonian, where $2h$ is the level splitting. The spin Hamiltonian (13.34) therefore becomes, for the two cases $\alpha = x$ and $\alpha = z$,

$$H = -h \sum_i \sigma_i^z - \sum_{ij} J_{ij} \sigma_i^x \sigma_j^x \tag{13.35}$$

$$H = -h \sum_i \sigma_i^z - \sum_{ij} J_{ij} \sigma_i^z \sigma_j^z \tag{13.36}$$

Expression (13.35) is actually only approximate since the canonical transformation implies a change in the operator σ^z for the case when $\alpha = x$; (13.35) is the "Ising model in a transverse field" and can have a transition for a range of fields h. In (13.36), the usual Ising model in a (longitudinal) field, the field couples to the spin component, which would exhibit the long-range order and this destroys the phase transition, as described in Section 13.1. Equation (13.35) describes the Jahn–Teller transition in $DyVO_4$, where $2h$ is the splitting, and in $TmVO_4$ where h is actually an applied magnetic field. Equation (13.36) applies to systems like $CeEthSO_4$, which do not have a sharp transition.

13.2.10. Ferroelectric Transitions

In ferroelectric transitions the long-range order is a spontaneous electric polarization, though this can also be accompanied by (or triggered by) structural reordering. There are again displacive types (such as $KTaO_4$) and order–disorder types.[124] The hydrogen-bonded ferroelectrics, such as KH_2PO_4 (KDP) and RbH_2AsO_4, are considered to be of the second type, though a proper description of their excitations requires the addition of phonon terms to a pseudospin model[125,126] The simplest pseudospin model for these "tunneling" ferroelectrics[12] is (13.35) above. There, the interaction term represents the coupling of displacements of hydrogen atoms on neighboring O—H—O bonds (this displacement triggers the ferroelectric distortion of the complex). The transverse field term in (13.35) represents the tunneling of the hydrogen atom between the two equivalent positions it may occupy on the O—H—O bond.

13.2.11. Percolation

We conclude with a rather different system exhibiting a transition, the percolation model.[127-133]

An illustration is the square lattice of Figure 13.8, which has been diluted by removal of bonds at random until a concentration p of bonds remains. In Figure 13.8a, the bond concentration is still sufficiently large that the system has linkage paths between pairs of opposite sides. In Figure 13.8b the bond concentration is now small enough that no such linkage paths occur: there only appear "finite connected clusters" of bonds. Figure 13.8a includes such finite clusters, and also an "infinite" cluster that provides the extended linkage paths. The two situations (a), (b) occur respectively for $p > p_c$, $p < p_c$, where p_c is a limiting (critical) concentration at which the infinite cluster appears or disappears. An appropriately defined correlation length diverges at $p = p_c$: a percolation transition is said

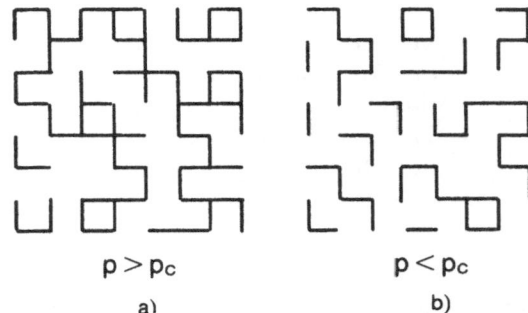

Figure 13.8. Bond-diluted square lattice: (a) infinite and finite clusters ($p > p_c$), (b) only finite clusters ($p < p_c$).

$p > p_c$ $p < p_c$

a)   b)

to occur, which is of second order. A possible order parameter is the probability that any selected bond is a member of the infinite cluster. This quantity goes to zero like a power of $p - p_c$ as p_c is approached from above, and that power is one of a set of critical exponents (cf equation (13.3)); another (cf equation (13.19)) is the correlation length exponent ν defined by

$$\xi \propto |p - p_c|^{-\nu}, \qquad p \sim p_c \qquad (13.37)$$

Another exponent is related to the divergence, at p_c, of the mean cluster size, which plays a role analogous to the susceptibility in magnetic systems.

The percolation transition is not caused by the competition between thermal fluctuations and some interaction tending to give cooperative effects, but simply by configurational aspects. The cooperative character is still present however, since linkages across a large specimen depend on the combined efforts of short linkages to provide such a path. It is clear that, while in the Ising chain there is no long-range order if $T \neq 0$, in the one-dimensional percolation problem there is no infinite cluster if $p \neq 1$.

The percolation problem may appear in various generalizations or disguises. One can have site dilution, where a site with all its incident bonds is removed: the same ideas apply, the critical concentration is different, but the critical exponents are the same as in bond percolation (for a given dimensionality). The percolation transition is important in, for example, dilute magnets, which even at $T = 0$ cannot have an ordered phase if $p < p_c$.

13.2.12. Summary

The above examples are intended to show that

1. Phase transitions always involve some cooperative aspects which tend to give an ordered state.

2. The ordered state can be characterized by an order parameter.
3. The order can be long-range order or something much more subtle (as, for example, in the spin glass or Kosterlitz–Thouless transition).
4. The order can be removed by fluctuations (usually thermal fluctuations but possibly, for example, configurational fluctuations).
5. Critical effects can occur at the transition, such as the divergence of a correlation length in a second-order phase transition.

We conclude with Table 13.2, which lists, for the various systems exhibiting phase transitions, the following quantities: parameters occurring in their description, the associated order parameter, any special features.

Appendix 13.2. Bose Condensation

This is a simple, exactly soluble case[101,134,135] in which there are no interactions present. The effect is a consequence of the Bose statistics. The occupation of the single-particle state k with energy ε_k is

$$n_k = [e^{\beta(\varepsilon_k - \mu)} - 1]^{-1}, \qquad \varepsilon_k = \hbar^2 k^2 / 2m \qquad (13.38)$$

The chemical potential μ is determined by the number of particles:

$$N = \sum_k n_k \qquad (13.39)$$

For a macroscopic system, since the level spacing is much less than the thermal energy per particle, for any reasonable temperature it is usual to replace the sum over k by an integral. However, for $\mu \leq 0$,

$$\frac{(2s+1)}{(2\pi)^3} V \int d^3\mathbf{k} \, \frac{1}{e^{\beta(\varepsilon_k - \mu)} - 1} \leq V \left(\frac{2mKT}{\hbar^2}\right)^{3/2} \gamma \qquad (13.40)$$

where the equality holds for $\mu = 0$ and γ is a dimensionless constant given by

$$\gamma = \frac{(2s+1)}{2\pi^2} \int_0^\infty \frac{x^2 dx}{e^{x^2} - 1} \qquad (13.41)$$

Thus as the temperature is reduced to T_c defined by

$$\frac{N}{V} = \gamma \left(\frac{2mKT_c}{\hbar^2}\right)^{3/2} \qquad (13.42)$$

Table 13.2. Summary of Selected Aspects of Some Systems Exhibiting Phase Transitions

System	Parameters[a]	Order parameter	Special features
Bose gas	KT, N/V	Fraction in single-particle ground state	No interaction
Ferromagnet	KT, J, h	Magnetization	Often directly related to simple models
Antiferro-magnet	KT, J, h	Sublattice magnetization	Often directly related to simple models
Spin glass	KT, J_1, J_2 concentration	—	Thermal and configurational (frustration, disorder) aspects
Alloys (binary)	KT, J, concentration	Density of one kind of sublattice	May be mapped onto spin model
Liquid gas	KT, P	$\rho_L - \rho_g$	May be mapped onto spin model
He⁴ λ-transition	KT, P	Single-particle ground state amplitude	Two-dimensional case has special features
Melting	KT, P	Fourier components of density	Two-dimensional case has special features
Supercon-ductivity	T	Electron pair amplitude, or energy gap	Long-range interaction
Peierls' lattice		Lattice distortion, or energy gap	Effect of electron–lattice interaction
Mott insulator/conductor	P	Free-carrier density	Effect of Coulomb interaction
Anderson model	Strength of disorder	Free-carrier density	No interaction. Disorder
Jahn–Teller	KT, h, strain	Distortion	Spin model
Other structural transitions	KT, h, strain	Distortion	Order–disorder and displacive types
Ferro-electrics	KT, ε, strain	Electric polarization	Order–disorder and displacive types
Percolation	Bond (or site) concentration	Percolation probability	Configurational. Otherwise normal second-order transition

[a] T = temperature, N/V = number density, J = exchange interation, h = magnetic field, P = pressure, ε = electric field.

μ approaches zero and the ground-state occupation becomes very large. For $T < T_c$ the ground state has to accommodate a macroscopic number N_0 of particles

$$N_0 = N - V\gamma \left(\frac{2mKT}{\hbar^2}\right)^{3/2} = N[1 - (T/T_c)^{3/2}] \tag{13.43}$$

where N_0 is an order parameter distinguishing the condensed phase ($N_0 \neq 0$, $T < T_c$) from the high-temperature phase. Thus for T just less than T_c the order parameter has the behavior

$$N_0 \propto (T_c - T)^\beta \qquad (13.44)$$

[cf (13.3)] with critical exponent $\beta = 1$.

The internal energy $\Sigma_k \varepsilon_k n_k$ can be evaluated easily for $T < T_c$ (where, by the above argument, μ can be taken as zero in the integral over k) to yield the following result for the specific heat C:

$$C \propto T^{3/2}, \qquad T < T_c \qquad (13.45)$$

The calculation of C for $T > T_c$ involves the calculation of μ, which is not now zero. But it is straightforward to show that the specific heat is continuous at $T = T_c$, where it has the cusp behavior illustrated in Figure 13.9. This is somewhat similar to the behavior of He^4 near the λ-point. Indeed, the λ-point is close to the Bose T_c for the helium mass and density. However, as remarked earlier, liquid helium is not an ideal Bose gas because of interparticle forces.

13.3. Mean-Field Theory

13.3.1. Mean-Field Theory for Magnetic Systems[136]

Mean-field theory was first developed for magnetic systems, and perhaps its simplest application is to the Ising model. The Hamiltonian for this model is given in equation (13.22). The terms in the Hamiltonian that

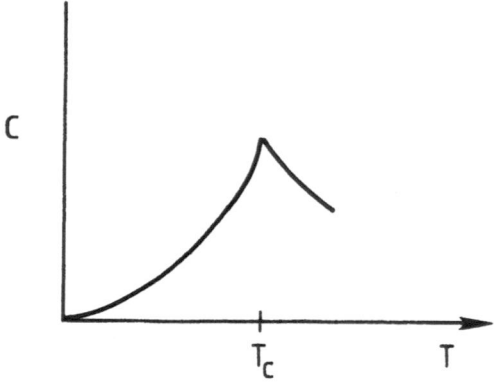

Figure 13.9. Specific heat of a system of noninteracting bosons.

involve the spin at a particular site l are

$$H(l) = + \left(h - 2 \sum_{j \neq l} J_{lj} S_j^z \right) S_l^z \qquad (J_{lj} = J_{jl}) \qquad (13.46)$$

If many spins j are linked by J_{lj} to l, the individual states of those spins are probably not of significance; then we can replace the second term in parentheses in (13.46) by its mean. This is the same as replacing the effect of all other spins on l by the mean field they produce:

$$H(l) \rightarrow H_{\text{eff}}(l) = - \gamma S_l^z \qquad (13.47)$$

$$\gamma = h + 2 \sum_{j \neq l} J_{lj} \langle S_j^z \rangle \qquad (13.48)$$

Since the (thermal) average $\langle S_j^z \rangle$ is independent of j for a translationally invariant system and J_{lj} normally only depends on the distance between l and j, the effective field γ on spin l is usually independent of l. Then $\langle S^z \rangle$ can be worked out self-consistently: for spin $\frac{1}{2}$

$$\langle S^z \rangle = \tanh \tfrac{1}{2} \beta \gamma \qquad (13.49)$$

$$\gamma = h + 2J(0) \langle S^z \rangle \qquad (13.50)$$

$J(0)$ is here the $q = 0$ form of a general Fourier transform that will be useful later:

$$J(q) = \sum_j \exp[i\mathbf{q} \cdot (\mathbf{r}_l - \mathbf{r}_j)] J_{lj} \qquad (13.51)$$

The above results (13.49) and (13.50) apply also for the Heisenberg model. Though in this case the Hamiltonian involves the scalar product of the spin vectors (see equation (13.21)), only the z-component of the spin has a nonzero average and all the results from equation (13.47) follow.

The argument leading to (13.47) requires long-range interactions. In that case, $J(q)$ is only nonzero for very small q and all the wave-vector dependences arising in the system from $J(q)$ (in the excitation frequencies, for example) are concentrated into the small q-regime. The mean-field theory actually neglects altogether such dependences. For this reason it gives an inadequate account of the excitation frequencies of the Heisenberg model (see Sections 13.3.6(2) and 13.4.2). For both Ising and Heisenberg systems the theory only allows excitations corresponding to spin flips in the mean field. Such single spin-flip excitations have energy 2γ.

The self-consistency condition, (13.49) and (13.50), gives spontaneous magnetization ($\langle S^z \rangle \neq 0$ in zero external field) at temperatures lower than a transition temperature that can be located as follows. With $h = 0$, put $x = J(0) \langle S^z \rangle / KT$. The self-consistency condition then becomes

$$\frac{KT}{J(0)} x = \tanh x \qquad (13.52)$$

This can be solved graphically[69] (Figure 13.10) by plotting each side against x. Whether there is a nonzero solution for x depends on whether $KT/J(0)$ (the slope of the straight line representing the left-hand side) is less than 1. If it is, there are three solutions, labeled A, B, C in the figure. A and C are the stable solutions (least free energy or, equivalently, positive susceptibility) and which one occurs depends on symmetry-breaking effects. Spontaneous magnetization therefore occurs for $T < T_c$ where

$$KT_c = J(0) \qquad (13.53)$$

The magnetization curve resulting from the theory has the form shown in Figure 13.3. The susceptibility diverges (like $|T - T_c|^{-1}$) near T_c. The magnetic specific heat has a discontinuity at $T = T_c$, and vanishes for $T > T_c$ due to the complete absence of short-range order within the theory: (13.47) leads to the spin–spin correlation function (13.9) reducing to

$$\Gamma_{ij}^0 = \delta_{ij} [\langle (S_i^z)^2 \rangle - \langle S_i^z \rangle^2] \equiv \delta_{ij} \Gamma^0 \qquad (13.54)$$

(the effective Hamiltonian is of single-particle type; it nevertheless contains cooperative effects through the self-consistent field).

13.3.2. Ornstein–Zernicke Theory[136]

An improved description of the correlation function Γ_{ij} is given by the Ornstein–Zernicke theory; the basic idea is illustrated qualitatively in

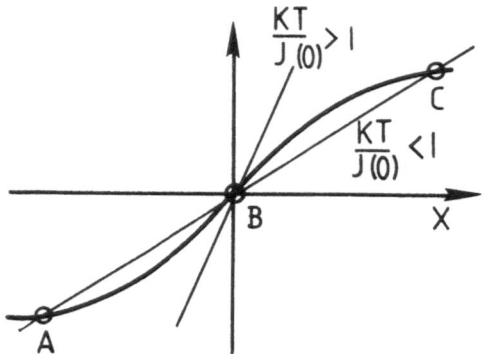

Figure 13.10. Graphical solution of mean field equation (13.52).

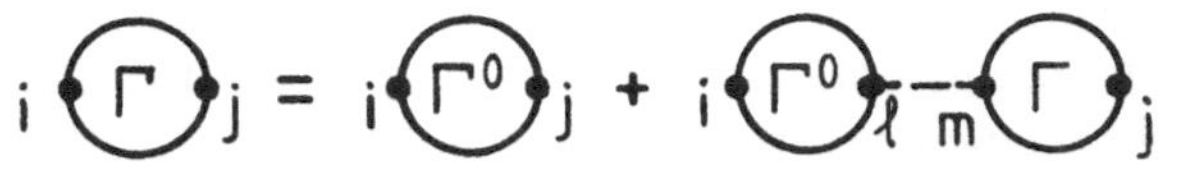

Figure 13.11. Graphical relationship between the Ornstein–Zernicke and mean field correlation functions (Γ and Γ^0). The dashed line is the interaction βJ_{lm}.

Figure 13.11, where a finite-range correlation Γ_{ij} is generalized self-consistently from a zero range one (Γ^0) by the mediating effects of the interactions represented schematically by a dashed line. This graphical equation has a precise meaning in many-body theory for classical or quantum systems, and in the Ising model takes the form[136]

$$\Gamma_{ij} = \Gamma_{ij}^0 + \sum_{lm} \Gamma_{il}^0 (\beta J_{lm}) \Gamma_{mj} \tag{13.55}$$

Thus, Fourier transforming and using (13.54) and (13.12),

$$\chi(q) = \frac{\Gamma^0}{1 - \beta J(q) \Gamma^0} \tag{13.56}$$

In zero field $\chi(0)$ diverges at the mean-field transition temperature (13.53) (where $\beta J(0) = 1$ and $\Gamma_0 = 1$). Thus the Ornstein–Zernicke theory gives the same transition temperature as mean-field theory. The small-q form of $\chi(q)$ can be obtained by inserting into equation (13.56) the following expansion, which is appropriate for lattices with inversion symmetry:

$$J(q) = J(0)[1 - \alpha q^2 + \cdots] \tag{13.57}$$

This leads to

$$\chi(q) = \frac{A}{q^2 + 1/\xi^2}, \qquad q \sim 0 \tag{13.58}$$

where

$$\frac{1}{\xi^2} = \frac{1}{\alpha} \left(\frac{KT}{J(0)\Gamma^0} - 1 \right) \tag{13.59}$$

The long-range correlations are thus of screened Coulomb form with correlation length ξ given by (13.59).

13.3.3. Critical Exponents of Mean-Field and Ornstein–Zernicke Theories

The Ornstein–Zernicke result (13.58) is of the form (13.17) with critical exponent $\eta = 0$. Also, the correlation length given by (13.59) diverges as in (13.19), with $\nu = \frac{1}{2}$, close to the transition on either side. The divergence of the susceptibility $\chi(0)$ is like $|T - T_c|^{-1}$. This is the same behavior as is obtained by differentiating (13.49) and (13.50) with respect to h. These mean-field equations can also be used to obtain the critical behavior of the magnetization. The last row of Table 13.1 gives the full set of mean-field/Ornstein–Zernicke critical exponents[79] as defined by equations (13.2)–(13.5), (13.16), and (13.19).

The corresponding theory (same exponents) for the liquid–gas transition is van-der-Waals/Ornstein–Zernicke theory. Mean-field treatments can also be given of superconductivity, structural, ferroelectric, and many other transitions.

13.3.4. Mean-Field Theory for Superconductivity

As described qualitatively in Section 13.2.6, the exchange of a phonon causes an attractive interaction between electrons of nearly equal Bloch energy. Cooper[106] showed that the Fermi surface is then unstable: electrons of opposite spin and wave vector at the Fermi surface tend to form bound pairs. Since the resulting Cooper pair is of large extent one cannot regard the system as composed of independent Cooper pairs, but must allow for a correlated state. The large extent helps to validate a mean-field approximation.

The essential features are provided by the reduced Hamiltonian of Bardeen et al.[105]

$$H_{\text{red}} = \sum_K \varepsilon_K (a_K^+ a_K + a_{-K}^+ a_{-K}) - V_0 \sum_{}^{'} a_{K'}^+ a_{-K'}^+ a_{-K} a_K \qquad (13.60)$$

We here use a shorthand notation in which K specifies both wave vector and spin, and $-K$ is the opposite assignment:

$$K \equiv k\uparrow, \qquad -K \equiv -k\downarrow \qquad (13.61)$$

The electron Bloch energy ε_K is measured from the Fermi energy. Since the phonon-mediated attractive interaction V_0 is between electrons near the Fermi surface, the interaction term involves a sum Σ' going only over states in an energy shell, at the Fermi surface, of width $2\omega_D$, where ω_D is a typical phonon frequency.

The Hamiltonian can be rewritten as

$$H_{\text{red}} = -\sum_K \varepsilon_K (1 - n_K - n_{-K}) - V_0 \sum_{KK'}^{'} b_{K'}^+ b_K + \text{const} \qquad (13.62)$$

where

$$n_K = a_K^+ a_K, \qquad b_K = a_{-K} a_K, \qquad b_K^+ = a_K^+ a_{-K}^+ \qquad (13.63)$$

are, respectively, number operators and pair annihilation and creation operators.

The quantity H_{red} operates in the subspace in which the single electron states are occupied in pairs, K, $-K$. This pair subspace is spanned by column vectors (for each K)

$$\begin{pmatrix} 1 \\ 0 \end{pmatrix}, \begin{pmatrix} 0 \\ 1 \end{pmatrix} \qquad (13.64)$$

meaning, respectively, pair $(K, -K)$ empty ($n_K = n_{-K} = 0$) or occupied ($n_K = n_{-K} = 1$).

By their action on these empty and full pair states (13.64) it is easy to check that the operators in (13.62) can be represented as follows in terms of matrices, or Pauli operators, in the pair subspace:

$$1 - n_K - n_{-K} = \begin{pmatrix} 1 & 0 \\ 0 & -1 \end{pmatrix} \equiv \sigma_K^z$$

$$b_K^+ = \begin{pmatrix} 0 & 0 \\ 1 & 0 \end{pmatrix} \equiv \sigma_K^x + i\sigma_K^y$$

$$b_K = \begin{pmatrix} 0 & 1 \\ 0 & 0 \end{pmatrix} \equiv \sigma_K^x - i\sigma_K^y \qquad (13.65)$$

The Pauli operators carry the label, K, of the column vectors on which they act. The Hamiltonian may therefore be rewritten as[108]

$$H_{red} = -\sum_K \varepsilon_K \sigma_K^z - V_0 \sum_{KK'}' (\sigma_K^x \cdot \sigma_K^x + \sigma_{K'}^y \cdot \sigma_K^y) + \text{const} \qquad (13.66)$$

The form is that for an XY model in an "inhomogeneous" (K-dependent) transverse field.

The kinetic energy, or transverse-field term, tends to align the spins σ_K along the $+z$ direction (pair state unoccupied) or along the $-z$ direction (pair state occupied) for single-particle energies greater or less than the Fermi energy. This is the normal Fermi sea. The interaction term tends to align spins in the $x-y$ plane and at low temperatures will always succeed in pulling spins out of the $\pm z$ direction for single-particle energies close enough to the Fermi energy (ε_K sufficiently small). This corresponds to a qualitative change in the Fermi surface, producing the transition and the gap in the excitation spectrum.

To see this, consider mean-field theory, which replaces H_{red} by

$$H_{eff} = -\sum_K \gamma_K \cdot \sigma_K \tag{13.67}$$

$$\gamma_K = (\Delta, 0, \varepsilon_K) \tag{13.68}$$

Here we chose the x, y axes so that there is no projection of the average spin onto the y-axis, and we defined

$$\Delta \equiv 2V_0 \sum_{K'}' \langle \sigma_{K'}^x \rangle \tag{13.69}$$

The average spin $\langle \sigma_K \rangle$ must then point along the mean field γ_K, and have magnitude $\tanh(\tfrac{1}{2}\beta\gamma_K)$. Thus

$$\langle \sigma_K \rangle = (\sin\theta_K, 0, \cos\theta_K)\tanh\tfrac{1}{2}\beta\gamma_K \tag{13.70}$$

$$\tan\theta_K = \Delta/\varepsilon_K \tag{13.71}$$

Δ will turn out to be the energy gap (the basic order parameter). It is determined, from the last three equations, by

$$\Delta = 2V_0 \sum_{K'}' \frac{\Delta}{\sqrt{\varepsilon_{K'}^2 + \Delta^2}} \tanh\left(\tfrac{1}{2}\beta\sqrt{\varepsilon_{K'}^2 + \Delta^2}\right) \tag{13.72}$$

This is the BCS gap equation.

The transition is where the gap goes to zero and the system becomes normal. Taking $\Delta \to 0$, we obtain the following equation for $\beta_c \equiv (1/KT_c)$:

$$1 = 2V_0 \sum_{K'}' \frac{\tanh\tfrac{1}{2}\beta_c\varepsilon_{K'}}{\varepsilon_{K'}} \sim 2V_0\rho_F \int_{-\omega_D}^{\omega_D} \frac{d\varepsilon}{\varepsilon} \tanh\tfrac{1}{2}\beta_c\varepsilon \tag{13.73}$$

where ρ_F is the density of states at the Fermi energy. The integral is difficult, but for $\rho_F V_0 \ll 1$, $\omega_D\beta_c$ will have to be large so that the integral is then dominated by the region where the hyperbolic tangent is close to ± 1. Then we obtain

$$KT_c \sim \omega_D \exp\left(-\frac{1}{V_0\rho_F}\right) \tag{13.74}$$

This expression shows that the transition temperature does not have a weak-coupling expansion. It is not difficult to show, by similar methods, that the zero temperature value $\Delta(0)$ of the energy gap is comparable to KT_c.

A possible excitation of the system is a spin flip in the fictitious field γ_K. This has energy

$$2\gamma_K = 2\sqrt{\varepsilon_K^2 + \Delta^2} \tag{13.75}$$

showing that 2Δ is indeed the energy gap for pair creation in this approach.

In the discussion we have looked at simple properties in a simple approach to a simple model. For a more general view, see Chapter 16. The limitations of the mean-field approximation will be discussed in Section 13.3.6.

13.3.5. Mean-Field Theory for Jahn–Teller Transitions

In Section 13.2.9 we gave two pseudospin Hamiltonians for Jahn–Teller systems. One of these was the Ising model, to which mean-field theory has already been applied in Section 13.3.1. The other is the transverse Ising model, for which we now briefly discuss the mean-field approach.[120,137] The Hamiltonian and its mean-field reduction are

$$H = -h \sum_i \sigma_i^z - \sum_{ij} J_{ij}\sigma_i^x\sigma_j^x \tag{13.76}$$

$$\rightarrow -\sum_i \boldsymbol{\gamma} \cdot \boldsymbol{\sigma}_i \tag{13.77}$$

where

$$\boldsymbol{\gamma} = (2J(0)\langle \sigma^x \rangle, 0, h) \equiv (\Delta, 0, h) \tag{13.78}$$

This is very similar to equations (13.67)–(13.69) of the previous section, but without the K-dependence. So, in the same way, we obtain

$$\langle \boldsymbol{\sigma} \rangle = (\sin\theta, 0, \cos\theta)\tanh\tfrac{1}{2}\beta\gamma \tag{13.79}$$

$$\tan\theta = \Delta/h \tag{13.80}$$

where

$$\Delta = \frac{2J(0)\Delta}{\sqrt{h^2 + \Delta^2}}\tanh(\tfrac{1}{2}\beta\sqrt{h^2 + \Delta^2}) \tag{13.81}$$

The ordered state is where $\langle \sigma^x \rangle$ (and therefore Δ) is nonzero. The transition temperature is found by taking $\Delta \rightarrow 0$, with the result

$$\tanh\tfrac{1}{2}\beta_c h = h/2J(0) \qquad (h \leqslant 2J(0)) \tag{13.82}$$

For transverse fields h greater than $2J(0)$, Δ is always zero: no transition takes place.

The mean-field relationship (13.82) between critical temperature and transverse field is shown in Figure 13.12. The phases are separated by a critical line (on which the correlation length diverges).

The specific heat resulting from mean-field theory is shown by the solid line of Figure 13.13, here drawn for the case $h = 0$, which applies to TmVO$_4$ in zero field for which the experimental data[138] are also shown in the figure. This system is one of the best mean-field systems known (apart from superconductors), as is indicated by the very small specific heat for $T > T_c$; this suggests that short-range order effects, neglected by mean-field theory, are here unimportant.

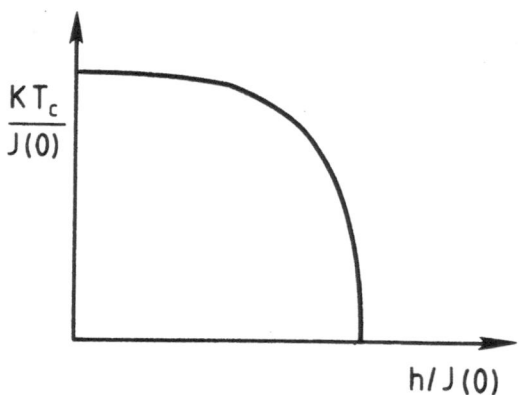

Figure 13.12. Mean field relationship between critical temperature and field, for Ising model in a transverse field.

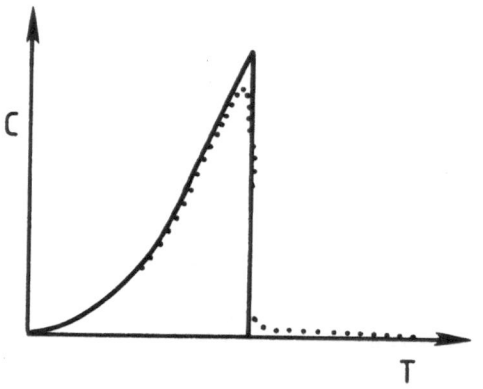

Figure 13.13. Experimental and mean field (solid line) values for the specific heat of TmVO$_4$ ($h = 0$).

13.3.6. Critique of Mean-Field Approaches

Mean-field theory can be very useful in indicating the possbility of a transition to a condensed state with long-range order. It can give a qualitative and sometimes quantitative understanding when such a condensation should occur. But it can have several major drawbacks:

1. It is wrong in low dimensions, where considerations absent from mean-field theory lead to instabilities in the mean-field ordered state. For example, mean-field theory predicts that the Ising model should order for $T < T_c = J(0)/K$, *whatever the dimensionality*. But entropy versus internal energy considerations (Section 13.1.1) showed that such ordering should not occur in the one-dimensional case except at $T = 0$. More generally, mean-field theory neglects the possibility that fluctuations can break up the mean-field ordered state. There is normally a least (integer) dimensionality d_1 for the mean-field ordering to occur.

2. It can lead to qualitatively wrong forms for the energies of collective excitations. For example, the simple mean-field theory (in which the Hamiltonian is approximated) leads, for the Heisenberg model, to (spin-flip) excitations with wave-vector-independent energy 2γ. The proper low-lying excitations (spin waves) can have arbitrarily low energy for small wave vector (Section 13.4.2). (This has important consequences for the stability of the ordered state in low dimensions (Section 13.5.1).) The proper description of dynamics is obtained by applying mean-field theory to the equations of motion, rather than to the Hamiltonian.

3. Even when a mean-field-predicted transition occurs, the mean-field values of the critical exponents will normally be wrong below a so-called upper critical dimensionality d_2.

We have introduced for each class of systems the lower and upper critical dimensionalities d_1, d_2 ($d_1 \le d_2$). There are then three possibilities for the actual dimensionality d of a particular system being considered:

(a) $d < d_1 \le d_2$: mean-field theory predicts a transition to a long-range ordered state, but the system is of too low a dimensionality to sustain it.

(b) $d_1 \le d < d_2$: the transition predicted by mean-field theory occurs, but the theory gives the wrong critical exponents.

(c) $d_1 \le d_2 < d$: mean-field theory correctly predicts both the transition and its associated critical exponents.

In the next section we discuss ground states, low-lying excitations, and fluctuations, for simple systems. Such considerations illustrate (2) above and will also be useful for the later discussion (Section 13.5) of the breakdown of long-range order below the lower critical dimensionality ((1), above).

13.4. Excitations

13.4.1. Ground States

We consider first the ground states of the Ising and Heisenberg ferromagnets, which have the Hamiltonians given in equations (13.22) and (13.21) with $J_{ij} \geq 0$. In both cases the first (Zeeman) term of the Hamiltonian tends to align spins parallel to the external field (i.e., to the z-direction) and the interaction term favors the alignment of spins that it couples. The ground state in both cases is therefore the state $|0\rangle$ of all spins "up" (Figure 13.14a). This clearly has the lowest energy, and is an eigenstate. Its long-range order remains even if the field h is taken to zero.

13.4.2. Low-Lying Excitations

The simplest possible excited state of the ferromagnet is obtained by flipping a single spin, at site i say, in the ground state (Figure 13.14b). This state is

$$S_i^- |0\rangle \tag{13.83}$$

where S_i^- is the usual lowering operator, for the spin on site i. This is the excitation allowed by mean-field theory. It is an eigenstate of the Ising Hamiltonian, but *not* however of the Heisenberg Hamiltonian (which, for the remainder of this section, we denote by H_H). To see this we use[136]

$$[H_\text{H}, S_i^-] = hS_i^- + 2 \sum_j J_{ij} [S_i^- S_j^z - S_j^- S_i^z] \tag{13.84}$$

and the ground-state properties

$$S_j^z |0\rangle = \tfrac{1}{2} |0\rangle \tag{13.85}$$

$$H_\text{H} |0\rangle = E_0 |0\rangle \tag{13.86}$$

↑ ↑ ↑ ↑. . . .↑   ↑ ↑ ↑ ↑. . . .↑
↑ ↑ ↑ ↑. . . .↑   ↑ ↓ ↑ ↑. . . .↑
↑ ↑ ↑ ↑. . . .↑   ↑ ↑ ↑ ↑. . . .↑

(a)   (b)

Figure 13.14. Ground state (a) and single spin-flip state (b) of a ferromagnet.

where E_0 is the ground-state energy. Thus

$$H_H S_i^- |0\rangle = (E_0 + h + J(0))S_i^- |0\rangle - \sum_j J_{ij} S_j^- |0\rangle \qquad (13.87)$$

The right-hand side is not a multiple of $S_i^- |0\rangle$, so this state is not an eigenstate of H_H. But the form of the equation makes it clear that particular linear combinations of the $S_j^- |0\rangle$, for all j, will be eigenstates. The translational invariance of the system suggests that the appropriate linear combinations are

$$\sum_j e^{i\mathbf{q}\cdot\mathbf{r}_j} S_j^- |0\rangle \equiv S_q^- |0\rangle \qquad (13.88)$$

and it is straightforward to check that equation (13.87) leads to

$$H_H S_q^- |0\rangle = (E_0 + \omega_q) S_q^- |0\rangle \qquad (13.89)$$

where

$$\omega_q = h + J(0) - J(q) \qquad (13.90)$$

and $J(q)$ is as defined in (13.51). $S_q^- |0\rangle$ is therefore an eigenstate, and it is the "spin-wave" state, with excitation energy ω_q.

However, $S_q^- S_{q'}^- |0\rangle$ is not an eigenstate of H_H; thus spin waves are not linearly superposable. This effect can be represented, at low temperatures, by an interaction (the kinematic interaction of Dyson[139]).

Nevertheless, when few spin waves are excited they behave like bosons. This is easily seen by considering the commutator

$$[S_i^+, S_j^-] = 2S_i^z \delta_{ij} \qquad (13.91)$$

When applied to the ground state $|0\rangle$ this becomes, in terms of the Fourier-transformed operators,

$$[S_q^+, S_{q'}^-]|0\rangle = \delta_{qq'}|0\rangle \qquad (13.92)$$

But S_q^- is the creation operator for a spin wave. So to the extent to which there are few spin waves present (use of $|0\rangle$ and of linear superposition despite what was said above) the spin waves behave like bosons. We shall use and generalize these results in the next sections.

13.4.3. Random-Phase Approximation

There are approximations that retain the essential idea of the mean-field approach while giving an adequate treatment of the collective excitations (objection (2) of Section 13.3.6). An example is the so-called "random-phase" approximation, first introduced by Bohm and Pines[140,141] for the plasmon problem (for examples, see Chapter 3 in the book by Pines[162]). It arises in somewhat modified forms in the treatment of such diverse situations as, for example, nuclear collective motion, highly anharmonic solids, and magnetic and pseudospin models. We illustrate the form of the method for the Heisenberg model[142,143] using some results of the previous subsection.

Here, the method gives a generalization of spin-wave theory to arbitrary temperatures. In spin-wave theory, the operator S^z in equations (13.84) and (13.91) is replaced by its (expectation) value in the ground state $|0\rangle$. An approximate generalization is obtained by replacing, in those equations, S^z by its *thermal* average:

$$S^z \to \langle S^z \rangle \tag{13.93}$$

We then recover from (13.91) a boson commutation relation, without extra factors, by defining

$$\alpha_q^+ = S_q^-(2\langle S^z \rangle)^{-1/2}, \qquad \alpha_q = S_q^+(2\langle S^z \rangle)^{-1/2} \tag{13.94}$$

With the replacement (13.93), equations (13.84) and (13.91) become

$$[H_{\mathrm{H}}, \alpha_q^+] = \varepsilon_q \alpha_q^+ \tag{13.95}$$

$$[\alpha_q, \alpha_{q'}^+] = \delta_{qq'} \tag{13.96}$$

where

$$\varepsilon_q = h + 2\langle S^z \rangle [J(0) - J(q)] \tag{13.97}$$

Thus the approximation has reduced the system to an assembly of free bosons with single-particle energies ε_a. The approximation correctly reduces to spin-wave theory in the low-temperature regime, where $\langle S^z \rangle$ approaches its ground-state (saturation) value.

$\langle S^z \rangle$ can be evaluated self-consistently by using the following result of the spin $\frac{1}{2}$ commutation relations:

$$S_i^z = \frac{1}{2} - S_i^- S_i^+ = \frac{1}{2} - \frac{1}{N} \sum_q S_q^- S_q^+ \tag{13.98}$$

The average of this now becomes, using (13.94)–(13.97),

$$\langle S^z \rangle = \frac{1}{2} - \frac{1}{N} \sum_q \frac{2\langle S^z \rangle}{e^{\beta \varepsilon_q} - 1} \tag{13.99}$$

This is the random-phase equation of state giving $\langle S^z \rangle$ (which also occurs in ε_q) as a function of β and h.

The above results are usually arrived at using Green-function equation-of-motion methods (see Zubarev[144] and Stinchcombe[80]).

The result (13.97) exhibits a crude type of "critical slowing down": for zero field h, where a transition can occur, the characteristic frequency of the renormalized spin waves disappears in the neighborhood of the transition like the order parameter $\langle S^z \rangle$. Critical slowing down and mode softening are important in the interpretation of dynamic behavior of systems at a phase transition (see Van Hove,[145] Cochran,[146,147] and Anderson[148]).

13.4.4. Broken Symmetry

The discussion in Section 13.4.2 emphasizes the difference between the low-lying excitations in the Ising model (localized spin flips) and in the Heisenberg model (spin waves). The difference arises because the transverse interaction terms of the Heisenberg model ($\Sigma_{i \neq j} J_{ij} S_i^+ S_j^-$) transfer the spin flip from one site to its neighbor.

The low-temperature excitation spectra of typical Ising and Heisenberg ferromagnets at $h = 0$, $T = 0$ are shown in Figure 13.15. For zero field the spin-wave energies form a band while the Ising spin-flip energy is a constant. In particular, the $q = 0$ spin-wave excitations of the zero-field Heisenberg model have zero frequency:

$$\omega_q \xrightarrow[q \to 0]{} 0, \qquad h = 0 \qquad\qquad (13.100)$$

The fundamental reason for the gapless nature of the spin-wave spectrum is the symmetry of the zero-field Heisenberg-model Hamiltonian: this is

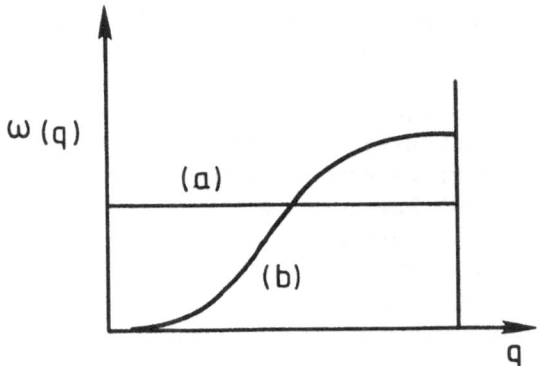

Figure 13.15. Dispersion relations for (a) spin–flip excitations in Ising ferromagnets and (b) spin-wave excitations in Heisenberg ferromagnets, at $h = 0$, $T = 0$.

invariant under a rotation (about any axis $\alpha = x$, y, or z) of all the spins:

$$\left[H_H, \sum_i S_i^\alpha\right] = 0 \qquad (h = 0) \qquad (13.101)$$

Equivalently, each component of total spin is a constant of motion. This implies, among other things, that

$$[H_H, S_{q=0}^-] = 0 \qquad (h = 0) \qquad (13.102)$$

so that the spectrum must be gapless, since S_q^- is the creation operator for the spin wave.

This is an illustration (one of the simplest) of the "Goldstone theorem."[102] The theorem states that a (boson) excitation with $\omega_q \to 0$ occurs whenever the ground state of a system breaks a (continuous) symmetry of the Hamiltonian. The "broken-symmetry" situation in which the theorem applies obviously occurs in the isotropic Heisenberg model, where the ground state does not have the rotational symmetry of the Hamiltonian. The symmetry breaking is required in order that the constant of motion related to the symmetry of the Hamiltonian (e.g., $S_{q=0}^-$, above) should produce a new state when applied to the ground state. The theorem strictly applies to systems with short-range potentials.[103,149]

Another example of the broken-symmetry situation is a perfect lattice with periodic boundary conditions, and forces dependent only on atomic separation. The Hamiltonian is invariant with respect to translation of all atoms, but the lattice has broken the longitudinal and translational symmetry. The "Goldstone bosons" are, in this case, the transverse acoustic phonons which, for $q \to 0$, become a uniform displacement of all the atoms.

Other examples of broken symmetry are the superfluid and superconducting states related to nonzero values of the order parameters given in equations (13.28) and (13.29). Here a gauge symmetry is broken by the appearance of a gauge angle ϕ giving the phase of the complex order parameter. The original Hamiltonians are independent of the gauge angle (their symmetry is gauge invariance). The Goldstone bosons are here oscillations in the phase parameter (the phonons of the superfluid and the collective excitations of the superconductor).

We have seen that broken continuous symmetries imply the appearance of gapless modes; but then, as discussed in the next sections, such gapless modes may, in low dimensionalities, destroy the broken symmetry states. Such considerations exclude the possibility of stable broken-symmetry states in low dimensionalities.

13.5. Instabilities and Fluctuations

13.5.1. Instabilities in Heisenberg Magnets

We continue to use the Heisenberg ferromagnet as an illustration. At low temperatures, where the spin-wave picture applies, the "sum rule" (13.98) yields

$$\langle S^z \rangle = \frac{1}{2} - \frac{1}{N} \sum_q n_q \qquad (13.103)$$

$$n_q \equiv \langle S_q^- S_q^+ \rangle = \frac{1}{e^{\beta \omega_q} - 1} \qquad (13.104)$$

using the boson character and the spin-wave frequency (13.90).

Now, in d-dimensions,

$$\frac{1}{N} \sum_q \propto \int q^{d-1} dq \qquad (13.105)$$

writing a form appropriate to summands dependent only on $q = |\mathbf{q}|$. Also, in zero field and at small q, for lattices with inversion symmetry

$$\omega_q \propto q^2 \qquad (h = 0) \qquad (13.106)$$

Thus for the three-dimensional isotropic Heisenberg model at low temperatures the difference of the magnetization $\langle S^z \rangle$ from its saturation value is

$$\frac{1}{N} \sum_q n_q \propto T^{3/2}, \qquad d = 3, \quad h = 0 \qquad (13.107)$$

This is the usual Bloch "$T^{3/2}$-law" for the low-temperature magnetization decrease.[150] In the Ising model the gap in the excitation spectrum leads to a magnetization decrease proportional to $\exp[-2\gamma/KT]$.

A crude estimate of the transition temperature of the three-dimensional Heisenberg model can be obtained by using equations (13.103) and (13.104) to find where the boson excitations reduce $\langle S^z \rangle$ to zero. The resulting transition temperature is of order $J(0)/K$ (as in mean-field theory). A more satisfactory discussion with essentially the same result is obtained by carrying out the same procedure but using the random-phase equation of state (13.99).[143]

A very slight extension of the above discussion indicates an instability in low-dimensional Heisenberg systems: the number Σn_q of excited spin waves diverges for $d \leq 2$ because the small-wave-vector part of the integral is proportional to

$$\int_0^{\cdots} \frac{q^{d-1}dq}{q^2} = [\ln q]_0^{\cdots} \qquad d = 2$$

$$= \left[-\frac{1}{q} \right]_0^{\cdots} \qquad d = 1 \qquad (13.108)$$

The ground state is therefore unstable with respect to the formation of spin waves, at any finite temperature. This shows that the isotropic Heisenberg ferromagnet cannot retain the long-range order of its ground state at nonzero temperatures for dimensionalities $d \leq 2$.

Since the system orders for $d \geq 3$, we can identify the lower critical dimensionality of the isotropic Heisenberg ferromagnet as $d_1 = 3$.

The isotropic Heisenberg antiferromagnet has a linear spin-wave dispersion ($\omega_q \propto q$, q small). This might suggest that it has long-range order in two dimensions. This however is not the case: a more detailed investigation than we have room for here shows that (for nearest-neighbor coupling between opposite sublattices only) there are two branches of spin waves with frequencies $\pm \omega_q$, where (in zero field)

$$\omega_q^2 = J(0)^2 - J(q)^2 \qquad (13.109)$$

and the contribution of the two branches to the sublattice magnetization decrease is

$$\frac{1}{2N} \sum_q \left[\left(1 + \frac{J(0)}{\omega_q} \right) n(\omega_q) + \left(1 - \frac{J(0)}{\omega_q} \right) n(-\omega_q) \right] \qquad (13.110)$$

where $n(\omega_q)$ is the Bose distribution function.[151] The resulting integral is in low dimensionalities dominated by a small-q part proportional to

$$\int \frac{q^{d-1}dq}{\omega_q^2} \qquad (13.111)$$

which shows the same divergences for $d = 1, 2$ as in the ferromagnetic case, (13.108). A more rigorous discussion of the absence of long-range order in ferro- and antiferromagnetic Heisenberg models in one and two dimensions has been given by Mermin and Wagner.[152]

13.5.2. Instabilities in Ising Magnets

In Section 13.1, free-energy considerations were used to show that long-range order in the one-dimensional Ising ferromagnet is destroyed by a single reverse coupling. In this argument, the excitation is the wall or soliton recently receiving much attention in anisotropic spin systems and elsewhere.[153] It is easy to show similarly that the Ising antiferromagnetic chain also has no long-range order at finite temperatures. Whether it, like the Ising ferromagnet, has long-range order in two dimensions can depend on whether "frustration" effects[154–156] are present: the square-lattice Ising antiferromagnet with nearest-neighbor coupling, for example, is not frustrated and it is possible to divide the spins into two sublattices each directly coupled by the antiferromagnetic exchange only to the other. Thus it can be mapped into the square-lattice ferromagnet by mapping the spins of one sublattice into minus themselves:

$$\sigma_i \rightarrow \sigma'_i = -\sigma_i, \quad i \text{ on one sublattice} \tag{13.112}$$

So it must have the same lower critical dimensionality and critical properties (in zero field) as the Ising ferromagnet.

A similar argument does not apply to the triangular lattice with antiferromagnetic nearest-neighbor coupling, and other frustrated lattices with closed paths containing odd numbers of nearest-neighbor bonds. For a discussion of frustration effects, see Chapter 18.

13.5.3. Lower Critical Dimensionality for Bose–Einstein Condensation

In the discussion of Appendix 13.2, Bose condensation was necessary in three dimensions because the single-particle states of nonzero energy could not accommodate a sufficient number of particles. In general dimensionality d, and with the chemical potential again necessarily nonpositive, the total occupation of the states with nonzero energy is proportional to

$$V \int_0^\infty \frac{k^{d-1} dk}{\exp[\beta(\hbar^2 k^2/2m - \mu)] - 1} \tag{13.113}$$

For $d \le 2$ this integral diverges as $\mu \rightarrow 0$ so that at any low but nonzero temperature the N particles can be accommodated with an appropriate negative μ. The macroscopic occupation of the ground state, characteristic of Bose condensation, does not therefore occur for $d < d_1 = 3$.

13.5.4. Instability of Low-Dimensional Crystal Lattices

We now consider, following Peierls[157,158] and Landau[159] (see particularly Peierls' book[160]), whether the long-range order of low-dimensional crystal lattices should persist at nonzero temperatures. We only consider the degrees of freedom related to atomic displacements (discarding, for example, electron–phonon interactions and, with them, such possible instabilities as the Jahn–Teller or Peierls transition).

Consider first a chain of atoms. Suppose u_n is the displacement of the nth atom from its equilibrium position, and a is the equilibrium spacing. The criterion for long-range crystalline order is

$$u_n \lesssim a \tag{13.114}$$

The separation of neighboring atoms will normally be close to a. We call the difference δ_n:

$$\delta_n = u_n - u_{n-1} \tag{13.115}$$

The mean-square difference is given in terms of the curvature of the interatomic interaction potential U by, roughly,

$$\langle \delta_n^2 \rangle = KT \Big/ \frac{d^2 U}{dx^2} \tag{13.116}$$

In real systems, even at the melting temperature, δ_n is a good deal smaller than a (as will be seen in Section 13.5.5). It is then reasonable to use a harmonic interaction U. Nevertheless, the displacement of the nth atom from its equilibrium position may be large, since

$$u_n - u_0 = \sum_{r=1}^{n} \delta_r \tag{13.117}$$

This superposition of n independent δ_r's, each normally distributed (in the harmonic approximation), increases with n like $n^{1/2}$. So, far enough along the chain the criterion (13.114) for long-range order is violated at any nonzero temperature. Thus the atomic chain does not have long-range order for $T > 0$.

For later generalization, this argument can be rewritten as follows in terms of the spectrum of fluctuations. The displacements u_n can be rewritten in terms of the normal-mode coordinates q_k:

$$u_n = \sum_k q_k e^{ikna}, \qquad q_{-k} = q_k^* \tag{13.118}$$

The wave vectors k are integer multiples of $2\pi/L$ lying in the first Brillouin zone $(-\pi/a, \pi/a)$.

For any mode k with frequency ω_k significantly less than KT, the equipartition law can be applied to yield

$$\langle q_k q_{-k'} \rangle = \delta_{kk'} KT / M\omega_k^2 \tag{13.119}$$

where M is the mass of the (whole) chain. Thus the contribution to $\langle (u_n - u_0)^2 \rangle$ from long-wavelength modes is

$$2 \sum_k{}' \langle q_k q_{-k} \rangle (1 - \cos kna) = \frac{KTa}{\pi mc^2} \int{}' dk \left(\frac{1 - \cos kna}{k^2} \right) \tag{13.120}$$

where the primes on the sum and integral indicate the region of the Brillouin zone in which the equipartition result is applicable. In the conversion of the sum to the integral the weight factor $(L/2\pi)dk$ has been used. We also used the long-wavelength form of the phonon frequency

$$\omega_k = ck \tag{13.121}$$

where c is the sound velocity and the atomic mass $m = Ma/L$. Using the change of variables $x = kna$, the integral becomes, for large n,

$$na \int_{-\infty}^{\infty} \frac{dx}{x^2} (1 - \cos x) \tag{13.122}$$

The dimensionless integral is of order 1. Thus, for large n, the dominant contribution to $\langle (u_n - u_0)^2 \rangle$ is

$$\langle (u_n - u_0)^2 \rangle \sim \frac{KT}{mc^2} na^2 \qquad \text{for large } n \tag{13.123}$$

This result is of the same form as that obtained from the simple considerations with which we started this section. It confirms that the fluctuation in atomic displacements is enough to remove the coherence of positions of sufficiently distant atoms in a linear chain at any finite temperature.

For the two-dimensional case the dominant contribution to $\langle (u_n - u_0)^2 \rangle$ at large n is proportional to

$$\int d\theta \int{}' kdk \frac{[1 - \cos(kna \cos \theta)]}{k^2} \tag{13.124}$$

where θ is the angle between $\mathbf{k}$ and the line joining the two atoms. The same change of variable takes this to the form

$$\int d\theta \int{}' dx \frac{[1 - \cos(x \cos \theta)]}{x} \tag{13.125}$$

The integral over x would diverge logarithmically at large $|x|$, so it is necessary to consider the upper limit which (as before) is of order $anKT/c$ and therefore large in magnitude for large n. The logarithmic divergence is thus cut off at these limits, so that for large n the integral gives a result proportional to $\ln n$ and, finally,

$$\langle (u_n - u_0)^2 \rangle \sim \frac{KT}{mc^2} a^2 \ln n \qquad (13.126)$$

So the two-dimensional lattice is also unstable with respect to such fluctuations.[157–161]

Monolayers of adsorbed helium do nevertheless appear to show a fairly well-defined transition between a solid and liquid phase. The reason appears to be that the two-dimensional crystal lattice can exist over distances up to about n^*a where n^* is such that

$$\frac{KT}{mc^2} \ln n^* \sim 1 \qquad (13.127)$$

At low temperatures this distance, $a \exp(mc^2/KT)$, can exceed the dimensions of the actual monolayers, and the effect we have been considering is then masked by size effects.

The type of discussion given above, considering fluctuations, can also be applied to the Heisenberg magnets, for example, to recover the same critical dimensionality as obtained from the earlier discussion of the magnetization decrease. The fluctuation to be considered is

$$\langle (S_i^z - \langle S_i^z \rangle)^2 \rangle \sim \left(\sum_q \langle S_q^+ S_{-q}^- \rangle \right) \left(\sum_{q'} \langle S_{q'}^- S_{q'}^+ \rangle \right) \qquad (13.128)$$

The approximate form, on the right-hand side, applies within the spin-wave description.

13.5.5. Melting

The three-dimensional crystal lattice does not show the same complete instability as the one- and two-dimensional lattices. Thus, as we know, three-dimensional crystal lattices can exist, and those structures occur that give the least free energy.

But, of course, any such structure melts at some finite temperature. The melting temperature can be estimated, following Lindemann,[95] by finding where the mean-square atomic displacement becomes comparable to the square of the atomic spacing. At the melting temperature T_m their ratio is given, from the analysis of the preceding section, by

$$\gamma \equiv \frac{\langle u_n^2 \rangle}{a^2} \doteq b_0 \frac{KT_m}{mc^2} \tag{13.129}$$

provided the equipartition theorem can be applied; b_0 is a numerical constant, about 0.4 for a simple cubic lattice. In the previous calculations, of $\langle (u_n - u_0)^2 \rangle$ with n large, the $[1 - \cos(kna\ldots)]$ factor caused the integral to be dominated by modes with small k, and the equipartition theorem was always applicable. But for the calculation of $\langle u_n^2 \rangle$, where no such factor occurs, all modes contribute and the equipartition theorem, and hence the result (13.129), only applies for melting temperatures well above the Debye temperature. This condition is satisfied for all the alkali metals except lithium, for the noble metals, and for calcium, and in all these cases melting occurs at a value of γ in the range

$$\gamma = 1/64\text{–}1/100 \tag{13.130}$$

This corresponds to a root-mean-square atomic vibration roughly $\frac{1}{8}$ to $\frac{1}{10}$ of the interactomic spacing.[162]

The main conclusion of the last two sections is that the lowest dimensionality for long-range crystalline order is given by

$$d_1 = 3 \tag{13.131}$$

13.6. Peierls' Transition, Charge-Density Waves, Solitons, and Phasons

We now revisit the Peierls' transition. We discussed it earlier for a linear chain, making no reference to the fluctuation effects, which cause there to be no crystalline chain in the first place. However, real solid-state systems are not normally one-dimensional, but at best composed of weakly coupled chains. The weak interchain coupling can be enough to stabilize the "quasi-one-dimensional" lattice at some low temperature. The Peierls' transition can then occur.

This transition can be considered as the instability of the lattice with respect to the formation of a static periodic distortion. It is driven by the electron–phonon interaction and the coupling of the two systems implies that the electron density, as well as the lattice, normally distorts in a similar way. (An exception is the half-filled band case for tight binding "alternate" structures: these are structures in which the sites can, as in the nonfrustrated antiferromagnet, be divided into two interpenetrating sublattices so that any site of one sublattice has as nearest neighbors only sites of the other. Any atomic distortion preserving the alternate topology can be shown to leave the same number of electrons at each site.[110] Except in such special cases the electron density shows a periodic distortion, the charge density wave (CDW).[163-167] The CDW has an associated Coulomb energy, which should properly be considered in discussing whether the Peierls' distortion occurs.

The above discussion is oversimplified in several respects,[110,168-171] of which we mention only two. The lattice will itself respond to the CDW distortion; a possible consequence is that the phase of the initial periodic lattice distortion is modulated. This leads to consideration of the so-called "soliton" solutions of the sine-Gordon equation.[153] Whether the lattice distortion can be translated (with respect to the lattice) without change of energy also needs to be considered. In Figure 13.16a, b, c we represent three situations, using an open circle for the atoms of the undistorted lattice and a line for the profile of the *lattice* distortion (the peaks are also indicated). In curves a and b, the wavelength of the distortion is commensurate with the lattice, and translating the distortion (a to b) gives different configurations and different energies. The distortion will lock in to give the lowest energy. In the incommensurate case c, all possible placements of the distortion are equivalent (of equal energy): any placement of the distortion simply puts a given peak in the same configuration relative to its nearest undistorted-lattice site, as could have been found further down the line with another placement. Thus no energy change is needed to translate the distortion. The associated zero-energy mode is the *phason*.

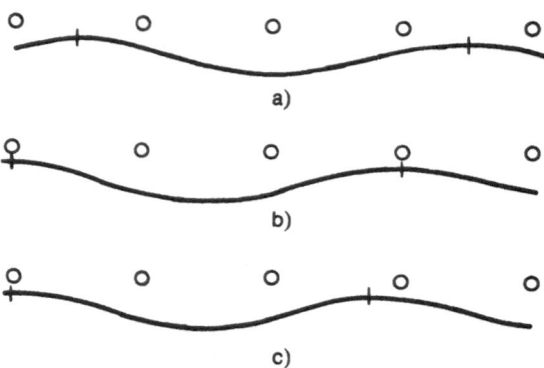

Figure 13.16. Placements (a), (b) of a commensurate lattice distortion, and (c) of an incommensurate distortion.

13.7. Quasi-Low-Dimensional Systems

13.7.1. Introduction

As remarked above, most "one-dimensional systems" are in actual practice weakly coupled chains — hereafter called "quasi-one-dimensional." Examples are KCP,[110,168] TMMC,[172] and TTF (TCNQ).[110,168] Likewise, real layer systems normally have weak coupling between the layers; examples are Rb_2MnF_4[173,174] and transition metal dichalcogenides.[175] We use Rb_2MnF_4 to illustrate some general points.

This system is an almost ideal two-dimensional isotropic Heisenberg (antiferro) magnet. Despite what was proved earlier about there being no long-range order in that model, Rb_2MnF_4 nevertheless exhibits a transition to an antiferromagnetic state at very low but nonzero temperatures. The weak coupling between layers is a possible reason for this. The other reason is a small departure from isotropy in the exchange that adds a small, essentially Ising-like term to the Hamiltonian. Even in the absence of layer coupling, this anisotropy would make the system order like a (two-dimensional) Ising model at very low temperatures. In the treatment of low-dimensional systems one often has to consider small couplings, which increase the effective dimensionality or change the symmetry of the system. We shall discuss the two effects in turn, using the weakly anisotropic, quasi-two-dimensional Heisenberg ferromagnet as our example.

13.7.2. Weakly-Coupled-Layer Heisenberg Ferromagnet

In the first case we ignore the anisotropy and consider only the weak coupling between layers. We use equations (13.103) and (13.104) to obtain a crude estimate of the transition temperature T_c by finding where the total number of excited spin waves is $N/2$:

$$\frac{1}{2} = \frac{1}{N} \sum_q n_q = \left(\frac{a}{2\pi}\right)^3 \int_{-\pi/a}^{\pi/a} dq_x \int_{-\pi/a}^{\pi/a} dq_y \int_{-\pi/a}^{\pi/a} dq_z \frac{1}{e^{\beta_c \omega_q} - 1} \quad (13.132)$$

Here, a is the lattice spacing, $\beta_c = 1/KT_c$, and

$$\omega_q = J(0) - J(q) = \sum_j J_{ij}(1 - e^{i\mathbf{q} \cdot \mathbf{r}_{ij}}) \quad (13.133)$$

For simplicity consider a cubic lattice with interaction J_{ij} coupling only nearest-neighbor spins. The essential feature of the weakly coupled layer system is that J_{ij} is much smaller (by a factor δ, say) when the nearest-neighbor vector $\mathbf{r}_{ij}$ is perpendicular to the layer than when $\mathbf{r}_{ij}$ is in the layer

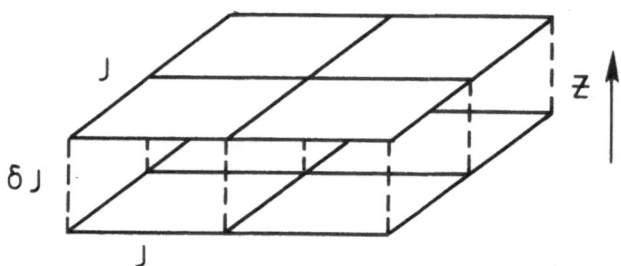

Figure 13.17. Exchange couplings for a weakly-coupled-layer system.

(Figure 13.17). Thus

$$\omega_q = J[2 - (\cos aq_x + \cos aq_y) + \delta(1 - \cos aq_z)] \qquad (13.134)$$

where we took the z-direction perpendicular to the layers, and the nearest-neighbor interaction strength as J. In the strictly two-dimensional case ($\delta = 0$) the integral (13.132) has a divergence from the regime in which q_x and q_y are small. For δ small but nonzero the small q_x, q_y regime still dominates the integral, which can therefore be approximated by using the small q_x, small q_y, expansion of the denominator:

$$\frac{1}{2} \sim \frac{1}{2\pi^2} \int_{-\pi}^{\pi} d\zeta \int_0^{\sim \pi^2} \frac{d(\xi^2 + \eta^2)}{\beta_c J[(\xi^2 + \eta^2) + 2\delta(1 - \cos \zeta)]}$$

$$\sim -\frac{1}{4\pi^2} \frac{1}{\beta_c J} \int_{-\pi}^{\pi} d\zeta \ln\left[\frac{2\delta}{\pi^2}(1 - \cos \zeta)\right]$$

$$\sim \frac{\ln(1/\delta)}{2\pi\beta_c J} \qquad (13.135)$$

In the first integral we used $\xi = aq_x$, etc. We dropped terms of order one, which are small in comparison with $\ln(1/\delta)$. Thus

$$KT_c \sim \frac{2\pi J}{\ln(1/\delta)} \qquad (13.136)$$

The transition temperature is thus smaller than the mean-field result by a factor proportional to $1/\ln(1/\delta)$, where δ is the ratio of the interlayer coupling to that within layers.

13.7.3. Two-Dimensional Weakly Anisotropic Heisenberg Ferromagnet

If the Hamiltonian is of the form

$$H = - \sum_{i \neq j} J_{ij} [\alpha \mathbf{S}_i . \mathbf{S}_j + (1 - \alpha) S_i^z S_j^z] \qquad (13.137)$$

the system is the isotropic Heisenberg model for $\alpha = 1$ and the Ising model for $\alpha = 0$. By reconsidering the analysis of Section 13.4.2 it is easy to see that the spin wave is still the low-lying excitation, but its energy is now

$$\omega_q = J(0) - \alpha J(q) \qquad (13.138)$$

Estimating the transition temperature by again counting spin waves now gives, for a two-dimensional (square) lattice with *small* anisotropy $(1 - \alpha)$,

$$\frac{1}{2} \doteq \frac{1}{\pi} \int_0^{\sim \pi^2} \frac{d(\xi^2 + \eta^2)}{\beta_c J(0)[(\xi^2 + \eta^2) + 4(1 - \alpha)]} \qquad (13.139)$$

Similarly to the previous section this gives

$$KT_c \sim \frac{\pi}{2} \frac{J(0)}{\ln[1/(1 - \alpha)]} \qquad (13.140)$$

Thus weak anisotropy $(1 - \alpha)$ would be sufficient to give a strictly two-dimensional Heisenberg system a finite, but low ordering temperature, smaller than the mean-field value by a factor of order $1/\ln[1/(1 - \alpha)]$.

Effects like those discussed in this and the previous section are seen in many other real low-dimensional systems, like those referred to earlier and, for example, the Ising chain systems (like $CoCl_2 2NC_5H_5$), which can show low but finite ordering temperatures because of interchain couplings.[92,93,176]

13.8. Transitions without Usual Long-Range Order; Kosterlitz–Thouless Transition

13.8.1. Introduction

In the previous sections we have concentrated on transitions to (or the instability of) condensed states with long-range order. We now discuss transitions to states without long-range order.

An intriguing paradox was suggested by high-temperature series-expansion results of Stanley and Kaplan,[177] which indicated the existence of a phase transition for two-dimensional isotropic spin systems for which the Mermin–Wagner theorem[152] proves that there cannot exist long-range order. The resolution would be to have order of a different type. One possibility is where the order-parameter correlation function decays at long distances like

$$\langle \phi\phi \rangle \propto 1/r^{n} \qquad (d=2) \qquad (13.141)$$

(Rice[178]). This is qualitatively distinct from the form (13.16) (exponential decay) and a transition from the phase having only short-range order to one having the above power-law decay (quasi-long-range order) does not violate the Mermin–Wagner theorem.[152]

Various two-dimensional systems seem to behave in this way (the XY spin model in two dimensions, superfluid and superconducting films, two-dimensional melting, and liquid-crystal films). The Hamiltonians of these systems all have a type of continuous symmetry (related in the XY model to the rotation of all spins about the z-direction, or related to a gauge symmetry in the case of superfluids or superconductors). The symmetry breaking (Section 13.4.4) occurs as the appearance of a definite spin angle, or gauge, through the phase angle of a multicomponent order parameter.

We briefly consider this aspect for the case of two-dimensional melting. It was argued (Section 13.5.4) that long-range positional order (the periodicity of atomic positions), which would show up in sharp X-ray or neutron diffraction lines, is removed at any finite temperature by fluctuations in the two-dimensional solid. There may still remain local order (near-neighbor atomic separations not differing much from those in the perfect lattice). If that is the case, then it is possible to show[89–91,161] that there must remain long-range directional order, an atom, and its neighbors forming a local set of basis vectors, which are nearly the same throughout the system. This long-range directional order would show up as sets of parallel crystal planes, with fixed angles between the sets. Consider the angle $\theta(\mathbf{r})$ describing the orientation (in two dimensions) of a nearest-neighbor bond at $\mathbf{r}$. In a perfect lattice $\theta(r)$ would be independent of $\mathbf{r}$ (modulo $\pi/2$ for a square lattice, or $\pi/3$ for a triangular lattice, etc.). Even though fluctuations cause the mean-square displacement to diverge, they leave $\theta(\mathbf{r})$ essentially independent of $\mathbf{r}$.

In this discussion we have ignored the possibility of dislocations, or some other strong disturbance of local order, whose occurrence could cause the long-range directional order to be broken down. This is the mechanism for the Kosterlitz–Thouless transition. Consider a two-dimensional system with short-range but no long-range positional order, the latter being excluded by the Peierls–Landau arguments. The system will have long-range directional order. That can be removed by the formation of dislocations. A finite energy, K_0 say, is required to form a single dislocation, so their density is exponentially small at low temperatures. But, as for spin-flip excitations in the Ising model, as more and more dislocations are formed it may become easier to form others. There is then a temperature T_c at which a macroscopic number of dislocations is formed and the long-range directional order disappears; T_c is the critical temperature for the Kosterlitz–Thouless transition.

13.8.2. Two-Dimensional XY Model and Superfluid

To make these considerations more quantitative, we briefly consider the two-dimensional XY model or the two-dimensional superfluid, which are more easily described than two-dimensional melting.[104]

The XY model can be characterized by the two-component ($n = 2$) order parameter (S^x, S^y) at each point $\mathbf{r}$ or, equivalently, by

$$\psi(r) = S^x - iS^y = |\psi| e^{i\phi} \tag{13.142}$$

It is well known that such a complex order parameter is also appropriate for the description of the superfluid.[179,180] In this case

$$\mathbf{v}(\mathbf{r}) = \frac{\hbar}{m} \nabla \phi(\mathbf{r}) \tag{13.143}$$

is the superfluid velocity and $|\psi|^2$ is directly proportional to the superfluid density ρ_s.

If we consider only the contribution of fluctuations in the phase ϕ the correlation function becomes

$$\langle \psi^*(\mathbf{r})\psi(0)\rangle \sim |\psi|^2 \langle e^{-i\phi(\mathbf{r})} e^{i\phi(0)}\rangle \tag{13.144}$$

If, furthermore, the fluctuations in ϕ are determined by a Boltzmann probability distribution with energy density

$$H = \tfrac{1}{2}K_0 \int |\nabla \phi|^2 d^2\mathbf{r} \tag{13.145}$$

the Gaussian nature of the distribution for ϕ allows us to write

$$\langle e^{-i\phi(r)} e^{i\phi(0)}\rangle = \exp\{-\tfrac{1}{2}\langle [\phi(\mathbf{r}) - \phi(0)]^2\rangle\} \tag{13.146}$$

The average in the final exponent can be evaluated using the equipartition law and other steps similar to our treatment of the Landau–Peierls lattice instability to yield [cf equations (13.120)–(13.126)]

$$\langle [\phi(\mathbf{r}) - \phi(0)]^2 \rangle \sim 2\gamma \frac{KT}{K_0} \ln(r/a) \tag{13.147}$$

for large r (which now replaces the site separation na of the earlier discussion); γ is a constant of order unity ($\gamma = 1/(2\pi)$ for a square lattice, if evaluated carefully). Then for large r,

$$\langle \psi^*(\mathbf{r})\psi(0) \rangle \propto (1/r)^\eta \tag{13.148}$$

where

$$\eta = \gamma KT/K_0 \tag{13.149}$$

The result (13.148) shows the quasi-long-range order introduced in (13.141).

The form of energy density given in (13.145) is appropriate in continuum descriptions of the XY ferromagnet and the superfluid for the following reasons. In the XY ferromagnet the exchange Hamiltonian can be expressed in the form (13.145), with $K_0 \sim Ja^2S^2$, for slowly varying spin fields for which a continuum description is possible ($S_i \to S(\mathbf{r}_i)$):

$$-\sum_{i \neq j} J_{ij}\mathbf{S}_i \cdot \mathbf{S}_j + \text{const} = \frac{1}{2} \sum_{i \neq j} J_{ij}(\mathbf{S}_i - \mathbf{S}_j)^2$$

$$\to \tfrac{1}{2} \int Ja^2(\nabla S)^2 d^2\mathbf{r} \to \tfrac{1}{2}Ja^2|\psi|^2 \int |\nabla\phi|^2 d^2\mathbf{r} \tag{13.150}$$

In the superfluid, where $(\hbar/m)\phi$ is the superfluid velocity potential, (13.145) would be an appropriate kinetic-energy density if the constant K_0 is related to the superfluid density by

$$K_0 = \left(\frac{\hbar}{m}\right)^2 \rho_s \tag{13.151}$$

Hereafter we use the superfluid as our illustration, but the discussion has a rather obvious rephrasing for the XY model.

In the discussion leading to (13.148) fluctuations in the magnitude of ψ were ignored. Now let us suppose that fluctuations may occur that are large enough to make $|\psi|$ vanish at certain points. If that happens, the phase ϕ is not single-valued at these points. The gradient of ϕ is still single-valued, but its line integral around a point where $|\psi|$ vanishes will be a nonzero integer multiple of 2π:

$$\oint_C \mathbf{v}(r) \cdot d\mathbf{r} = 2\pi N\hbar/m \qquad (13.152)$$

This corresponds to a vortex (or equivalently to N elementary vortices), or to a disclination in the vector field $\mathbf{S}$ in the case of the XY model. For a vortex (in an incompressible fluid)

$$|\mathbf{v}(r)| \propto 1/r \qquad (13.153)$$

where r is measured from the axis of the vortex. The energy of an elementary vortex in a system of radius R is therefore

$$\tfrac{1}{2}K_0 \int_a^R \frac{1}{r^2} 2\pi r \, dr + E_a = \pi K_0 \ln\frac{R}{a} + E_a \equiv E \qquad (13.154)$$

where the "core contribution" from the region $r < a$ has been separated off into the constant E_a. The Boltzmann probability density for a vortex is therefore proportional to

$$e^{-E/KT} \qquad (13.155)$$

Thus the probability of finding a vortex in the system of radius R is proportional to

$$R^2 e^{-E/KT} \propto R^{[2-(\pi K_0/KT)]} \qquad (13.156)$$

Thus vortices will form freely in a large system unless

$$KT < \frac{\pi K_0}{2} \qquad (13.157)$$

(which actually implies $\eta < \tfrac{1}{4}$).

By considering, for example the decay of persistent currents in a cylindrical shell it is possible to see that vortex formation can cause the decay of the current.[104] The condition (13.157) is therefore a requirement for superfluidity, and gives the transition temperature. A more complete account must also consider interactions between vortices; these lead to a renormalized constant in place of the bare value K_0 in equations like (13.157), but do not change the essential points.

The order is destroyed by the formation of topological defects (vortices, disclinations) when the entropy gain overcomes the energy increase (just as in the earliest, spin chain, example of Section 13.1.1).

13.9. Critical Behavior

13.9.1. Exponents and Universality

We now consider, in more detail, the behavior of systems in the vicinity of a phase transition. This behavior can be, as we saw earlier, characterized by critical exponents. Descriptions such as mean-field theory or the random-phase approximation normally give incorrect values for these exponents when tested against any known results. Such exact results are few, but those for the two-dimensional Ising model[83,84] provide a test for approximate theories.

Series-expansion results for various systems suggested that critical exponents were dependent on the dimensionality of the lattice but otherwise independent of lattice structure, dependent on the number of components of the order parameter but not on spin magnitude, and independent of strength and range of forces provided they were of finite range. Experiments indicated that simple fluids had the same critical exponents (see Table 13.1), and that these were the same as the exponents of the three-dimensional Ising-like magnets. Such "universality" of critical behavior was not understood, nor was any method known of reliably obtaining exponents except by exactly solving models or by the laborious method of series expansions.

13.9.2. Upper Critical Dimensionality, Power Counting

It was realized early on that the mean-field theories failed because of their neglect of fluctuations,[136] and that the effect of fluctuations on critical behavior (as on the ordered state itself) decreases with increasing dimensionality. It is now known that mean-field theory (in the form of the "Gaussian model," in which the correlation functions correpond to those of Ornstein–Zernicke theory) correctly gives the critical exponents in dimensions greater than or equal to the upper critical dimensionality d_2 referred to earlier. In many cases (such as simple pure magnets) $d_2 = 4$, but we shall give an exception to this below.

The upper critical dimensionality can be obtained by power-counting methods, which we now briefly describe. The (free) Gaussian propagator, or correlation function, is equation (13.17) (i.e., $\chi_q \propto 1/q^2$ at the transition). Most systems (e.g., the magnetic systems, helium, liquid–gas) have interactions that can be represented graphically by vertices with four incident Gaussian propagators (ϕ^4 field theories).[181,182] The interaction effects modify the Gaussian propagator by adding to its denominator self-energy terms like those represented graphically in Figure 13.18a, b. The self-energy graph of Figure 13.18a has three internal propagators and two vertices (assumed to be virtually wave-vector independent for small wave vectors). Wave-vector conservation at the vertices leaves two unrestricted wave-vector labels on internal propagators and, in a d-dimensional system, each of these wave vectors gives rise to a d-dimensional integral. The resulting self-energy contribution is therefore

$$\Sigma^{(a)} \propto \frac{(q^d)^2}{(q^2)^3} \propto q^{2d-6} \tag{13.158}$$

This graph does not modify the small-q Gaussian denominator ($\propto q^2$) provided

$$2d - 6 \geqslant 2 \qquad (d \geqslant 4) \tag{13.159}$$

A similar reasoning applies to the graph of Figure 13.18b and indeed in general. A general graph with n vertices has n unrestricted internal wave-vector labels, each to be integrated over, and $(2n - 1)$ internal propagators. The Gaussian q^2-form is therefore preserved as long as, for any (positive) n,

$$nd - 2(2n - 1) \geqslant 2 \tag{13.160}$$

which again implies $d \geqslant 4$. Thus $d_2 = 4$ is the upper critical dimensionality for systems representable as ϕ^4 field theories.

One of the rather few systems which is not a ϕ^4 field theory is percolation. Here, vertices with three incident lines occur (it is a ϕ^3 field theory).[183,184] Self-energy graphs, such as Figure 13.18c, with n vertices (n

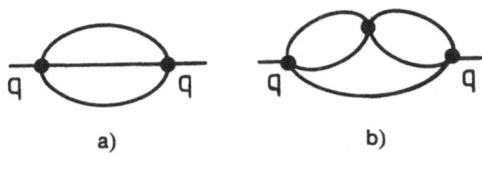

a)   b)

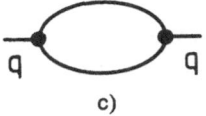

Figure 13.18. (a), (b): Self-energy terms in a ϕ^4 field theory; (c) a self-energy term for a ϕ^3 field theory.

c)

is necessarily even) have $n/2$ unrestricted internal wave-vector labels and $\frac{3}{2}n - 1$ propagators. The condition for keeping the Gaussian form is therefore (for all positive even n)

$$d\frac{n}{2} - 2(\tfrac{3}{2}n - 1) \geqslant 2 \tag{13.161}$$

giving $d_2 = 6$ as the upper critical dimensionality for ϕ^3 theories (including percolation).

This gives a partial explanation of universality, showing that the exponents in the last row of Table 13.1 should occur in all systems for $d \geqslant d_2$ (though, strictly speaking, there can be logarithmic corrections for $d = d_2$). Unfortunately the explanation does not apply to realizable dimensionalities, except in special cases (like tricritical points, where $d_2 = 3$ because a ϕ^6 field theory arises).

13.9.3. Homogeneity and Scaling

Two important advances led up to the breakthrough provided by the development of renormalization group theory. The first advance was the so-called homogeneity hypothesis,[185,186] which proposed that the critical correlation functions, and hence the free energy, were homogeneous functions of the basic variables. This had the consequence that the various critical exponents should be related as follows (we do not here distinguish between the two possibly different exponents on either side of the transition):

$$\alpha + 2\beta + \gamma = 2 \tag{13.162}$$

$$\gamma = \beta(\delta - 1) \tag{13.163}$$

$$\gamma = (2 - \eta)\nu \tag{13.164}$$

$$\alpha = 2 - \nu d \tag{13.165}$$

These relationships are satisfied by the Onsager solution of the two-dimensional Ising model, and [except for equation (13.165)] by the mean-field exponents; series results give very sensitive tests of the exponent relationships.

The second advance came through Kadanoff's scaling approach,[187] which for a simple magnetic model showed how homogeneity could be understood by considering the explicit transformations of the basic parameters (temperature and field) under a length scaling. At the same time this illustrated the importance of dimensionality, and the irrelevance of lattice structure, spin magnitude, and range and strength of forces. The basic idea is that at the transition, provided it is of second order, the correlation length ξ diverges. Near the transition it is therefore the largest and, presumably, the only characteristic length associated with critical phenomena. All features on the scale of, say, the atomic spacing become irrelevant. But the correlation length diverges in a temperature-dependent way

$$\xi \propto |T - T_c|^{-\nu} \tag{13.166}$$

[given previously in equation (13.19)]. This temperature dependence of ξ then leads to the temperature dependence of all the singular properties in the critical regime.

Let us consider a dilatation (increase of length scale) of the system by an arbitrary factor b. On the new length scale the correlation length appears shrunk by the same factor

$$\xi \to \xi' = \xi/b \tag{13.167}$$

Now look at a second quantity whose behavior under a change of scale is known. An example is the free energy, which is extensive:

$$F \to F' = b^{-d}F \tag{13.168}$$

where d is the dimensionality. But F' should be the same function of ξ' as F is of ξ, if the correlation length is the only relevant length. Therefore

$$F(\xi) = F'(\xi') = b^{-d}F(\xi/b) \tag{13.169}$$

Since b is arbitrary, we can put it equal to ξ to obtain

$$F(\xi) = \xi^{-d}F(1) \propto |T - T_c|^{\nu d} \tag{13.170}$$

Thus, for example, taking two derivatives with respect to temperature we obtain the singular part of the specific heat C:

$$C \propto |T - T_c|^{-\alpha} \tag{13.171}$$

with

$$\alpha = 2 - \nu d \tag{13.172}$$

This is the exponent relationship (13.165). Equation (13.169) is a reduced from of the homogeneity assumption of Widom and Griffiths, and the steps from (13.169) to (13.172) are typical of those used in homogeneity arguments.

Kadanoff's discussion identifies the homogeneity parameter b with dilatation. The discussion also shows that instead of obtaining just exponent *relationships*, the exponents could be found if it was known how the parameters describing the system (temperature, field, etc.) transformed under the dilatation of the system. Suppose, for example, that under the dilatation by b the scale of temperature (measured from the critical temperature T_c) transforms as

$$\delta T \equiv T - T_c \rightarrow \delta T' = \lambda_b \delta T \tag{13.173}$$

where λ_b is some number dependent on b. Then, using (13.166) and (13.167),

$$\frac{1}{b} = \frac{\xi'}{\xi} = \frac{(\delta T')^{-\nu}}{(\delta T)^{-\nu}} = \lambda_b^{-\nu} \tag{13.174}$$

Thus, if the "eigenvalue" λ_b is known for the transformation of the temperature scale, we can obtain the critical exponent ν from

$$\nu = \ln b / \ln \lambda_b \tag{13.175}$$

13.9.4. Renormalization Group Method

The renormalization group method[188–192] ties these ideas together in a neat way, and at the same time provides methods of obtaining the transformation of parameters under a dilatation, and hence the critical exponents, and of properly answering the question of universality.

The idea is to explicitly construct a transformation of basic parameters such that under a dilatation the system looks exactly as it did before the dilatation (in the sense of preserving the same absolute correlation length, etc.). This is normally done by removing degrees of freedom and finding the transformation of parameters that leaves unchanged the Boltzmann probabilities for the remaining degrees of freedom. The original, and still the most controlled, methods of doing this use field-theoretic techniques in momentum space, where the degrees of freedom removed are the short-wavelength components of the fields. Perturbation graphs are usually selected by classifying their order with respect to $\varepsilon \equiv d_2 - d$, or sometimes with respect to $d - d_1$ or $1/n$ (where d_2 and d_1 are upper and lower critical dimensionalities, respectively, and n is the number of components of the order parameter). In the first case, the perturbation analysis (ε-expansion) is an expansion from the mean-field (Gaussian) results. The $1/n$ expansion[193] is from the spherical-model results, which are exact for $n \to \infty$. The second principal method of explicitly constructing the renormalization group transformation is the direct position space method.[192] This is based on grouping degrees of freedom[191] (block spin methods), or on deleting degrees of freedom[194,195] (e.g., decimation methods removing every other spin or atom). The decimation methods will be briefly illustrated in Section 13.10.

We suppose that some such method has been applied to construct a transformation. If we denote by μ all the parameters (temperature, field, etc.) the resulting transformation can be written formally as

$$\mu \to \mu' = R_b \mu \qquad (13.176)$$

where R_b is the operator giving the transformation resulting from dilatation by the factor b.

At the (second-order) transition the only relevant length ξ diverges and the system becomes length-invariant. So it should not notice the dilatation. One way this situation may occur is through the basic parameters μ having particular values given by μ^*, where

$$\mu^* = R_b \mu^* \qquad (13.177)$$

so that they do not change under the dilatation; μ^* is one of the equation's so-called fixed points (in the multidimensional parameter space). Another way of having the system critical (divergent ξ) is to have μ one of, for example, a "critical surface" of points such that repeated applications of R_b to any point on the surface takes the point into a fixed point. The transformation of points near the fixed point then determines the asymptotic critical behavior (see below). This concept of the transformations effecting "flow" in the parameter space is illustrated in Figure 13.19. It

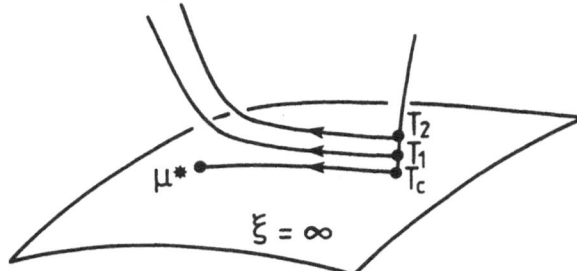

Figure 13.19. Critical surface ($\xi = \infty$) and flow in parameter space: under the renormalization group transformations the set μ of parameters flows to the fixed point μ^* if the initial μ is critical ($T = T_c$).

gives an explanation of universality and of crossover. Universality is the irrelevance of many parameters (such as spin and strength of forces), which simply scale away as the system approaches the fixed point, whose character only basically depends on the spatial dimensionality d and the number n of components of the order parameter. Crossover is the change of critical behavior from that characteristic of one canonical system to that of another as the point flows between the fixed points associated with the two characteristic behaviors. Thus a weakly anisotropic Heisenberg system crosses over ultimately to the critical behavior of an Ising model as the flow goes from the vicinity of the isotropic Heisenberg fixed point ($n = 3$) to the Ising fixed point ($n = 1$).

We shall now consider the transformation in a little more detail to make contact with the scaling discussion given previously. Near the fixed point we can work with $\delta\mu \equiv \mu - \mu^*$ and $\delta\mu' \equiv \mu' - \mu^*$, which are related through a linearized form of (13.176):

$$\delta\mu' = R_b^L \delta\mu \tag{13.178}$$

Since two successive dilatations are equivalent to a single dilatation we obtain the (semi-) group property

$$R_b^L R_{b'}^L = R_{bb'}^L \tag{13.179}$$

This implies that the eigenvalues λ_b of R_b^L satisfy a similar equation for any b, b', so that

$$\lambda_b = b^y \tag{13.180}$$

where y is independent of b. In a simple one-parameter case (where typically $\delta\mu = T - T_c$)

$$\lambda_b = dR(\mu^*)/d\mu^* \tag{13.181}$$

and the same arguments (see 13.174) used to relate (13.166) and (13.173) can be applied to yield

$$\xi \propto (\delta\mu)^{-\nu} \tag{13.182}$$

where

$$\nu = 1/y = \ln b/\ln \lambda_b \tag{13.183}$$

Thus equation (13.180) is equivalent to (13.173) and (13.174) and we have rederived (13.175).

The eigenvalues of the linearized transformations therefore give the critical exponents. Further, whether a parameter is relevant or irrelevant (scales away under the transformation) depends on whether its associated eigenvalue is greater or less than 1. The critical surface referred to earlier is the surface through the fixed point on which all relevant parameters have their fixed point value, and so includes all points flowing into the fixed point.

Thus, the transformation of parameters under a change of length scale determines the critical behavior of a system. Simple examples of renormalization group transformations will be given in Section 13.10.

13.10. Critical Behavior of Low-Dimensional Systems

13.10.1. One-Dimensional Magnetic Systems

One-dimensional systems (with finite range forces) do not have transitions at finite temperatures. However, they can have ordered ground states (as in the Ising and Heisenberg models) and critical behavior in the region near $T = 0$.

We shall first give a transfer-matrix discussion[196] of the Ising chain, and then discuss by simple renormalization group methods the critical behavior of Ising and Heisenberg chains in turn (considering for simplicity only zero-field cases).

13.10.1.1. Ising Chain: Transfer Matrix

The chain with Hamiltonian (13.22), where $\beta J \equiv K$ and $h = 0$, can be discussed by considering the possible configurations of a pair of adjacent spins, σ_1 and σ_2 say (Figure 13.20). The unnormalized Boltzmann probability for various configurations of the spin pair is

$$
\begin{aligned}
\exp(-\beta H_{12}) &= \exp K, && (\sigma_1, \sigma_2) = (1, 1), (-1, -1) \\
&= \exp(-K), && (\sigma_1, \sigma_2) = (1, -1), (-1, 1)
\end{aligned}
\tag{13.184}
$$

This can be written as a 2×2 transfer matrix T whose row and column

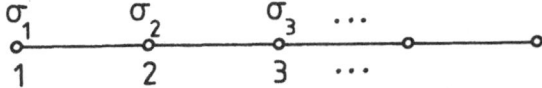

Figure 13.20. Chain of spins $\sigma_1, \sigma_2, \ldots$ at lattice sites $1, 2, \ldots$.

labels are the states of σ_1 and σ_2, respectively

$$T = \begin{pmatrix} e^K & e^{-K} \\ e^{-K} & e^K \end{pmatrix} \tag{13.185}$$

Since the multiplication of two such matrices involves a sum over states, the total partition function for a (closed) chain of N spins is

$$Z(N) = \operatorname{Tr} T^N = \lambda_1^N + \lambda_2^N \tag{13.186}$$

where λ_1 and λ_2 are the eigenvalues of T:

$$\lambda_1 = 2 \cosh K \tag{13.187}$$

$$\lambda_2 = 2 \sinh K \tag{13.188}$$

For a large system the partition function is thus the Nth power of the larger eigenvalue (λ_1). From the partition function we can in principle obtain all the thermodynamic quantities of interest (of course, in order to get the magnetic moment or susceptibility we should have kept the field-dependence). It is easy to check from our result that the specific heat, for example, has no singularity at finite temperature (no transition).

One can also consider in a similar way the correlation function, which is obtained from

$$\langle \sigma_1 \sigma_{n+1} \rangle = \frac{1}{Z(N)} \operatorname{Tr} \left[\begin{pmatrix} 1 & 0 \\ 0 & -1 \end{pmatrix} T^n \begin{pmatrix} 1 & 0 \\ 0 & -1 \end{pmatrix} T^{N-n} \right] \tag{13.189}$$

$$= \frac{\lambda_1^n \lambda_2^{N-n} + \lambda_2^n \lambda_1^{N-n}}{\lambda_1^N + \lambda_2^N} \tag{13.190}$$

$$\xrightarrow[N \text{ large}]{} \left(\frac{\lambda_2}{\lambda_1} \right)^n = (\tanh K)^n \tag{13.191}$$

Comparing this with (13.16) we see that, with a the spin spacing,

$$\xi = \frac{a}{\ln(\lambda_1/\lambda_2)} \tag{13.192}$$

$$\sim \tfrac{1}{2} a e^{2K} \qquad (K \gg 1) \tag{13.193}$$

The last result is the well-known exponential divergence of the Ising chain correlation length as the temperature goes to zero.

13.10.1.2. Ising Chain Renormalization Group Method (Decimation)

We shall now apply a simple renormalization group method to the above problem. Of course the method only comes into its own with problems which are *not* exactly soluble by simple means.

In the decimation method, every other spin (for example) is removed to take the lattice into one with twice the original lattice spacing; this achieves a dilatation by the factor $b = 2$. We seek the transformation R_b of parameters K that leaves the properties of the system unchanged. These properties are determined by the Boltzmann probability factor, which for the spin pair σ_1, σ_2 takes the form

$$e^{-\beta H_{12} + K_0} = e^{K\sigma_1\sigma_2 + K_0} = e^{K_0} \cosh K[1 + \sigma_1\sigma_2 \tanh K] \qquad (13.194)$$

$$\equiv u[1 + \sigma_1\sigma_2 t] \qquad (13.195)$$

Here we used the fact that $\sigma_1\sigma_2$ can only be ± 1, and to normalize the probability we introduced the constant K_0 (which is actually the free energy per spin pair). Finally, for convenience we introduced the new parameters

$$t = \tanh K \qquad (13.196)$$

$$u = e^{K_0} \cosh K \qquad (13.197)$$

If we remove spin σ_2 (and all other even-site spins) σ_1, σ_3 now form a nearest-neighbor pair on the decimated lattice, and their Boltzmann factor is

$$\text{Tr}_{\sigma_2} u^2[1 + \sigma_1\sigma_2 t][1 + \sigma_2\sigma_3 t] = u'[1 + \sigma_1\sigma_3 t'] \qquad (13.198)$$

where

$$t' = t^2 \qquad (13.199)$$

$$u' = 2u^2 \qquad (13.200)$$

The new probability factor (13.198) has the same form as the original one (13.195), but with new (transformed) values t', u' of the parameters t, u.

Since the new lattice is equivalent to the old one but with a length change by $b = 2$, (13.199) and (13.200) are the renormalization group transformation equations. We do not consider further here the second transformation equation, though it can be very useful since it gives the transformation of the free energy.

The first equation (13.199) gives the transformation of the thermal parameter t. The fixed points of this equation are those values of t that yield the same value for t':

$$t^* = 0, 1 \tag{13.201}$$

We now linearize equation (13.199) in the vicinity of the fixed points by working to first order in $\delta t' = t' - t^*$, $\delta t = t - t^*$:

$$\delta t' = 2\delta t \qquad (t^* = 1) \tag{13.202}$$

$$\delta t' = 0\delta t \qquad (t^* = 0) \tag{13.203}$$

These are linearized equations [cf (13.178)] with eigenvalues $\lambda_b = 2, 0$, respectively. Since the eigenvalue (2) at $t^* = 1$ is greater than one, $t^* = 1$ is the nontrivial fixed point associated with (low-temperature) criticality. Any point with $t = 1$ will flow away to the trivial fixed point $t^* = 0$ associated with the noncritical high-temperature behavior.

In the critical region (low temperatures) where (13.202) applies we therefore have, using (13.180)–(13.183),

$$\xi \propto (\delta t)^{-\nu} \tag{13.204}$$

where

$$\nu = \ln b / \ln \lambda_b = \ln 2 / \ln 2 = 1 \tag{13.205}$$

It can be easily seen using the low-temperature form of

$$\delta t = \tanh K - 1 \propto e^{-2K} \tag{13.206}$$

that the results (13.204) and (13.205) are equivalent to (13.193).

13.10.1.3. Heisenberg Chain, Decimation Method

We shall now apply the method to the classical Heisenberg chain: here the spin magnitude is sufficiently great that commutation problems can be ignored, and the spins behave like classical unit vectors. Fisher[197] has shown that the ratio of the two largest eigenvalues of the transfer "matrix" for this system is, at low temperatures $(K \gg 1)$,

$$\frac{\lambda_2}{\lambda_1} = 1 - \frac{1}{K} \equiv f(K) \tag{13.207}$$

where K is, once again, βJ. It is then possible to show[198] that the renormalization group transformation of the parameter K, for dilatation by factor b, is $K \to K'$ where

$$f(K') = [f(K)]^b \tag{13.208}$$

For low temperatures this becomes

$$\frac{1}{K'} = b \frac{1}{K} \tag{13.209}$$

This can be regarded as a transformation linearized in the neighborhood of the zero-temperature fixed point. Again using (13.180)–(13.183) we thus find

$$\xi \propto \left(\frac{1}{K} \right)^{-\nu} \tag{13.210}$$

where $\nu = \ln b / \ln b = 1$. Thus ξ diverges like $1/T$ in the low-temperature Heisenberg chain. This dependence has been observed in neutron scattering experiments on the Heisenberg chain compound TMMC.[172]

Such methods have been generalized to deal with the much more difficult problem of weakly anisotropic Heisenberg chains, where soliton effects occur and where at low temperatures the critical behavior crosses over between the dependences (13.210) and (13.193) characteristic of the Heisenberg and Ising models.[199]

13.10.2. Two-Dimensional Magnetic Systems

Once again an exact solution exists (for the two-dimensional Ising model). The two-dimensional Heisenberg model has not been exactly solved, but it and the Ising model are equally amenable to treatment by, for example, the ε-expansion method. A decimation method has also been constructed for both systems by Migdal.[200] We illustrate an earlier decimation procedure[194,195] for the Ising model.

Let us consider the square lattice shown by full lines in Figure 13.21. If the spins $1, 2, \ldots$ at every other vertex are decimated, only the spins $A, B, \ldots$ remain and these lie on a new square lattice, shown by dashed lines, that is larger by a scale factor $b = \sqrt{2}$. The interaction between a nearest-neighbor pair of spins (A, B, say) on the new lattice can be evaluated approximately by considering the cluster A1B2 of spins and taking the partial trace over spins 1, 2 [compare equation (13.198)]. The resulting transformed parameter K' is related to that (K) for the original lattice by

$$K' = 2\tilde{K} \tag{13.211}$$

$$\tanh \tilde{K} = \tanh^2 K \tag{13.212}$$

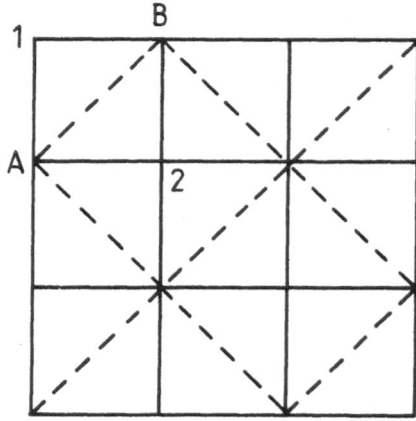

Figure 13.21. Decimation transformation of a square lattice (full lines) to one (dashed lines) larger by a factor $b = \sqrt{2}$ by removing spins $1, 2, \ldots$.

($\tilde{K}$ is the contribution to the exchange coupling between A and B resulting from decimating 1 from the "linear chain" A1B. The chain A2B gives an equal contribution). The resulting transformation

$$t' = \frac{2t^2}{1 + t^4} \tag{13.213}$$

(where $t' = \tanh K'$, etc.) has high- and zero-temperature fixed points, and one at

$$t^* = 0.54 \tag{13.214}$$

This can be compared with the critical condition given by Onsager's exact solution $t^* = \sqrt{2} - 1 = 0.41$.

Linearization of the transformation about the fixed point gives

$$\delta t' = \lambda \delta t \tag{13.215}$$

where $\lambda = 1.68$. Inserting this, with $b = \sqrt{2}$, into (13.182) and (13.183) gives

$$\xi \propto (T - T_c)^{-\nu} \tag{13.216}$$

where $\nu = 0.67$. Onsager's exact value is $\nu = 1$. The errors arise from having restricted our consideration to a small cluster (four spins, A1B2). Larger-cluster calculations of this type[201] give critical exponents with about the accuracy of series-expansion methods.

13.10.3. Percolation and Other Disorder Problems

Procedures like those just illustrated for the low-dimensional magnets also apply to the percolation problem.[202-204] In this case the parameter is the bond concentration p. In place of the Boltzmann probability we can use the probability of a nearest-neighbor connection. Let us consider decimation of a linear chain, as illustrated in Figure 13.22, by removing every other site ($b = 2$). In the original chain the probability of nearest neighbors (such as 1, 2) being linked is the bond concentration p. In the decimated chain the probability of nearest neighbors (1, 3) being linked is

$$p' = p^2 \tag{13.217}$$

This is the renormalization group transformation, and by the same methods as before yields

$$p^* = 0, 1 \tag{13.218}$$

$$\nu = 1 \tag{13.219}$$

These results are exact (and trivial).

Percolation in the two-dimensional square lattice can also be discussed in a similar manner. Again let us consider decimation of sites $1, 2, \ldots$ in Figure 13.21 to yield the dashed-line lattice. The probability p' of a nearest-neighbor linkage AB on the new lattice arising from paths via 1, 2 is the sum of the probabilities of the three independent events (a) linkage AB via 1 and not via 2, (b) linkage via 2 and not via 1, (c) linkage via 1 and via 2:

$$p' = p^2(1 - p^2) + p^2(1 - p^2) + p^4 \equiv R(p) \tag{13.220}$$

The nontrivial fixed point is

$$p^* = 0.62 \tag{13.221}$$

and this is our result for the critical concentration (the exact value from

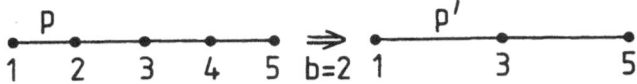

Figure 13.22. Decimation of a linear chain. The probability of a nearest-neighbor linkage is the same as the bond concentration p.

duality arguments is 0.5). The eigenvalue of the linearized equation is

$$\lambda = \frac{dR(p^*)}{dp^*} = 1.53 \qquad (13.222)$$

which yields, using (13.183) and (13.182),

$$\xi \propto (p - p_c)^{-\nu} \qquad (13.223)$$

with $\nu = 0.82$. Calculations of this sort, with larger clusters,[202] give $\nu = 1.34 \pm 0.01$, in agreement with series results.[205]

Decimation methods have also been used to obtain the phase diagrams and critical properties of dilute magnets[206–209] and mixed magnets.[210] They have also been applied to the spin glass[206,211,212] and to other systems. For a review of this area see the lectures by Thouless,[213] Kirkpatrick,[214] and Anderson.[156]

13.11. Concluding Remarks

In this chapter our aim has been to provide some general coverage at an elementary level, together with some more detailed discussion of selected areas. The emphasis has been on simple methods. We have generalized or replaced them where they fail. Particular attention has been given to mean-field aspects and, because of our interest in low dimensionality, to fluctuation effects. Some recent developments have been discussed. Though reference has been made to a wide range of systems, the emphasis has been on those (magnetic, structural, and superfluid systems) that most easily illustrate the main theme of the chapter.

Many-Electron Effects

S. Lundqvist

14.1. Response to Electric and Magnetic Fields

A basic problem in understanding the behavior of a many-electron system is the study of the response of the system when various external fields are applied. The field may vary in space and time in an arbitrary way. A particularly simple case is when the field is a harmonic wave in space and time, characterized by its wave number q and frequency ω. The strength of the external field is usually assumed to be small so that the response is linear. The response of the system to such a field will give important information about the eigenmodes or resonances of the system. These will appear as peaks in the response of the system plotted as a function of the frequency of the probe.

The response to an external electric field leads to the introduction of a generalized dielectric function. This function describes, for example, the optical properties, collective excitations, and the linear response to fast charged particles. It is essential in the study of most many-electron properties because it depends crucially on the internal dynamic correlations in the system. The analogous role in the response to external magnetic fields is taken by the magnetic susceptibility. We shall limit ourselves to give the formulas for the spin susceptibility of a paramagnetic system.

S. Lundqvist · Institute of Theoretical Physics, Chalmers University of Technology, 41296 Göteborg, Sweden.

14.1.1. Some General Formulas for the Dielectric Function

We shall start by writing down some general standard relations for later use. Let us consider the linear response of the system to a small external potential $V_{ext}(\mathbf{r}, t)$. We obtain the following formula for the induced charge density:

$$\rho_{ind}(\mathbf{r}, t) = \int d\mathbf{r}' dt' R(\mathbf{r}, \mathbf{r}', t - t') V_{ext}(\mathbf{r}', t') \qquad (14.1)$$

The response function $R(\mathbf{r}, \mathbf{r}', t - t')$ is nonzero only for $t > t'$. The induced charges give rise to an internal screening field and we can calculate the effective electrostatic potential:

$$V_{eff}(\mathbf{r}, t) = \int v(\mathbf{r}, \mathbf{r}') \rho_{ind}(\mathbf{r}', t) d\mathbf{r}' + V_{ext}(\mathbf{r}, t) \qquad (14.2)$$

where $v(\mathbf{r}, \mathbf{r}') = e^2 / |\mathbf{r} - \mathbf{r}'|$ is the Coulomb potential. The relation between V_{eff} and V_{ext} defines the inverse dielectric function through the formula

$$V_{eff}(\mathbf{r}, t) = \int d\mathbf{r}' dt' \varepsilon^{-1}(\mathbf{r}, \mathbf{r}', t - t') V_{ext}(\mathbf{r}', t') \qquad (14.3)$$

The effective field is what we associate with the total electrostatic field acting on a small test charge. It is convenient to introduce a new response function P which relates ρ_{ind} to the effective field V_{eff} rather than to the external field V_{ext}

$$\rho_{ind}(\mathbf{r}, t) = \int d\mathbf{r}' dt' P(\mathbf{r}, \mathbf{r}', t - t') V_{eff}(\mathbf{r}', t') \qquad (14.4)$$

To proceed, it is suitable to simplify the notation by considering the quantities as vectors and matrices in the variables $\mathbf{r}$, t and r', t'. We can then write

$$\rho_{ind} = P V_{eff} = P(v \rho_{ind} + V_{ext})$$

$$\rho_{ind} = \frac{1}{1 - Pv} P V_{ext} \qquad (14.5)$$

$$V_{eff} = \left(1 + v \frac{1}{1 - Pv} P\right) V_{ext} = \frac{1}{1 - vP} V_{ext} = \frac{1}{\varepsilon} V_{ext}$$

The last line shows that $\varepsilon = 1 - vP$, or written out in full

$$\varepsilon(\mathbf{r}, \mathbf{r}', t - t') = \delta(\mathbf{r}, \mathbf{r}') - \int d\mathbf{r}'' v(\mathbf{r}, \mathbf{r}'') P(\mathbf{r}'', \mathbf{r}', t - t') \qquad (14.6)$$

where P is called the irreducible polarization propagator and, as just mentioned, gives the response of an electron to the total effective electric field that the electron sees. It is therefore a somewhat simpler quantity to consider than the response function itself and is the key quantity to calculate the dielectric function.

What we calculate using linear response theory is the retarded dielectric function, which is obtained by averaging the commutator of the density of two points $\mathbf{r}$, t and $\mathbf{r}'$, t':

$$R(\mathbf{r}, \mathbf{r}', t - t') = -\frac{i}{\hbar} \, \Theta(t - t') \langle [\rho(\mathbf{r}, t); \rho(\mathbf{r}', t')] \rangle \qquad (14.7)$$

When using many-body perturbation techniques such as diagrammatic methods, one encounters a dielectric function of the same form, where R is replaced by the average of the time-ordered product replacing R by Q, and

$$Q(\mathbf{r}, \mathbf{r}', t - t') = -\frac{i}{\hbar} \langle T[\rho(\mathbf{r}, t)\rho(\mathbf{r}', t')] \rangle \qquad (14.8)$$

The time-ordered dielectric function and the retarded dielectric function are closely related and differ only with respect to their different causality properties. The difference is easiest to see if we take the Fourier transform with respect to the time variable $t - t'$ and consider the functions as functions of the variables $(\mathbf{r}, \mathbf{r}', \omega)$. The two functions are equal for positive frequencies (energies). The time-ordered dielectric function is an even function of frequency, whereas the retarded dielectric function has an even real part and an odd imaginary part with respect to frequency (energy). Knowing these properties, one can use any suitable technique to start the construction of such functions and then put in the proper boundary conditions. When it is necessary to distinguish them, we shall use the notation ε_r (for retarded) and ε_c (for causal or time-ordered). Similarly one can define the irreducible polarization propagator P either as a retarded or as a time-ordered quantity. One often uses diagrammatic methods to obtain approximate formulas for P. We refer to standard textbooks[215,216] for discussions of techniques to calculate P.

14.1.2. The Dielectric Function of a Uniform System

Next we shall give some general formulas related to the dielectric function. The formulas will be given for a uniform system in order to keep the notation simple. However, most of the relations have analogues for periodic systems, solids with a surface or small finite systems. For a uniform system, the response functions do not depend on $\mathbf{r}$ and $\mathbf{r}'$ separately but only on the difference $\mathbf{r} - \mathbf{r}'$. It is simplest to consider the complete Fourier transform with respect to both space and time and to use as variables q and the frequency ω. In this section we shall merely collect a number of general relations, approximations to the dielectric function discussed in the following section dealing with the electron liquid.

Using the notation $R(q, \omega)$ for the Fourier transform of $R(\mathbf{r} - \mathbf{r}', \omega)$ and analogous notations for the other quantities, we obtain the following relations:

$$\rho_{\text{ind}}(\mathbf{q}, \omega) = R(\mathbf{q}, \omega) V_{\text{ext}}(\mathbf{q}, \omega) = \frac{4\pi e^2}{q^2} R(\mathbf{q}, \omega) \rho_{\text{ext}}(\mathbf{q}, \omega) \qquad (14.9)$$

This gives an alternative way to define the generalized dielectric function $\varepsilon(\mathbf{q}, \omega)$ by using the relation $\rho_{\text{ind}} = (1/\varepsilon - 1)\rho_{\text{ext}}$, and we obtain

$$\frac{1}{\varepsilon(\mathbf{q}, \omega)} = 1 + \frac{4\pi e^2}{q^2} R(\mathbf{q}, \omega) \qquad (14.10)$$

The effective field $V_{\text{eff}}(\mathbf{q}, \omega)$ is given as the sum of the external and the induced field, hence

$$V_{\text{eff}}(\mathbf{q}, \omega) = V_{\text{ext}}(\mathbf{q}, \omega) + \frac{4\pi e^2}{q^2} \rho_{\text{ind}}(\mathbf{q}, \omega) \qquad (14.11)$$

and the response to V_{eff} is given by the formula

$$\rho_{\text{ind}}(\mathbf{q}, \omega) = P(\mathbf{q}, \omega) V_{\text{eff}}(\mathbf{q}, \omega) \qquad (14.12)$$

Now V_{eff} is the field due to both external and induced charges, and thus

$$V_{\text{eff}}(\mathbf{q}, \omega) = \frac{4\pi e^2}{q^2} (\rho_{\text{ext}}(\mathbf{q}, \omega) + \rho_{\text{ind}}(\mathbf{q}, \omega))$$

$$= \frac{4\pi e^2}{q^2} \frac{1}{1 - \varepsilon(\mathbf{q}, \omega)} \rho_{\text{ext}}(\mathbf{q}, \omega)$$

so we obtain

$$\varepsilon(\mathbf{q}, \omega) = 1 - \frac{4\pi e^2}{q^2} P(\mathbf{q}, \omega) \qquad (14.13)$$

Formulas (14.10) and (14.13) are important relations, which hold in analogous forms also for nonuniform systems. We notice that the inverse dielectric function is defined in terms of the response to the external field while the dielectric function itself is directly related to the response to the effective field. Since the response function P is much easier to calculate, one usually first finds a result for ε analogous to equation (14.13). The problem to find ε^{-1}, which is a key quantity for dynamic screening properties, is generally nontrivial, leading to the inversion of a discrete matrix for a periodic system (but still a continuous function of the frequency) and the inversion of a continuous matrix in most other cases.

For a uniform system, the transformation from ε to ε^{-1} or from P to R is trivial and we obtain the following explicit relation between $R(\mathbf{q}, \omega)$ and $P(\mathbf{q}, \omega)$:

$$R(\mathbf{q}, \omega) = \frac{P(\mathbf{q}, \omega)}{1 - (4\pi e^2/q^2)P(\mathbf{q}, \omega)} \tag{14.14}$$

The response function can be directly related to the physical properties of a system studied in experiments, through some important formulas called spectral representations. Using the definition of the retarded response or the time-ordered function, one obtains after a short calculation

$$R(\mathbf{q}, \omega) = \sum_n |\langle n | \rho_{-q} | 0 \rangle|^2 \left(\frac{1}{\omega - \omega_n + i\delta} - \frac{1}{\omega + \omega_n \pm i\delta} \right) \tag{14.15}$$

where $\hbar\omega_n = E_n - E_0 \geq 0$, E_0 being the ground-state energy and E_n an excited state of the system. The upper sign in equation (14.15) holds for the retarded function and the lower sign for the time-ordered function. Introducing nonexplicitly the spectral weight functions, or simply spectral functions, the one related to equation (14.15) is defined by

$$B(\mathbf{q}, \omega) = \sum_n |\langle n | \rho_{-q} | 0 \rangle|^2 \delta(\omega - \omega_n)$$

and

$$B(\mathbf{q}, \omega) = -B(\mathbf{q}, -\omega) \tag{14.16}$$

Using these relations we can write down the spectral representations for ε_r^{-1} and ε_c^{-1} as follows:

$$\left.\begin{array}{l} \dfrac{1}{\varepsilon_r(\mathbf{q}, \omega)} - 1 = \dfrac{4\pi e^2}{q^2} \displaystyle\int_{-\infty}^{\infty} \dfrac{B(\mathbf{q}, \omega')d\omega'}{\omega - \omega' + i\delta} \\[4mm] \dfrac{1}{\varepsilon_c(\mathbf{q}, \omega)} - 1 = \dfrac{4\pi e^2}{q^2} \displaystyle\int_{-\infty}^{\infty} \dfrac{B(\mathbf{q}, \omega')d\omega'}{\omega - \omega' + i\omega'\delta} \end{array}\right\} \tag{14.17}$$

These formulas show explicitly that the real parts of ε_r^{-1} and ε_c^{-1} are identical but that the imaginary parts differ as in equations (14.17). The spectral function defined by equation (14.16) gives the weighted density of states for exciting the system to the state n. The full density of states of excited levels is given by the formula

$$N(\omega) = \sum_n \delta(\omega - \omega_n) \tag{14.18}$$

The spectral function $B(\omega)$ in equation (14.16) weights each level with the probability factor $|\langle n | \rho_{-q} | 0 \rangle|^2$ that the level n can be excited through a probe, coupling to the density operator of the system. In this sense it is appropriate to refer to $|\langle n | \rho_{-q} | 0 \rangle|^2$ as a generalized oscillator strength for this kind of excitation. We stress again that the results given are just special cases of very general relations for retarded and time-ordered response functions.

In the case of the dielectric function the spectral weight function $B(\mathbf{q}, \omega)$ for $\omega > 0$ is usually referred to as the dynamic form factor $S(\mathbf{q}, \omega)$. The dynamic form factor contains all information that can be obtained from an experiment, which couples to the density of the system. From the spectral formulas (14.17) we obtain a useful relation between $S(\mathbf{q}, \omega)$ and $\operatorname{Im} \varepsilon^{-1}(\mathbf{q}, \omega)$, namely

$$S(\mathbf{q}, \omega) = -\frac{q^2}{4\pi e^2} \operatorname{Im} \varepsilon^{-1}(\mathbf{q}, \omega) \tag{14.19}$$

The usual static structure factor is obtained by integrating according to

$$S(\mathbf{q}) = \frac{1}{n} \int d\omega S(\mathbf{q}, \omega) = -\frac{q^2}{4\pi^2 n e^2} \int_0^\infty d\omega \operatorname{Im} \varepsilon^{-1}(\mathbf{q}, \omega) \tag{14.20}$$

where n is the number of electrons per unit volume. The static structure factor gives the instantaneous density–density correlations of wave number q. The Fourier transform is related to a particularly important quantity, the pair correlation function $g(r)$, defined as

$$g(r) = 1 + \frac{1}{n} \sum_q [S(q) - 1] \exp(i\mathbf{q} \cdot \mathbf{r}) \tag{14.21}$$

From the formulas for $\varepsilon(\mathbf{q}, \omega)$, $S(\mathbf{q})$, and $g(r)$ we can write down different but equivalent expressions for the total Coulomb interaction energy between the electrons. Using the pair correlation function $g(r)$ we have that

$$E_{int} \sim \int d\mathbf{r} g(r)/r$$

Going into Fourier space we obtain

$$\left. \begin{aligned} E_{int} &= \left\langle 0 \left| \sum_q \frac{2\pi e^2}{q^2} (\rho_q^* \rho_q - 1) \right| 0 \right\rangle \\ &= n \sum_q \frac{2\pi e^2}{q^2} [S(q) - 1] \end{aligned} \right\} \qquad (14.22)$$

Expressing $S(\mathbf{q})$ in terms of the dielectric function according to equation (14.20) we find the equivalent formula

$$E_{int} = - \sum_q \left[\frac{2\pi n e^2}{q^2} + \int_0^\infty \frac{d\omega}{2\pi} \, \mathrm{Im}\, \varepsilon^{-1}(\mathbf{q}, \omega) \right] \qquad (14.23)$$

We can go one step further and express the whole correlation energy in terms of $\varepsilon(\mathbf{q}, \omega)$. This makes use of an integration over the coupling constant, due to Pauli. We denote by $\varepsilon_\lambda(\mathbf{q}, \omega)$ the dielectric function, where we have replaced e^2 by an interaction strength λ. The result for the total ground-state energy of the system can be written as

$$E = E_0 - \sum_q \left[\frac{2\pi n e^2}{q^2} + \int_0^{e^2} \frac{d\lambda}{\lambda} \int_0^\infty \frac{d\omega}{2\pi} \, \mathrm{Im}\, \varepsilon_\lambda^{-1}(\mathbf{q}, \omega) \right] \qquad (14.24)$$

E_0 being the energy of the noninteracting system. This frequently used formula was first derived by Nozières and Pines.[217]

We shall conclude this presentation of general formulas by mentioning some very useful invariants called sum rules, which have to be satisfied by the dielectric function. For the derivation we refer to standard textbooks, a particularly extensive discussion being found in the book by Pines and Nozières.[218] The response to a field of sufficiently high frequency cannot be influenced by the Coulomb interaction between the particles. Therefore it must be the same as that for an ideal gas, i.e.,

$$\lim_{\omega \to \infty} R(\mathbf{q}, \omega) = nq^2/m\omega^2 \qquad (14.25)$$

and we obtain

$$\lim_{\omega \to \infty} \frac{1}{\varepsilon(\mathbf{q}, \omega)} = 1 + \frac{\omega_p^2}{\omega^2} \qquad (14.26)$$

where $\omega_p = (4\pi n e^2/m)^{1/2}$ is the plasma frequency. At very high frequency the response to the effective field must be the same as for free particles and there relation (14.25) must also hold for $P(\mathbf{q}, \omega)$. We therefore have

$$\lim_{\omega \to \infty} \varepsilon(\mathbf{q}, \omega) = 1 - \frac{\omega_p^2}{\omega^2} \tag{14.27}$$

From this equation and the analytic properties of $\varepsilon(\mathbf{q}, \omega)$ (which are the same as for ε^{-1}), one can show that

$$\int_0^\infty d\omega \, \omega \, \mathrm{Im} \, \varepsilon(\mathbf{q}, \omega) = \frac{\pi}{2} \omega_p^2 \tag{14.28}$$

The zero-frequency properties can be related to the hydrodynamics of the system. Two important results follow. The reader who is interested in the proofs will find them elsewhere.[218] The relation

$$\lim_{q \to 0} \frac{1}{\varepsilon(\mathbf{q}, 0)} = 0 \tag{14.29}$$

guarantees that any long-wavelength disturbance is completely screened out. The other relation takes the form

$$\varepsilon(\mathbf{q}, 0) = 1 + \frac{\omega_p^2}{s^2 q^2} \tag{14.30}$$

where s is the velocity of sound. Sum rules are usually expressed in terms of the spectral weight function or, in this case, the dynamic form factor $S(\mathbf{q}, \omega)$. The sum rules just presented can be expressed as conditions on $S(\mathbf{q}, \omega)$ as follows:

$$\left.\begin{aligned}
\int_0^\infty d\omega \, \omega S(\mathbf{q}, \omega) &= nq^2/2m \\[2mm]
\int_0^\infty d\omega \, \frac{S(\mathbf{q}, \omega)}{\omega} |\varepsilon(\mathbf{q}, \omega)|^2 &= \frac{nq^2}{2m} \\[2mm]
\lim_{q \to 0} \int_0^\infty d\omega \, \frac{S(\mathbf{q}, \omega)}{\omega} &= \frac{q}{8\pi e^2} \\[2mm]
\lim_{q \to 0} \int_0^\infty d\omega \, \frac{S(\mathbf{q}, \omega)}{\omega} |\varepsilon(q, \omega)|^2 &= \frac{n}{2ms^2}
\end{aligned}\right\} \tag{14.31}$$

We refer again to the aforementioned book[218] for the derivation and a detailed discussion of these formulas. The first is nothing but the well-known f sum rule. Using the definition of $S(\mathbf{q}, \omega)$ we can perform the integration over ω and write the result in the usual form of the Thomas–Kuhn–Reich sum rule:

$$\sum_n f_n = N$$

where N is the total number of electrons in the system and

$$f_n = \frac{2m}{q^2} \omega_n |\langle n | \rho_{-q} | 0 \rangle|^2 \qquad (14.32)$$

is the generalized oscillator strength.

14.1.3. The Magnetic Susceptibility

Analogous relations to those presented for the dielectric function hold in other situations, a particularly important case being the response to an external magnetic field. We shall content ourselves with a few definitions and refer to the standard textbooks for a detailed discussion. For simplicity we assume the medium to be uniform. A small external field $H(\mathbf{r}, t)$ will give rise to a magnetization

$$M(\mathbf{r}, t) = \int d\mathbf{r}' dt' \chi(\mathbf{r} - \mathbf{r}', t - t') H(\mathbf{r}', t') \qquad (14.33)$$

which defines the magnetic susceptibility $\chi(\mathbf{r} - \mathbf{r}', t - t')$. We have suppressed the vectorial character of H and M in the notation, χ being a tensor of second order in an anisotropic system. Taking the Fourier transform we obtain

$$M(\mathbf{q}, \omega) = \chi(\mathbf{q}, \omega) H(\mathbf{q}, \omega) \qquad (14.34)$$

The retarded susceptibility is obtained from the retarded commutator:

$$\chi(\mathbf{r} - \mathbf{r}', \omega) = i \int_0^\infty dt \, \exp(i\omega t) \langle [M(r, t), M(r, 0)] \rangle \qquad (14.35)$$

In complete analogy with the dielectric response we can also introduce a time-ordered response function, which is convenient to use in connection with diagrammatic perturbation theory. The analytic properties are the same as for the dielectric function. A case of particular interest is that of the spin susceptibility. The interaction energy between the magnetic moment due to the spin and a magnetic field is given by

$$- \boldsymbol{\mu} \cdot \mathbf{H} = g\mu_B \mathbf{S} \cdot \mathbf{H} \tag{14.36}$$

where g is the gyromagnetic ratio. The dynamic spin susceptibility can then be written as

$$\chi_{\alpha\beta}(\mathbf{r}, \mathbf{r}', \omega) = ig^2\mu_B^2 \int dt \exp(-i\omega t) \langle S_\alpha(\mathbf{r}, t) S_\beta(\mathbf{r}', 0) \rangle_r \tag{14.37}$$

where $\langle \ \rangle_r$ means that the retarded commutator is taken. The spin density operator $\mathbf{S}(\mathbf{r}, t)$ is given by the formula ($\hbar = 1$)

$$\mathbf{S}(\mathbf{r}, t) = \tfrac{1}{2}\Psi(\mathbf{r}, t)\boldsymbol{\sigma}\Psi^*(\mathbf{r}, t) \tag{14.38}$$

where Ψ and Ψ^* are two-component vectors and $\boldsymbol{\sigma} = (\sigma_x, \sigma_y, \sigma_z)$ are the Pauli spin matrices. The components of the spin-density operator are conveniently expressed as

$$\left.\begin{aligned}
S^+(\mathbf{r}, t) &= S_x + iS_y = \Psi_\uparrow^*(\mathbf{r}, t)\Psi_\downarrow(\mathbf{r}, t) \\
S^-(\mathbf{r}, t) &= S_x - iS_y = \Psi_\downarrow(\mathbf{r}, t)\Psi_\uparrow(\mathbf{r}, t) \\
S_z(\mathbf{r}, t) &= \tfrac{1}{2}[\Psi_\uparrow^*(\mathbf{r}, t)\Psi_\uparrow(\mathbf{r}, t) - \Psi_\downarrow^*(\mathbf{r}, t)\Psi_\downarrow(\mathbf{r}, t)]
\end{aligned}\right\} \tag{14.39}$$

where $\Psi_\uparrow^*$ and $\Psi_\downarrow^*$ are creation operators for an up or down spin electron, $\Psi_\uparrow$ and $\Psi_\downarrow$ are the corresponding annihilation operators, S^+ describes a spin-flip process in which we flip a down spin into an up spin, and S^- describes the reverse process. The density operator $\rho_\uparrow$ for up-spin electrons is evidently $\Psi_\uparrow^*\Psi_\uparrow$ while $\rho_\downarrow = \Psi_\downarrow^*\Psi_\downarrow$ is the density of down-spin electrons. Thus

$$S_z(\mathbf{r}, t) = \tfrac{1}{2}(\rho_\uparrow - \rho_\downarrow) \tag{14.40}$$

measures the spin density in the system. The total charge density is given by the operator

$$\rho(\mathbf{r}, t) = \Psi_\uparrow^*\Psi_\uparrow + \Psi_\downarrow^*\Psi_\downarrow = \rho_\uparrow + \rho_\downarrow \tag{14.41}$$

These results afford a starting point for discussing spin and charge-density phenomena. We note that the spin-flip operators do not couple to the direct Coulomb interaction and that there are no induced charge densities or induced field associated with them.

14.2. Self-Consistent Independent-Particle Models

The concept of one-electron states forms a fundamental basis in practically all theories of many-electron systems and the notion of independent particle motion is essential in all discussions of many-body assemblies. A key point is that each electron must then be viewed as moving in some average field, which is determined by the motion of all the other electrons. This introduces the concept of self-consistency, which is of central importance in all realistic independent-particle models.

There is a wide variety of different independent particle descriptions and it is outside the scope of this chapter to try to describe them in quantitative detail. Historically the Hartree model is the first. In this model each electron moves in the average electrostatic potential arising from all the other electrons. The Hartree model does not satisfy the Pauli principle requirement of antisymmetry of the total wave function and was soon generalized to include the effect of exchange, characteristic of the antisymmetry requirement. The corresponding theory is the well-known Hartree–Fock method. The potential acting on an electron is the total average electrostatic potential plus the exchange potential. This latter potential describes the repulsion between electrons having parallel spins that arises because of the exclusion principle. With the development of systematic many-body theory it has been possible to include also the effects of correlations between electrons and appropriate potentials can be constructed by diagrammatic or other methods. The interaction between an electron and the rest of the system is described by a nonlocal and energy-dependent quantity called the self-energy. The self-energy is a complex quantity, where the imaginary part is related to the decay of a single-particle state because of the interactions with the rest of the system. Thus the single-particle states are no longer states with a well-defined energy but decaying states, where one conveniently introduces the appropriate spectral weight function.

The Hartree and Hartree–Fock theories are generally used to describe the ground state of the system. However, the same approach can equally well be applied to time-dependent situations to consider either the response of the system to an external perturbation, or the free oscillations of the system in an excited state. By studying this time-dependent problem, some insight can be gained into the classical aspects of dynamic correlations. One can thereby see how collective modes can arise and also how to calculate in an approximate way the dielectric response function. The time-dependent Hartree theory is equivalent to the well-known random-phase approximation (RPA) and the time-dependent Hartree–Fock theory yields the random-phase approximation with exchange (RPAE).

An alternative approach to the correlation problem, density functional theory, developed by Kohn and co-workers in the midsixties[219,220] (a general presentation is given elsewhere[221]), has already been referred to in Chapter 12. The essential point for present purposes is that the ground-state properties of an N-particle system can be expressed in terms of exactly N one-electron orbitals where each satisfies a Schrödinger-like one-electron wave equation, in which all exchange and correlation contributions appear through a local, energy-independent exchange-correlation potential (which can also depend on spin). This gives a tremendous simplification in the explicit calculation of the physical properties and applications made to date have been extremely successful. But exact knowledge of the exchange-correlation potential, of course, would require an exact solution of the many-body problem, which is currently not feasible.

As the above single-particle methods are widely discussed in the literature we shall not give the basic equations here, but rather turn immediately to approximate dielectric functions of the electron liquids. In the context of this book, there will be considerable interest in the role of Coulomb interactions in d-dimensions, d being 1, 2, or 3.

14.3. The Electron Liquid

The electron liquid and the jellium model have been extensively studied to demonstrate the effects of electron correlation. In the jellium model, the electrons move in a continuous distribution of charge obtained by smearing out the ions. The jellium is in equilibrium at a density close to that of sodium metal.

The noninteracting electron gas is described below in three dimensions, to be definite. The generalization to d-dimensions is straightforward. In the paramagnetic phase, the ground state is obtained by filling all electron states up to a maximum momentum k_F, the Fermi momentum, which defines a Fermi surface in k-space. For the three-dimensional electron gas the Fermi surface is a sphere, in two dimensions it becomes a circle, and in one dimension it degenerates to the two points $\pm k_F$.

The only parameter needed to characterize the electron gas is the electron density. One often characterizes the density by means of a dimensionless parameter r_s, which is the radius (in units of the Bohr radius a_0) of a sphere that an electron occupies on average:

$$\frac{4\pi (r_s a_0)^3}{3} = \frac{V}{N} = \rho^{-1} \tag{14.42}$$

Since the density is related to the Fermi momentum k_F through $\rho = k_F^2/3\pi^2$

we have the relation

$$k_F = \frac{1}{\alpha r_s a_0} \; ; \qquad \alpha = (4/9\pi)^{1/3} = 0.521 \qquad (14.43)$$

Except in the surface layer, the sum of the charge densities of the electrons and of the uniform background is zero. Therefore the electrostatic potential is constant in the interior. In the Hartree model, we have only the average electrostatic interaction. The one-electron energy measured from the average potential is simply

$$\mathscr{E}_k = \frac{h^2 k^2}{2m} \qquad (14.44)$$

The average energy per electron is $\frac{3}{5}\mathscr{E}_F$, where $\mathscr{E}_F = h^2 k_F^2/2m = 1/(\alpha r_s)^2$.

14.3.1. Approximate Dielectric Functions

There is a vast literature with different methods and approximate results for an electron liquid. We shall not enter into a detailed discussion but only mention some of the most important developments. Some physical properties depend only on the gross features of the dielectric function while other properties, particularly screening, are quite sensitive to details. The general formulas for the dielectric function were collected in Section 14.1.2.

14.3.1.1. The Lindhard Dielectric Function (RPA)

The dielectric function is given by equation (14.13), where $P(q, \omega)$ is seen from equation (14.12) to be the response function to the effective field. Lindhard, in a famous paper,[222] treated the response of an electron gas to an external electromagnetic field. He used the time-dependent Hartree approach, briefly referred to in Section 14.2. This is equivalent to the calculation of $P(q, \omega)$ using first-order perturbation theory, or, alternatively, to the calculation of the Fourier transform of the retarded density–density commutator for the noninteracting electron gas:

$$P_0(q, \omega) = -2 \sum_k \frac{n^0(\mathbf{k}+\mathbf{q}) - n^0(\mathbf{k})}{\omega - \mathscr{E}(\mathbf{k}+\mathbf{q}) + \mathscr{E}(\mathbf{k}) + i\delta} \qquad (14.45)$$

n^0 denoting the Fermi factors for the noninteracting system. For the dielectric function itself we obtain

$$\varepsilon(q, \omega) = 1 + \frac{8\pi e^2}{q^2} \sum_k \frac{n^0(\mathbf{k}+\mathbf{q}) - n^0(\mathbf{k})}{\omega - \mathscr{E}(\mathbf{k}+\mathbf{q}) + \mathscr{E}(\mathbf{k}) + i\delta} \qquad (14.46)$$

The integral has been performed analytically, but the result is a rather lengthy expression.[222,223]

The Lindhard formula describes the gross properties of the system quite well. It is less accurate at shorter wavelengths. The pair correlation function $g(r)$ takes on large negative values at small distances, which reflects the inaccuracy in the response function at intermediate and large values of q. It is to be expected that the Hartree approximation fails to describe things well at distances shorter than the average distance between particles. At such distances, the average-field description becomes insufficient and exchange and correlation effects play an important role.

The function $P_0(q, \omega)$ determines the screening properties of the system. In the static case and long-wavelength limit, the screening is the same as in the Thomas–Fermi approach and one obtains the limiting value

$$\lim_{q \to 0} \varepsilon(q, 0) = 1 + k_{\text{TF}}^2/q^2 \qquad (14.47)$$

where $k_{\text{TF}} = 6\pi m e^2/\mathscr{E}_F$ is the Thomas–Fermi screening wave vector. We note in passing that equation (14.47) violates the exact relation usually referred to as the compressibility sum rule:

$$\lim_{q \to 0} \varepsilon(q, 0) = 1 + \frac{k_{\text{TF}}^2}{q^2} \frac{K}{K_0} \qquad (14.48)$$

where K is the compressibility of the interacting electron assembly while K_0 is that of the noninteracting gas.

In the static case, $P^0(q, 0)$ from equation (14.45) has denominator $\mathscr{E}(\mathbf{k} + \mathbf{q}) - \mathscr{E}(\mathbf{k})$ and of particular interest in its calculation are the terms for which this energy denominator vanishes. Since one of the terms must represent an occupied state and the other an empty state, it follows that both $\mathbf{k}$ and $\mathbf{k} + \mathbf{q}$ must be at the Fermi surface. In the three-dimensional electron liquid, such contributions occur from $q = 0$ up to the limiting value $q = 2k_F$; for larger values of q the energy denominator is always nonzero. As a result there will be a logarithmic infinity in the derivative of $\varepsilon(q, 0)$ at $q = 2k_F$. This kink in $\varepsilon(q, 0)$ gives rise to the so-called Kohn anomalies in the phonon dispersion curves of metals, and to the Friedel oscillations in the spatial screening around a point charge.

Similar considerations apply also in the two- and one-dimensional electron liquid, but the singular behavior becomes enhanced because of the reduced dimensionality. In two dimensions, the response function has a cusp and a discontinuous derivative at $q = 2k_F$. In one dimension, we obtain a logarithmic singularity when $q \to 2k_F$. It is convenient to represent the q-dependence of the screening by setting

$$\varepsilon(q) = 1 + Q(q)/q^2 \qquad (14.49)$$

where $Q(q) = 4\pi e^2 P(q, 0)$. We can refer to $Q(q)$ as the screening function, describing how the screening reduces from the Thomas–Fermi screening with increasing q. As an illustration of the behavior mentioned above, Figure 14.1 shows the screening function $Q(q)$ in the Lindhard approximation for one-, two-, and three-dimensional cases. It should be noted that the singular behavior in one dimension is very sensitive to temperature and is smeared out at finite T when it is replaced by a structure with a pronounced hump. The three-dimensional case exhibits a continuous behavior in the screening properties, while the singular properties become more and more pronounced with reduced dimensionality. It is a very general feature that the effects of correlations become much more pronounced in lower dimensions. This also puts much higher demands on the accuracy of the methods of solution and explains why so much effort has been expended on finding exact solutions to certain one-dimensional models.

14.3.1.2. Random Phase Approximation with Exchange (RPAE): The Hubbard Modification

A serious oversimplification of the Lindhard approximation is that each electron is assumed to respond to the total electrostatic potential. If we take into account the exclusion principle by using time-dependent

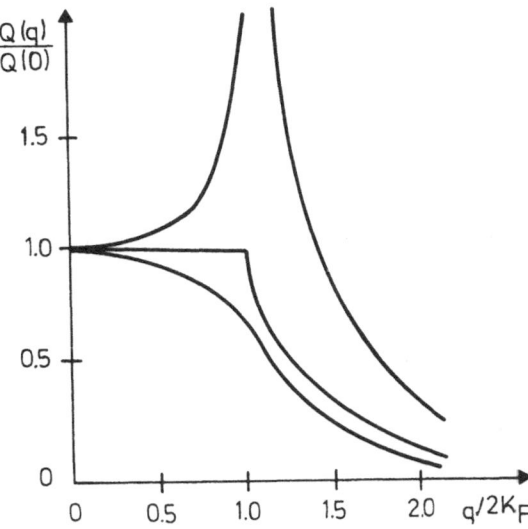

Figure 14.1. Screening function $Q(q)$ in the Lindhard approximation for one-, two-, and three-dimensional electron gas.

Hartree–Fock equations or some equivalent formalism, we include the important effect that each electron is surrounded by an exchange or Fermi hole. This is a local depletion in the charge density around each electron in the distribution of electrons with parallel spins. In more accurate theories, where we include correlations explicitly in the description, we obtain the corresponding effect, an exchange-correlation hole. This changes the total effective field acting on an electron. The depletion in the charge density around the electron gives rise to what one refers to as a local-field correction. There are many different ways of incorporating such effects and we shall only exemplify the concept in this section.

In the work of Hubbard,[224] he used a suitable average in the exchange term of Hartree–Fock theory that yielded a new approximation to $P(q, \omega)$:

$$P_H(q, \omega) = \frac{P_0(q, \omega)}{1 + (4\pi e^2/q^2)G(q)P_0(q, \omega)} \tag{14.50}$$

where $G(q)$ is defined as

$$G(q) = \tfrac{1}{2}v_s(q)/v(q) \tag{14.51}$$

In equation (14.51), $v(q)$ represents the bare Coulomb potential, but the averaging referred to above implies that the average potential in the theory becomes a short-range interaction, which we can represent approximately by the formula

$$v_s(q) = \frac{4\pi e^2}{q^2 + q_s^2} \tag{15.52}$$

where q_s is a screening constant of the order of k_F or k_{TF}. Thus the range of $v_s(q)$ will be typically of the order 1 Å in a metal. One can refer to $G(q)$ as the local-field correction factor. The actual local field itself, which modifies the potential that an electron sees, is given by

$$V_{\text{local field}} = -\frac{4\pi e^2}{q^2} G(q) \tag{14.53}$$

showing the local field itself to be short-range.

14.3.1.3. Self-Consistent Approach Including Local-Field Corrections

The available, formal many-body techniques lead to a formidable problem to include local-field effects beyond a simplified time-dependent

Hartree–Fock theory as just outlined above. A significant step forward was taken by Singwi *et al.*[225] (see also the review paper by Tosi[226] and the more recent one by Singwi and Tosi[227]). A similar approach was outlined by Hubbard.[228] The key idea in this development was that of local-field corrections as introduced in the previous section, but now extended to the general case of an exchange-correlation hole and connected to a self-consistent procedure.

We write the potential felt by an electron from the induced charges modified by the exchange-correlation hole as

$$V_{pol}(q, \omega) = \frac{4\pi e^2}{q^2} [1 - G(q)]\langle \rho(q, \omega)\rangle = U(q)\langle \rho(q, \omega)\rangle \quad (14.54)$$

The local reduction in density around an electron is described by the pair correlation function $g(\mathbf{r} - \mathbf{r}')$. We can write the total polarization field from the induced charges as

$$\mathbf{E}_{pol} = -\nabla_r \int U(\mathbf{r} - \mathbf{r}')\langle \rho(\mathbf{r}')\rangle d\mathbf{r}'$$

$$= -\int g(\mathbf{r} - \mathbf{r}')\nabla_r V(\mathbf{r} - \mathbf{r}')\langle \rho(\mathbf{r}')\rangle d\mathbf{r}' \quad (14.55)$$

which defines the effective interaction $U(\mathbf{r} - \mathbf{r}')$ in terms of $g(\mathbf{r} - \mathbf{r}')$ and the bare Coulomb interaction $v(\mathbf{r} - \mathbf{r}')$. Going back to q-space this gives the formula for the local-field correction factor in the form

$$G(q) = 1 - \int \frac{d\mathbf{q}'}{(2\pi)^3} g(\mathbf{q} - \mathbf{q}') \frac{\mathbf{q} \cdot \mathbf{q}'}{|q'|^2} \quad (14.56)$$

where $g(k)$ is the Fourier transform of $g(r)$.

The total field felt by an electron is $V_{ext}(q, \omega) + V_{pol}(q, \omega)$. We calculate the induced density due to this response, and calculate the response function using first-order perturbation theory for a noninteracting electron gas. This leads to the dielectric constant in the form

$$e(q, \omega) = 1 - \frac{4\pi e^2}{q^2} \frac{P_0(q, \omega)}{1 + (4\pi e^2/q^2)G(q)P_0(q, \omega)} \quad (14.57)$$

which is of the same form as the one derived by Hubbard.[224] However, the important difference is that we have an expression for $G(q)$ in terms of the pair correlation function. Thus, the dielectric function will depend on the pair distribution function $g(r)$. However, $g(r)$ is related to the density–density correlation at equal times and can, referring to the discussion in Section 14.2, be expressed as an integral $\int d\omega(1/\varepsilon - 1)$. Therefore a self-consistent procedure can be set up. The simplest approximation is obtained by assuming that $g(r)$ is given by the Fermi hole due to exchange only. This choice will give essentially the approximation discussed in the previous section. Singwi and co-workers have made extensive self-consistent calculations using various modifications of the approach briefly outlined above. The results are generally in excellent agreement with experiment. However, it is clear that the notion of a static exchange-correlation hole has its limited validity and work has been done to include dynamic effects.

14.3.1.4. Static Dielectric Screening in the Density Functional Scheme

We shall briefly summarize now how the ideas referred to in the previous section apply to the density functional approach. Essentially, we must replace the electrostatic field $V(q)$ in the Hartree scheme by the potential $V(q) + \mu_{xc}(q)$ where $\mu_{xc}(q)$ is the exchange-correlation contribution to the chemical potential of the uniform electron gas. In this way we include a local-field correction in analogy with the discussion in the previous section, the result for $\varepsilon(q)$ being

$$\varepsilon(q) = 1 - \frac{v(q)P_0(q)}{1 + v(q)G(q)P_0(q)} \tag{14.58}$$

where $G(q)$ is now given by

$$G(q) = -\frac{1}{v(q)}\frac{d\mu_{xc}}{d\rho} = \tfrac{1}{4}\gamma q^2 \tag{14.59}$$

The theory outlined here was developed by Hedin and B. Lundqvist.[229] Their paper contains tables of γ for an interacting electron liquid. We note that the actual local field is given by

$$v(q)G(q) = \pi e^2 \gamma \tag{14.60}$$

i.e., the field is constant in q-space. This is closely connected with the fact that the local-density approximation employed just represents an average over the local depletion in the charge density around the electron.

14.3.2. Collective Modes and Screening Effects

We recall next that the imaginary part of a response function or the equivalent to the corresponding spectral weight function describes the creation of real excitations of the system. In the dielectric response $\mathrm{Im}\, 1/\varepsilon$ describes the density fluctuation spectrum of the system.

The long-wavelength response of a three-dimensional electron liquid is dominated by a strong resonance at the plasma frequency ω_p. The system will oscillate with this frequency without a driving external field. The restoring force is given by the internal field set up by the disturbance in the charge distribution. At shorter wavelengths, the plasma oscillations show disperison, i.e., their frequency depends on the wave number q. The condition for having a plasma oscillation is given by $\varepsilon(q, \omega) = 0$, which determines the dispersion relation $\omega_p = \omega_p(q)$ of the collective oscillation. At not too short wavelengths, the dispersion relation can be expanded as follows:

$$\omega_p(q) = \omega_p + \alpha q^2 + \cdots \qquad (14.61)$$

where the dispersion coefficient is given by $\alpha = 3k_F^2/10\, m^2 \omega_p$ in the Lindhard approximation (RPA).

Besides the collective oscillations, we have also a continuous spectrum of screened particle-hole excitations. For wave numbers larger than a characteristic q_c, the plasma modes will be degenerate with the particle-hole excitations and a considerable damping of the plasmons will set in. The spectrum of excitations in the RPA is shown schematically in Figure 14.2. At small and moderate q the plasmons will nearly exhaust the sum

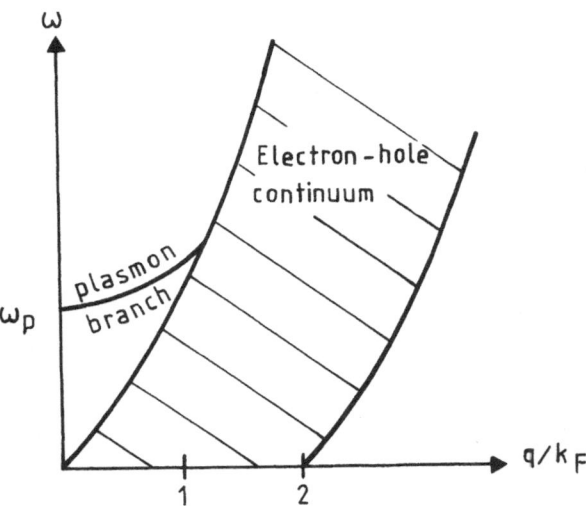

Figure 14.2. Spectrum of excitations in the RPA.

rule

$$\int_0^\infty d\omega \, \omega \, \mathrm{Im} \, \varepsilon^{-1}(q, \omega) = -\frac{\pi}{2} \, \omega_p^2 \tag{14.62}$$

The qualitative behavior of $\mathrm{Im} \, \varepsilon^{-1}(q, \omega)$ appears as in Figure 14.3. The energy to excite plasmons is quite high in free-electron metals (5–15 eV). In doped semiconductors the plasma frequency is a small fraction of an eV and may mix strongly with longitudinal phonons.

14.3.3. Comparison between Different Approximations to the Dielectric Function

In Section 14.3.1 we described briefly three different theoretical approaches to deal with correlations in the dielectric properties. Below we shall merely give a brief discussion of some of the numerical results of these approaches.

We first consider the correlation energy $\mathscr{E}_c$, defined as the difference between the total energy and the Hartree–Fock energy (given per electron in Rydbergs). We regard $\mathscr{E}_c$ as a function of the dimensionless parameter r_s, and curves L, H, and S in Figure 14.4 represent, respectively, the approaches by Lindhard, Hubbard, and Singwi with co-workers. There are also many other alternatives and interpolation formulas, of which we mention a frequently used formula given by Pines and Nozières.[218] The general conclusion is that the various approximations beyond the RPA give results that are consistent within 10^{-2} Rydbergs over the range of metallic densities.

Results for the pair correlation function are similarly given in Figure 14.5. These curves show that the approach by Singwi and co-workers

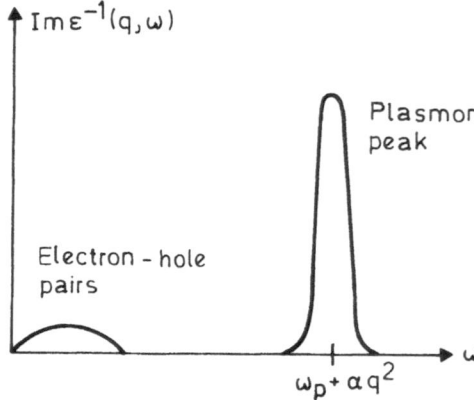

Figure 14.3. Typical behavior of $\mathrm{Im} \, \varepsilon^{-1}(q, \omega)$ for a small q. The energy to excite plasmons is quite high in free-electron metals (5–15 eV). In doped semiconductors the plasma frequency is a small fraction of an eV and may mix strongly with longitudinal phonons.

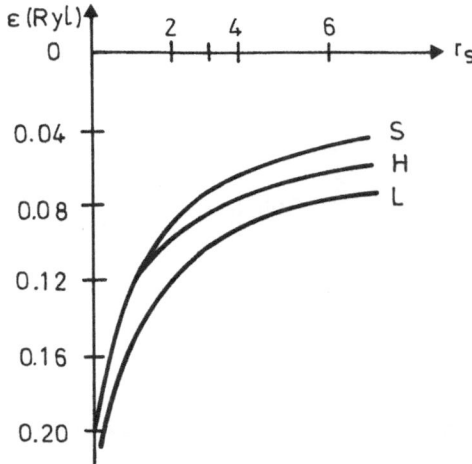

Figure 14.4. Some different results for the correlation energy.

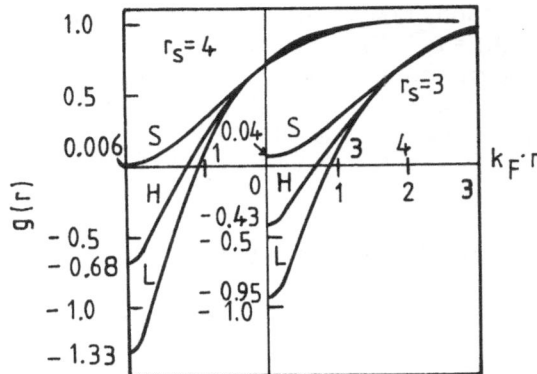

Figure 14.5. The pair correlation function $g(r)$ for $r_s = 3$ and $r_s = 4$. S = Singwi *et al.*, H = Hubbard, L = Lindhard (after).

represents a dramatic improvement compared with H and L, being positive for all r.

As a final example we show in Figure 14.6 the screening of a fixed point charge. The overall features are very similar in the three cases. The Friedel oscillations,[230] deriving from the singular nature of the screening at $q = 2k_F$, are clearly shown in the inset to Figure 14.6.

14.3.4. The One-Electron Spectrum

The common picture of one-electron spectra in solids is obtained from energy-band calculations, in which the Schrödinger equation is solved

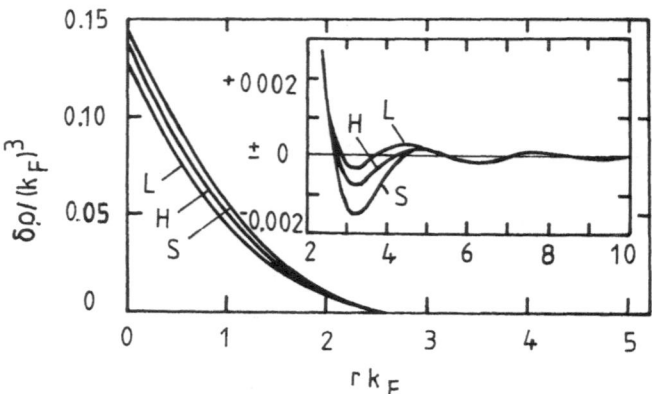

Figure 14.6. Screening charge distribution round a point charge (after Hedin–Lundqvist[223]).

using some suitable average potential. This gives the electron energy as a function of the wave vector, $\mathscr{E}(\mathbf{k})$. To transcend the Hartree approximation $\mathscr{E}(\mathbf{k}) = \hbar^2 k^2 / 2m$, relative to the average electrostatic potential, the step referred to in Section 14.2 is to introduce the nonlocal energy-dependent potential called the self-energy, $\Sigma(\mathbf{k}, E)$. To find the electron states we have now to solve

$$E = \mathscr{E}(\mathbf{k}) + \Sigma(\mathbf{k}, E) \qquad (14.63)$$

The Green function is given by the formula

$$G(k, E) = \frac{1}{E - \mathscr{E}(\mathbf{k}) - \Sigma(\mathbf{k}, E)} \qquad (14.64)$$

and the spectral function by

$$A(\mathbf{k}, E) = \frac{1}{\pi} |\mathrm{Im}\ G(\mathbf{k}, E)| \qquad (14.65)$$

Because of the imaginary part of $\Sigma(\mathbf{k}, E)$, we have now one-electron states that decay and we talk about quasi-particles and quasi-electrons. The real part of Σ will shift the energy so that $\mathscr{E}(\mathbf{k}) \to \tilde{\mathscr{E}}(\mathbf{k})$. For quasi-particle states of long lifetime, we may have poles in $G(\mathscr{E})$ close to the real axis. In the neighborhood of the pole we may write

$$G(\mathscr{E}) = \frac{Z}{\mathscr{E} - \mathscr{E}_*} + \phi(\mathscr{E}) \qquad (14.66)$$

where $\mathscr{E}^* = \tilde{\mathscr{E}} + i\Gamma$, $Z = Z_1 + iZ_2$, and $\phi(\mathscr{E})$ gives a smooth background. The damping term Γ is very small, close to the Fermi level, and goes to zero when $\tilde{\mathscr{E}} \to \mathscr{E}_F$. For a further discussion of the quasi-particle properties, the reader may consult elsewhere.[230]

If, by way of illustration, we take the Hartree–Fock case, the self-energy is given by the formula

$$\Sigma_{HF}(\mathbf{k}) = -\int \frac{d\mathbf{k}'}{(2\pi)^3} \frac{4\pi e^2}{|\mathbf{k} - \mathbf{k}'|^2} = -\frac{e^2}{2\pi}\left(\frac{k_F^2 - k^2}{k} \ln\left|\frac{k + k_F}{k - k_F}\right| + 2k_F\right) \quad (14.67)$$

which is plotted schematically in Figure 14.7. There is substantial variation in the exchange energy with k. In particular there is an infinite slope at $k = k_F$, which implies rather unpleasant properties, such as that the density of states $N(\mathscr{E}) \to 0$ when $\mathscr{E} \to \mathscr{E}_F$. This makes the Hartree–Fock theory quite unacceptable as a theory for metals. These properties of the Hartree–Fock exchange have to do with the long range of the Coulomb interaction in equation (14.67). Going further to include correlations, the Coulomb interaction will be strongly screened and the self-energy from exchange will behave in a completely satisfactory manner. While discussing the Hartree–Fock self-energy, we can easily calculate the total exchange energy by summing the Coulomb interaction over all pairs of electrons with parallel spins, when we find the exchange energy per electron as $0.916/r_s$ Rydberg. We have already mentioned that the Coulomb interaction energy

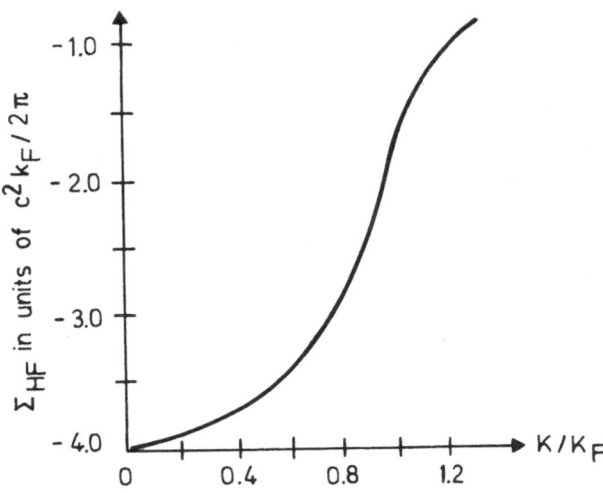

Figure 14.7. $\Sigma_{HF}(k)$ for an electron gas.

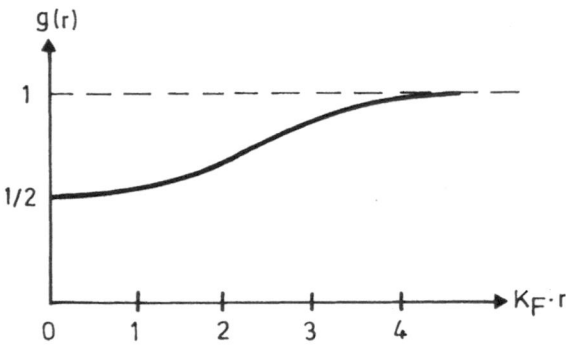

Figure 14.8. Function $g(r)$ in the Hartree–Fock approximation.

can be calculated from the pair distribution function $g(r)$, and we merely state the result[230]:

$$g(r) = 1 - F^2(k_F r) \tag{14.68}$$

where

$$F(x) = 3\, \frac{\sin x - x \cos x}{x^3} \tag{14.69}$$

This is illustrated in Figure 14.8. The density of electrons having the same spin is reduced in the vicinity of the electron, while the electrons of opposite spin are uniformly distributed. One says that the electron is surrounded by an exchange hole with total charge $+e$.

In order to avoid the difficulties with the Hartree–Fock exchange, one has frequently introduced the concept of screened exchange. In fact, there will also be a Coulomb hole due to the actual Coulomb repulsions, as seen in Figure 14.5, as well as the exchange hole that is in fact of the same structure for all Fermion systems irrespective of the nature of the interaction. Hedin,[231] using a diagrammatic approach, has worked out classically the self-energy for a charged heavy particle passing through the electron gas. Using in his formulation a plasmon-pole approximation, one obtains the self-energy as

$$\Sigma(\mathbf{k}, \mathscr{E}) = -\int \frac{d\mathbf{q}}{(2\pi)^3} \frac{v(\mathbf{q}) n^0(\mathbf{k}+\mathbf{q})}{\varepsilon(\mathbf{q}, \mathscr{E}(\mathbf{k}+\mathbf{q}) - \mathscr{E}(\mathbf{k}))}$$
$$+ \omega_p^2 \int \frac{d\mathbf{q}}{(2\pi)^3} \frac{v(\mathbf{q})}{\omega(\mathbf{q})} \frac{1}{\mathscr{E} - \mathscr{E}(\mathbf{k}+\mathbf{q}) - \omega} \tag{14.70}$$

The first term represents an energy-dependent screened exchange potential while the second term describes the Coulomb hole.

Equation (14.70), as well as all more elaborate formulas, gives rise to a rather complex spectrum. The spectrum usually shows a typical quasi-particle peak with a strength corresponding to 50–60% of the total strength. There are broad wings due to particle-hole excitations and there is usually a characteristic satellite structure due to plasmon excitations coupled to the one-electron motion. We cite an earlier reference[223] and other references in that review for a more extensive discussion of the one-electron spectrum.

14.3.5. Effect of Correlations on Spin Susceptibility

Let us start by getting a qualitative understanding of the role of electron–electron interactions for the static susceptibility, using reasoning first developed by Landau in the theory of a Fermi liquid.

Hartree–Fock theory, to be explicit, has a one-electron energy

$$\mathcal{E}(\mathbf{k}) = \frac{\hbar^2 k^2}{2n} - \int \frac{d\mathbf{k}'}{(2\pi)^3} \frac{4\pi e^2}{|\mathbf{k} - \mathbf{k}'|^2} \tag{14.71}$$

We shall here be concerned only with processes in which we change the spin of an electron from a down-spin to an up-spin, or *vice versa*. We know from the previous section that screening of the exchange is important. For this calculation we only need the integral over momentum states in a narrow shell round the Fermi surface. For simplicity we account for the exchange interaction by considering an average interaction $\langle v \rangle$.

If a magnetic field is applied, we might first imagine that the field raises the energy of the up-spins and lowers the energy of the down-spins, as shown in Figure 14.9. The distribution in Figure 14.9a cannot be stable; the electrons will adjust so that the chemical potential of up-spin and down-spin electrons will be the same. This results in a decrease in the number of

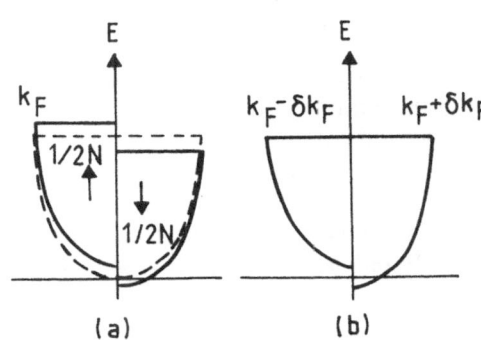

Figure 14.9. Distribution of electrons in the presence of a magnetic field.

up-spins and an equally large increase in the number of down-spin electrons. Therefore we shall obtain a magnetic moment equal to

$$M = - g\mu_B[+\tfrac{1}{2}(- \delta N) - \tfrac{1}{2}(+ \delta N)] = g\mu_B \delta N \qquad (14.72)$$

The change in energy of an up-spin electron at the Fermi level is given by

$$\delta E_{k_F - \delta k_F \uparrow} = g\mu_B \tfrac{1}{2} H - \frac{\partial E}{\partial k} \delta k_F + \delta N \langle v \rangle \qquad (14.73)$$

The first term gives the interaction with the external field, while the second gives the reduction in energy due to the change of the Fermi momentum from k_F to $k_F - \delta k_F$. The last term is the change in the exchange interaction due to the decrease $- \delta N$ in the number of spin-up electrons. For electrons with down-spins we obtain by the same argument

$$\delta E_{k_F + \delta k_F \downarrow} = - g\mu_B \tfrac{1}{2} H + \frac{\partial E}{\partial k} \delta k_F - \delta N \langle v \rangle \qquad (14.74)$$

The last two formulas should give the chemical potential, so putting them equal and using the formula

$$\frac{\partial E}{\partial k} \delta k_F = \frac{\partial E}{\partial k} \frac{\partial k_F}{\partial N} \delta N = \frac{\delta N}{N_0} \qquad (14.75)$$

we obtain

$$2\delta N \left(\frac{1}{N_0} - \langle v \rangle \right) = g\mu_B H \qquad (14.76)$$

and finally the spin susceptibility χ in the form

$$\chi = \frac{M}{H} = \frac{g\mu_B \delta N}{H} = \chi_P \frac{1}{1 - N_0 \langle v \rangle} \qquad (14.77)$$

where χ_P is the Pauli susceptibility $\chi_P = \tfrac{1}{2} g\mu_B N_0$. We note that $\langle v \rangle$ is positive and therefore we have an enhancement of the susceptibility that is usually 20–40% above the Pauli value. Thus the Coulomb interaction modifies the static susceptibility and we point out again that it is the exchange part of the Coulomb interaction that gives the enhancement.

We next consider the frequency-dependent susceptibility of an electron liquid. For a uniform system we have

$$\chi_{xx} = \chi_{yy} = \chi_{zz} \qquad (14.78)$$

and that all cross-terms are zero. This can be used to express the susceptibility in terms of the spin-flip operators S^+ and S^- introduced in equations (14.39) and we find

$$\chi(\mathbf{q}, \omega) = \frac{i}{2} g^2 \mu_B^2 \int dt \exp(-i\omega t) \langle S^+(\mathbf{q}, t) S^-(-\mathbf{q}, 0) \rangle \quad (14.79)$$

We note that there is a fundamental difference in comparison with the dielectric response. The spin-flip operators will not couple to the direct Coulomb field and there are no induced charge densities or induced fields.

The spin-flip operators S^+ and S^- are expressed in terms of creation and annihilation operators for one-electron states through the formulas

$$\left. \begin{array}{l} S^+(\mathbf{q}) = \sum_{\mathbf{k}} a_{\uparrow}^*(\mathbf{k} + \mathbf{q}) a_{\downarrow}(\mathbf{k}) \\[2mm] S^-(\mathbf{q}) = \sum_{\mathbf{k}} a_{\downarrow}^*(\mathbf{k} + \mathbf{q}) a_{\uparrow}(\mathbf{k}) \end{array} \right\} \quad (14.80)$$

Except for the spin flip, these formulas are the same as for the density-fluctuation operator $\rho(\mathbf{q})$:

$$\rho(\mathbf{q}) = \int d\mathbf{r}\, \Psi^*(\mathbf{r}) \exp(i\mathbf{q} \cdot \mathbf{r}) \Psi(\mathbf{r}) = \sum_{\mathbf{k}} a^*(\mathbf{k} + \mathbf{q}) a(\mathbf{k}) \quad (14.81)$$

When the external field only acts on the spins, we have no net change in the charge distribution, i.e., $\langle \rho_{\uparrow} + \rho_{\downarrow} \rangle = 0$. Thus there is no net induced charge density but only a spin polarization.

We first calculate the susceptibility $\chi_0(\mathbf{q}, \omega)$ in the independent particle approximation. Applying formulas (14.80) to the Fermi sea, one obtains in units of $g^2 \mu_B^2$

$$\chi_0(\mathbf{q}, \omega) = \sum_{\mathbf{k}} \frac{n^0(\mathbf{k}) - n^0(\mathbf{k} + \mathbf{q})}{\mathscr{E}(\mathbf{k} + \mathbf{q}) - \mathscr{E}(\mathbf{k}) - \omega} \quad (14.82)$$

Except for a factor of 2 due to spin summation, $\chi_0(\mathbf{q}, \omega)$ is the same as the zero-order free-electron polarizability $P_0(\mathbf{q}, \omega)$. Specializing to $\omega = 0$, we see immediately that the static susceptibility shows the same singularity at $q = 2k_F$ as the formula for the dielectric function. Equation (14.82) with $\omega = 0$ is equivalent to the Ruderman–Kittel–Yoshida formula, which gives the interaction between localized spins via the conduction electrons. Without going into details, we note that there is spin polarization of the conduction electrons round a localized spin of oscillatory character akin to the Friedel oscillations in the charge density shown in Figure 14.6.

In the RPA we have that density fluctuations couple via the Coulomb interaction. In the case of a magnetic interaction we have that the particle and the hole will be coupled via a screened or unscreened Coulomb interaction. All these diagrams will be beyond the RPA. The first-order approach is to attempt to include the particle-hole interaction in the way we discussed, following Hubbard,[224] the effect of exchange in the case of dielectric response. Using the same procedure in the present case one finds

$$\chi(\mathbf{q}, \omega) = -g^2 \mu_B^2 \frac{\chi_0(\mathbf{q}, \omega)}{1 - I(\mathbf{q})\chi_0(\mathbf{q}, \omega)} \tag{14.83}$$

The average exchange interaction in the first paper by Hubbard just included an average over the Fermi sea and suggested the choice

$$I(q) = -\tfrac{1}{2} \frac{4\pi e^2}{q^2 + k_F^2} \tag{14.84}$$

In later work by Hubbard supported by, for example, the work of Hedin[231] he suggested replacing k_F^2 in equation (14.84) by $k_F^2 + q_s^2$, where q_s is an appropriate screening wave number.

It will be clear to the reader that the theory just sketched has many features in common with the treatment of dielectric screening. The method outlined here was first given by Wolff.[232] What we were discussing by his method was the inclusion of exchange in the dynamic response. We can continue the discussion in analogy with how we proceeded in the case of the dielectric response. In that case we introduced the function $G(q)$ to account for the local fields due to exchange and correlation. When applying the same idea to the spin response, we are no longer intereasted in the pair correlation function but rather in the spin correlation function round a given spin (e.g., ↑) defined as

$$\tilde{g}(r) = \tfrac{1}{2}[g_{\uparrow\uparrow}(r) - g_{\uparrow\downarrow}(r)] \tag{14.85}$$

The corresponding structure factor is given by

$$\tilde{S}(\mathbf{q}) = 1 + n \int d\mathbf{r}\tilde{g}(r)\exp(-i\mathbf{q}.\mathbf{r}) \tag{14.86}$$

In the Hubbard approximation, the spin correlation is due to exchange alone and therefore only parallel spins correlate. Going beyond that, we can proceed as in the theory of Singwi and co-workers. The function $I(q)$ is related to the form factor $S(q)$ by

$$I(q) \sim \int \frac{d\mathbf{q}'}{(2\pi)^3} \frac{\mathbf{q}.\mathbf{q}'}{q'^2} [\tilde{S}(\mathbf{q}-\mathbf{q}') - 1] \tag{14.87}$$

The form factor $\bar{S}(q)$, however, is in turn related to the spin susceptibility through the exact formula

$$\bar{S}(q) = \frac{1}{\pi n g^2 \mu_B^2} \int_0^\infty d\omega \operatorname{Im} \chi(\mathbf{q}, \omega) \tag{14.88}$$

These formulas impose a self-consistency condition on $\bar{S}(\mathbf{q})$ and $\chi(\mathbf{q}, \omega)$. The equations have been solved by Lobo *et al.*[233,234] We mention here that the interaction parameter I can also be obtained from the density functional approach.[235]

14.3.6. Different Types of Ground States: Ferromagnetic, Wigner Lattice, Spin Density Waves, and Charge Density Waves

Traditionally, it was believed that an electron liquid can have either a paramagnetic or a ferromagnetic Hartree–Fock ground state depending on the density. The Hartree–Fock energy per electron in the paramagnetic state is obtained by adding the exchange energy given above to the Fermi energy. This yields

$$\mathscr{E}_p = \frac{2.21}{r_s^2} - \frac{0.916}{r_s} \tag{14.89}$$

in Rydbergs.

The ferromagnetic state of an electron liquid having the same density has a Fermi sphere occupied with only one spin direction and a radius therefore equal to $2^{1/3} k_F$. The Hartree–Fock energy per electron is now

$$\mathscr{E}_f = 2^{2/3} \frac{2.21}{r_s^2} - 2^{1/3} \frac{0.916}{r_s} \tag{14.90}$$

Consequently a transition from the paramagnetic to the ferromagnetic state should take place at $r_s = 5.45$. This criterion was first derived by Bloch.[236] The Hartree–Fock exchange with the unscreened Coulomb interaction gives too much weight to the tendency to become ferromagnetic. We know now from the Monte-Carlo calculations of Ceperley and Alder[237] that the ferromagnetic ground state only lies below the paramagnetic ground state for $r_s \gtrsim 80$.

In the high-density limit the kinetic energy dominates the potential energy and the Hartree theory becomes exact as $r_s \to 0$. At the opposite extreme of very low density, the electrostatic interaction dominates and the electrons tend to condense on a Wigner lattice.[238] The electrons will oscillate around their equilibrium lattice sites and one obtains to leading order the formula[239,240]

$$\mathscr{E}_w = \left(-\frac{1.792}{r_s} + \frac{2.65}{r_s^{3/2}} - \frac{0.73}{r_s^2} + \cdots \right) \text{ Rydbergs} \tag{14.91}$$

The transition to the Wigner lattice is found by Ceperley and Alder[237] to take place at $r_s \sim 100$.

Besides the ferromagnetic state and the Wigner lattice, there are other types of states that are of considerable current interest. These are the spin-density waves (SDW) and the charge-density waves (CDW), which were first proposed by Overhauser.[241]

Let us consider first the Hartree–Fock theory. The conventional solution for an electron gas in a box of volume V consists of doubly occupied plane waves filling the Fermi sphere:

$$\left. \begin{array}{l} \phi_{\mathbf{k}\uparrow}(\mathbf{r}) = V^{-1/2} \exp(i\mathbf{k}.\mathbf{r}) \alpha(\xi) \\ \phi_{\mathbf{k}\downarrow}(\mathbf{r}) = V^{-1/2} \exp(i\mathbf{k}.\mathbf{r}) \beta(\xi) \end{array} \right\} \tag{14.92}$$

Overhauser considered superpositions of, for example, the form

$$\left. \begin{array}{l} \phi_{\mathbf{k}}^{(+)} = \phi_{\mathbf{k}\uparrow} \cos\theta(\mathbf{k}) + \phi_{\mathbf{k}+\mathbf{Q}\downarrow} \sin\theta(\mathbf{k}) \\ \phi_{\mathbf{k}}^{(-)} = -\phi_{\mathbf{k}-\mathbf{Q}\uparrow} \sin\theta(\mathbf{k}) + \phi_{\mathbf{k}\downarrow} \cos\theta(\mathbf{k}) \end{array} \right\} \tag{14.93}$$

where $\theta(\mathbf{k})$ is a periodic function of k with period $\mathbf{Q}$. These functions form a complete orthonormal set and reduce to the conventional set of plane waves for $\theta(\mathbf{k}) \equiv 0$. These functions represent the so-called spin-density waves. The spin density associated with a function forms a spiral wave characterized by the wave vector $\mathbf{Q}$.

Overhauser found that the paramagnetic Hartree–Fock ground state is always unstable against formation of spin-density waves. This result is again a consequence of the unscreened exchange potential. Model calculations with screened interactions do not seem to give any stable SDW solution. Nevertheless the spin-density waves are of great interest in connection with other systems and in particular low-dimensional systems.

These ideas were extended by Overhauser to include also the case in which the charge density in the ground state forms a wave, a charge-density wave. Consider a supposed ground state for which the spin-up and spin-down electron densities are given by

$$\left. \begin{array}{l} \rho^+(r) = \tfrac{1}{2}\rho_0(1 + p\cos(\mathbf{Q}.\mathbf{r} + \phi)) \\ \rho^-(r) = \tfrac{1}{2}\rho_0(1 + p\cos(\mathbf{Q}.\mathbf{r} - \phi)) \end{array} \right\} \tag{14.94}$$

The mean electron density is ρ_0 and the modulation amplitude is characterized by p. This gives the following three possible types of waves:

$$\left.\begin{array}{ll} \phi = 0 & \text{pure CDW} \\ \phi = \tfrac{1}{2}\pi & \text{pure SDW} \\ 0 < \phi < \tfrac{1}{2}\pi & \text{mixed CDW–SDW} \end{array}\right\} \qquad (14.95)$$

We note that a CDW ground state is highly unlikely for an electron gas because of the high Coulomb energy involved. However, the primary concern is the real system with discrete positive ions, whose equilibrium positions can adjust to cancel out most of the Coulomb energy mentioned above.

The existence of a wave-like ground state described by equation (14.94) requires a corresponding variation in the self-consistent potential. It will have the form

$$V(\mathbf{r}) = A\sigma_z \sin(\mathbf{Q}.\mathbf{r}) - C\cos(\mathbf{Q}.\mathbf{r}) \qquad (14.96)$$

where σ_z is the Pauli matrix. This periodic potential will introduce energy gaps in the one-electron energy spectrum. This will occur on planes perpendicular to the wave vector $\mathbf{Q}$ at a distance $\tfrac{1}{2}\mathbf{Q}$ from the origin.

14.4. The Two-Dimensional Electron Liquid

The two-dimensional electron liquid is frequently used as a model for the electrons in an inversion layer. Since the properties of space charges will be covered in detail in Chapter 15, we shall only give a brief discussion of how some important physical quantities are affected by the electron–electron interaction. We shall present results for the plasmon dispersion, pair correlation function, exchange and correlation energy, and the self-energy at the Fermi level. We shall use the same three main approximations to the many-body problem as for the three-dimensional liquid, namely:

1. The RPA.
2. The Hubbard approximation.
3. The self-consistent approach of Singwi and co-workers.

The results show that the first two schemes are less satisfactory in two dimensions than in three. This is a typical example that the effects of interactions become more important with lower dimensionality and require a more accurate approximation scheme.

In practice, a system like an inversion layer is not a truly two-dimensional system. We shall therefore comment on the effect of a finite width of the layer and give the results for the quasi-two-dimensional electron liquid. We choose the coordinate axes such that the xy plane defines the two-dimensional electron liquid and the z-axis is perpendicular to it. The z-dependence of the wave functions is taken to be a delta function $\delta(z)$ for the two-dimensional liquid. In the quasi-two-dimensional case, we use an approximate form[242] of the Hartree solution for an inversion layer calculated by Stern,[243] of the form

$$\xi_0(z) = \sqrt{\frac{b^2}{2}}\, z \exp(-bz/2) \tag{14.97}$$

The average of z weighted by the charge distribution is given by

$$\langle z \rangle = \int_0^\infty dz\, z\, |\xi_0(z)|^2 = 3/b \tag{14.98}$$

A typical value in connection with the MOSFET device (cf Chapter 15) is $\langle z \rangle \sim 25$ Å.

In the following, we always integrate over the z-dependence. All equations will have the same form as in the three-dimensional case with the difference that $\mathbf{r}$, $\mathbf{r}'$ will now refer to the two-dimensional vectors in the xy plane and q will be the two-dimensional wave vector with components $q = (q_x, q_y)$. The two-dimensional Coulomb interaction is given by

$$v(\mathbf{r} - \mathbf{r}') = \frac{e^2}{\bar\varepsilon\, |\mathbf{r} - \mathbf{r}'|} \tag{14.99}$$

where $\bar\varepsilon$ is an appropriately defined dielectric constant. For the MOSFET device $\bar\varepsilon = (\varepsilon_{sc} + \varepsilon_{ox})/2$ (cf Chapter 15). In the quasi-two-dimensional case we average over the z-dependence and obtain an effective two-dimensional Coulomb interaction, which for the MOSFET device has the form

$$v(\mathbf{r} - \mathbf{r}') = \frac{e^2}{\varepsilon_{sc}} \int_0^\infty dz \int_0^\infty dz' \frac{|\xi_0(z)|^2 |\xi_0(z')|^2}{[(\mathbf{r} - \mathbf{r}')^2 + (z - z')^2]^{1/2}}$$
$$+ e^2 \frac{\varepsilon_{sc} - \varepsilon_{ox}}{\varepsilon_{sc} + \varepsilon_{ox}} \int_0^\infty dz \int_0^\infty dz' \frac{|\xi_0(z)|^2 |\xi_0(z')|^2}{[(\mathbf{r} - \mathbf{r}')^2 + (z - z')^2]^{1/2}} \tag{14.100}$$

The last term is due to the image chrage distribution in the oxide.

In applications to real inversion layers, one would have to consider the effects of the degeneracies in the bands, the quantized levels in the

z-direction, the different effective masses, parallel and perpendicular to the plane of the dielectric properties of the system, etc. Such aspects will be referred to in Chapter 15 and we only mention a couple of general points. To establish the connection, we introduce the appropriate notation. We shall use effective atomic units, $a_B^* = \bar{\varepsilon}\hbar^2/m^*\varepsilon^2$ is the effective Bohr radius determined by the effective mass m^* in the xy plane, $\bar{\varepsilon}$ being the appropriate dielectric constant. The effective energy unit becomes Ryd* = $m^*e^4/2\bar{\varepsilon}^2\hbar^2$. It is convenient to introduce the electron-liquid parameter r_s related to the density n through the form $n^{-2} = \pi r_s^2 a_{B2}^*$. As in the three-dimensional case, this gives the possibility to express the results in terms of r_s, for example,

$$k_F = (2\pi n)^{1/2} = 2^{1/2}/r_s a_B^*$$

As an example we give the numbers for an Si(100)–SiO$_2$ inversion, where we have double degeneracy. The parameters are $\varepsilon_{sc} = 11.8$, $\varepsilon_{ox} = 3.8$, $m^*/m = 0.19$, which gives $a_B^* = 22$ Å, and Ryd* = 42 MeV.

One can vary the electron density within wide ranges. This means that by varying the density one can cover a wide range of situations. At high densities we have the ordinary metallic phase. At lower densities we may have a transition to the Mott–Anderson localized phase to be described in the next section, and at very low densities we may even reach a situation where the strong electron correlations induce a crystallization of the inversion layer electrons so that a Wigner lattice is formed. As already pointed out, the effects of electron–electron correlations come out much stronger in two than in three dimensions.

We shall follow closely the discussion in the previous sections, and take over all the formulas and general results, but applied now to the two-dimensional electron liquid. For the three approximate theories the response function R, defined through the relation $\rho_{ind} = R V_{ext}$, is given by

$$R(\mathbf{q}, \omega) = \frac{P^0(\mathbf{q}, \omega)}{1 - v(\mathbf{q})[1 - G(\mathbf{q})]P^0(\mathbf{q}, \omega)} \tag{14.101}$$

Here $P^0(\mathbf{q}, \omega)$ is the first-order perturbation response and $G(\mathbf{q})$ is the local-field correction factor; $G(q) = 0$ corresponds to the RPA. The Hubbard approximation is well represented by the approximate formula[244]

$$G_H(q) = \frac{1}{2} \frac{q}{(q^2 + k_F^2)^{1/2}} \tag{14.102}$$

In the self-consistent theory by Singwi et al., $G(q)$ has to be calculated numerically.

We first discuss some results for the pair correlation function.[244] Figure 14.10 shows the pair correlation function $g(R)$ for a density corresponding to $r_s = 4$. We first compare the RPA results in two and three dimensions. The outstanding difference is that the two-dimensional results are more negative (about a factor of two) for small R. The conclusion is that the RPA is a poorer approximation in two dimensions than in three. We now turn to the quasi-two-dimensional electron liquid. The RPA still gives an unphysical result, but the approximation is somewhat better than in the strictly two-dimensional model. We can understand this as follows. In the quasi-two-dimensional model the effective interaction is less singular at small distances than the $1/r$ potential of the two-dimensional electron liquid. Because of the finite width, the potential behaves rather like $\ln(1/r)$ for small r. This makes the Coulomb repulsion weaker and the exchange-correlation hole will be more shallow.

We next discuss the energy of the electron liquid. The kinetic energy is given by

$$E_{\text{kin}} = \frac{1}{r_s^2} \text{Ryd}^* \tag{14.103}$$

while the exchange energy was calculated by Stern[245] as

$$E_x = -\frac{8.2^{1/2}}{3\pi} \frac{1}{r_s} \text{Ryd}^* \tag{14.104}$$

It should be noted that the dependence on r_s of the kinetic and exchange terms is the same as in three dimensions. However, the coefficient of the kinetic term is smaller than that of the exchange term, in contrast to the three-dimensional case.

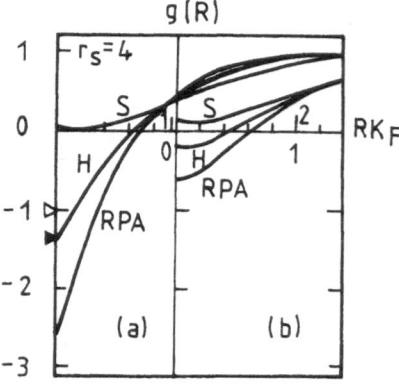

Figure 14.10. Pair correlation function $g(R)$ for (a) a two-dimensional electron gas and (b) a quasi-two-dimensional gas with $r_s = 4$. The full arrow indicates $g(0)$ in three dimensions and the RPA and the open arrow gives $g(0)$ in the Hubbard approximation.

The correlation energy in the high-density limit of a three-dimensional electron assembly was calculated by Gell-Mann and Brueckner in a famous paper in 1957.[246] Their result, which includes the second-order exchange contribution in addition to the RPA, is to leading order in r_s given by

$$E_c = (0.062 \ln r_s - 0.096) \text{ Ryd}^* \tag{14.105}$$

The corresponding calculation in two dimensions by Rajagopal and Kimball[247] gave the result

$$E_c = (-0.38 - 0.172 r_s \ln r_s) \text{ Ryd}^* \tag{14.106}$$

in the high-density limit. However, it should be remarked that these results only apply to densities much higher than those of physical interest.

In the opposite extreme of low densities, namely large values of r_s, the electrons will form a two-dimensional lattice. Srinivasan and Jonson[248] studied a two-dimensional Einstein solid and obtained an energy of $-1.103/r_s$ Ryd* per electron. By interpolating with the result of the Singwi approach at $r_s = 0.5$ they obtained an interpolation formula for the correlation energy, similar to that given by Wigner[238] for the three-dimensional case:

$$E_c = -\frac{1.103}{r_s + 4.41} \text{ Ryd}^* \tag{14.107}$$

This formula gives energies which stay within 25% of those by the Singwi approach over a wide range of densities ($r_s < 16$). The RPA correlation is off by a factor of 2–4 in the same range and the Hubbard approximation is not substantially better. The conclusion is that a self-consistent solution of the type developed by Singwi *et al.* is needed to describe the correlation effects in the two-dimensional electron liquid.

The self-energy at the Fermi level shows the same characteristics. The RPA and the Hubbard approach give very similar results and one has to go to the Singwi approach or some equivalent or better formulation to improve the accuracy.

The dielectric function $\varepsilon(q, \omega)$ contains information about the excitation spectrum. We shall leave out the electron-hole excitations and focus on the collective plasmon excitations, which are found as the solutions to the equation $\varepsilon(q, \omega) = 0$. In the limit $q \to 0$ the dispersion is given by the formula

$$\omega_p(q) = \omega_s q^{1/2} \tag{14.108}$$

where $\omega_s = (2\pi n e^2/\bar{\varepsilon} m^*)^{1/2}$ is the surface plasmon frequency. The special feature of the two-dimensional liquid is that the plasmon dispersion at long wavelengths is proportional to $q^{1/2}$ rather than being constant as in three dimensions. The reason becomes clear as one recalls that the long-wavelength plasma oscillations can be considered as set up by charges on the surface of the sample. In three dimensions the electrostatic field corresponding to such a charge is independent of the size of the sample, while in two dimensions the field goes to zero when we increase the size of the sample. The plasmon dispersion for larger values of q will tell us the effect of how we treat the short-range correlations. It turns out that for high densities the Hubbard approximation seems to give a good description of the plasmon dispersion, while for lower densities the approach by Singwi *et al.* gives results that differ considerably from the Hubbard and RPA approximation.

All the examples mentioned here indicate the need to treat correlations accurately by some self-consistent procedure. If we go to the quasi-two-dimensional case we find smaller deviations, which tend to increase with increasing values of r_s. We refer to Jonson's paper[244] for further discussion.

We mentioned in Section 14.3 that charge-density waves were proposed by Overhauser[241] in 1968. He emphasized the correlation energy as the source of the instability, but there are also other sources. Consider a charge-density wave of the form

$$\rho(\mathbf{r}) = \rho_0(\mathbf{r})[1 + \phi \cos \mathbf{q} \cdot \mathbf{r}] \qquad (14.109)$$

Such an oscillating charge distribution in an electron gas would, if we keep the positive background fixed, produce a net charge that oscillates from negative to positive at maxima and minima in the wave. So far we have assumed that the ground state of the two-dimensional electron gas is the usual paramagnetic state. However there are many quasi-two-dimensional systems where the ground state is different and we mention particularly layered compounds (cf Chapter 12) where we often find a ground state characterized by charge-density waves. As examples we can mention the layered transition-metal dichalcogenides, which form a sandwich structure in which we focus on the layers formed by the metal atoms. This will give rise to a Coulomb energy typically of the order of 10 eV. However, if the positive ions are allowed to move toward the maxima and away from the minima, the Coulomb energy will be much smaller and such a coupled distortion may be the stable ground state of the system. Thus the charge-density waves (CDW) to be considered in the layered compounds and also for one-dimensional systems correspond to a static, coupled periodic distortion of the conduction electron density and the lattice.

It is clear that lattice waves having the same wave vector $\mathbf{q}$ and with displacements parallel to $\mathbf{q}$ (longitudinal phonons) will be the important ones. It is also clear that the formation of a CDW is favored by a strong electron–phonon interaction. Besides a large electron–phonon interaction the shape of the Fermi surface is also important. Consider the static case and the first-order perturbation formula for $P^0(\mathbf{q})$:

$$P^0(\mathbf{q}) = \sum_{\mathbf{k}} \frac{n^0(\mathbf{k}) - n^0(\mathbf{k}+\mathbf{q})}{\mathscr{E}(\mathbf{k}+\mathbf{q}) - \mathscr{E}(\mathbf{k})} \qquad (14.110)$$

It is clear that $P^0(\mathbf{q})$ will be large for a $\mathbf{q}$ that connects many filled states to empty states having the same energy. This happens, for example, if the Fermi surface has plane-parallel sections, if there is a nesting of electron and hole surfaces, or if there are saddle points in the band structure at the Fermi level.

In order to have some idea how instabilities in the electron liquid appear due to electron–phonon or electron–electron interactions, we may use the RPA formula for the static susceptibility $R(\mathbf{q})$:

$$\mathbf{R}(\mathbf{q}) = \frac{P^0(\mathbf{q})}{1 - v(\mathbf{q})P^0(\mathbf{q})} \qquad (14.111)$$

where $v(\mathbf{q})$ represents all the interactions. Now we see that an instability will occur at the q where $v(\mathbf{q})P^0(\mathbf{q}) = 1$. Overhauser used the three-dimensional electron-gas model and showed that the correlation energy for unscreened Coulomb potentials in a Hartree–Fock calculation diverges at $q = 2k_F$. If $v(\mathbf{q})$ instead is dominated by exchange, a spin-density wave (SDW) is formed as in chromium. If $v(q)$ is dominated by the electron–phonon interaction, we expect a CDW instability. If $\mathbf{q}$ is determined by the Fermi surface, then the wavelength of the CDW will usually be incommensurate with the lattice: the wavelength will not equal a lattice translation. Such an incommensurate CDW (ICDW) tells us that the instability is "driven" by the Fermi surface. However, there are also cases with a commensurate state (CCDW) and then it is difficult to assess the role of the Fermi-surface properties.

14.5. Partially Localized Electrons

So far we have only considered electron–electron interactions in systems where the electrons behave more or less as free electrons. The fact that the solid has been built up from atoms has only been incorporated in a schematic way through effective masses and the average dielectric constant of the background medium. The full effects of band structure could formally be taken into account (see, for example, the work of Hedin and Lundqvist[223]). This leads to rather formidable numerical analysis and would bring in a number of new aspects on the detailed nature of the correlations but hardly anything that is conceptually new. What we wish to introduce here is the atomic aspect of the correlation problem.

Particularly for heavier elements we have very strong intra-atomic correlations, which give properties rather different from the Hartree–Fock model, such as collective resonances. However, even in small systems the Coulomb interaction between a pair of electrons gives rise to a large change in the total energy if we change the number of electrons on the atom. That means that we could not take the corresponding Hartree–Fock orbital energy as a measure of the energy change. We have also to correct explicitly for the corresponding change in the interaction energy.

Consider a system of closed-shell atoms or molecules on a lattice with a lattice parameter large enough that the electron wave functions are completely localized on the atoms and with no overlap between them. This is the idealized picture of a van der Waals crystal. The correlation between electrons on different atoms will arise through the pairwise interaction between dipolar fluctuations in the charge density of the atoms. The van der Waals energy is conventionally calculated for a pair of atoms using second-order perturbation theory. Here we shall consider the correlations in the same spirit as developed in the earlier sections. The dipolar excitation between any pair of atomic levels will be coupled through the dipolar interactions and will be broadened into an exciton band of polarization waves. We can calculate the frequency-dependent polarizability and the total energy of the crystal in the RPA. This discussion emphasizes the collective aspect of the interaction rather than the two-body aspect.

Let us now consider the case when the atoms have moved closer so that the wave functions on neighboring atoms have a small overlap. This makes it possible for electrons to tunnel from one atom to another, and hence gives rise to a coupling matrix element to transfer an electron between neighboring atoms. Such an element is usually referred to as the hopping matrix element. This is, in short, the background for the tight-binding approximation in band theory (cf Chapter 12). The atomic aspect is then introduced by taking account of the intra-atomic interaction. The transfer of an electron from one atom to the next depends on the number of electron states already occupied. The intra-atomic correlation plays a decisive role in determining what kind of state will be formed. Thus the problem involves both the intra-atomic correlation, the hopping problem, and of course also the interatomic correlations, which are usually ignored. These problems are normally discussed using a simplified theoretical model developed by Hubbard[249-251] and extensions of this model. There is an enormous literature dealing with the Hubbard model, its extensions, and its applications to a wide variety of problems. We shall only give a brief introduction to this topic, restricting ourselves to the presentation of the basic ideas.

One of the most important aspects is the possibility of a metal-insulator transition due to correlation. The physical ideas were put forward already around 1950 in some famous papers by Mott[252,253] and started what is now a very active field of research. We shall briefly present the intuitive physical arguments put forward by Mott. Extensions of the Hubbard model also give a theoretical description of the metal-insulator transition, but it seems that the connection with the original Mott model is not at all obvious.

14.5.1. Electron–Electron Interactions in van der Waals Crystals

Let us consider an idealized model of a van der Waals crystal and assume the size of the atoms to be small compared with the distance between them. We consider the case of a crystal with one kind of atom. Since all the atoms are neutral and nonoverlapping, there will be no interaction energy in lowest order. The leading term in the interaction will be due to the coupling between dipolar fluctuations in the charge density. The total dipolar interaction is given by

$$V = -\tfrac{1}{2} \sum_{\mathbf{RR'}} \mathbf{p_R} \cdot \mathbf{T}(\mathbf{R} - \mathbf{R'}) \cdot \mathbf{p_{R'}} \tag{14.112}$$

where $\mathbf{p_R}$ is the operator representing the total dipole moment of the atom at $\mathbf{R}$; $\mathbf{T}(\mathbf{R} - \mathbf{R'})$ is the interaction tensor given by

$$\mathbf{T}(\mathbf{R} - \mathbf{R'}) = \frac{\partial^2}{\partial \mathbf{R} \partial \mathbf{R'}} \frac{1}{|\mathbf{R} - \mathbf{R'}|} \tag{14.113}$$

We wish to treat the interaction in RPA and can use the equations discussed in the earlier sections of this chapter. However, here we can simplify the problem since we do not need the full density response but only the dipole moment. We therefore use instead the dipolar response function defined by

$$\alpha(\mathbf{R} - \mathbf{R'}, t) = i\langle [\mathbf{p_R}(t); \mathbf{p_{R'}}(0)]\rangle \tag{14.114}$$

For a free atom, this is just the atomic polarizability

$$\alpha_0(t) = i\langle [\mathbf{p}(t); \mathbf{p}(0)]\rangle_0 \tag{14.115}$$

where the average is taken with respect to the ground state of the atom. Expressed in terms of eigenstates and excitation frequencies of the free atom, we have the well-known formula

$$\alpha_0(\omega) = \frac{e^2}{m} \sum_n \frac{f_n}{\omega_n^2 - \omega^2} \tag{14.116}$$

where

$$f_n = \frac{2m}{e^2} \omega_n |\mathbf{p}_n|^2 \tag{14.117}$$

is the oscillator strength, $\mathbf{p}_n$ being the dipole matrix element for the atomic excitation n.

Let us first look at the eigenmodes of the system of coupled dipoles. The field on the atom R is given by the equation

$$\mathbf{E_R}(t) = \sum_{\mathbf{R}'} \mathbf{T}(\mathbf{R} - \mathbf{R}') \cdot \mathbf{p_{R'}}(t) \tag{14.118}$$

and the induced dipole is

$$\mathbf{p_R} = \sum_{\mathbf{R}'} \alpha_0(\omega) \mathbf{T}(\mathbf{R} - \mathbf{R}') \cdot \mathbf{p_{R'}} \tag{14.119}$$

The coupled equations (14.119) have solutions of wave form:

$$\mathbf{p_R} = \mathbf{A} \exp(i\mathbf{q} \cdot \mathbf{R}) \tag{14.120}$$

and we obtain

$$[1 - \alpha_0(\omega)\mathbf{T}(\mathbf{q})] \cdot \mathbf{A} = 0 \tag{14.121}$$

with

$$\mathbf{T}(\mathbf{q}) = \sum_{\mathbf{R}} \mathbf{T}(\mathbf{R}) \exp(-i\mathbf{q} \cdot \mathbf{R}) \tag{14.122}$$

The possible values for ω are obtained from the corresponding determinantal equation

$$\text{Det}(1 - \alpha_0 \mathbf{T}(\mathbf{q})) = 0 \tag{14.123}$$

For each wave vector $\mathbf{q}$ we obtain in general three different values for ω. Therefore each atomic excitation ω_n will be broadened into three exciton bands characterized by their dispersion relations $\omega_{ns}(q)$, $s = 1, 2, 3$, and the corresponding polarization vectors. Thus the dipolar atomic excitations will couple and form collective modes in the form of polarization waves.

Let us now consider the response to an external field and calculate the polarizability $\alpha(\mathbf{q}, \omega)$ of the system. In the RPA this gives, in complete analogy with the electron-gas case, the equation

$$\alpha(\mathbf{q}, \omega) = \alpha_0(\omega)\mathbf{1} + \alpha_0(\omega)\mathbf{T}(\mathbf{q}) \cdot \alpha(\mathbf{q}, \omega) \qquad (14.124)$$

with solution

$$\alpha(\mathbf{q}, \omega) = \frac{\alpha_0(\omega)}{1 - \alpha_0(\omega)\mathbf{T}(\mathbf{q})} \qquad (14.125)$$

The physical significance of $\alpha(\mathbf{q}, \omega)$ should be clear. Consider an external field with a wavelength large compared to the dimensions of a single atom. The average induced dipole moment of an atom is then given by

$$\mathbf{p}(\mathbf{q}) = \alpha(\mathbf{q}, \omega) \cdot \mathbf{E}_{ext}(\mathbf{q}, \omega) \qquad (14.126)$$

The effect of the internal field is now included in $\alpha(\mathbf{q}, \omega)$ through the denominator in equation (14.125). Thus resonances will occur at the frequencies corresponding to the internal excitations of the systems. They correspond to the zeros of the denominator in equation (14.125), i.e., to the solutions of the homogeneous problem just discussed.

Since we are considering a system of interacting dipoles, we would expect to get the well-known Lorentz–Lorenz results in the limit q tends to zero. Indeed, for a cubic lattice the values of $T(q)$ give the classical result for longitudinal and transverse waves and we obtain for the dielectric function when $\mathbf{q} \rightarrow 0$

$$\varepsilon(\omega) = \frac{1 + (8\pi/3)n\alpha_0(\omega)}{1 - (4\pi/3)n\alpha_0(\omega)} \qquad (14.127)$$

with n denoting the number of atoms per unit volume. The poles of the dielectric constant for $q = 0$ give the transverse excitations, while the zeros of the dielectric constant give the longitudinal excitations. It is only at $q = 0$ that the dielectric properties are described by a scalar function. For a general q, we have a dielectric tensor with nondiagonal elements and the physics is not as simple as for $q = 0$. Further information on this can be obtained elsewhere.[254]

14.5.2. The Hubbard Hamiltonian

We shall conclude this chapter by referring briefly to the treatment of intra-atomic correlation using the Hubbard Hamiltonian. We start by discussing the following example. Let us consider a monovalent metal. Each atom has one valence electron, which is an s-electron. The conduction band of the metal is then half-filled. For simplicity we do not consider the possbility of overlapping bands. The conduction electrons are free to move around in the crystal and the almost-free electron model is adequate for the description of the metallic properties.

Let us now increase the lattice constant and thereby reduce the width of the s-band. If we increase the lattice constant so there will be no interaction between the atoms, the band finally reduces to the sharp s-level of the free atom. In this limit we shall have exactly one valence electron localized round every atom. Metallic conduction is no longer possible although in the band model we still have a half-filled band. The meaning of this is simply that somewhere the band model breaks down and we have a transition from delocalized to localized electron states. We notice that the localization implies a strong correlation between electrons. In the intermediate situation with narrow bands, we have a competition between the terms, which tend to keep the electrons itinerant, and the correlations, which will bring about localization on individual atoms at sufficiently large interatomic distances.

We notice that what we have just discussed had its analogue in the earlier discussion of the electron liquid. At high densities, the kinetic energy dominates over the potential energy and the behavior is like a gas of free electrons. As the density is lowered the importance of the Coulomb repulsion between electrons grows rapidly, and for sufficiently large r_s there is a phase transition from the electron liquid to an electron crystal, the resulting Wigner lattice being a consequence of strong localization of the electrons through electron–electron correlation.

We briefly outline the framework within which a quantitative formulation of these ideas is possible. We shall use the above model of an s-band which is just half-filled with N electrons, N being the number of atoms. The Hamiltonian is divided into two parts, the first being an effective one-electron part h, the "band Hamiltonian," and the electron–electron interaction contribution v. The Hamiltonian is conveniently expressed in terms of creation and annihilation operators, $c_{k\sigma}^*$ and $c_{k\sigma}$, respectively, for an electron of wave vector $\mathbf{k}$ and spin σ. The full Hamiltonian then takes the form

$$
\begin{aligned}
H = &\sum_{\mathbf{k}\sigma} E(\mathbf{k}) c_{\mathbf{k}\sigma}^* c_{\mathbf{k}\sigma} \\
&+ \tfrac{1}{2} \sum_{\substack{\mathbf{k}_1 \mathbf{k}_2 \\ \mathbf{k}_1' \mathbf{k}_2' \\ \sigma_1 \sigma_2}} \langle \mathbf{k}_1 \mathbf{k}_2 | v | \mathbf{k}_1' \mathbf{k}_2' \rangle c^* \mathbf{k}_{1\sigma_1} c_{\mathbf{k}_2\sigma_2}^* c_{\mathbf{k}_1'\sigma_1} c_{\mathbf{k}_2'\sigma_2}
\end{aligned}
\tag{14.128}
$$

Since we wish to discuss how correlation effects appear when we reduce the bandwidth and pass over to a localized description, it is appropriate to transform the Hamiltonian from the Bloch wave picture to the Wannier representation. The Wannier functions are expressed in terms of the Bloch functions $\psi(\mathbf{k}, \mathbf{r})$ through

$$\psi(\mathbf{k}, \mathbf{r}) = \frac{1}{N^{1/2}} \sum_n a(\mathbf{R}_n, \mathbf{r}) \exp(i\mathbf{k} \cdot \mathbf{R}_n) \qquad (14.129)$$

or

$$a(\mathbf{R}_m, \mathbf{r}) = \frac{1}{N^{1/2}} \sum_{\mathbf{k}} \exp(-i\mathbf{k} \cdot \mathbf{R}_n) \psi(\mathbf{k}, \mathbf{r}) \qquad (14.130)$$

The summation over $\mathbf{k}$ extends over the entire Brillouin zone. The Wannier function $a(\mathbf{R}_n, \mathbf{r})$ is localized round the atom n and Wannier functions centered on different atoms are orthogonal. In the Wannier picture the Hamiltonian becomes

$$H = \sum_{lk} t_{lk} c_{i\sigma}^* c_{k\sigma}$$

$$+ \tfrac{1}{2} \sum_{ijkl} \sum_{\sigma\sigma'} \langle ij \,|\, v \,|\, kl \rangle c_{i\sigma}^* c_{j\sigma'}^* c_{k\sigma} c_{l\sigma'} \qquad (14.131)$$

where t_{ij} is given by

$$t_{ij} = \frac{1}{N} \sum_{\mathbf{k}} E(\mathbf{k}) \exp(i\mathbf{k} \cdot (\mathbf{R}_i - \mathbf{R}_j)) \qquad (14.132)$$

The operators and matrix elements refer to the Wannier functions. The matrix elements describe interactions between electrons on different lattice sites. When we pass to the limit of large separation it is obvious that the interactions between electrons on the same site will be the most important ones, in order to end up with a state possessing precisely one electron per atom. As a first step, one therefore neglects all matrix elements except those for which $i = j = k = l$. Putting $\langle ii \,|\, v \,|\, ii \rangle = U$ we obtain the approximate Hamiltonian

$$H = \sum_{lk\sigma} t_{lk} c_{i\sigma}^* c_{k\sigma} + \frac{U}{2} \sum_{i\sigma\sigma'} c_{i\sigma}^* c_{i\sigma'}^* c_{i\sigma} c_{i\sigma} \qquad (14.133)$$

One now introduces the number operator $n_{i\sigma} = c_{i\sigma} c_{i\sigma}$ for an electron of spin σ on site i, in which case the Hamiltonian (14.133) becomes

$$H = \sum_{lk\sigma} t_{lk} c_{i\sigma}^* c_{k\sigma} + U \sum_i n_{i\uparrow} H_{i\downarrow} \qquad (14.134)$$

This Hamiltonian was first discussed by Hubbard[249-251] and is usually referred to as the Hubbard Hamiltonian. It contains essentially three parameters: t_0, $t_1 = t_{ik}$ (i and k nearest neighbors), and U. Below we shall briefly discuss the meaning of the model and refer to approximate solutions. Since there are many extensions of the Hubbard model and an enormous literature exists on approximate solutions and applications, we shall not even be able to summarize these further developments but just limit the summary below to the very basic aspects of the model.

14.5.3. Discussion of the Hubbard Model

We shall first discuss the atomic limit. Hubbard's main observation was that, in the limit of increasing lattice constant, the coupling matrix elements $t_{ij} \to 0$ while U remains large. Therefore one should first solve the problem for

$$H_{atom} = \sum_i (t_0 n_i + U n_{i\uparrow} n_{i\downarrow})$$
(14.135)

For each atom we have the following states:

$$|0\rangle \quad \text{vacuum}$$
$$c_i^* |0\rangle \quad \text{one electron}$$
$$c_{i\uparrow} c_{i\downarrow} |0\rangle \quad \text{two electrons}$$

with energies 0, t_0, and $2t_0 + U$. Thus t_0 is the energy needed to bind one electron on an isolated atom and $t_0 + U$ is the energy needed to bind a second electron of opposite spin; U is therefore the interaction energy of two electrons located on the same atom.

If we now have N_e electrons in a system of N atoms, we can form product wave functions in which N_0 atoms are unoccupied and N_1 have two electrons. The total energy of the system becomes

$$E = N_1 t_0 + (N_e - N_1)(2t_0 + U)$$
(14.136)

Though we shall not give much detail, the atomic limit requires closer study with a view to later inclusion of the "hopping terms" t_{ij}, which permit an electron to move from one atom to the next. Suppose we insert an extra electron. The propagation of this electron will depend on whether it goes into an empty site or on a singly occupied site (the exclusion principle prevents it going into a doubly occupied site). This is a matter of probability so that, on average, an extra electron will propagate in a mixture of states. In order to develop these ideas in a quantitative way, one finds it helpful to formulate the theory in terms of "propagators" or Green's functions, as we discuss briefly below.

In particular, we wish to consider the one-electron Green's function in the Wannier representation, namely

$$G^\sigma_{ij}(t) = \langle a_{i\sigma}(t); a^*_{j\sigma}(0) \rangle \qquad (14.137)$$

The physical interpretation is that we insert an extra electron on site j at $t = 0$ and require the probability to find it at site i at time t. We now apply this technique to the limit of zero bandwidth so that $t_{ij} = t_0 \delta_{ij}$. Working with the Fourier-transform variable ω instead of t in equation (14.137), this limit can then be solved. We shall omit the details as the result

$$G^\sigma_{ij}(\omega) = \frac{1}{2}\,\delta_{ij} \left(\frac{1 - \frac{1}{2}n}{\omega - t_0} - \frac{\frac{1}{2}n}{\omega - t_0 - U} \right) \qquad (14.138)$$

is easy to understand. In equation (14.138) we have written $\langle n_{i\sigma} \rangle = \frac{1}{2}n$, for having assumed a translationally invariant system the average $\langle n_{i\sigma} \rangle$ must be independent of both i and σ. The quantity n is the average number of electrons per atom. The formula (14.138) corresponds exactly to the intuitive picture before we started to discuss propagators. If the system is in its ground state and $n = N_e/N < 1$, then we shall have a fraction $n/2$ of occupied sites with a given spin and a fraction $(1 - n/2)$ of the sites not occupied by electrons with this spin. The first term in equation (14.138) represents the situation in which there was no electron before, therefore the probability factor is $1 - \frac{1}{2}n$ and the Green's function has a pole at $\omega = t_0$, which means that an energy t_0 is needed to land an extra electron on the site. If the state is already occupied, the probability factor is $n/2$ and the energy required to put the extra electron on site i is $t_0 + U$. Thus G represents a mixed amplitude for the electron to propagate in either empty or singly occupied states. The density of states (spectral density) is given by

$$N(\omega) = \frac{1}{\pi}\,\mathrm{Im}\,G(\omega) \qquad (14.139)$$

from which we obtain

$$N(\omega) = (1 - \tfrac{1}{2}n)\delta(\omega - t_0) + \tfrac{1}{2}n\delta(\omega - t_0 - U) \qquad (14.140)$$

Thus the system behaves as if it had two energy levels t_0 and $t_0 + U$ having $(1 - \frac{1}{2}n)$ and $\frac{1}{2}n$ states per atom.

We conclude this discussion of the correlation aspects near the atomic limit by now allowing the electrons to tunnel between neighboring atoms;

in this case $t_{ij} = 0$ for $i = j$ and the atomic levels are broadened into a band. To start with, we can usefully consider the case $U = 0$, which for small overlap then gives us the usual tight-binding model. The Green's function is then simply

$$G_{\mathbf{k}}(\omega) = \frac{1}{\omega - E(\mathbf{k}) + i\delta} \qquad (14.141)$$

Next we recall that in the extreme atomic limit the propagator for an electron going into an empty state of an atom is given by

$$G_e(\omega) = \frac{1}{\omega - t_0 + i\delta} \qquad (14.142)$$

The Green's function G_{ij} can now be written, with the inclusion of t_{ij}, as an infinite series in t_{ij}. Hubbard's approximate solution for the full Hamiltonian with $U = 0$ corresponds to replacing this infinite series G_e in equation (14.142) by the solution in the atomic limit $G_0^\sigma(\omega)$. Reverting to the $\mathbf{k}$ representation, and writing $E(\mathbf{k}) = t_0 + t(\mathbf{k})$, the infinite series can then be summed to yield the approximate solution

$$G_{\mathbf{k}}^\sigma(\omega) = \frac{1}{G_{\mathbf{k}}^0(\omega)^{-1} - t(\mathbf{k})} \qquad (14.143)$$

By substituting the explicit expression for the solution in the atomic limit we obtain

$$G_{\mathbf{k}}^\sigma(\omega) = \frac{1}{(\omega - t_0)\left[\dfrac{\omega - t_0 - U}{\omega - t_0 - U(1 - n/2)}\right] - t(\mathbf{k})} \qquad (14.144)$$

For each value of $\mathbf{k}$, this function has two discrete poles given as the solutions of the quadratic equation

$$(\omega - t_0)(\omega - t_0 - U) = t(\mathbf{k})[\omega + t_0 - U(1 - n/2)] \qquad (14.145)$$

In the limit of vanishing bandwidth, $t(\mathbf{k}) \to 0$, these solutions reduce to the energies t_0 and $t_0 + U$ of the atomic states. Corresponding to these atomic levels we now have two bands. This splitting has nothing to do with the periodic structure but follows from the intra-atomic correlations. This behavior is illustrated in Figure 14.11. We note that a gap can exist in the spectrum provided the two bands do not overlap. The corresponding densities of states in these two Hubbard bands are shown in Figure 14.12.

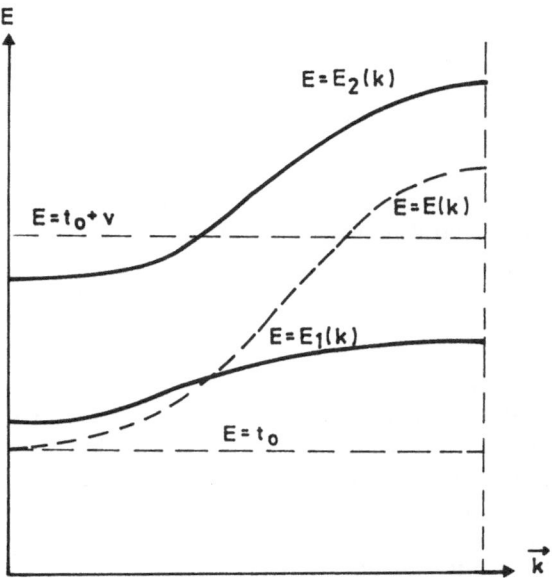

Figure 14.11. Schematic illustration of energy bands in the Hubbard model; $E_1(\mathbf{k})$ and $E_2(\mathbf{k})$ are the two Hubbard bands and $E(\mathbf{k})$ is the case when $v = 0$.

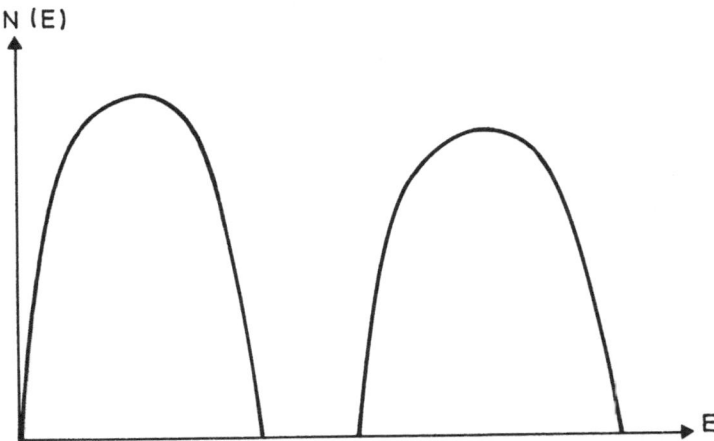

Figure 14.12. Schematic illustration of the densities of states in the Hubbard bands.

We finally calculate the number of electrons per atom in the ground state:

$$n = \frac{1}{N} \sum_{i,\sigma} \langle n_{i,\sigma} \rangle \tag{14.146}$$

This can be expressed in terms of the imaginary part of the Green's function, or the spectral density, as follows:

$$n = \frac{1}{N} \sum_{i,\sigma} \frac{1}{\pi} \int_{-\infty}^{\mathscr{E}_F} dE \, \text{Im} \, G_{ii}^{\sigma}(E) \tag{14.147}$$

In the atomic limit and just one electron per atom ($n = 1$), we have obviously that the lower level is full and the upper level is empty. The Fermi level lies between t_0 and $t_0 + U$. However, when we consider a finite bandwidth, we find that the ground state will not correspond to having the lower band filled and the upper band empty. This is a deficiency of the approximate solution we have just discussed and can be remedied only by a more accurate solution of the model.

Space-Charge Layers

F. Stern

Space-charge layers can arise in many different kinds of physical systems. In this chapter, we shall limit the consideration to a very simple category of space-charge layers, namely those that arise in semiconductors in thermodynamic equilibrium or quasi-equilibrium. This category includes systems of considerable interest in research and technology, including semiconductor–insulator systems, p–n junctions, and Schottky barriers. In some cases, the most interesting aspect of the problem is not the space-charge layer as such but the nature of the physical mechanisms that lead to its formation. While a discussion of such mechanisms in detail is outside the scope of this chapter, we shall refer at least briefly to work bearing on such problems. Space charge layers can be detected experimentally in many different ways: by capacitance, conductance, and optical measurements, for example. The experimental methods will be described briefly in relation to particular systems.

Some simple theoretical ideas are outlined first, because they provide a common framework for the remaining discussion. Then we consider a number of different physical systems and experiments, to illustrate the ways in which space-charge layers arise and how they are studied. Inversion layers at semiconductor–insulator interfaces are considered in some detail, because they are a particularly rich system both experimentally and theoretically.

An excellent description of space-charge layers in the context of semiconductor surfaces is given in the book by Many et al.[255] and also in the book by Frankl.[256] These books provide additional coverage of many

F. Stern · IBM Thomas J. Watson Research Center, Yorktown Heights, NY 10598, USA.

of the subjects considered here. Properties of Schottky barriers are described in the book by Rhoderick.[257]

The space-charge layers considered in this chapter generally arise from the response of mobile charges in a semiconductor to a fixed charge, usually near a surface or interface. Thus we deal with static systems in thermal equilibrium, although the response of the system to weak electric or electromagnetic fields will also be considered. The subject of space-charge layers can include much more complicated systems, like those that arise when nonequilibrium electrons and holes are injected into a semiconductor[258]; this latter system will not be discussed here.

A knowledge of basic semiconductor physics, as found in standard textbooks,[259] is assumed in the discussion that follows. The chapter is intended to give the reader a glimpse into the extensive subject of space-charge layers, but references are given that should allow the interested reader to pursue particular aspects of the subject in greater depth.

15.1. Screening in One- and Three-Dimensional Systems

15.1.1. Screened Point Charge in Three Dimensions

Perhaps the most familiar system with a space charge is the screened point charge in three dimensions. When a fixed charge is introduced at the origin in a semiconductor, mobile carriers of opposite sign are attracted toward it and mobile carriers of like sign are repelled. In the linear screening approximation, which will be described in detail below in connection with one-dimensional systems, the electrostatic potential at a distance r from the fixed charge is

$$\phi(r) = \frac{Ze}{4\pi\varepsilon r} e^{-r/L} \tag{15.1}$$

where the screening length L is given at elevated temperatures by

$$L^{-2} = \frac{(n+p)e^2}{\varepsilon k_B T} \tag{15.2}$$

n and p are the concentrations of electrons and holes, $\varepsilon \equiv \kappa\varepsilon_0$ is the permittivity and κ is the dielectric constant, k_B is Boltzmann's constant, and T is the absolute temperature. We shall derive this result for L below as a special case of a more general result, but this example is found in many textbooks.[255,256,259] The space charge density associated with this familiar example is

$$\rho(r) = -\varepsilon \nabla^2 \phi,$$

$$= Ze \left[\delta^{(3)}(r) - \frac{e^{-r/L}}{4\pi L^2 r} \right] \tag{15.3}$$

whose integral over all space vanishes.

Throughout this chapter we use the Système International (SI) system of units. To convert to cgs units, replace ε_0 by $1/4\pi$.

It is important to recognize that equation (15.1) results from a linear-screening approximation: the potential energy $-e\phi(r)$ must be small compared to a characteristic energy of carriers in the system for this approximation to be valid. In the high-temperature regime that leads to equation (15.2), this means that $-e\phi(r)$ must be small compared to $k_B T$, which breaks down at small distances from the point charge. Many authors have studied nonlinear screening. A familiar nonlinear theory is the Thomas–Fermi approximation.[260] A study that includes many-body effects was carried out by Vinter.[261]

15.1.2. Screening with One Spatial Variable

The physical systems in which we shall primarily be interested, such as Schottky barriers and semiconductor–insulator interfaces, can usually be regarded as uniform in two dimensions, say x and y, so that physical quantities vary only in the z direction. Thus the equations governing the systems have only one spatial variable and become one-dimensional in that sense.

Perhaps the simplest example is the response of a semiconductor to a sheet of charge with charge density Q per unit area. The response of the system is given by Poisson's equation

$$\phi''(z) = -\frac{\rho(z)}{\varepsilon} \tag{15.4}$$

$$\rho(z) = Q\delta(z) + \rho_{ind}(z) \tag{15.5}$$

where ρ_{ind} is the induced-charge density, which represents the response of the mobile charges in the semiconductor to the perturbing potential ϕ, and the primes indicate derivatives with respect to z.

For ρ_{ind} in a uniform medium in thermal equilibrium we can write

$$\rho_{ind}(z) = g(E_F + e\phi(z)) - g(E_F) \tag{15.6}$$

where g is a function that represents the total-charge density as a function of the Fermi energy E_F and the local electrostatic potential $\phi(z)$. The second term on the right subtracts out any compensating background charge and assures that the material is neutral far from any disturbance, where we take $\phi = 0$.

For a weak perturbation we can expand (15.6) in powers of ϕ and find, to lowest nonvanishing order,

$$\rho_{\text{ind}}(z) = e\phi(z)\frac{dg}{dE_F} \tag{15.7}$$

If we let

$$L^{-2} = -\frac{e}{\varepsilon}\frac{dg}{dE_F} \tag{15.8}$$

then equation (15.4) becomes

$$\phi''(z) - L^{-2}\phi(z) = -Q\delta(z)/\varepsilon \tag{15.9}$$

whose solution is

$$\phi(z) = \frac{QL}{2\varepsilon}\exp(-|z|/L) \qquad \text{(two-sided)} \tag{15.10}$$

if the medium extends on both sides of the charge sheet and

$$\phi(z) = \frac{QL}{\varepsilon}\exp(-z/L) \qquad \text{(one-sided)} \tag{15.11}$$

if the medium extends only on one side ($z > 0$) of the charge sheet. Thus the screening length L is the characteristic length for the exponential decay of a weak disturbance. The total-charge density per unit area in the screening layers is $-Q$ for both equations (15.10) and (15.11).

Now consider a simple example for the function g. For an intrinsic, nondegenerate semiconductor,

$$g(E_F) = -en_i\left[\exp\left(\frac{E_F - E_i}{k_BT}\right) - \exp\left(\frac{E_i - E_F}{k_BT}\right)\right] = e(p - n) \tag{15.12}$$

where E_i is the intrinsic Fermi level,

$$n_i = 2\left(\frac{k_BT}{2\pi\hbar^2}\right)^{3/2}(m_c m_v)^{3/4}\exp(-E_g/2k_BT) \tag{15.13}$$

is the intrinsic carrier concentration, E_g is the energy gap, and m_c and m_v are the density-of-states effective masses for the conduction and valence bands, respectively. Then

$$\frac{dg}{dE_F} = -\frac{en_i}{k_B T}\left[\exp\left(\frac{E_F - E_i}{k_B T}\right) + \exp\left(\frac{E_i - E_F}{k_B T}\right)\right] = -\frac{e}{k_B T}(n + p) \quad (15.14)$$

Combining equations (15.8) and (15.14) leads to the conventional result for the screening length already used in equation (15.2).

Other results for screening obtain in other situations. For example, for a degenerate n-type semiconductor with conduction-band edge at E_c we have

$$L^{-2} = \frac{3ne^2}{2(E_F - E_c)\varepsilon} \quad (15.15)$$

The concept of linear screening breaks down when the band bending is comparable to or larger than a characteristic energy, such as $k_B T$. Then a different approximation for band bending can be used if the direction of band bending is such as to deplete the majority carrier. One can assume that the space-charge layer, with thickness z_d, is free of mobile charge and that all impurities are ionized. For a p-type semiconductor with permittivity ε_{sc} and acceptor concentration N_A, bounded at $z = 0$ by a sheet of charge density Q, we find in this approximation that

$$\phi(z) = \frac{(z - z_d)^2 N_A e}{2\varepsilon_{sc}}, \qquad 0 < z < z_d$$

$$= 0, \qquad\qquad z > z_d \quad (15.16)$$

The total-charge density in the space-charge layer is $-eN_A z_d$, which must equal $-Q$ to maintain charge neutrality for the entire system. This condition fixes the value of the depletion-layer thickness z_d and of the surface potential

$$\phi_s = \frac{eN_A z_d^2}{2\varepsilon_{sc}} \quad (15.17)$$

This nonlinear approximation assumes that $e\phi_s \gg k_B T$. However, $e\phi_s$ must remain smaller than $(E_c - E_F)_b$, the energy separation between the conduction-band edge and the Fermi level in the bulk, or minority carriers will affect the charge density. The approximation given in equations (15.16) and (15.17) is widely used and will be referred to again below.

15.2. Physical Examples

The previous section contains some basic results for screening and space charge in systems that are varying in only one space dimension. Here we give a few simple examples of physical systems to which these results apply.

15.2.1. p–n Junctions

The familiar example of a p–n junction is a simple illustration of the results of the previous section. Suppose that the junction has N_D donors per unit volume in the half-space $z < 0$ and N_A acceptors per unit volume in the half-space $z > 0$. Then electrons will flow from the left to the right until a double layer is established to equalize the Fermi levels on the two sides. The nonlinear approximation of equation (15.16) applies in the usual case for which the band bending is much larger than $k_B T$. Thus the sum of the band bendings on the two sides of the junction is

$$\frac{e^2 N_D d_n^2}{2\varepsilon_{sc}} + \frac{e^2 N_A d_p^2}{2\varepsilon_{sc}} = (E_F - E_v)_n - (E_F - E_v)_p \tag{15.18}$$

where the terms on the right give the Fermi level relative to the top of the valence band in the n- and p-type bulk, respectively. The parameters d_n and d_p are the thicknesses of the space-charge layers on the two sides. They are determined by the condition of charge neutrality, that the total space charge in the double layer must vanish:

$$e d_n N_D - e d_p N_A = 0 \tag{15.19}$$

This is just the standard textbook example for a p–n junction.[259,262] We will not pursue it further here.

15.2.2. The Jellium–Vacuum Interface

Jellium is a theoretical substance with a fixed uniform positive charge density and a corresponding density of free electrons that is often used as a model for simple metals. When the positive charge density of jellium is abruptly terminated, the electrons will spill out slightly, creating a space-charge double layer. Because of the high charge densities present in metals, the characteristic distances are of order 0.1 nm. Many-body effects are essential for a correct treatment of the jellium–vacuum double layer and work function. This subject has been very well reviewed by Lang[263] and will not be considered further here.

15.2.3. Semiconductor Surfaces

The surface of a semiconductor is different from that of a metal because most of the electrons in a semiconductor are firmly attached to bonding orbitals, leaving only a relatively low density of free carriers. When a semiconductor surface is formed, there is generally a rearrangement of the outermost layers of atoms and there is often a reconstruction, which changes the symmetry of the outer layer or layers. The characterization of the semiconductor–vacuum surface is one of the central problems of surface science, and has a vast literature. One powerful experimental technique is photoemission, which has been reviewed elsewhere.[264]

The property of a semiconductor surface which we shall consider here is its tendency to become charged. This has traditionally been modeled by introducing surface states, energy levels for states localized near the surface whose charge state depends on the position of the Fermi level. Intrinsic surface states arise in an ideal semiconductor–vacuum interface with no impurity atoms or structural defects. These states form bands that reflect the symmetry of the surface, and therefore depend on the type of surface reconstruction present. A review of surface structures and surface states is given, for example, by Appelbaum and Hamann.[265] Spectroscopic methods of detecting surface states have been reviewed by Chiarotti.[266]

Intrinsic surface states have a density of order 10^{15} cm^{-2}, roughly the same as the density of surface atoms. Extrinsic surface states, which arise from impurities or structural defects, will usually have densities smaller by two or more orders of magnitude. Many of the early experiments on semiconductor space-charge layers used gas ambients and photoeffects to vary the surface charge.[255,256,267] In other cases, such as ZnO, both gas ambients and photoeffects can be used, and interface-charge densities of order 10^{14} cm^{-2} can be achieved.[268,269]

A classic example of intrinsic surface states is found near the top of the valence band in silicon, as demonstrated in the photoemission experiments of Eastman and Grobman[270] and Wagner and Spicer.[271] These states disappear when the silicon surface is oxidized. Other states appear, but they are not near the band edges. Only states within the band gap or near the band edges affect space-charge layers because only these states can have their occupations changed by band bending. The absence of intrinsic surface states in or near the band gap of silicon at the Si–SiO$_2$ interface helps to make it so important technologically.

15.2.4. Schottky Barriers

A very important semiconductor space-charge layer is the one that often arises when a metal is deposited on a semiconductor surface. It is called a Schottky barrier and leads to nonohmic current flow in the direction perpendicular to the surface. For a detailed discussion, the book by Rhoderick[257] can be consulted. References to early work can be found in the book by Mott and Gurney.[272]

The simplest model of a Schottky barrier, illustrated in Figure 15.1, leads to a band bending at the surface given by

$$- e\phi_s = \Phi_m - \chi_{sc} - (E_c - E_F)_b \qquad (15.20)$$

where Φ_m is the work function of the metal, the energy to remove an electron from the bulk to a point outside the surface; χ_{sc} is the electron affinity of the semiconductor, the energy required to take an electron from the bottom of the conduction band to a point outside the semiconductor; and $(E_c - E_F)_b$ is the energy separation between the bottom of the conduction band and the Fermi level in the bulk. The Schottky barrier height, as distinguished from the band bending given in equation (15.20), is usually defined as the magnitude of the energy difference between the majority-carrier-band edge at the surface and the Fermi level, and therefore omits the last term in equation (15.20).

Equation (15.20) describes the band bending in ionic semiconductors rather well, but less well for more covalent semiconductors.[273] The barrier height has also been related to the chemical heat of formation.[274] Theoretical models to explain departures from equation (15.20) involve

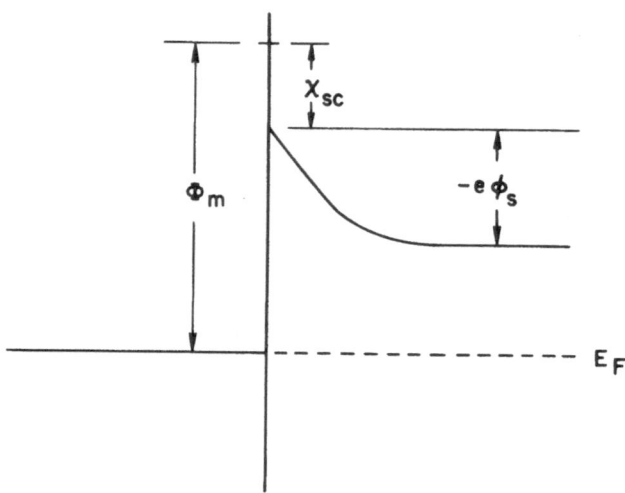

Figure 15.1. Idealized model of a Schottky barrier without interface states showing the work function Φ_m of the metal, and the electron affinity χ_{sc} and band bending $-e\phi_s$ of the semiconductor.

surface states,[275] the penetration of charge from the metal into the semiconductor,[276,277] or other mechanisms.

The structural basis of the simple models of metal–semiconductor interfaces have been questioned. The ideal, abrupt interface visualized in some models is being superseded by interfaces in which defects, chemical reactions, or transition layers are invoked as important factors in determining the barrier height. This also implies that the barrier height can depend on the method of formation of the metal–semiconductor contact, and requires that detailed structural information be available before a realistic theory can be constructed. Work on the mechanisms that determine the interface structure and presumably also the Schottky barrier height is being actively pursued,[278–282] and should lead to new insights and to a better experimental base of knowledge. The Proceedings of the annual conferences on the Physics of Compound Semiconductor Interfaces, published in the *Journal of Vacuum Science and Technology*, are a good guide to this rapidly evolving subject.

15.2.5. Metal–Insulator–Semiconductor Systems

A metal–insulator–semiconductor structure is basically a capacitor in which a semiconductor is one of the plates. This simple structure has taken on great technical importance because SiO_2 on Si is a particularly useful insulator. It has, when carefully made, a very high dielectric breakdown strength, a low density of interface states, and adequate stability. The charge density on the capacitor plates is given approximately by

$$Q = CV \sim \frac{\varepsilon_{ox}}{d_{ox}} V \sim \varepsilon_{ox} F_{ox} \qquad (15.21)$$

where V is the applied voltage, C is the capacitance per unit area, ε_{ox} is the permittivity of the oxide, d_{ox} is its thickness, and F_{ox} is the electric field in the oxide. Some correction terms have been ignored here for simplicity. If we take $\varepsilon_{ox} = 3.5 \times 10^{-13}$ F/cm and limit the electric field to 10^7 V/cm to avoid dielectric breakdown, we find that 2×10^{13} electrons can be induced per cm^2.

The effect of applying a field at the semiconductor–insulator interface is best understood from an energy-band diagram. Figure 15.2 shows the band bending that results when the semiconductor is p-type and a positive voltage is applied to the metal. As the applied voltage is increased, the band bending increases and the thickness

$$z_d = \left(\frac{2\varepsilon_{sc}\phi_s}{N_A e} \right)^{1/2} \qquad (15.22)$$

of the depletion layer, so called because it is largely depleted of mobile carriers, also increases.

As the applied voltage continues to increase, the conduction band at the surface will approach and eventually pass through the Fermi level. Then there will be an inversion layer at the surface, so called because the sign of the surface charge carriers is opposite to that of charge carriers in the bulk. Because the depletion layer provides an effective electric barrier between the inversion layer and the bulk, the transport properties of the inversion layer can be measured separately, at least at low frequencies.

Figure 15.3 shows the structure used to study the properties of the inversion layer. Contacts called source and drain contacts are introduced, usually by implanting or diffusing a high concentration of impurities with the same conductivity type as the inversion layer, here n-type. The charge density, and therefore the conductance, of the inversion layer is controlled by varying the voltage on the metallic electrode, called the gate. This is the basic structure of a metal-oxide-semiconductor field-effect transistor (MOSFET), sometimes called an insulated gate field-effect transistor. It is widely used in semiconductor memories and in other device applications.

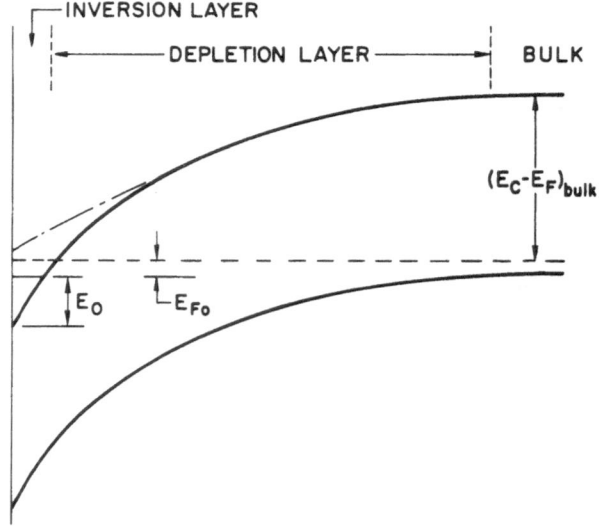

Figure 15.2. Schematic band bending near a semiconductor surface showing the conduction and valence band edges (solid lines) and the Fermi energy (dashed line). The dash–dot curve indicates the part of the band bending near the surface due to the fixed space charge alone. The additional band bending originates from the inversion layer charge density. E_0 is the lowest-inversion-layer energy level.

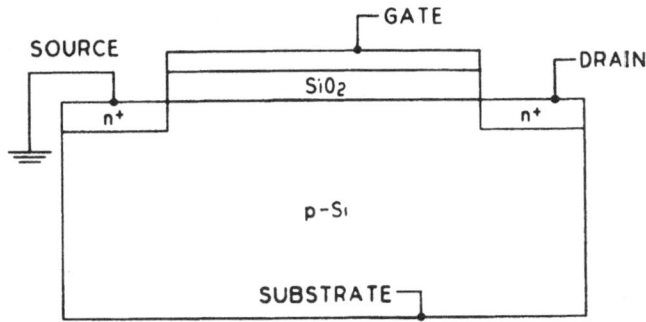

Figure 15.3. Schematic MOSFET structure showing the contacts to the source and drain for measuring the surface conductance, to the gate electrode for inducing charge near the silicon surface, and to the substrate for changing the potential drop in the depletion layer, used to control the surface electric field independently of the surface carrier concentration.

For a discussion of device properties, see for example the books of Grove[283] and Sze.[262]

In addition to voltages between the source–drain contacts and the gate, which control the inversion and depletion layer charge, and the voltage between source and drain that induces currents along the surface, one can also apply a voltage between the source–drain contacts and the semiconductor bulk. This voltage, called substrate bias, provides an independent control over the depletion layer charge and the electric field at the semiconductor–insulator interface.

The example given here is one in which a positive gate voltage is applied to an MOS system with a p-type bulk. If the gate voltage has the opposite sign, the bands bend the other way, the concentration of majority carriers at the surface is enhanced, and we speak of an accumulation layer.

15.2.6. Electrons on Liquid Helium

A space-charge layer of a rather different nature is formed when a layer of electrons is placed on the surface of liquid helium and is held there by the weak image potential and, in most cases, an applied electric field. This system has been well reviewed[284–289] and will not be discussed here. We note only that the long-expected electron crystal was first observed in this dynamically two-dimensional system.[290]

15.3. Experimental Probes of Space-Charge Layers

This section will contain simple accounts of some of the standard methods for characterizing space-charge layers and systems that have space-charge layers. More extensive accounts are given elsewhere.[255-257] Reference may also be made to the lectures of Many in the 1974 Surface Science course at Trieste.[291]

15.3.1. Capacitance

One of the most direct methods of studying a space-charge layer is to measure its capacitance. An idealized space-charge layer is simply an insulating region bounded by conducting electrodes, whose capacitance per unit area is simply the dielectric constant divided by the thickness, as in equation (15.21). More generally the capacitance — strictly speaking, the differential capacitance — is given by

$$C = \left| \frac{dQ}{dV} \right| \tag{15.23}$$

For the depletion-layer system described in equations (15.16) and (15.17), the charge density per unit area is given by the fixed bulk charge density $-eN_A$ times the depletion-layer thickness $z_d = (2\varepsilon_{sc}\phi_s/eN_A)^{1/2}$. Thus the space-charge capacitance is

$$C_{sc} = \left| \frac{dQ}{d\phi_s} \right| = \left| \frac{Q}{2\phi_s} \right| = \frac{\varepsilon_{sc}}{z_d} \tag{15.24}$$

which applies, for example, to a Schottky barrier, to the depletion layer of a metal-oxide-semiconductor system, or to each side of a p–n junction, all with uniform doping.

In a metal-oxide-semiconductor structure, the space-charge capacitance (15.24) is in series with the oxide capacitance ε_{ox}/d_{ox} and the combined capacitance is

$$C = \frac{1}{1/C_{ox} + 1/C_{sc}} \tag{15.25}$$

In a space-charge layer that is strongly accumulated or strongly inverted, the space charge is very close to the interface and the space-charge capacitance is very large. In those limits therefore, the capacitance approaches the insulator capacitance. As the bands are bent through the flat-band position towards inversion, the space-charge thickness reaches its maximum value, and the capacitance reaches its minimum value, just before the onset of inversion. In this case the value of the surface potential is approximately equal to the band gap of the semiconductor. If a substrate bias is applied in the structure of Figure 15.3, then the value of the substrate bias must be included in calculating the total band bending in the depletion layer.

Several experimental considerations enter in actual measurements. One is the dynamic response of the system. The expressions we have given assume that the system remains in quasi-equilibrium during the course of a measurement. Normally, capacitance is measured by slowly sweeping the gate voltage while applying a small ac voltage that is used to measure the capacitance. If carriers do not respond fast enough, then the measured value will not correspond to the steady-state capacitance. This can be particularly serious when a capacitor is measured in inversion and there are no source–drain contacts present. Then the minority carriers that form the inversion layer must be thermally generated, a process that can take hours at low temperatures. One can deal with this problem in several ways, one of which is to shine light on the sample to generate minority carriers.

A second complication of capacitance measurements in real MOS systems is the presence of interface charge. As the surface potential changes, the charge in the interface region changes, which gives an additional contribution to the capacitance (15.23). If the states charge rapidly they are called fast states and can give structure to the capacitance–voltage curve. If the states charge so slowly that they do not keep up with the slow dc voltage sweep, they are called slow states. The presence of slow states can lead to hysteresis and other time-dependent effects in the capacitance. One of the reasons the Si–SiO$_2$ system is so useful in device applications is that ways have been found to reduce the density of interface states to very low values, of order 10^{10} per cm^2 or even less, which means that almost all the induced charge in an MOS device goes into the space-charge layer and very little goes into interface charge. This was achieved only after a massive technological effort involving many laboratories. Much of the scientific work on inversion layers described in Section 15.4 below has been made possible by these advances in technology.

For additional information on capacitance measurements, the reader may consult, for example, some of the works already cited.[255–257,262,283] Detailed analysis of MOS capacitors has been published by McNutt and Sah,[292] and modeling of the transient response has been described by Collins et al.[293]

15.3.2. Conductance

A space-charge layer can also be studied by measuring the conductance either through the layer, as commonly used to study Schottky barriers,[257] or along the layer. Here we only discuss the latter method.

Field-effect measurements, in which the conductance of a sample is measured under the influence of an external field or of some other external variable that changes the surface potential, constitute one of the classical tools for studying surface space-charge layers. The theory of surface effects on conduction, with and without band bending, developed for many years along lines suggested by Fuchs[294] that invoke specular and diffuse scattering of carriers incident on the surface. There are many good reviews of this work.[255,256,295]

For MOS structures, it is easy to use an external voltage to change the surface potential. To provide an easy way of measuring the surface conductance, source and drain electrodes of conductivity type opposite to that of the bulk are diffused or implanted into the semiconductor, as shown in Figure 15.3. A p–n junction barrier then prevents current from flowing between the contacts through the bulk. The conductance of the surface space-charge layer can be measured alone, a much simpler experimental task. There is appreciable surface conductance only when the surface is inverted. In a relatively ideal system with few interface states, the inversion-layer-charge density is given approximately by

$$eN_s \simeq C_{ox}(V - V_{th}) \tag{15.26}$$

where V_{th} is the nominal threshold voltage for introducing charge in the inversion layer. The effective inversion-layer mobility is then given by

$$\mu_{eff} = \frac{g_D L}{eN_s W} \tag{15.27}$$

where g_D is the source–drain conductance, and L and W are the length and width, respectively, of the conducting channel on the surface. Some of the experimental problems in determining the threshold voltage, and other experimental limitations on the accuracy of equation (15.26) have been discussed by Fowler and Hartstein.[296]

An extensive analysis of the conductance in an MOSFET that includes interface states and potential fluctuations is given by Muls et al.[297] Goodman and Fritzsche[298] have described a detailed application of field effect and capacitance measurements to space-charge layers in hydrogenated amorphous silicon to determine the density of electronic states in the pseudogap.

A number of authors[299,300] have considered the possibility that an inversion layer might be superconducting, but the predicted transition temperature is well below 0.1 °K and does not appear to have been reached.

15.3.3. Photoeffects

Optical measurements can also be used to characterize space-charge layers. Surface-photovoltage measurement, an old technique,[255] has been recently used by Maltby et al.[301] to characterize the space-charge layer and the surface of CdS, and by Goldstein and Szostak[302] for similar characterization of amorphous hydrogenated silicon. Two other optical methods, which we shall describe briefly, are internal photoemission as used to measure semiconductor–insulator barrier heights, and the photocurrent-voltage method used to characterize space charges in the insulator of an MOS structure.

The energy gap of insulators used in MOS devices is generally considerably larger than the energy gap of the semiconductor. The bands generally align in such a way that there is a barrier both for electron injection and for hole injection from the semiconductor to the insulator. One way to measure this barrier height is to excite electron–hole pairs optically in the presence of an applied field of suitable polarity and measure the current as a function of the photon energy. The photon energy at which the extrapolated photocurrent goes to zero is equal to the barrier height of the injected carrier plus the energy gap of the semiconductor. This method was used by Williams[303] to determine that there is a barrier of 3.1 to 3.2 eV from the conduction band of Si to the conduction band of SiO_2 in a Si–SiO_2 interface. Note that electrons in the conduction band of SiO_2 have a mobility of 20 cm^2/V s at room temperature.[304]

Optical excitation is also used in the photocurrent–voltage or photo-I–V method to inject charge into the insulator of an MOS structure and determine its location. Powell and Berglund[305] and DiMaria[306] have shown that some average properties of fixed charge in the oxide can be determined from the current–voltage characteristics of an MOS structure in which carriers are excited optically in one of the electrodes. The oxide charges influence the electric field near the electrode–oxide interface and therefore modify the contact resistance. DiMaria[306] showed that by studying the current–voltage characteristics for both polarities and comparing the curves with those of a nominally uncharged oxide, he could determine both the amount of oxide charge and its centroid. The mechanisms that trap charge in SiO_2 have been reviewed by Young.[307]

15.3.4. Interface Characterization

Simple treatments of space-charge layers regard the surface or interface as a sharp boundary that terminates the space-charge layer. We have already noted that the Schottky barrier can be a rather complex interface and has been attracting considerable experimental attention. The semiconductor–insulator interface also generally differs from the ideal, as demonstrated by the presence of interface charges and by the surface recombination velocity, which is an interface effect on carrier lifetime.[255,256,259] In this section we note some of the methods that have been used to characterize the physical structure of the Si–SiO_2 interface.

A simple model shows that even a crystallographically perfect interface between two materials like Si and SiO_2 must have at least one layer of Si atoms whose bonding is intermediate between that of Si and SiO_2. Evidence for the presence of such a layer has been seen in photoemission[308,309] through the change in the core levels produced by the different local environment of the atoms in the interface layer, and its thickness has been deduced to be of order 1 nm or less.

Another powerful analytic method that has been applied to the characterization of the $Si-SiO_2$ interface is nuclear backscattering. This method uses simple conservation of energy and momentum to deduce the mass of a target nucleus that scatters an incident nucleus, say a 1 MeV alpha particle, into a particular direction with a particular energy loss. Additional energy losses due to low-energy excitations define a depth scale in the target material. This method therefore allows the composition profile of a suitable target to be determined rather directly. By such means, the $Si-SiO_2$ interface can be shown to be approximately 1 to 2 atomic layers thick.[310] This method gives perhaps the most convincing evidence that the interface between Si and its thermally grown oxide is quite sharp. Other evidence[311-313] shows that the surface has slight deviations from flatness, as had already been deduced from analysis of the mobility in inversion layers, to be described below.

15.4. Semiconductor Inversion Layers

Semiconductor inversion layers are a particular example of space-charge layers that have been widely studied because they are involved in an important class of semiconductor devices widely used in computers and other applications, and because their quasi-two-dimensional character and the easily varied carrier concentration make them a model system for many types of experiments and theories. There have been several reviews[314-320] of this field, and the Proceedings of the International Conferences on Electronic Properties of Two-Dimensional Systems — published in Volumes 58, 73, 98, 113, and 142 of *Surface Science* — are good guides to its development.

15.4.1. Quantum Effects

When a strong electric field attracts electrons to a semiconductor surface to form an inversion layer, the electrons are confined in an asymmetric potential well bounded on one side by the barrier that keeps electrons in the semiconductor and on the other side by the sloping conduction-band edge. It is customary to represent the former by an abrupt, infinite barrier because that simplifies the calculations without introducing substantial errors in most cases. The latter is sometimes approximated by the linear potential of a constant electric field, but this turns out to be only qualitatively correct. The states in the inversion layer can be represented approximately by wave functions of the form

$$\psi = \zeta(z)\exp(i\mathbf{k} \cdot \mathbf{r})u(x, y, z) \tag{15.28}$$

where u is a three-dimensional Bloch function for the bottom of the conduction band, $\zeta(z)$ is an envelope function, and $\mathbf{k}$ and $\mathbf{r}$ are two-dimensional vectors in the directions along the surface. The separation of the $\mathbf{r}$ and z coordinates is exact only for an abrupt, infinite surface barrier and for bulk ellipsoids whose major axes are parallel and perpendicular to the surface.[242] For silicon, which is the prototypical semiconductor for inversion-layer studies because of its desirable oxide and interface properties, the sixfold conduction-band degeneracy means that there are six different Bloch functions. We shall see that the surface field splits this degeneracy in most cases.

The energy levels for the states (15.28) are given in a one-electron approximation by

$$E = E_i + \frac{\hbar^2 k_x^2}{2m_x} + \frac{\hbar^2 k_y^2}{2m_y} \tag{15.29}$$

where m_x and m_y are the principal effective masses for motion parallel to the surface and where the E_i are the eigenvalues of the Schrödinger equation

$$-\frac{\hbar^2}{2m_z}\frac{d^2\zeta_i(z)}{dz^2} + [V(z) - E_i]\zeta_i(z) = 0 \tag{15.30}$$

Each value E_i is the beginning of a two-dimensional continuum of energies called a subband, and all the values E_i for a given value of m_z, the effective mass for motion perpendicular to the surface, form a so-called ladder of subbands.

For inversion layers at a (001) Si–SiO₂ interface, the most widely studied case, the six conduction bands divide into a twofold ladder of subbands conventionally called $0, 1, 2, \ldots$ and a fourfold ladder of subbands conventionally called $0', 1', 2', \ldots$. The twofold ladder is associated with the valleys having their heavy mass perpendicular to the surface, giving lower-energy solutions of equation (15.30). Energy levels for a particular case are shown in Figure 15.4, taken from Stern's work.[243]

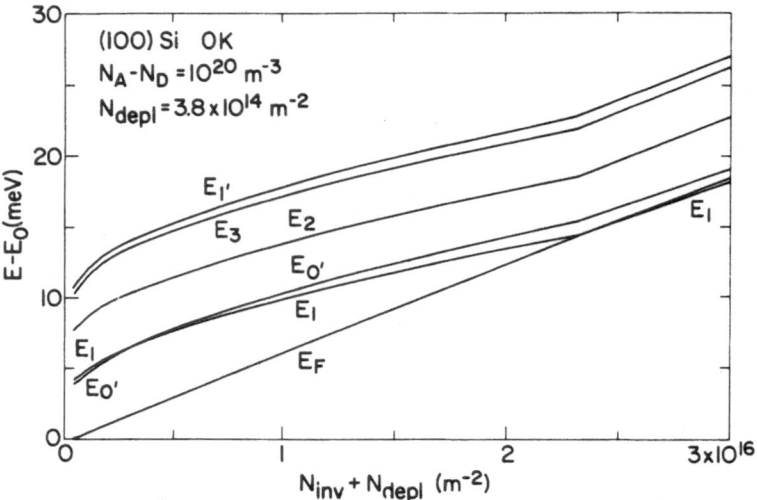

Figure 15.4. Energy levels for electrons in an inversion layer on a silicon (100) surface vs total density of charges in the space-charge layer, calculated from a self-consistent Hartree scheme without image or exchange and correlation effects. Note the change in slope as the Fermi energy E_F crosses the bottom of the first excited subband. All energies are measured from the bottom of the lowest subband (after Stern[243]).

The potential energy is

$$V(z) = V_d(z) + V_i(z) + V_{im}(z) + V_{xc}(z) \tag{15.31}$$

Here V_d is the potential energy associated with the fixed charges and V_i is the potential energy associated with the inversion-layer charge distribution

$$\rho(z) = -e \sum_i N_i \zeta_i^2(z) \tag{15.32}$$

where

$$N_i = \frac{m_{di} g_{vi}}{\pi \hbar^2} k_B T \ln\left[1 + \exp\left(\frac{E_F - E_i}{k_B T}\right)\right] \tag{15.33}$$

gives the number of electrons per unit area in the ith subband, $\dot{m}_d = (m_x m_y)^{1/2}$ is the density-of-states effective mass per valley, g_v is the valley degeneracy, and E_F is the Fermi energy. Conventionally V_d and V_i are taken to vanish at the interface, where $z = 0$, so V_d thus differs by a constant from $-e\phi(z)$ with $\phi(z)$ given by equation (15.16), which was chosen to vanish in the bulk. The image potential energy seen by an electron in the semiconductor because of the different dielectric constants of the semiconductor and the oxide is

$$V_{im}(z) = \frac{\varepsilon_{sc} - \varepsilon_{ox}}{\varepsilon_{sc} + \varepsilon_{ox}} \frac{e^2}{16\pi\varepsilon_{sc}z} \tag{15.34}$$

The remaining term in the potential energy, if included, is an effective one-electron potential energy V_{xc} to represent many-body exchange and correlation effects as discussed, for example, by Kalia et al.[321]

Because V_i is the solution of Poisson's equation for the charge density (15.32) and because the wave functions ζ_i are the solutions of equation (15.30), which includes V_i in the potential, the solution of the Schrödinger equation becomes a nonlinear eigenvalue problem that is usually solved iteratively. Several methods of solution have been described by Stern.[322]

Calculations that omit exchange and correlation effects are usually called Hartree calculations. They give results that are in substantial disagreement with the optical measurements discussed below. Accounts of many-body theory as applied to the energy levels in inversion layers can be found in the works of Ando,[323] Jonson,[244] and Vinter,[324] among others.

When only one subband is occupied, the inversion layer electrons are free to move only in two dimensions. The system is then said to be quasi-two-dimensional or dynamically two-dimensional although the wave functions have a finite extension in the third direction (typically 1 to 10 nm).

15.4.2. Transport Properties

The transport and magnetotransport properties of semiconductor inversion layers have been actively studied for many years, and are discussed in a copious body of literature. Here we shall only very briefly describe a few key features of this work and give references to the literature. Measurements using the basic-metal-oxide-semiconductor field-effect transistor structure of Figure 15.3 have been carried out for samples with a wide range of bulk doping and interface charge, and over a wide range of temperatures. The transport has been studied with a magnetic field perpendicular to the surface and at various angles to the surface and to the current and has also been studied as a function of applied uniaxial stress.

Mobility of electrons in a silicon (001) inversion layer at several temperatures is shown in Figure 15.5, taken from a review of transport by Fowler.[325] Comprehensive mobility measurements were published by Fang and Fowler.[326] The general features of the data can be explained by invoking phonon scattering, Coulomb scattering from charges near the Si–SiO$_2$ interface, and interface roughness scattering. Note in particular the decrease of mobility in Figure 15.5 as the oxide-charge density increases.

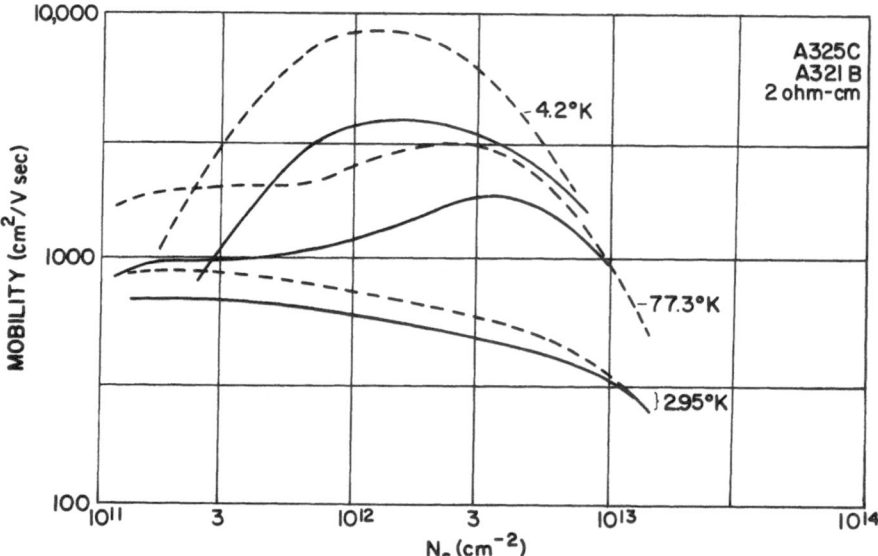

Figure 15.5. Mobility vs temperature and inversion-layer carrier concentration for an inversion layer of electrons on the (100) surface of silicon. The two sets of curves are for samples with different amounts of oxide charge, with the dashed curves corresponding to the sample with lower oxide-charge density near the silicon–silicon dioxide interface (after Fowler[325]).

Scattering from acoustic[327] and optical[328] phonons accounts in a general way for the temperature dependence of the mobility, although there are some quantitative problems. Recent work by Cham and Wheeler,[329] Stern,[330] and Kawaguchi et al.[331] suggests that at low temperatures a large part of the observed temperature dependence of mobility may be a result of the temperature dependence of screening rather than a result of phonon scattering.

Measurements of mobility in samples in which the interface charge is varied, as carried out by Hartstein et al.,[332] can be accounted for at least grossly in terms of Coulomb scattering from interface charges, as treated by Stern and Howard[242] for a simple case and by Mori and Ando[333] for a case in which more than one subband is occupied. The rapid fall-off in mobility near threshold at low temperatures is attributed to localization effects, as discussed in Section 15.4.4, to carrier-density fluctuations as discussed by Brews,[334] and to changes in screening near threshold as discussed by Shiue and Sah[335] and by Yagi and Nakai.[336]

The fall-off in mobility for high values of the surface electron density is attributed to interface roughness, a deviation of the Si–SiO₂ interface from flatness, as first discussed for electrons in magnetic-field-induced surface states by Prange and Nee[337] and as discussed more fully for inversion layers by Matsumoto and Uemura[338] and by Ando.[339] The existence of interface roughness has been verified by independent means, as discussed for example by Sugano et al.[313]

A magnetic field perpendicular to the surface quantizes the motion along the surface into a series of Landau levels and therefore removes all the remaining degrees of freedom of the carriers in an inversion layer. The density of states becomes a series of delta functions, apart from broadening effects.[340,341] Fowler et al.[342] first showed that by increasing the carrier concentration and sweeping the Fermi level through the ladder of Landau levels one can see a uniformly spaced series of oscillations in the conductance, which is a clear demonstration of the two-dimensional character of the system. Tilting the magnetic field changes the ratio of spin splitting and Landau splitting, because the latter depends only on the normal component of the magnetic field, and has been used to show that the g-factor is substantially different from the free-electron value 2.0 and increases as the electron density N_s decreases.[343] This has been explained as a consequence of electron–electron interaction.[344,345] A similar but weaker dependence of the electron effective mass on N_s, determined from the temperature dependence of the magnetoconductance oscillations,[346] is also attributed to many-body effects.[347-349] Some interesting questions about details of many-body theory arise in this connection (see, for example, papers by Vinter[348] and Lee et al.[350]) and are still unresolved at the time of writing.

Application of uniaxial stress changes the relative energies of the six conduction-band minima of silicon, and therefore changes the relative energies and populations of the subbands. Of particular interest is the behavior of the conductance, magnetoconductance, and cyclotron resonance when two subbands associated with different valleys approach each other in energy and are both populated by electrons. For experiments using stress see, for example, Dorda,[351] Tsui and Kaminsky,[352] Eisele,[353] and Stallhofer et al.[354] Related calculations have been carried out by Takada and Ando[355] and by Vinter.[356]

Magnetotransport properties of inversion layers have been widely studied and given detailed information about level splittings and level broadening. Klitzing et al.[357] showed that the Hall conductance measured when the Fermi level lies in the density-of-states minimum between two Landau levels is a precise integer multiple of e^2/h and may prove to be another way to measure fundamental constants.

15.4.3. Optical Properties

In this section we shall briefly mention a few of the optical techniques that have been used to study semiconductor inversion layers. Of these the optical absorption associated with transitions between subbands is perhaps the most direct confirmation of the theory of quantization in the surface potential well. Figure 15.6 is a sample absorption curve, taken from the work of Kneschaurek and Koch[358] on (100) silicon. The curve shows a transition from the lowest subband of the twofold set of valleys to the first excited subband associated with the same conduction-band valleys. At higher temperatures, a line attributed to transitions from the lowest subband of the other set of valleys to the first excited state of those valleys appears as the ground state for that transition becomes thermally populated. The matrix element for dipole transitions is very small unless the

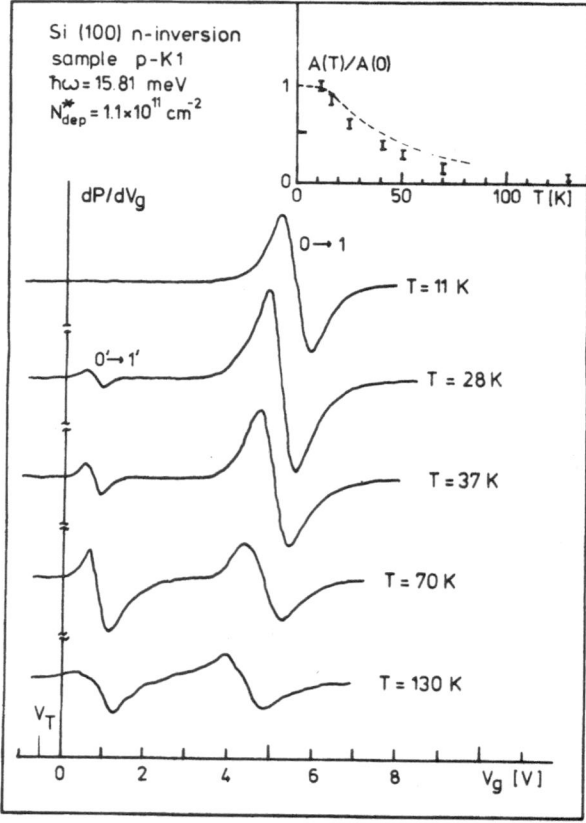

Figure 15.6. Optical absorption of a silicon (100) inversion layer showing the transition from the lowest subband (0) to the first excited subband (1) of the same valleys and, at higher temperatures, the appearance of a transition between subbands 0' and 1' associated with the higher-lying valleys. The photon energy is fixed at 15.8 meV. A change of 1 V in gate voltage changes the inversion-layer electron concentration by about 10^{11} cm^{-2}. The inset shows the temperature dependence of the $0 \rightarrow 1$ transition amplitude (after Kneschaurek and Koch[358]).

initial and final subbands are attached to the same set of valleys. Optical transitions between subbands of different valleys have not been observed.

The measurements are most commonly carried out at a fixed frequency, using a far-infrared gas laser, varying the gate voltage on the sample to change the energy splittings between subbands. They can also be done at fixed surface conditions by using a Fourier transform spectrometer.[359] The dependence of the energy levels on inversion-layer electron concentration can only be understood if many-body effects are included. Several other corrections to the measured spectra also enter (see, for example, Ando[360]) but do not alter this conclusion. Theoretical treatments of the generalized nonlocal dielectric function of surface space-charge layers have been given by Dahl and Sham[361] and by Eguiluz and Maradudin.[362]

Cyclotron resonance is another optical experiment that verifies the quantum description of electrons in inversion layers, although there are still many unresolved matters of detail. Discussions of this subject have been given by Kotthaus[363] and by Wagner *et al.*[364]

Collective or plasma modes of a two-dimensional electron gas have been known for many years[365] to give an excitation spectrum that starts at zero energy, rather than at a finite energy as for bulk or surface plasmons. This was verified first for electrons on liquid helium[366] and later for electrons in inversion layers.[367] A good review has been given by Theis.[368]

The inelastic light-scattering methods so successfully used to study the electronic subbands in superlattices and heterojunctions[369,370] have not yet been successfully applied to silicon. The theoretical considerations that enter in the Raman scattering by two-dimensional electron systems have been reviewed by Burstein *et al.*[371]

15.4.4. Impurity Bands and Localization

The low-temperature conductivity of an inversion layer near threshold is activated, as shown by Fang and Fowler.[326] Many of the observed effects[372] can be qualitatively understood in terms of a mobility edge[373] separating localized from delocalized states. (For different views see, for example, Arnold[374] and Adkins.[375]) The theoretical basis for the existence of a mobility edge in a two-dimensional system is in question. There is now a large body of theoretical[376-387] and experimental[388-394] work on quasi-metallic two-dimensional electron systems which is consistent with a logarithmic increase in resistance with decreasing temperature at low temperatures. The size of the increase is rather small, often of the order of 10% or less, and the effect has been called weak localization to distinguish it from the strong localization — accompanied by exponentially activated conductance — that occurs for carriers deep in a band tail. Some authors[377] attribute the logarithmic terms at low temperature to electron–electron interaction rather than to the effect of potential fluctuations. The subject is presently being actively pursued both experimentally and theoretically.

One of the concepts that had been used to categorize the data until more recent work on weak localization is the concept of minimum metallic conductivity. In some[395] but not all[396,397] measurements, the division between activated and quasi-metallic conduction occurred near a conductivity of $0.1\ e^2/\hbar$,[398] which equals about 3×10^{-5} siemens. Theoretical support for such a value had come from the work of Licciardello and Thouless[399] (see also Thouless and Elzain[400] and Stein and Krey[401]). Pepper[402] speculated that the samples which exhibited the expected minimum metallic conductivity had potential fluctuations of rather short range, while those that deviated from it had longer range fluctuations. He found that he could influence the behavior by pulling electrons away from the surface with a substrate bias.

The simplest example of activated conduction is that associated with a shallow impurity level. Such levels in inversion layers were first studied by Harstein and Fowler,[403] who drifted sodium ions to the Si–SiO$_2$ interface and measured the conductance near threshold. The sodium is believed to be the surface analog of a conventional shallow impurity in a bulk semiconductor. Their most detailed experiments[404] found that the conductance has a low activation energy — attributed to hopping within the impurity band — which has a minimum near a half-filled impurity band and a second, larger activation energy — attributed to activation to the mobility edge — which also changes with impurity-band occupation. These effects have been considered theoretically by Hayden and Butcher,[405,406] who obtain qualitative agreement with the experimental results.

The binding energy of electrons to interface charges has been studied by Wallis and co-workers,[407,408] Lipari,[409] Vinter,[410] and Hipolito and Campos.[411] Only the last of these calculations considered screening effects under conditions that can be compared with the experiments, and found a moderate reduction from the values calculated without screening. There is at least qualitative agreement between calculated and measured binding energies.

15.5. Two-Dimensional Physics: Some Examples

In this section we briefly describe some results for dynamically two-dimensional systems embedded in a three-dimensional space. An inversion layer at a semiconductor–insulator interface is one example of such a system. The systems are two-dimensional in the sense that motion can only take place in the xy plane. Electromagnetic fields, on the other hand, are free to penetrate outside the plane. Such systems, in which charges interact with a $1/r$ potential, are to be distinguished from the two-dimensional Coulomb gas,[412] in which fields do not leave the plane and charges interact with logarithmic potentials. We briefly discuss first scattering by an unscreened Coulomb potential and then some properties of bound states, both of which differ in interesting ways from the corresponding three-dimensional results. Another difference between a two-dimensional electron gas and its three-dimensional counterpart is the spectrum of plasma oscillations, mentioned briefly in Section 15.4.3.

15.5.1. Coulomb Scattering

In three dimensions, the differential cross section for scattering by an unscreened Coulomb potential $Ze/4\pi\varepsilon r$ is the same whether determined classically, from the Born approximation, or from the solution of the Schrödinger equation. For an electron in two dimensions, the results of the three cases differ. The Schrödinger equation solution gives[242]

$$\sigma(\vartheta) = \frac{G \tanh \pi G}{2k \sin^2 \vartheta/2} \tag{15.35}$$

where $\sigma(\vartheta)$ is the differential cross section at scattering angle ϑ, k is the electron wave vector, $G = mZe^2/4\pi\varepsilon k\hbar^2$, and m is the electron effective mass.

For high velocities, where the Born approximation is expected to be valid, G is small and the cross section becomes

$$\sigma(\vartheta) = \frac{\pi(Ze^2/4\pi\varepsilon\hbar v)^2}{2k \sin^2 \vartheta/2} \tag{15.36}$$

where $v = \hbar k/m$, which is the Born approximation result.[242]

For low velocities, for which the classical approximation is valid,[413] G is large and the cross section becomes

$$\sigma(\vartheta) = \frac{|Z|e^2}{8\pi\varepsilon mv^2 \sin^2 \vartheta/2} \tag{15.37}$$

which is the classical result.[414]

15.5.2. Bound States

We call attention here to a few ways in which bound states in two dimensions differ from their three-dimensional counterparts. One is the binding to an unscreened attractive Coulomb potential, which is 4 Ry in two dimensions as against 1 Ry in three dimensions, where $Ry = me^4/32\pi^2\varepsilon^2\hbar^2$ is the effective rydberg and m is the effective mass. Note that the dielectric constant in the two-dimensional case is the average dielectric constant of the media, assumed homogeneous, surrounding the electron layer. Another difference is the fact that an arbitrarily weak attractive potential has a bound state in two dimensions[415] while a minimum strength is required to bind a state in three dimensions. These and other aspects of binding in two dimensions are discussed elsewhere.[242] The transition from two-dimensional to three-dimensional behavior for the binding of an electron to a fixed attractive charge in a layer of finite thickness has been discussed by Keldysh.[416]

Summary

Space-charge layers constitute a rich field with important physical and technological aspects, a few of which have been briefly mentioned here. An inversion layer at a semiconductor–insulator interface is an example of a space-charge layer that has a dynamically two-dimensional character and — because of its easily variable carrier concentration — lends itself well to a number of experiments not easily carried out in bulk systems. Only a glimpse at the content of this large and active field could be given here.

Acknowledgments

I am indebted to D. J. DiMaria, J. L. Freeouf, R. Ludeke, D. R. Young, and other colleagues for helpful comments; to A. B. Fowler for providing Figure 15.5; and to M. P. Tosi, E. Tosatti, G. Srinivasan, H. G. Reik, and the participants in the 1980 Spring College for their hospitality and interest.

Note Added in Proof. For more recent developments in the physics of Schottky barriers and related interface problems, see the Proceedings of the Second Trieste Semiconductor Symposium (1982), *Surfaces and Interfaces: Physics and Electronics* (R. S. Bauer, ed.), published as Vol. 132 of *Surface Science*. A review of electronic properties of two-dimensional systems has been given by T. Ando, A. B. Fowler, and F. Stern, *Rev. Mod. Phys.* **54**, 437–672 (1982).

Superconductivity via Electron–Phonon and Electron–Exciton Interactions

S. Strässler and P. Wyder

In this chapter an attempt will be made to present some basic ideas in the theory of conductivity. We start immediately below in Section 16.1 with a discussion of the electron–phonon Hamiltonian in the jellium approximation. The effective electron–electron interaction is derived using a canonical transformation and the "Cooper pair" is then discussed in Section 16.2. The Bardeen–Cooper–Schrieffer ground-state wave function is derived in Section 16.3 while the general theory of the inhomogeneous superconductor is discussed in Section 16.4 on the basis of the Bogoliubov equation. In Section 16.5 this equation is solved for the homogeneous case and, in particular, is used to discuss some aspects of the transition temperature. In Section 16.6 the derivation of the Landau–Ginzburg theory is given. In the last two sections, this background is utilized to give short accounts on the excitonic mechanism and the effects of limited dimensionality.

A number of books and articles deal with some of the topics considered in the text below. We mention, as examples, references to second quantization,[417,418] and to the problem of the electron–phonon interaction.[419] A rather complete treatment of superconductivity can be found in the book by Parks.[420] Eilenberger and Jacobs[421] give the

S. Strässler · Cerberus A6, CH-8708 Männedorf, Switzerland, **P. Wyder** · Research Institute for Materials, University of Nijmegen, Toernooiveld, Nijmegen, The Netherlands.

derivation of the Landau–Ginzburg theory mentioned in Section 16.4, while Götzel *et al.*[422] discuss the possibility of deriving transition temperatures from first-principle calculations. Without more ado we turn to the electron–phonon interaction.

16.1. Electron–Phonon Hamiltonian

In this section we discuss the interaction between electrons and lattice vibrations in a crystal within the jellium model (cf Chapter 14). In this model the ions of charge Ze are described by a homogeneous density n_i. If we consider a general displacement field $\psi_\nu(\mathbf{r})$ for the ions but leave the electronic system unchanged, the system will become locally charged. The charge density is given by

$$q(\mathbf{r}) = -Zn_i e \nabla \psi \tag{16.1}$$

This charge leads to an electrostatic potential $\phi(\mathbf{r})$, which can be computed from Poisson's equation

$$-\nabla^2 \phi(\mathbf{r}) = +4\pi q(\mathbf{r}) \tag{16.2}$$

The potential energy for the jellium model then becomes

$$U = \tfrac{1}{2} \int d^3 r \, q(\mathbf{r}) \phi(\mathbf{r}) \tag{16.3}$$

Note that U is harmonic in ψ, therefore the Hamiltonian can be diagonalized directly.

The Hamiltonian of the ionic jellium of mass density ρ is

$$H = \frac{1}{2\rho} \int \boldsymbol{\pi}(\mathbf{r})\boldsymbol{\pi}(\mathbf{r}) d^3 r + U \tag{16.4}$$

where the field operators satisfy the commutation relation

$$[\psi_\mu(\mathbf{r}), \pi_{\mu'}(\mathbf{r})] = \delta_{\mu\mu'} \delta(\mathbf{r} - \mathbf{r}') \tag{16.5}$$

Here $\boldsymbol{\pi}(\mathbf{r})$ is the momentum density.

Let us consider a bulk of volume V and assume cyclic boundary conditions. We now make the ansatz

$$\psi(\mathbf{r}) = \frac{1}{\sqrt{V}} \sum_{\mathbf{k}} \mathbf{Q}(\mathbf{k}) e^{i\mathbf{k}\mathbf{r}} \tag{16.6}$$

$$\pi(\mathbf{r}) = \sqrt{\frac{1}{V}} \sum_{\mathbf{k}} P(\mathbf{k}) e^{i\mathbf{k}\mathbf{r}} \tag{16.7}$$

$$\phi(\mathbf{r}) = \frac{1}{\sqrt{V}} \sum_{\mathbf{k}} \phi(\mathbf{k}) e^{i\mathbf{k}\mathbf{r}} \tag{16.8}$$

$$q(\mathbf{r}) = \frac{1}{\sqrt{V}} \sum_{\mathbf{k}} q(\mathbf{k}) e^{i\mathbf{k}\mathbf{r}} \tag{16.9}$$

From equation (16.1)

$$q(\mathbf{k}) = -Zn_i e i \mathbf{k} \cdot \mathbf{Q}(\mathbf{k}) \tag{16.10}$$

and from equation (16.2)

$$k^2 \phi(\mathbf{k}) = 4\pi q(\mathbf{k}) \tag{16.11}$$

The potential energy thus becomes

$$U = \frac{1}{2} \sum_{\mathbf{k}} q(\mathbf{k})\phi(-\mathbf{k}) = \frac{4\pi}{2} \sum_{\mathbf{k}} \frac{1}{k^2} q(\mathbf{k})q(-\mathbf{k})$$

$$= 4\pi n_i^2 e^2 Z^2 \sum_{\mathbf{k}} \frac{1}{k^2} [\mathbf{k}\mathbf{Q}(\mathbf{k})] \cdot [\mathbf{k}\mathbf{Q}(-\mathbf{k})] \tag{16.12}$$

The problem separates in longitudinal modes $\mathbf{Q} \parallel \mathbf{k}$ ($\kappa = 1$) and transverse modes $\mathbf{Q} \perp \mathbf{k}$ ($\kappa = 2, 3$):

$$H = \frac{1}{2} \sum_{\mathbf{k},\kappa} \left[\frac{1}{\rho} P_\kappa(\mathbf{k}) P_\kappa(-\mathbf{k}) + \frac{1}{2} \rho \omega_\kappa^2(\mathbf{k}) Q_\kappa(\mathbf{k}) Q_\kappa(-\mathbf{k}) \right] \tag{16.13}$$

where

$$\omega_\kappa(\mathbf{k}) = \begin{cases} \sqrt{\dfrac{4\pi n_i e^2 Z^2}{M}} = \omega_p & \kappa = 1 \\[2mm] 0 & \kappa = 2, 3 \end{cases} \tag{16.14}$$

Here we have introduced the ion-mass density $\rho = n_i M$ where M is the mass of the ions. It is noteworthy that for the jellium model there is no restoring force for transverse modes and that the dispersionless longitudinal frequency is equal to the so-called plasma frequency ω_p. Diagonalization of (16.13) proceeds as usual:

$$Q_\kappa(\mathbf{k}) = \sqrt{\frac{1}{2\omega_\kappa(\mathbf{k})\rho}} [a_\kappa(\mathbf{k}) + a_\kappa^+(-\mathbf{k})]$$

$$\tag{16.15}$$

$$P_\kappa(\mathbf{k}) = +\sqrt{\frac{\omega_\kappa(\mathbf{k})\rho}{2}} i[a_\kappa^+(\mathbf{k}) - a_\kappa(-\mathbf{k})]$$

where from (16.5) it follows that the operators $a_\kappa^{(+)}(\mathbf{k})$ are Boson operators:

$$[a_\kappa(\mathbf{k}), a_{\kappa'}^+(\mathbf{k}')] = \delta_{\mathbf{kk}'}\delta_{\kappa\kappa'} \tag{16.16}$$

The phonon Hamiltonian thus obtains its standard diagonal form

$$H = \sum_{\kappa\mathbf{k}} \omega_\kappa(\mathbf{k})[a_\kappa^+(\mathbf{k})a_\kappa(\mathbf{k}) + \tfrac{1}{2}] \tag{16.17}$$

In the derivation of (16.17) we have assumed that the electron gas does not react to the perturbation caused by the vibrations of the ions. In other words we must now consider how the electronic system reacts to the perturbation represented by the potential $\phi(\mathbf{r})$. This will lead to an additional contribution q_i due to the induced charge of the electrons in (16.1).

Now we have to deal with an inhomogeneous gas. This is an extremely difficult problem, which we can only deal with approximately. In the following we shall use the so-called Thomas–Fermi approximation. The basic approximation of this method consists in:

1. A local relation between charge density and energy density.
2. Neglecting exchange and correlation.
3. Assuming that the potential $\phi(\mathbf{r})$ varies only weakly.

Let us first assume that ϕ is a constant. The energy of a particle in state $\mathbf{k}$ is then given by

$$\varepsilon(\mathbf{k}) = \frac{k^2}{2m} - e\phi = \frac{k^2}{2m} + U \tag{16.18}$$

The density of electrons at zero temperature is obtained from

$$n = \frac{N}{V} = \frac{2}{V}\sum_{\mathbf{k}} \frac{1}{e^{[\varepsilon(\mathbf{k})-\mu]/T}-1} = \frac{2V}{(2\pi)^3}\frac{4\pi}{3}k_F^3 \tag{16.19}$$

where k_F is defined by

$$\mu = \frac{k_F^2}{2m} + U \tag{16.20}$$

When equation (16.20) is substituted into (16.19), the electron density is given by

$$n = \frac{[2m(\mu - U)]^{3/2}}{3\pi^2} \tag{16.21}$$

Suppose now the potential varies as a function of position in space, but sufficiently slowly. Then equation (16.21) still applies, where U and n are now functions of position. In the following we also assume that $U \ll \mu$ so that (16.21) can be expanded in the form

$$n = n_0 - \tfrac{3}{2}n_0(U/\varepsilon_F) \qquad (16.22)$$

where

$$\varepsilon_F = k_F^2/2m \qquad (16.23)$$

with

$$n_0 = \frac{k_F}{3\pi^2} \qquad (16.24)$$

The induced charge in the electron gas is thus given by

$$q_i = -e(n - n_0) = -\tfrac{3}{2}n_0\frac{\phi e^2}{\varepsilon_F} = -e^2 N_0 \phi \qquad (16.25)$$

where we have introduced the density of states N_0.

Equation (16.2) for the potential now becomes

$$\nabla^2 \phi(\mathbf{r}) = -4\pi(q + q_i) \qquad (16.26)$$

or

$$(\nabla^2 - k_s^2)\phi(\mathbf{r}) = -4\pi q(\mathbf{r}) \qquad (16.27)$$

where we have introduced

$$k_s^2 = \frac{6\pi n_0 e^2}{\varepsilon_F} = 4\pi N_0 e^2 \qquad (16.28)$$

The analysis leading to equation (16.17) can now be carried through again. The main difference is that equation (16.11) assumes the form

$$-(k^2 + k_s^2)\phi(\mathbf{k}) = 4\pi q(\mathbf{k}) \qquad (16.29)$$

and therefore

$$\omega_\kappa(\mathbf{k}) = \begin{cases} \omega_P \dfrac{k}{k_s} \dfrac{1}{\sqrt{1 + k^2/k_s^2}} & \kappa = 1 \\[2ex] 0 & \kappa = 2, 3 \end{cases} \qquad (16.30)$$

Hence the longitudinal phonons satisfy a dispersion relation

$$\omega_\kappa(\mathbf{k}) = ck \qquad \text{for } \kappa = 1, \quad k \to 0 \tag{16.31}$$

where the sound velocity is given by

$$c^2 = \frac{\omega_p^2}{k_s^2} = \frac{4\pi n_i e^2 Z^2 \varepsilon_F}{6\pi n_0} = \tfrac{1}{3} Z \frac{m}{M} v_F^2 \tag{16.32}$$

Here we have used $n_0 = Zn_i$ and $\varepsilon_F = mv_F^2/2$, where

$$v_F = k_F/m \tag{16.33}$$

is the Fermi velocity.

Since the electron–ion mass ratio is typically of order 10^{-4}–10^{-5}, equation (16.32) predicts a sound velocity of about 1/100 of the Fermi velocity, or of order 10^6 cm/s. For Na, $v_F = 10^8$, $Z = 1$, and $M = 23m_p$ so that $c = 2.81 \times 10^5$ cm/s in reasonable agreement with $c_{exp} = 2.3 \times 10^5$ cm/s.

Within the jellium model and the Thomas–Fermi approximation the electron–phonon interaction is given by

$$H_{el-ph} = -e \int d^3r n(\mathbf{r})\phi(\mathbf{r}) \tag{16.34}$$

where

$$n(\mathbf{r}) = \psi^+(\mathbf{r})\psi(\mathbf{r}) \tag{16.35}$$

is the electron-density operator and $\phi(\mathbf{r})$ is the potential induced by the phonons as described by equation (16.27). In the long-wavelength limit $k \ll k_s$ we have

$$\phi(\mathbf{r}) = -\frac{4\pi}{k_s^2} q(\mathbf{r}) \tag{16.36}$$

where

$$q(\mathbf{r}) = -\frac{Zen_i}{\sqrt{V\rho}} \sum_{\mathbf{k}} e^{i\mathbf{k}\cdot\mathbf{r}} q \frac{k}{\sqrt{2\omega_1(\mathbf{k})}} [a_1(\mathbf{k}) + a_1^+(-\mathbf{k})] \tag{16.37}$$

Because

$$\int e^{i\mathbf{k}\cdot\mathbf{r}} \psi^+(\mathbf{r})\psi(\mathbf{r}) d^3r = \sum_{\mathbf{k},\mathbf{k}'} c_{\mathbf{k}'}^+ c_{\mathbf{k}} \delta_{\mathbf{k}'-\mathbf{k},\mathbf{q}} \tag{16.38}$$

the electron–phonon interaction can be written as

$$H_{\text{el-ph}} = \sum_{\substack{k,k' \\ \kappa}} M_{kk',\kappa} c_{k'}^+ c_k a_\kappa (k'-k) + \sum_{\substack{k,k' \\ \kappa}} M_{kk',\kappa}^* c_k^+ c_{k'} a_\kappa^+ (k'-k) \quad (16.39)$$

where

$$M_{kk',\kappa} = i \frac{Ze^2 n_i |k-k'|}{\sqrt{2 V \rho \omega_1 (k-k')}} \frac{4\pi}{k_s^2} \delta_{k,1} \quad (16.40)$$

Sometimes one finds equation (16.39) written as

$$H_{\text{el-ph}} = \frac{1}{\sqrt{V}} \sum_{\substack{k,k' \\ \kappa}} g_{kk',\kappa} c_{k'}^+ c_k \phi_\kappa (k'-k) \quad (16.41)$$

where

$$\frac{1}{\sqrt{V}} g_{kk',\kappa} = M_{kk',\kappa} \sqrt{\frac{2}{\omega_\kappa (k'-k)}} \quad (16.42)$$

and

$$\phi_\kappa (q) = \sqrt{\frac{\omega(q)}{2}} [a_\kappa (q) + a_\kappa^+ (-q)] \quad (16.43)$$

In particular, for the jellium model in the limit $k \to 0$

$$g_{kk',\kappa} = \frac{1}{\sqrt{N_0}} \delta_{1,\kappa} \quad (16.44)$$

If we go beyond the jellium approximation, electrons will scatter with all types of phonons. Equation (16.39) will still be valid but $M_{kk',\kappa}$ is no longer given by equation (16.40).

16.2. Canonical Transformation and the Cooper Pair

In Section 16.1 we derived a Hamiltonian that describes the electron–phonon interaction for the jellium model. In this section we wish to show that such a Hamiltonian leads to an instability of the "free-electron state," which is then the starting point for the construction of the "superconducting state."

In this section we work with the model Hamiltonian given by

$$H = H_{\text{el}} + H_{\text{ph}} + H_{\text{el-ph}} \quad (16.45)$$

Here

$$H_{el} = \sum_{k} \varepsilon(\mathbf{k})c_k^+ c_k \tag{16.46}$$

describes the electrons with the "band" energy $\varepsilon(\mathbf{k})$, and for convenience we drop the band and spin index;

$$H_{ph} = \sum_{q} \omega(\mathbf{q})(a_q^+ a_q + \tfrac{1}{2}) \tag{16.47}$$

describes phonons with dispersion $\omega(\mathbf{q})$, and again we drop the band index. Finally, for the electron–phonon interaction we take for simplicity $M_{kk'} = iD$ (independent of wave vectors) and

$$H_{el-ph} = iD \sum_{k,q} c_{k+q}^+ c_k (a_q - a_{-q}^+) \tag{16.48}$$

In the following the energy $\varepsilon(\mathbf{k})$ is measured relative to the Fermi energy ε_F ($\varepsilon(\mathbf{k}) \to \varepsilon(\mathbf{k}) - \varepsilon_F$), and H_{el-ph} induces many changes in the behavior of the free particle, such as finite conductivity, finite lifetime, and changes in phonon dispersion. In the following we are interested in a process in which an electron of momentum $\mathbf{k}_1$ scatters into a state $\mathbf{k}_1'$ and emits a phonon of wave vector $\mathbf{q} = \mathbf{k}_1 - \mathbf{k}_1'$. This phonon is then absorbed in a scattering process $\mathbf{k}_2 \to \mathbf{k}_2 + \mathbf{q}$. As far as the electrons which concerned we can also start with $\mathbf{k}_2$ first. The two processes together are equivalent to a Hamiltonian with an effective electron–electron interaction.

In order to obtain an explicit expression we can use the concept of the canonical transformation, in which both the states and the operators are transformed such that all matrix elements remain invariant. Canonical transformations can always be written by the action of a unitary operator U satisfying

$$U^{-1} = U^+ \tag{16.49}$$

in which case

$$|\tilde{\psi}'\rangle = U|\psi'\rangle, \qquad \langle\tilde{\psi}| = \langle\psi|U^+ \tag{16.50}$$

$$\tilde{A} = UAU^{-1} \tag{16.51}$$

and

$$\langle\tilde{\psi}'|\tilde{A}|\tilde{\psi}\rangle = \langle\psi'|U^+ UAU^{-1}U|\psi\rangle \equiv \langle\psi'|A|\psi\rangle \tag{16.52}$$

A unitary transformation can be written as

$$U = e^{-S} \quad \text{where } S^+ \equiv -S \tag{16.53}$$

Consider now a Hamiltonian of the form

$$H = H_0 + H_1 \tag{16.54}$$

Then

$$\tilde{H} = e^{-S} H e^{S} = H + [H, S] + \tfrac{1}{2}[[H, S], S]$$
$$= H_0 + H_1 + [H_0, S] + [H_1, S] + \tfrac{1}{2}[[H_0, S], S] + \cdots \tag{16.55}$$

We note that it is possible to eliminate the linear term H_1 in $\tilde{H}$ if we choose S such that

$$H_1 + [H_0, S] = 0 \tag{16.56}$$

and therefore

$$\tilde{H} = H_0 + \tfrac{1}{2}[H_1, S] + \cdots \tag{16.57}$$

Suppose now that the eigenstates of H_0 are denoted by $|n\rangle, |m\rangle$. Then it follows from equation (16.56) that

$$\langle n | S | m \rangle = \frac{\langle n | H_1 | m \rangle}{E_m - E_n} \tag{16.58}$$

We are interested in the part of $\tilde{H}$ that is diagonal in the phonon occupation numbers, since these elements represent the effective electron–electron interaction. We set $H_1 = H_{\text{el-ph}}$ and consider the phonon system at absolute zero, so that n or m refers to the vacuum phonon state. In equation (16.58) we take the matrix elements with respect to the phonon system only. Then

$$\langle 1_q | S | 0 \rangle = - iD \sum_k c_{k-q}^+ c_k \frac{1}{\varepsilon(k) - \varepsilon(k-q) - \omega(q)} \tag{16.59}$$

$$\langle 0 | S | 1_q \rangle = + iD \sum_{k'} c_{k'+q}^+ c_{k'} \frac{1}{\varepsilon(k') - \varepsilon(k'+q) + \omega(q)} \tag{16.60}$$

Hence

$$\tilde{H} = H_0 + \tfrac{1}{2} \sum_q (\langle 0 | H_1 | 1_q \rangle \langle 1_q | S | 0 \rangle - \langle 0 | S | 1_q \rangle \langle 1_q | H_1 | 0 \rangle) \tag{16.61}$$

or

$$H_{el-el} = \tfrac{1}{2} D^2 \sum_{\mathbf{q}} \sum_{\mathbf{k,k'}} c^+_{\mathbf{k'+q}} c_{\mathbf{k'}} c^+_{\mathbf{k-q}} c_{\mathbf{k}}$$

$$\times \left(\frac{1}{\varepsilon(\mathbf{k}) - \varepsilon(\mathbf{k-q}) - \omega(\mathbf{q})} - \frac{1}{\varepsilon(\mathbf{k'}) - \varepsilon(\mathbf{k'+q}) + \omega(\mathbf{q})} \right) \qquad (16.62)$$

Rearrangement of the second term $(\mathbf{k'} \to \mathbf{k},\ \mathbf{q} \to -\mathbf{q})$ gives

$$H_{el-el} = \sum_{\mathbf{q}} \sum_{\mathbf{k,k'}} V_{\mathbf{kq}} c^+_{\mathbf{k'+q}} c_{\mathbf{k'}} c^+_{\mathbf{k-q}} c_{\mathbf{k}} \qquad (16.63)$$

where

$$V_{\mathbf{kq}} = \frac{D^2 \omega(\mathbf{q})}{[\varepsilon(\mathbf{k}) - \varepsilon(\mathbf{k-q})]^2 - \omega^2(\mathbf{q})} \qquad (16.64)$$

The interaction matrix element given by equation (16.64) can be either attractive or repulsive.

Cooper[106] was the first to point out that unusual properties would result from the attractive interaction. He considered two electrons with total momentum $\mathbf{K} = 0$ and opposite spin. A complete set of wave functions is given by

$$|\mathbf{k}\rangle = c^+_{\mathbf{k}\uparrow} c^+_{-\mathbf{k}\downarrow} |F\rangle \qquad (16.65)$$

where F is the filled Fermi sea.

The eigenvalue equation

$$(H_0 + H_{el-el}) |\psi\rangle = E |\psi\rangle \qquad (16.66)$$

becomes

$$\sum_{\mathbf{k'}} \langle \mathbf{k} | H_0 + H_{el-el} | \mathbf{k'} \rangle \langle \mathbf{k'} | \psi \rangle = E \langle \mathbf{k} | \psi \rangle \qquad (16.67)$$

Note that $|\mathbf{k}|, |\mathbf{k'}| > k_F$.

We have

$$\langle \mathbf{k'} | H_0 | \mathbf{k} \rangle = \delta_{\mathbf{kk'}} \varepsilon(\mathbf{k}) \qquad (16.68)$$

In order to evaluate $\langle \mathbf{k'} | H_{el-el} | \mathbf{k} \rangle$ it is convenient to rewrite H_{el-el}. Since the superconducting state as implied by equation (16.65) is formed through the association of electrons in pairs with opposite momentum and spin, we drop all terms in equation (16.63) that do not connect such pairs. Thus we replace H_{el-el} by H_{el-el}^{eff} where

$$H_{el-el}^{eff} = \frac{1}{2} \sum_{\substack{k,k' \\ \sigma,\sigma'}} U_{kk'} c_{k'\sigma}^+ c_{-k'-\sigma}^+ c_{-k-\sigma} c_{-k\sigma} \tag{16.69}$$

Here

$$U_{kk'} = \frac{2D^2 \omega(k-k')}{[\varepsilon(k)-\varepsilon(k')]^2 - \omega(k-k')^2} \tag{16.70}$$

Since $U_{kk'} = U_{-k-k'}$ we can write equation (16.69) as

$$H_{el-el}^{eff} = \sum_{k,k'} U_{kk'} c_{k'\uparrow}^+ c_{-k'\downarrow}^+ c_{-k\downarrow} c_{k\uparrow} \tag{16.71}$$

Hence

$$\langle k' | H_{el-el}^{eff} | k \rangle = U_{kk'} \tag{16.72}$$

and equation (16.67) becomes, with $\alpha_k = \langle k | \psi \rangle$,

$$[2\varepsilon(k) - E]\alpha_k + \sum_{|k|>k_F} U_{k,k'} \alpha_{k'} = 0 \tag{16.73}$$

If equation (16.73) has a solution $E < 0$, then the pair of particles will prefer to be in the state $|\psi\rangle$ rather than in the states on the Fermi surface for which $E = 0$.

For a general $U_{kk'}$ we still have a complicated integral equation to solve. Fortunately the essential physics survives if $U_{kk'}$ is replaced by

$$U_{kk'} = \begin{cases} -V & \text{for } -\omega_D < \varepsilon(k),\ \varepsilon(k') < \omega_D \\ 0 & \text{otherwise} \end{cases} \tag{16.74}$$

Here ω_D is the Debye frequency. Hence equation (16.73) becomes

$$[2\varepsilon(k) - E]\alpha_k - V \sum_{k,|k|>k_F}' \alpha_k = 0 \tag{16.75}$$

The prime in equation (16.75) takes care of equation (16.74).
Equation (16.75) has the solution

$$\alpha_k = \frac{A}{[2\varepsilon(k) - E]} \tag{16.76}$$

where A is some constant. Substituting this back into equation (16.75), we obtain

$$1 = V \sum_{\mathbf{k},|\mathbf{k}|>k_F}' \frac{1}{2\varepsilon(\mathbf{k}) - E} = N(0)V \int_0^{\omega_D} \frac{d\varepsilon}{2\varepsilon - E} \qquad (16.77)$$

Here we have introduced the density of states

$$N(\varepsilon) = \sum_{\mathbf{k}} \delta[\varepsilon(\mathbf{k}) - \varepsilon] \sim N(0) \qquad (16.78)$$

Evaluation of equation (16.77) gives

$$1 = \tfrac{1}{2}N(0)V \ln \frac{2\omega_D - E}{-E} \qquad (16.79)$$

Hence

$$E = -2\omega_D \frac{1}{e^{2/N(0)V} - 1} \sim -2\omega_D e^{-2/N(0)V} \qquad (16.80)$$

for $N(0)V \ll 1$.

The fact that we have been able to find a solution of equation (16.66) with $E < 0$ shows that pairs of fermions added to the Fermi sea would like to form a state that is bound relative to the Fermi sea. This suggests that one should now look for a formulation that allows one to describe the formation of a large number of bound pairs. This has been achieved in the BCS theory and will be discussed in the following section.

Before closing this section we would like to comment on the single Cooper-pair solution. The result $E < 0$ depends on the fact that the interaction is attractive. If the interaction is negative, equation (16.77) has no solution with $E < 0$. For an interaction that is partly attractive and partly repulsive (as is the case in practice) the equation must be solved with more care.

16.3. Superconductivity: The Ground State

In the previous section we considered a many-fermion system with an attractive electron–electron interaction. We have shown that in this case the Fermi sea is unstable with respect to the formation of a bound pair. In the BCS theory one constructs a variational solution, which allows the formation of a large number of pairs to be described.

The Hamiltonian is

$$H = \sum_{\mathbf{k},\sigma} \varepsilon(\mathbf{k})c_{\mathbf{k}\sigma}^+ c_{\mathbf{k}\sigma} + \sum_{\mathbf{k},\mathbf{k}'} U_{\mathbf{k},\mathbf{k}'} c_{\mathbf{k}\uparrow}^+ c_{-\mathbf{k}\downarrow}^+ c_{-\mathbf{k}'\downarrow} c_{\mathbf{k}'\uparrow} \qquad (16.81)$$

where, as before, we set

$$U_{\mathbf{k},\mathbf{k}'} = \begin{cases} -V & \text{for } -\omega_D < \varepsilon(\mathbf{k}),\ \varepsilon(\mathbf{k}') < \omega_D \\ 0 & \text{otherwise} \end{cases} \tag{16.82}$$

Many approximations were involved in deriving this so-called pair Hamiltonian. It is, of course, quite surprising that a theory based on equation (16.81) is in fact able to account for many of the observed facts about superconductivity.

We write the Hamiltonian (including the chemical potential) as

$$H = \sum_{\mathbf{k},\sigma} (\varepsilon(\mathbf{k}) - \mu) c^+_{\mathbf{k}\sigma} c_{\mathbf{k}\sigma} + \sum_{\mathbf{k},\mathbf{k}'} U_{\mathbf{k}\mathbf{k}'} \beta^+_{\mathbf{k}} \beta_{\mathbf{k}'} \tag{16.83}$$

where

$$\beta_{\mathbf{k}} = c_{-\mathbf{k}\downarrow} c_{\mathbf{k}\uparrow}, \qquad \beta^+_{\mathbf{k}} = c^+_{\mathbf{k}\uparrow} c^+_{-\mathbf{k}\downarrow} \tag{16.84}$$

In Section 16.2 we have shown that the energy of the filled Fermi sea $|F\rangle$ can be lowered by creating a bound electron pair of opposite spin:

$$|\psi\rangle = \sum_{|\mathbf{k}|>k_F} \alpha_{\mathbf{k}} \beta^+_{\mathbf{k}} |F\rangle \tag{16.85}$$

Equation (16.85) represents a wave function in which one artificially keeps all $N-2$ electrons fixed in the Fermi sphere and only solves the corresponding Schrödinger equation for two electrons outside the sphere. To do a proper job one must allow for the possibility that all electrons assume new configurations, not just two.

This suggests that one should construct the ground state by applying the pair creation operator

$$C^+ = \left(\sum_{\mathbf{k}} \alpha_{\mathbf{k}} b^+_{\mathbf{k}} \right) \tag{16.86}$$

$N/2$ times on the vacuum state

$$|\psi\rangle = C^{+N/2} |0\rangle \tag{16.87}$$

The coefficients α_k have to be determined by minimizing the total energy

$$\delta \langle \psi | H | \psi \rangle = 0 \tag{16.88}$$

In order to carry through the calculation, it is convenient to express equation (16.87) in the form

$$|\psi_{BCS}\rangle = Ae^{C^+}|0\rangle = A \prod_k e^{\alpha_k b_k^+}|0\rangle \tag{16.89}$$

Here A is a normalization factor.

The wave function (16.87) can be seen to be the component of the wave function (16.89) having exactly N particles. Equation (16.89) is the famous "BCS" wave function. It is an acceptable representation of the physical N-particle wave function if the mean-square-root variation satisfies

$$\frac{\overline{\Delta N}}{\bar{N}} \to 0 \qquad \text{for } \bar{N} \to \infty \tag{16.90}$$

Because $b_k b_k = 0$ the expansion of equation (16.89) gives

$$|\psi_{BCS}\rangle = A \prod_k (1 + \alpha_k b_k^+)|0\rangle \tag{16.91}$$

$$\equiv \prod_k (u_k + v_k b_k^+)|0\rangle \tag{16.92}$$

Here we have written

$$A = \prod_k u_{k'} \qquad v_k = \alpha_k u_k \tag{16.93}$$

The normalization condition becomes

$$\langle \psi_{BCS}|\psi_{BCS}\rangle = \prod_k \langle 0|u_k^2 + v_k^2|0\rangle = 1 \tag{16.94}$$

so that

$$u_k^2 + v_k^2 = 1 \tag{16.95}$$

The expectation value of equation (16.83) with respect to the BCS ground-state wave function is now easily calculated. In particular $(\phi = \psi_{BCS})$

$$\langle \phi | c_{k\sigma}^+ c_{k\sigma} | \phi \rangle = v_k^2 \tag{16.96}$$

$$\langle \phi | b_k^+ b_{k'} | \phi \rangle = u_k v_k u_{k'} v_{k'} \tag{16.97}$$

Therefore

$$\langle \phi | H | \phi \rangle = \sum_{\mathbf{k}} 2[\varepsilon(\mathbf{k}) - \mu]v_{\mathbf{k}}^2 + \sum_{\mathbf{kk'}} U_{\mathbf{kk'}} u_{\mathbf{k}} v_{\mathbf{k}} u_{\mathbf{k'}} v_{\mathbf{k'}} \qquad (16.98)$$

If this expectation value is minimized with respect to $v_{\mathbf{k}}$ one obtains using $u_{\mathbf{k}} = \sqrt{1 - v_{\mathbf{k}}^2}$

$$2 \cdot 2[\varepsilon(\mathbf{k}) - \mu]v_{\mathbf{k}} + 2 \sum_{\mathbf{k'}} U_{\mathbf{kk'}} u_{\mathbf{k'}} v_{\mathbf{k'}} \left(u_{\mathbf{k}} - v_{\mathbf{k}}^2 \frac{1}{u_{\mathbf{k}}} \right) = 0 \qquad (16.99)$$

or

$$2[\varepsilon(\mathbf{k}) - \mu]u_{\mathbf{k}} v_{\mathbf{k}} = \left(\sum_{\mathbf{k'}} U_{\mathbf{kk'}} u_{\mathbf{k'}} v_{\mathbf{k'}} \right)(u_{\mathbf{k}}^2 - v_{\mathbf{k}}^2) \qquad (16.100)$$

The requirement on the mean number of particles gives the condition

$$\bar{N} = \langle \phi | N | \phi \rangle = 2 \sum_{\mathbf{k}} v_{\mathbf{k}}^2 \qquad (16.101)$$

The set of equations (16.100) and (16.101) determines $v_{\mathbf{k}}$ and μ. In general these equations must be solved numerically, but for the particular model represented by equation (16.82) the equations can be solved analytically. For this case it is convenient to define

$$\Delta = V \sum_{\mathbf{k}} u_{\mathbf{k}} v_{\mathbf{k}} \qquad (16.102)$$

Equation (16.100) then becomes

$$2[\varepsilon(\mathbf{k}) - \mu]u_{\mathbf{k}} v_{\mathbf{k}} = \Delta(u_{\mathbf{k}}^2 - v_{\mathbf{k}}^2) \qquad (16.103)$$

or

$$4[\varepsilon(\mathbf{k}) - \mu]^2 u_{\mathbf{k}}^2 v_{\mathbf{k}}^2 = \Delta^2(u_{\mathbf{k}}^2 - v_{\mathbf{k}}^2)^2 \qquad (16.104)$$

Because

$$1 - (u_{\mathbf{k}}^2 - v_{\mathbf{k}}^2)^2 = (1 - u_{\mathbf{k}}^2 + v_{\mathbf{k}}^2)(1 + u_{\mathbf{k}}^2 - v_{\mathbf{k}}^2) = 2v_{\mathbf{k}}^2 2u_{\mathbf{k}}^2 \qquad (16.105)$$

equation (16.104) becomes

$$[\varepsilon(\mathbf{k}) - \mu]^2[1 - (u_{\mathbf{k}}^2 - v_{\mathbf{k}}^2)^2] = \Delta^2(u_{\mathbf{k}}^2 - v_{\mathbf{k}}^2)^2 \qquad (16.106)$$

from which we obtain

$$u_k^2 - v_k^2 = \frac{\varepsilon(\mathbf{k}) - \mu}{E}$$

$$u_k^2 = \frac{1}{2}\left[1 + \frac{\varepsilon(\mathbf{k}) - \mu}{E(\mathbf{k})}\right]$$

$$v_k^2 = \frac{1}{2}\left[1 - \frac{\varepsilon(\mathbf{k}) - \mu}{E(\mathbf{k})}\right] \tag{16.107}$$

where

$$E(\mathbf{k}) = \sqrt{[\varepsilon(\mathbf{k}) - \mu]^2 + \Delta^2} \tag{16.108}$$

The parameters Δ and μ are as yet undetermined.

Let us first evaluate the gap parameter Δ. From equations (16.102) and (16.103)

$$\Delta = \frac{V}{2}\sum_{\mathbf{k}}\frac{\Delta}{[\varepsilon(\mathbf{k}) - \mu]}(u_k^2 - v_k^2) = \frac{\Delta V}{2}\sum_{\mathbf{k}}\frac{1}{E(\mathbf{k})} \tag{16.109}$$

Equation (16.109) has the trivial solution

$$\Delta = 0 \tag{16.110}$$

which corresponds to the noninteracting Fermi sea. In this case $\mu = \varepsilon_F$. The nontrivial solution $\Delta \neq 0$ is given by the solution of the equation

$$1 = \frac{V}{2}\sum_{\mathbf{k}}\frac{1}{E(\mathbf{k})} \tag{16.111}$$

The chemical potential μ is determined by the equation

$$\bar{N} = 2\sum_{\mathbf{k}}v_k^2 = \sum_{\mathbf{k}}\left[1 - \frac{\varepsilon(\mathbf{k}) - \mu}{E(\mathbf{k})}\right] \tag{16.112}$$

For $\omega_D \ll \varepsilon_F$ and $\varepsilon = \varepsilon(\mathbf{k}) - \mu$

$$\sum_{\mathbf{k}}\cdots = N(0)\int d\varepsilon \cdots \tag{16.113}$$

where $N(0)$ is the density of states per spin. The two equations (16.111) and (16.112) then become

$$1 = \frac{VN(0)}{2}\int_{-\omega_D}^{+\omega_D}\frac{d\varepsilon}{\sqrt{\varepsilon^2 + \Delta^2}} \tag{16.114}$$

and

$$\int_{\mu-\varepsilon_F-\omega_D}^{\mu-\varepsilon_F-\omega_D} d\varepsilon \left(1 - \frac{\varepsilon}{\sqrt{E^2+\Delta^2}}\right) = 0 \qquad (16.115)$$

The second equation gives

$$\mu = \varepsilon_F \qquad (16.116)$$

The first equation becomes

$$1 = \tfrac{1}{2}VN(0)\sinh^{-1}\left(\frac{\omega_D}{\Delta}\right) \qquad (16.117)$$

Thus

$$\Delta = \frac{\omega_D}{\sinh[1/N(0)V]} \qquad (16.118)$$

Note that Δ is not an analytical function of V. Therefore the result (16.118) cannot be obtained by a perturbation series.

The condition (16.90) can now be checked. We have

$$\langle N^2 \rangle = \left\langle \phi \left| \sum_{\substack{kk' \\ \sigma\sigma'}} c_{k\sigma}^{+} c_{k\sigma} c_{k'\sigma'}^{+} c_{k\sigma} \right| \phi \right\rangle$$

$$= 4\left(\sum_{k\neq k'} v_k^2 v_{k'}^2 + \sum_k v_k^2\right)$$

$$= 4\left[\left(\sum_k v_k^2\right)^2 - \sum_k v_k^4 + \sum_k v_k^2\right] \qquad (16.119)$$

Furthermore

$$\bar{N} = 2\sum_k v_k^2 \qquad (16.120)$$

so that

$$\langle N^2 \rangle - \bar{N}^2 = 4\sum_k v_k^2(1-v_k^2) = 4\sum_k v_k^2 u_k^2 \sim \bar{N} \qquad (16.121)$$

Therefore equation (16.90) is satisfied. With this, an acceptable ground state has been determined and it remains to be shown how the various properties of a superconductor can be explained.

To summarize, the BCS ground state

$$|\phi\rangle = \prod_k (u_k + v_k c^+_{+k\uparrow} c^+_{k\downarrow})|0\rangle \tag{16.122}$$

has been constructed in analogy to the exactly solvable problem of binding two electrons above the Fermi sea. Equation (16.122) represents a natural generalization of this concept to describe a many-pair wave function.

The replacement of an N-particle wave function by a superposition is correct in the limit $N \to \infty$ and represents the "mathematical" breakthrough of the BCS theory. At this moment it may not be very clear what sort of approximation is involved in the variational ansatz. This point will be discussed in the following section.

16.4. The Bogoliubov Equation

Let us start again with a general Hamiltonian for an interacting electron gas in second quantized form:

$$H = \sum_\sigma \int \psi^+_\sigma(\mathbf{r}) h_0(\mathbf{r}, \mathbf{r}') \psi_\sigma(\mathbf{r}') d^3r \, d^3r'$$

$$+ \tfrac{1}{2} \sum_{\sigma,\sigma'} \int \psi^+_\sigma(\mathbf{r}) \psi^+_{\sigma'}(\mathbf{r}') U(\mathbf{r}, \mathbf{r}') \psi_{\sigma'}(\mathbf{r}') \psi_\sigma(\mathbf{r}) d^3r \, d^3r'$$

$$= H_0 + H_s \tag{16.123}$$

where

$$h_0(\mathbf{r}, \mathbf{r}') = \left\langle \mathbf{r} \left| \frac{\mathbf{p}^2}{2m} - \mu \right| \mathbf{r}' \right\rangle = \left(-\frac{\nabla^2_r}{2m} - \mu \right) \delta(\mathbf{r} - \mathbf{r}') \tag{16.124}$$

and $U(\mathbf{r}, \mathbf{r}') = -V(\mathbf{r})\delta(\mathbf{r} - \mathbf{r}')$ is an arbitrary electron–electron interaction. In case we want to include external fields $\mathbf{p} \to [\mathbf{p} - (e/c)\mathbf{A}]$.

In Section 16.3 we have shown that it is correct to represent the ground state by a state that does not correspond to a fixed number of particles. Therefore we can no longer assume that

$$\langle \psi^+ \psi^+ \rangle \equiv 0 \tag{16.125}$$

Hence terms like this must be included in the construction of any approximation.

Using the identity

$$\psi^+_\uparrow(\mathbf{r}) \psi^+_\downarrow(\mathbf{r}) = \langle \psi^+_\uparrow(\mathbf{r}) \psi^+_\downarrow(\mathbf{r}) \rangle + \{ \quad \}^+ \tag{16.126}$$

in equation (16.123) and neglecting quadratic terms in { } (the molecular-field approximation) we obtain

$$H = \sum_{\sigma} \int \psi_{\sigma}^{+}(\mathbf{r}) \left(-\frac{\nabla^2}{2m} - \mu \right) \psi_{\sigma}(\mathbf{r}) d^3 r$$

$$- \int \Delta(\mathbf{r}) \psi_{\downarrow}^{+}(\mathbf{r}) \psi_{\uparrow}^{+}(\mathbf{r}) d^3 r + cc + \int \frac{1}{V(\mathbf{r})} |\Delta|^2 d^3 r \quad (16.127)$$

where we have introduced the order parameter

$$\Delta(\mathbf{r}) = V(\mathbf{r}) \langle \psi_{\downarrow}(\mathbf{r}) \psi_{\uparrow}(\mathbf{r}) \rangle \quad (16.128)$$

If the new Fermi operators

$$\begin{pmatrix} \phi_1 \\ \phi_2 \end{pmatrix} = \begin{pmatrix} \psi_{\uparrow} \\ \psi_{\downarrow} \end{pmatrix}, \qquad (\phi_1^{+}, \phi_2^{+}) = (\psi_{\uparrow}^{+}, \psi_{\downarrow}) \quad (16.129)$$

are introduced, then the Hamiltonian can be written as

$$H = \sum_{s,s'} \int \phi_s^{+}(\mathbf{r}) \langle \mathbf{r}s | h | \mathbf{r}'s' \rangle \phi_{s'}(\mathbf{r}') d^3 r d^3 r' \quad (16.130)$$

where

$$\langle \mathbf{r}s | h | \mathbf{r}'s \rangle = \left\langle \mathbf{r} \left| \begin{pmatrix} h_0 & -\Delta \\ -\Delta^* & -h_0 \end{pmatrix} \right| \mathbf{r}' \right\rangle \quad (16.131)$$

With the eigenstates of h given by

$$h |\lambda\rangle = E_{\lambda} |\lambda\rangle \quad (16.132)$$

and the new Fermi operators

$$\zeta_{\lambda}^{+} = \int \sum_s d^3 r \phi_s^{+}(\mathbf{r}) \langle \mathbf{r}s | \lambda \rangle \quad (16.133)$$

the Hamiltonian becomes diagonal:

$$H = \int \sum_s \phi_s^{+}(\mathbf{r}) \langle \mathbf{r}s | \lambda \rangle E_{\lambda} \langle \lambda | \mathbf{r}'s \rangle \phi_s(\mathbf{r}') d^3 r d^3 r'$$

$$= \sum_{\lambda} \zeta_{\lambda}^{+} E_{\lambda} \zeta_{\lambda} \quad (16.134)$$

In the coordinate representation with

$$u_\lambda(\mathbf{r}) = \langle \mathbf{r}1 | \lambda \rangle, \qquad v_\lambda(\mathbf{r}) = \langle \mathbf{r}2 | \lambda \rangle \tag{16.135}$$

equation (16.132) can be written as

$$\begin{pmatrix} -h_0 & -\Delta(\mathbf{r}) \\ -\Delta^*(\mathbf{r}) & -h_0 \end{pmatrix} \begin{pmatrix} u_\lambda(\mathbf{r}) \\ v_\lambda(\mathbf{r}) \end{pmatrix} = E_\lambda \begin{pmatrix} u_\lambda(\mathbf{r}) \\ v_\lambda(\mathbf{r}) \end{pmatrix} \tag{16.136}$$

Equation (16.136) represents the so-called Bogoliubov equations and must be solved together with equation (16.128). Since

$$\phi_s^+ = \sum_\lambda \zeta_\lambda^+ \langle \lambda | \mathbf{r}s \rangle \tag{16.137}$$

equation (16.128) becomes

$$\Delta(\mathbf{r}) = V(\mathbf{r}) \sum_\lambda \langle \lambda | \mathbf{r}2 \rangle \langle \mathbf{r}1 | \lambda \rangle \langle \zeta_\lambda^+ \zeta_\lambda \rangle \tag{16.138}$$

We have

$$\langle \zeta_\lambda^+ \zeta_\lambda \rangle = \frac{1}{e^{E_\lambda/T} + 1} \tag{16.139}$$

so that finally the order-parameter equation is

$$\Delta(\mathbf{r}) = V(\mathbf{r}) \sum_\lambda \frac{u_\lambda(\mathbf{r}) v_\lambda^*(\mathbf{r})}{e^{E_\lambda/T} + 1} \tag{16.140}$$

The free energy is given by

$$F = -T \sum_\lambda \log(1 + e^{-E_\lambda/T}) + \int \frac{1}{V(\mathbf{r})} |\Delta(\mathbf{r})|^2 d^3r \tag{16.141}$$

16.5. The Transition Temperature

Equations (16.136), (16.140), and (16.141) have been used to obtain a formal solution for the thermodynamics of an inhomogeneous superconductor. Explicit solutions still remain to be worked out. Let us again consider the homogeneous case. Now we can set

$$\begin{pmatrix} u_\lambda \\ v_\lambda \end{pmatrix} = \frac{1}{\sqrt{V}} e^{i\mathbf{k}\cdot\mathbf{r}} \begin{pmatrix} u_\mathbf{k} \\ v_\mathbf{k} \end{pmatrix} \tag{16.142}$$

Equation (16.136) becomes

$$\begin{pmatrix} \varepsilon(k) & -\Delta \\ -\Delta & -\varepsilon(k) \end{pmatrix} \begin{pmatrix} u(\mathbf{k}) \\ v(\mathbf{k}) \end{pmatrix} = E(\mathbf{k}) \begin{pmatrix} u(\mathbf{k}) \\ v(\mathbf{k}) \end{pmatrix} \tag{16.143}$$

which has the eigenvalues

$$E_\pm(\mathbf{k}) = \pm \sqrt{\varepsilon^2(\mathbf{k}) + \Delta^2} = \pm E(k) \tag{16.144}$$

and where

$$u_\pm(\mathbf{k}) = \left[\frac{1}{2}\left(1 + \frac{\varepsilon(\mathbf{k})}{E_\pm(k)}\right)\right]^{1/2} \tag{16.145}$$

$$v_\pm(\mathbf{k}) = \mp \left[\frac{1}{2}\left(1 - \frac{\varepsilon(\mathbf{k})}{E_\pm(k)}\right)\right]^{1/2} \tag{16.146}$$

$$\varepsilon(\mathbf{k}) = k^2/2m - \mu \tag{16.147}$$

The new Fermi operators become

$$\zeta_{\mathbf{k}\pm}^+ = c_{\mathbf{k}\uparrow}^+ u_\pm(\mathbf{k}) + c_{-\mathbf{k}\downarrow} v_\pm(\mathbf{k})$$
$$\zeta_{\mathbf{k}\pm} = c_{\mathbf{k}\uparrow} u_\pm(\mathbf{k}) + c_{-\mathbf{k}\downarrow}^+ v_\pm(\mathbf{k}) \tag{16.148}$$

The Hamiltonian is then given by

$$H = \sum_{\substack{\mathbf{k} \\ s=\pm1}} E_s \zeta_{\mathbf{k}s}^+ \zeta_{\mathbf{k}s} \tag{16.149}$$

For the homogeneous case the "gap" equation becomes

$$\Delta = V \sum_{\mathbf{k}s} \frac{u_s(\mathbf{k}) v_s^*(\mathbf{k})}{e^{E_s/T} + 1} \tag{16.150}$$

From equations (16.145) and (16.146)

$$u_s(\mathbf{k}) v_s^*(\mathbf{k}) = -s \frac{1}{2} \frac{\Delta}{|E|} \tag{16.151}$$

Furthermore

$$\frac{1}{e^{-E/T} + 1} - \frac{1}{e^{E/T} + 1} = \tanh(E/2T) \tag{16.152}$$

The gap equation thus becomes

$$\Delta = \frac{\Delta V}{2} \sum_{\mathbf{k}}{}' \frac{\tanh[E(\mathbf{k})/2T]}{E(\mathbf{k})} \qquad (16.153)$$

Note that if $V < 0$ (repulsive interaction) the only solution is $\Delta = 0$. The prime in equation (16.153) indicates a cutoff when

$$|\varepsilon(\mathbf{k})| > \omega_D \qquad (16.154)$$

As usual, the sum over $\mathbf{k}$ is converted into an integral by introducing the density of states $N(0)$ per spin.

At the transition temperature T_c, $E = |\varepsilon|$ so that the equation for T_c becomes

$$\frac{1}{N(0)V} = \int_0^{\omega_D} \frac{\tanh(\varepsilon/2T)}{\varepsilon} \, d\varepsilon$$

$$= \int_0^{\omega_D/2T_c} \frac{\tanh x}{x} \, dx = \ln\left(\frac{2\gamma}{\pi} \frac{\omega_D}{T_c}\right) \qquad (16.155)$$

where γ is Euler's constant. Thus

$$T_c = \frac{2\gamma}{\pi} \omega_D e^{-1/N(0)V} \qquad (16.156)$$

where

$$2\gamma/\pi = 1.13 \qquad (16.157)$$

At $T = 0$ equation (16.153) becomes

$$\frac{1}{N(0)V} = \int_0^{\omega_D} d\varepsilon \, (\varepsilon^2 + \Delta_0^2)^{-1/2}$$

$$= \ln[x + (x^2 + \Delta_0^2)^{1/2}]_0^{\omega_D}$$

$$= \ln(2\omega_D/\Delta_0) \qquad (16.158)$$

Therefore

$$\Delta_0 = 2\omega_D \exp[-1/N(0)V] \qquad (16.159)$$

or

$$2\Delta_0 = 3.5 T_c \qquad (16.160)$$

Experimental values of $2\Delta_0/T_c$ are given in Table 16.1.

Table 16.1. Experimental Values of $2\Delta_0/T_c$

Sn	Al	Pb	Cd
3.5	3.4	4.1	3.3

The question whether there is a natural well-defined temperature limit to the occurrence of superconductivity is still open. Nevertheless some predictions can be made if the attractive interaction originating from the phonons is known. The important parameter is

$$\lambda = N(0)V = \eta/(M\omega_D^2) \tag{16.161}$$

Among transition metals η is surprisingly constant (approximately 5). From equation (16.161) we see that a decrease in ω_D will increase λ. On the other hand, the variation in η for different classes of materials is more important in causing T_c to change. This is illustrated, for example, by Nb and Nb_3Sn in Table 16.2.

The failure to find higher T_cs despite their theoretical possibility seems to be connected with metastability. When a trend is found in which superconducting interactions are increased by chemical substitution [such as $(Tl—Pb)_{1-x}Bi_x$] the phase suddenly becomes metastable; Nb_3Ge with $T_c \sim 23$ °K is such a metastable system.

In the derivation of the equation for T_c [equation (16.156)] we have simplified the effective electron–electron interaction and have neglected Coulomb interaction. An improved solution has been worked out by McMillan whose numerical results can be well approximated by

$$T_c = \frac{\omega_D}{1.45} \exp\left[-\frac{1.04(1+\lambda)}{\lambda - \mu^*(1+0.62\lambda)} \right] \tag{16.162}$$

enabling the transition temperature to be estimated from knowledge of the

Table 16.2. Coupling Strength η as Measured in Units of a Characteristic Debye Energy

	T_c	ω_D	η
Nb	9.2	175	4.7
Nb_3Sn	18.1	146	7.9

Debye frequency, the coupling strength, and μ^* defined by

$$\mu^* = \frac{\mu}{1 + \mu \ln(\varepsilon_F/\omega_D)} \tag{16.163}$$

where

$$\mu = \left\langle \frac{4\pi e^2}{q^2 + q_s^2} \right\rangle \tag{16.164}$$

is the averaged screened Coulomb interaction with

$$q_s^2 = 8\pi N(0) e^2 \tag{16.165}$$

Let us now make an estimate of T_c based on the jellium model. Adding the screened Coulomb interaction to the effective phonon-induced interaction gives

$$U_{\mathbf{k},\mathbf{k}'} = \frac{4\pi e^2}{q_s^2} \frac{\omega_{\mathbf{k},\mathbf{k}'}^2}{\omega_{\mathbf{k},\mathbf{k}'}^2 - \omega_D^2} \tag{16.166}$$

where

$$\omega_{\mathbf{k},\mathbf{k}'} = \varepsilon_{\mathbf{k}} - \varepsilon_{\mathbf{k}'} \tag{16.167}$$

If we replace equation (16.166) by an appropriate average we obtain

$$N(0)V = \tfrac{1}{6} \tag{16.168}$$

Therefore

$$T_c = \frac{2\gamma}{\pi} \omega_D e^{-6} \tag{16.169}$$

which is of the order of a few degrees.

16.6. Landau–Ginzburg Theory

The Ginzburg–Landau equations were presented phenomenologically before the microscopic theory. They describe spatial variations of superconductivity. The expression assumed for the free energy is

$$F = \int d^3r \{ f(|\Delta|^2) + \gamma \, |[-i\nabla + eA(\mathbf{r})]\Delta(\mathbf{r})|^2 \} \tag{16.170}$$

This expression has subsequently been derived on the basis of the BCS theory by Gorkov with the use of Green's functions. Here we shall present a derivation of equation (16.170) based directly on the Bogoliubov equations.

In the following we concentrate on the case where Δ depends on one coordinate only. Furthermore we set the vector potential $\mathbf{A} = 0$. The general expression for the free energy as given by the Bogoliubov equations (see the preceding section) is

$$F = -T \sum_\lambda \log[1 + \exp(-E_\lambda/T)] + \int \frac{1}{V(\mathbf{r})} \Delta(\mathbf{r}) d^3r \quad (16.171)$$

The eigenvalues E are obtained from

$$\begin{pmatrix} -\dfrac{1}{2m} \nabla^2 - \mu & -\Delta(x) \\[2mm] -\Delta(x) & \dfrac{1}{2m} \nabla^2 + \mu \end{pmatrix} \begin{pmatrix} u_\lambda \\ v_\lambda \end{pmatrix} = E_\lambda \begin{pmatrix} u_\lambda \\ v_\lambda \end{pmatrix} \quad (16.172)$$

In order that equation (16.172) should define the eigenvalues E, one must also define boundary conditions, e.g.,

$$\phi_\lambda = \begin{pmatrix} u_\lambda \\ v_\lambda \end{pmatrix} \quad \phi(0) = \phi(L) = 0 \quad (16.173)$$

for a slab of width L, or

$$\phi(0) = \phi(L) \quad \text{and} \quad \frac{d}{dx} \phi(0) = \frac{d}{dx} \phi(L) \quad (16.174)$$

for a ring of length L. In either case the boundary conditions will lead to a dispersion relation

$$D(E) = 0 \quad (16.175)$$

from which the eigenvalues E_λ can be computed. In the complex ω-plane the first term in equation (16.171) can be written as

$$F_1 = -\frac{T}{2\pi i} \int_C d\omega \log(1 + e^{-\omega/T}) \frac{d}{d\omega} \log \frac{D(\omega)}{D_0(\omega)} + F_0 \quad (16.176)$$

where the path of integration goes around the real axis. For ω close to E_λ

$$D(\omega) = a(\omega - E_\lambda) \quad (16.177)$$

so that

$$\frac{d}{d\omega} \log D(\omega) = \frac{1}{\omega - E_\lambda} \qquad (16.178)$$

F_0 being the free energy of the normal state and D_0 the corresponding dispersion relation. We assume that $\log[D(\omega)/D_0(\omega)] \to 0$ for $\omega \to \pm\infty$ so that the boundary terms do not contribute when equation (16.176) is partially integrated. We then obtain

$$\Delta F = F_1 - F_0 = \frac{1}{2\pi i} \int_C d\omega \, \frac{1}{e^{\omega/T} + 1} \log \frac{D(\omega)}{D_0(\omega)} \qquad (16.179)$$

The function $(e^{\omega/T} + 1)^{-1}$ has poles at

$$\omega = i\omega_n = 2\pi T(n + \tfrac{1}{2})i, \qquad n = 0, \pm 1, \ldots \qquad (16.180)$$

If it is again assumed that $\log[D(\omega)/D_0(\omega)] \to 0$ sufficiently fast so that the path C can be deformed around the poles ω_n, then

$$\Delta F = -T \sum_{n=-\infty}^{+\infty} \log \left[\frac{D(-i\omega_n)}{D_0(-i\omega_n)} \right] \qquad (16.181)$$

We now assume periodic boundary conditions in the y, z direction and set

$$\phi(\mathbf{r}) = [\exp(ik_y y + ik_z z)] w(x) \qquad (16.182)$$

By introducing

$$\kappa^2 = k_y^2 + k_z^2, \qquad K^2 = k_F^2 - \kappa^2, \qquad \mu = k_F^2/2m \qquad (16.183)$$

and the Pauli matrix σ_i, the Bogoliubov equations become

$$\left[-\frac{1}{2m} \sigma_3 \left(\frac{\partial^2}{\partial x^2} + K^2 \right) - \Delta(x)\sigma_1 \right] w = Ew \qquad (16.184)$$

We now make the ansatz

$$w = e^{iKx} G^{(+)}(x) + e^{-iKx} G^{(-)}(x) \qquad (16.185)$$

Next we assume that

$$|K| \gg \left| \frac{d}{dx} G^{(\pm)} \right| \qquad (16.186)$$

which is reasonable because $|K|^{-1} \sim k_F^{-1}$ is of the order of atomic distances.

When equation (16.185) is substituted into equation (16.184) and use is made of equation (16.186), the Bogoliubov equation separates into two equations for $G^{(\pm)}$:

$$\left[-i\frac{K}{m}\sigma_3\frac{\partial}{\partial x} - \Delta(x)\sigma_1 \right] G^{(\pm)} = EG^{(\pm)} \tag{16.187}$$

If we set

$$G = \begin{pmatrix} u \\ iv \end{pmatrix} \tag{16.188}$$

the coupled set of equations for u, v becomes

$$\begin{aligned} \mp i\frac{K}{m} u' - \Delta iv &= Eu \\ -\Delta u \pm i\frac{K}{m} v' &= iEv \end{aligned} \tag{16.189}$$

and with

$$v_K = \frac{K}{m} \tag{16.190}$$

we finally obtain

$$\left. \begin{aligned} v_K^2 u'' + (E^2 - \Delta^2)u &= -\Delta' v v_K \\ v_K^2 v'' + (E^2 - \Delta^2)v &= -\Delta' u v_K \end{aligned} \right\} \tag{16.191}$$

These equations pertain to the region $0 < x < L$. We now consider the two independent special solutions

$$\left. \begin{aligned} G^{(+)} &= A_+^{(+)} \begin{pmatrix} f_- \\ ig_+ \end{pmatrix} - A_-^{(+)} \begin{pmatrix} ig_- \\ -f_+ \end{pmatrix} \\ G^{(-)} &= A_+^{(-)} \begin{pmatrix} -f_+ \\ ig_- \end{pmatrix} - A_-^{(-)} \begin{pmatrix} ig_+ \\ f_- \end{pmatrix} \end{aligned} \right\} \tag{16.192}$$

with initial conditions

$$\left. \begin{aligned} f_s(0) = 1, & \qquad v_K f_s' = -siE \\ g_s(0) = 0, & \qquad v_K g_s'(0) = -\Delta(0) \end{aligned} \right\} \tag{16.193}$$

For periodic boundary conditions $w(0) = w(L)$ and $w'(0) = w'(L)$, and because $KL \gg 1$, only the first of these conditions is relevant.

Let us consider first $G^{(+)}(0) = G^{(+)}(L)$. Then it follows that

$$A^{(+)} = A_+^{(+)}f_- iA_-^{(+)}g_-$$
$$A_-^{(+)} = iA_+^{(+)}g_+ + A_-^{(+)}f_+$$

$$(16.194)$$

These equations have a solution only if (we do not write explicitly $x = L$)

$$D_K = 1 - (f_+ - f_-) + f_+f_- - g_+g_- = 0 \qquad (16.195)$$

It follows directly from equation (16.189) that

$$\frac{d}{dx}(f_+f_- - g_+g_-) = 0 \Rightarrow f_+f_- - g_+g_- = 1 \qquad (16.196)$$

Hence we obtain

$$D_K = \tfrac{1}{2}(f_+ + f_-) - 1 = 0 \qquad (16.197)$$

The same result is obtained for $G^-(0) = G^-(L)$. Therefore the periodic boundary condition is satisfied with equation (16.197). In order to obtain decoupled equations we introduce

$$h_s = \tfrac{1}{2}[f_+ + f_- - s(g_+ + g_-)] \qquad (16.198)$$

which satisfies the initial conditions

$$h_s(0) = 1, \qquad v_K h_s'(0) = s\Delta(0) \qquad (16.199)$$

Function h_s satisfies the equation [see equation (16.191)]

$$v_K^2 h_s'' - \kappa^2 h_s = -s\Delta' v_K h_s \qquad (16.200)$$

Now we must remember that in the expression for the free energy, $E \to -i\omega_n$, so that

$$\kappa^2 = (\omega_n^2 + \Delta^2) \qquad (16.201)$$

The expression for D_K is now given by

$$D_K(-i\omega_n) = \tfrac{1}{2}[h_+(-i\omega_n) + h_-(-i\omega_n)] - 1 \qquad (16.202)$$

The equation for h_s can be put into the form of a Schrödinger equation

$$\left[-v_K^2 \frac{d^2}{dx^2} + V_s(x) \right] h_s(x) = -v_K^2 \omega_n^2 h_s(x) \qquad (16.203)$$

where

$$V_s(x) = \Delta^2 + s v_K \Delta' \qquad (16.204)$$

The free energy can now be computed from

$$\Delta F = -2T \sum_{n=-\infty}^{+\infty} \sum_K \log\left[\frac{D_K(i\omega_n)}{D_{0K}(-i\omega_n)}\right] \qquad (16.205)$$

For a slab of area L^2

$$\sum_K \cdots = \frac{1}{L^2} \frac{2\pi}{(2\pi)^2} \int_{K<k_F} K dK \cdots \qquad (16.206)$$

With equation (16.203) we have transformed the original Bogoliubov equations (16.172) into a new form. We now have two decoupled equations for the auxiliary functions h_s from which the free energy can be directly computed. One should note that this can be done only for given $\Delta(x)$. In order to avoid the self-consistency equation (16.150) it is convenient to use the variational principle. For a constant Δ the solution of equation (16.203) is

$$h_s(x) = \cosh qx + s v_K \Delta \sinh qx \qquad (16.207)$$

where

$$q = \sqrt{\omega_n^2 + \Delta^2/v} \qquad (16.208)$$

The term in the free energy that is proportional to $L^3 = 1$ then becomes

$$F = -T \sum_n \sum_K (\sqrt{\omega_n^2 + \Delta^2} - |\omega_n|) v_K^{-1} \qquad (16.209)$$

As is usual, we must introduce a cutoff in equation (16.209) for

$$|\omega_n| < \omega_D \qquad (16.210)$$

Furthermore

$$\sum_K v_K^{-1} = \frac{1}{2\pi} v_F^{-1} k_F^2 \qquad (16.211)$$

and finally

$$\Delta F = -\frac{T}{\pi} v_F^{-1} k_F^2 \sum_n' |\omega_n|(\sqrt{1 + \Delta^2/\omega_n^2} - 1) = f(\Delta^2) \qquad (16.212)$$

where f is the free-energy density appearing in equation (16.170).

According to the general principles of the Landau theory of second-order phase transitions, near the transition

$$f = -\alpha\Delta^2 + \tfrac{1}{2}\beta\Delta^4 \tag{16.213}$$

Comparison of equations (16.212) and (16.213) yields

$$\alpha = \frac{T}{\pi}\, v_F^{-1} k_F^2 \sum_{n=0}^{\omega_D/2\pi T} \frac{1}{2\pi T(n+\tfrac{1}{2})}$$

$$= \frac{k_F^2}{\pi^2 v_F} \sum_{n=0}^{\omega_D/2\pi T} \frac{1}{n+\tfrac{1}{2}} = \frac{k_F^2}{\pi^2 v_F} \ln\left(\frac{2\gamma\omega_D}{\pi T}\right) \tag{16.214}$$

$$\beta = \frac{T}{\pi}\, v_F^{-1} k_F^2 \frac{1}{4} \sum_{n=0}^{\infty} \frac{1}{[2\pi T(n+\tfrac{1}{2})]^3}$$

$$= \frac{k_F^2}{32\pi^4}\frac{1}{T^2 v_F} \sum_{n=0}^{\infty} \frac{1}{(n+\tfrac{1}{2})^3} = \frac{1}{32\pi^4}\frac{k_F^2}{T^2 v_F}\, 7\cdot\zeta(3) \tag{16.215}$$

The summation over n can be extended to infinity because $\omega_D/2\pi T \gg 1$. In order to obtain the term $|\Delta'|^2$ in equation (16.170) we set

$$h_s(x) = \exp\int^x K_s(x')dx \tag{16.216}$$

in equation (16.203). The equation for K_s becomes

$$K_s = \frac{1}{v_K}\sqrt{\omega_n^2 + \Delta^2 - v_K^2 K_s' + v_K s\Delta'} \tag{16.217}$$

In an expansion up to second order in Δ'^2

$$K_s = \frac{1}{v_K}|\omega_n|\left[1 + \frac{1}{8}\left(\frac{v_K\Delta'}{\omega_n^2}\right)^2 + \cdots\right] \tag{16.218}$$

The linear terms do not contribute to equation (16.218) and have been omitted. Hence

$$\gamma = 2T\sum_{n>0}\sum_{K}\frac{1}{8}v_K\frac{1}{\omega_n^3}$$

$$= \frac{7}{8}\frac{\zeta(3)}{24\pi^4}\frac{k_F^2}{T^2}v_F \tag{16.219}$$

16.7. Excitonic Mechanisms

In Section 16.2 it was shown how the electron–phonon interaction gives rise to an effective attractive interaction between the conduction electrons. There is no apparent reason why phonons should not be replaced by something else. Why not spin waves? In this case an electron scatters with a spin wave and the spin of the electrons is reversed. Carrying out the corresponding analysis of Section 16.2 leads to a sign change for the effective interaction.[423] In 1964 Little[424] suggested that the electron scattering with molecular excitations would lead to an effective attractive interaction. Because he argued that such an interaction would be most effective if the polarizable molecules were to form the side chains of long organic molecules, the field of one-dimensional conductors was strongly activated. Subsequently, Davis et al.[425] have given a specific model for an excitonic superconductor based on Little's ideas. Allender et al.[426] (ABB) considered the excitonic mechanism of superconductivity with respect to a particular model, a thin metal layer on a semiconductor surface. In this model, the metal electrons at the Fermi surface tunnel into the semiconductor gap where they interact with virtual excitons, producing a net attractive interaction among the electrons in direct analogy with the phonon mechanism of superconductivity. In such a case, therefore, the superconducting transition temperature is given by equation (16.156) as

$$T_c = 1.13\omega_{ex}\exp[-1/N(0)V_{ex}] \tag{16.220}$$

where ω_{ex} is the excitation energy for an electron–hole pair or exciton and V_{ex} is the effective interaction caused by the exciton exchange. An estimate of the size of V_{ex} based on equation (16.70) leads to

$$V_{ex} = 2|M^2|/\omega_{ex}^2 \tag{16.221}$$

where M represents the matrix element for the electron–exciton interaction. In analogy to equation (16.48) we have

$$H_{el-ex} = \sum_{k,q} M_{k,q}c_{k+q}^+ c_k(b_q + b_q^+) \tag{16.222}$$

where now b_q^+ and b_q are the creation operators of the excitons.

In his original estimate Little obtained a transition temperature of approximately 200 °K. At present there are no known superconductors in which pairing is excitonic. In fact the theoretical basis as discussed above is quite inadequate.

The most obvious difference between an electron–phonon and an electron–exciton interaction is that the exciton is an electron–hole pair, therefore it is made with the same entities for which it produces scattering. This results in exchange terms in the interaction between electrons and the excitons. In addition there is always the direct Coulomb repulsion, which in the Thomas–Fermi approximation is [see equation (16.29)]

$$U_c = \frac{4\pi e^2}{q^2 + k_s^2} \tag{16.223}$$

In fact equation (16.223) may be generalized to include the excitonic effects:

$$U = \frac{4\pi e^2}{q^2 \varepsilon(q, \omega = 0)} \tag{16.224}$$

where $\varepsilon(q, \omega)$ is the dielectric function of the medium at wave vector $\mathbf{q}$ and frequency ω. It contains the direct Coulomb interaction (16.223), but also the effect of all the excitations of the system that contribute to the effective interaction between conduction electrons. The formulation (16.224) gave rise to an interesting stability argument first noted by Cohen and Anderson.[427]

In the continuum approximation (no umklapp scattering or local-field corrections), stability requires that $\varepsilon(q, 0) \geqslant 0$. This stability argument is based on the fact that a static modulation of the background charge by $\delta\rho(q)$ would modify the total energy of the system by

$$\delta E = \frac{4\pi e^2}{q^2} \frac{|\delta\rho(q)|^2}{\varepsilon(q, 0)} \tag{16.225}$$

If umklapps are included, equation (16.225) has the form

$$\delta E = \sum_{\mathbf{K}} \frac{4\pi e^2}{q^2 + K^2} \frac{|\delta\rho(q)|^2}{\varepsilon(\mathbf{q}, \mathbf{K}, 0)} \tag{16.226}$$

and $\varepsilon(\mathbf{q}, 0, 0)$ may be less than zero. This means that the local-field correction becomes quite important when considering the excitonic mechanisms. This obviously makes it important to consider the details if one wants to make predictions.

It has been argued by Phillips[428] that the local-field corrections would be very ineffective in producing an attractive part, because the exciton, being an electronic excitation, occupies a large part of the unit cell and thus there is not much difference between the average field and the local field.

However, Gutfreund and Little[429] have pointed out that for molecular side-chain systems the unit cell is very large, and that the electrons are concentrated in a very small portion of it. Therefore local-field effects could be effective after all. In any case there is agreement that the excitonic effect, if at all possible, has the best prospects of success for "one"- and "two"-dimensional structures. This then leads us to the question of dimensionality and superconductivity discussed in the final section.

16.8. Effects of Limited Dimensionality

In the preceding section we discussed arguments apparently indicating that one- or two-dimensional structures have the best prospect of showing excitonic superconductivity. But here two additional difficulties have to be considered:

1. In a one-dimensional structure, thermal fluctuations prevent the formation of long-range order and there is no transition at all.[70] This argument has been extended by Rice[430] and later by Hohenberg,[431] who showed rigorously that long-range order could not occur in any infinite one- or two-dimensional system. Therefore if one wants to discuss real phase transitions, interchain coupling and interchain hopping must be included.

2. The competition of superconductivity against other transitions becomes more severe as the anisotropy of the system increases. For a one-dimensional system interacting with phonons, a charge-density wave (CDW) as proposed by Peierls[71] and Fröhlich[432] is the ground state. This state is of particular interest because the CDW can slide as a whole in an applied field. As Fröhlich noted a long time ago,[432] for an incommensurate charge-density wave in the absence of pinning to impurities, this leads to an alternative mechanism for superconductivity.

The first, very highly conducting anisotropic organic crystal was TTF-TCNQ, initially reported in 1973. After the suggestion that the conductivity peak observed above the Peierls transition observed at $T_c = 54$ °K is due to the Fröhlich conductivity from CDW fluctuations,[433] a semiphenomenological theory was derived by Allender et al.[434] A similar expression was derived by Strässler and Toombs[435] from an evaluation of a diagram corresponding to that which Aslimasov and Larkin used to describe paraconductivity from fluctuations above T_c in an ordinary superconductor. The conductivity comes out to be

$$\sigma = 0.2 \cdot ne^2\hbar/(mk_B T_c \varepsilon^{1/2}) \tag{16.227}$$

where n is the electron density, m the electron mass, and $\varepsilon = (T - T_c)/T_c$.

A slightly generalized expression, derived by Rice,[435] has been used by Heeger[436] to analyze the TTF-TCNQ data. As discussed by Bardeen,[437] the question whether the conductivity seen in TTF-TCNQ can be explained by the one-dimensional fluctuating Fröhlich superconductivity mechanism is not yet resolved.

Nevertheless the Fröhlich mechanism has been observed in several systems. The material KCP [$K_2Pt(CN)_4Br_{0.3} \cdot 3H_2O$] probably represents the best case for a comparison of theory and experiment. As a result of pinning to impurities, the Fröhlich CDW shows up as an enhancement of the far-infrared dielectric constant[438]:

$$\varepsilon(\omega) = 1 + \frac{\Omega_p^2}{\omega_p^2 - \omega^2 - i\Gamma\omega} \tag{16.228}$$

where $\Omega_p = 4\pi n e^2/M$ is the plasma frequency of the collective mode, M is the effective mass of the CDW, and Γ is the damping constant of the wave.

Another case where the Fröhlich mechanism plays an active role is NbSe$_3$. Although the CDW in NbSe$_3$ is pinned as in KCP, the CDW may be depinned by the application of extremely small fields (approximately 5 mV/cm). The depinned CDW provides an additional contribution to the electrical conductivity. The experimental evidence[439] for Fröhlich-mode conductivity in NbSe$_3$ as well as some other effects, including the noise, have been reviewed recently by Fleming.[440]

Until now, we have considered only the effect of the electron–phonon interaction on a quasi-one-dimensional system. If one includes also the interaction between electrons and explicitly takes into account the interchain interaction and interchain hopping, then there are many more possibilities. An excellent up-to-date review has been given by Rice.[441] Once these additional effects are included, the pairing superconductivity including the exciton mechanism again become theoretical possibilities.

At present it seems that interchain hopping is most essential in stabilizing the superconductors. In 1975 (SN)$_x$, believed to be a quasi-one-dimensional polymer, was found to be superconducting at about 0.3 °K. By now it has been shown to be not a one-dimensional material. Recently[442] superconductivity has been found in the chain compound (TMTSF)$_2$PF$_6$ under a hydrostatic pressure of 12 kbar. It is the first known organic superconductor. Transport properties and optical reflectance[442,443] had originally suggested that (TMTSF)$_2$PF$_6$ has quasi-one-dimensional electronic properties. More recent measurements indicate that the substance is one-dimensional near room temperature, but crosses over to become a two- or three-dimensional system at lower temperatures.[444] The transverse plasma frequency is about one-fifth of the parallel plasma frequency.

References for Part III

1. R. Huisman, R. DeJonge, C. Haas, and F. Jellinek, *J. Solid State Chem.* **3**, 56 (1971).
2. D. R. Salahub and R. P. Messmer, *Phys. Rev. B* **14**, 2592 (1976).
3. H. C. Longuet-Higgins and L. Salem, *Proc. R. Soc. London, Ser. A*, **251**, 172 (1959).
4. E. H. Lieb and C. Wu, *Phys. Rev. Lett.* **20**, 1445 (1968).
5. J. E. Lennard-Jones, *Proc. R. Soc. London, Ser. A* **158**, 280 (1937).
6. J. N. Murrell, *Theory of the Electronic Spectra of Organic Molecules*, p. 69, John Wiley and Sons, New York (1963).
7. L. Salem, *Molecular Orbital Theory of Conjugated Systems*, Benjamin, New York (1966).
8. H. Kuhn, *J. Chem. Phys.* **17**, 1198 (1949).
9. M. J. S. Dewar, *J. Chem. Soc.* 3544 (1952).
10. Y. Ooshika, *J. Phys. Soc. Jpn.* **12**, 1238, 1246 (1957).
11. See, for example, N. H. March, in: *Quantum Theory of Polymers* (J.-M. Andre, J. Delhalle, and J. Ladik, eds.), p. 49, D. Reidel Pub. Co., Dordrecht (1978).
12. A. Ferraz, P. J. Grout, and N. H. March, *Phys. Lett.* **66A**, 155 (1978).
13. B. Johansson and K. F. Berggren, *Phys. Rev.* **181**, 855 (1969).
14. D. R. Yarkony and R. Silbey, *Chem. Phys.* **20**, 183 (1977).
15. A. C. Laraga, R. J. Aerni, and M. Karplus, *J. Chem. Phys.* **73**, 5230 (1980).
16. M. Baldo, R. Pucci, and P. Tomasello, *J. Chem. Phys.* **70**, 4086 (1979).
17. A. A. Ovchinnikov, *Sov. Phys. JETP (Engl. Transl.)* **30**, 1160 (1970).
18. See, for example, *Highly Conducting One-Dimensional Solids* (J. T. Devreese, R. P. Evrard, and V. E. van Doren eds.), Plenum Press, New York (1979).
19. See, for instance, the article by A. J. Berlinsky in Ref. 18, p. 1.
20. B. Renker, H. Rietschel, L. Pintschovius, W. Gläser, P. Brüesch, D. Kuse, and M. J. Rice, *Phys. Rev. Lett.* **30**, 1144 (1973).
21. R. Comès, M. Lambert, H. Launois, and H. R. Zeller, *Phys. Rev. B* **8**, 571 (1973).
22. D. M. Whitmore, Ph.D. Thesis, Stanford University (1974).
23. R. P. Messmer and D. R. Salahub, *Phys. Rev. Lett.* **35**, 533 (1975).
24. L. V. Interrante and R. P. Messmer, in: *Extended Interactions between Metal Ions*, ACS Symp. Series No. 5, p. 382, ACS, Washington (1974).
25. J. A. Wilson and A. D. Yoffe, *Adv. Phys.* **18**, 193 (1969).
26. J. B. Goodenough, *Phys. Rev.* **171**, 466 (1968).
27. L. F. Mattheiss, *Phys. Rev. B* **8**, 3719 (1973).
28. C. C. J. Roothaan and R. S. Mulliken, *J. Chem. Phys.* **16**, 118 (1948).
29. C. A. Coulson and R. Taylor, *Proc. Phys. Soc.* **65A**, 815, 834 (1952).

30. Boron Nitride, *Phys. Rev.* (in press).
31. F. J. Corbato, *Proceedings of the Third Conference on Carbon*, p. 173, Pergamon Press, London (1959).
32. See, for example, the summary in W. Jones and N. H. March, *Theoretical Solid State Physics*, Vol. 2, p. 796, Wiley–Interscience, London (1973).
33. I. P. Batra, B. I. Bennett, and F. Herman, *Phys. Rev. B* **11**, 4927 (1975).
34. F. Herman, A. R. Williams, and K. H. Johnson, *J. Chem. Phys* **61**, 3508 (1974).
35. T. J. Kistenmacher, T. E. Phillips, and D. O. Cowan, *Acta Crystallogr., Sect. B* **30**, 763 (1974).
36. R. E. Long, R. A. Sparks, and K. N. Trueblood, *Acta Crystallogr.* **18**, 932 (1965).
37. K. H. Johnson, *J. Chem. Phys.* **45**, 3085 (1966); J. C. Slater, *Adv. Quantum Chem.* **6**, 1 (1972).
38. J. C. Slater and K. H. Johnson, *Phys. Rev. B* **5**, 844 (1972).
39. F. Herman and I. P. Batra, *Phys. Rev. Lett.* **33**, 94 (1974).
40. C. E. Klots, R. M. Compton, and V. F. Raaen, *J. Chem. Phys.* **60**, 1177 (1974).
41. H. Johansen (to appear).
42. W. D. Grobman, R. A. Pollak, D. E. Eastman, E. T. Maas, and B. A. Scott, *Phys. Rev. Lett.* **32**, 534 (1974).
43. M. Boudeulle, A. Douillard, P. Michel, and G. Vallet, *C. R. Acad Sci. Paris, Ser. C* **272**, 2137 (1971).
44. M. Boudeulle, Ph.D. Thesis, Université Claude-Bernard, Lyon (1974).
45. A. G. MacDiarmid, C. M. Mikulski, P. J. Russo, M. S. Saran, A. F. Garito, and A. J. Heeger, *J. Chem. Soc., Chem. Commun.* 477 (1975).
46. C. M. Mikulski, P. J. Russo, M. S. Saran, A. G. MacDiarmid, A. F. Garito, and A. J. Heeger, *J. Am. Chem. Soc.* **97**, 6358 (1975).
47. M. J. Cohen, A. F. Garito, A. J. Heeger, A. G. MacDiarmid, C. M. Mikulski, and M. S. Saran (to appear).
48. R. L. Greene, G. B. Street, and L. J. Suter, *Phys. Rev. Lett.* **34**, 577 (1975).
49. D. R. Salahub and R. P. Messmer, *J. Chem. Phys.* **64**, 2039 (1976).
50. P. Mengel, P. M. Grant, W. E. Rudge, B. H. Schechtman, and D. W. Rice, *Phys. Rev. Lett.* **35**, 1803 (1975).
51. L. Ley, *Phys. Rev. Lett.* **35**, 1796 (1975).
52. This method was used earlier in a one-dimensional band-structure calculation by D. E. Parry and J. M. Thomas, *J. Phys. C* **8**, L45 (1975).
53. N. H. March, *Adv. Phys.* **6**, 1 (1957).
54. A. S. Bamzai and B. M. Deb, *Rev. Mod. Phys.* **53**, 95 (1981).
55. N. H. March, Electron density description of atoms and molecules, in: *Theoretical Chemistry*, Vol. 4, p. 92, Specialist Periodical Reports, Royal Society of Chemistry, Burlington House, London (1981).
56. B. Stenhouse, P. J. Grout, N. H. March, and J. Wenzel, *Philos. Mag.* **36**, 129 (1977).
57. See R. G. Parr, R. A. Donnelly, M. Levy, and W. E. Palke, *J. Chem. Phys.* **68**, 3801 (1978) and earlier references cited therein.
58. R. Pucci and N. H. March, *J. Chem. Phys.* **74**, 2936 (1981).
59. N. H. March and J. S. Plaskett, *Proc. R. Soc. London, Ser. A* **235**, 419 (1956).
60. E. Teller, *Rev. Mod. Phys.* **34**, 627 (1962).
61. K. Ruedenberg, *J. Chem. Phys.* **66**, 375 (1977).
62. N. H. March, *J. Chem. Phys.* **67**, 4618 (1977).
63. M. Kertész, J. Koller, and A. Ažman, *J. Chem. Phys.* **69**, 2937 (1978).
64. A. D. Walsh, *J. Chem. Soc.* 2260 (1953).
65. N. H. March, *J. Chem. Phys.* **74**, 2973 (1981).
66. J. F. Mucci and N. H. March, *J. Chem. Phys.* **71**, 1495 (1979).

67. C. A. Coulson and H. C. Longuet-Higgins, *Proc. R. Soc. London, Ser. A* **191**, 39 (1947).
68. C. A. Coulson, *Proc. R. Soc. London, Ser. A* **169**, 413 (1939).
69. N. W. Ashcroft and N. D. Mermin, *Solid State Physics*, Holt, Rinehart, and Winston, New York (1976).
70. L. Landau and E. M. Lifshitz, *Statistical Physics*, Pergamon Press, Oxford (1969).
71. R. E. Peierls, *Quantum Theory of Solids*, Oxford University Press (1955).
72. C. N. R. Rao and K. J. Rao, *Phase Transitions in Solids*, McGraw-Hill, New York (1978).
73. T. Riste (ed.), *Electron–Phonon Interactions and Phase Transitions*, Plenum Press, New York (1977).
74. J. Woods Halley (ed.), *Correlation Functions and Quasi-Particle Interactions in Condensed Matter*, Plenum Press, New York (1978).
75. R. Balian, R. Maynard, and G. Toulouse (eds.), *Ill-Condensed Matter*, North-Holland Publ. Co., Amsterdam (1979).
76. G. H. Wannier, *Statistical Physics*, John Wiley and Sons, New York (1966).
77. R. E. Peierls, *Proc. Camb. Phil. Soc.* **32**, 477 (1936).
78. H. E. Stanley, *Introduction to Phase Transitions and Critical Phenomena*, Oxford University Press (1971).
79. M. E. Fisher, *Rep. Prog. Phys.* **30**, 615 (1967).
80. R. B. Stinchcombe, in: *Correlation Functions and Quasi-Particle Interactions in Condensed Matter* (J. Woods Halley, ed.), Plenum Press, New York (1978).
81. J. Als-Nielsen, O. W. Dietrich, and L. Passell, *Phys. Rev. B* **14**, 4908 (1977).
82. E. Ising, *Z. Phys.* **31**, 253 (1925).
83. L. Onsager, *Phys. Rev.* **65**, 117 (1944).
84. L. Onsager and B. Kaufman, *Phys. Rev.* **76**, 1232, 11244 (1949).
85. H. Ikeda and K. Hirakawa, *Solid State Commun.* **14**, 529 (1974).
86. H. Ikeda, I. Hatta, and M. Tanaka, *J. Phys. Soc. Jpn.* **40**, 334 (1976).
87. E. J. Samuelson, *Phys. Rev. Lett.* **31**, 936 (1973).
88. A. Tucciarone, H. Y. Lau, L. M. Corliss, A. Delapalme, and J. M. Hastings, *Phys. Rev. B* **4**, 3206 (1971).
89. J. M. Kosterlitz and D. J. Thouless, *J. Phys. C* **6**, 1181 (1973).
90. J. M. Kosterlitz and D. J. Thouless in: *Progress in Low Temperature Physics* (D. F. Brewer, ed.), Vol. VII B, North-Holland Publ. Co., Amsterdam (1978).
91. J. M. Kosterlitz, *J. Phys. C* **7**, 1046 (1974).
92. L. J. de Jongh and A. R. Miedema, *Experiments on Simple Model Magnetic Systems*, Taylor and Francis, London (1974).
93. L. J. de Jongh and A. R. Miedema, *Adv. Phys.* **23**, 1 (1974).
94. W. L. Bragg and E. J. Williams, *Proc. R. Soc. London, Ser. A* **145**, 699 (1934).
95. F. Lindemann, *Phys. Z.* **11**, 609 (1910).
96. D. R. Nelson and B. I. Halperin, *Phys. Rev. B* **19**, 2457 (1979).
97. B. I. Halperin and D. R. Nelson, *Phys. Rev. Lett.* **41**, 121 (1978).
98. D. R. Nelson, *Phys. Rev. B* **18**, 2318 (1978).
99. A. P. Young, *Phys. Rev. B* **19**, 1855 (1979).
100. L. D. Landau, *J. Phys. USSR* **5**, 71 (1944).
101. R. P. Feynman, *Statistical Physics*, Benjamin, New York (1972).
102. J. Goldstone, *Nuovo Cimento* **19**, 154 (1961).
103. T. W. Kibble, in: *Oxford International Conference on Elementary Particle Physics*, Rutherford High Energy Laboratory (1966).
104. B. I. Halperin, in: *Proc. Summer Institute on Low-Dimensional Systems*, Kyoto University Press (1979).
105. J. Bardeen, L. N. Cooper, and J. R. Schrieffer, *Phys. Rev.* **108**, 1175 (1957).
106. L. N. Cooper, *Phys. Rev.* **104**, 1189 (1956).

107. J. Bardeen and D. Pines, *Phys. Rev.* **99**, 1140 (1955).
108. P. W. Anderson, *Phys. Rev.* **112**, 900 (1958).
109. R. E. Peierls, *Quantum Theory of Solids*, Oxford University Press (1955).
110. J. Friedel, in: *Electron–Phonon Interactions and Phase Transitions* (T. Riste, ed.), Plenum Press, New York (1977).
111. N. F. Mott, *Adv. Phys.* **16**, 49 (1967).
112. N. F. Mott, *Metal–Insulator Transitions*, Taylor and Francis, London (1974).
113. J. Hubbard, *Proc. Roy. Soc. London, Ser. A* **281**, 401 (1964).
114. P. W. Anderson, *Phys. Rev.* **109**, 1492 (1958).
115. D. J. Thouless, *Phys. Rep. C* **13**, 94 (1974).
116. N. F. Mott and W. D. Twose, *Adv. Phys.* **10**, 107 (1960).
117. K. Ishii, *Suppl. Prog. Theor. Phys.* **53**, 77 (1973).
118. E. Abrahams, P. W. Anderson, D. C. Licciardello, and T. V. Ramakrishnan, *Phys. Rev. Lett.* **42**, 673 (1979).
119. J. F. Scott, *Rev. Mod. Phys.* **46**, 83 (1974).
120. R. B. Stinchcombe, in: *Electron–Phonon Interactions and Phase Transitions* (T. Riste, ed.), Plenum Press, New York (1977).
121. R. J. Elliott, R. T. Harley, W. Hayes, and S. R. P. Smith, *Proc. R. Soc. London, Ser. A* **328**, 217 (1972).
122. H. Thomas, in: *Electron–Phonon Interactions and Phase Transitions* (T. Riste, ed.), Plenum Press, New York (1977).
123. G. A. Gehring and K. A. Gehring, *Rep. Prog. Phys.* **38**, 1 (1975).
124. R. Blinc and B. Zeks, *Soft Modes in Ferroelectrics and Antiferroelectrics*, North-Holland Publ. Co., Amsterdam (1974).
125. K. K. Kobayashi, *J. Phys. Soc. Jpn.* **24**, 497 (1968).
126. R. J. Elliott and A. P. Young, *Ferroelectrics* **7**, 23 (1974).
127. S. R. Broadbent and J. M. Hammersley, *Proc. Camb. Phil. Soc.* **53**, 629 (1957).
128. S. Roach, *Theory of Clumping* Methuen (1968).
129. J. W. Essam, in: *Phase Transitions and Critical Phenomena*, (C. Domb and M. S. Green, eds.), Vol. 2, Academic Press, New York (1972).
130. V. K. S. Shante and S. Kirkpatrick, *Adv. Phys.* **20**, 325 (1971).
131. J. W. Essam, *Rep. Prog. Phys.* **43**, 833 (1980).
132. D. J. Stauffer, *Phys. Rep.* **54**, 1 (1979).
133. P. Pfeuty and E. Guyon (to appear).
134. L. D. Landau and E. M. Lifshitz, *Statistical Physics*, Pergamon Press, Oxford (1959).
135. K. Huang, *Statistical Mechanics*, John Wiley and Sons, New York (1963).
136. R. Brout, *Phase Transitions*, Benjamin, New York (1965).
137. R. B. Stinchcombe, *J. Phys. C* **6**, 2459 (1973).
138. A. H. Cooke, S. J. Swithenby, and M. R. Wells, *Solid State Commun.* **10**, 265 (1972).
139. F. Dyson, *Phys. Rev.* **102**, 1217, 1230 (1956).
140. D. Bohm and D. Pines, *Phys. Rev.* **92**, 609 (1952).
141. D. Pines and D. Bohm, *Phys. Rev.* **85**, 338 (1952).
142. N. N. Bogoliubov and S. V. Tyablikov, *Sov. Phys. Dokl.* **4**, 589 (1959).
143. F. Englert, *Phys. Rev. Lett.* **5**, 102 (1960).
144. D. N. Zubarev, *Sov. Phys. Usp.* **3**, 320 (1960).
145. L. Van Hove, *Phys. Rev.* **93**, 1374 (1954).
146. W. Cochran, *Phys. Rev. Lett.* **3**, 412 (1959).
147. W. Cochran, *Adv. Phys.* **10**, 401 (1962).
148. P. W. Anderson, in: *Physics of Dielectrics* (Russian) (G. I. Skanavi, ed.), Akad. Nauk SSSR, Moscow (1959).
149. R. V. Lange, *Phys. Rev. Lett.* **14**, 3 (1975).
150. F. Bloch, *Z. Phys.* **61**, 206 (1930).

151. C. Kittel, *Quantum Theory of Solids*, John Wiley and Sons, New York (1963).
152. N. D. Mermin and H. Wagner, *Phys. Rev. Lett.* **17**, 1133 (1966).
153. A. C. Scott, F. Y. F. Chu, and D. W. McLaughlin, *Proc. IEEE* **61**, 1443 (1973).
154. G. Toulouse, *Commun. Phys.* **2**, 115 (1977).
155. J. Villain, *J. Phys. C* **10**, 1717 (1977).
156. P. W. Anderson, in: *Ill-Condensed Matter* (R. Balian, R. Maynard, and G. Toulouse, eds.), North-Holland Publ. Co., Amsterdam (1979).
157. R. E. Peierls, *Helv. Phys. Acta* **7**, Suppl. **2**, 81 (1934).
158. R. E. Peierls, *Ann. Inst. Henri Poincaré* **5**, 177 (1935).
159. L. D. Landau, *Phys. Z. Sowjet Union* **11**, 26 (1937).
160. R. E. Peierls, *Surprises in Physics*, Princeton University Press (1979).
161. N. D. Mermin, *Phys. Rev.* **176**, 250 (1968).
162. D. Pines, *Elementary Excitations in Solids*, Benjamin, New York (1963).
163. A. W. Overhauser, *Phys. Rev.* **128**, 1437 (1962).
164. A. W. Overhauser, *Phys. Rev.* **167**, 691 (1968).
165. A. W. Overhauser, *Phys. Rev. B* **3**, 3173 (1971).
166. M. J. Rice and S. Strässler, *Solid State Commun.* **14**, 125 (1973).
167. S. K. Chan and V. Heine, *J. Phys. F* **3**, 795 (1973).
168. J. D. Axe, in: *Electron–Phonon Interactions and Phase Transitions* (T. Riste, ed.), Plenum Press, New York (1977).
169. A. Luther, in: *Electron–Phonon Interactions and Phase Transitions* (T. Riste, ed.), Plenum Press, New York (1977).
170. F. di Salvo, in: *Electron–Phonon Interactions and Phase Transitions* (T. Riste, ed.), Plenum Press, New York (1977).
171. W. L. McMillan, in: *Electron–Phonon Interactions and Phase Transitions* (T. Riste, ed.), Plenum Press, New York (1977).
172. R. J. Birgeneau, R. Dingle, M. T. Hutchings, G. Shirane, and S. L. Holt, *Phys. Rev. Lett.* **26**, 718 (1971).
173. R. J. Birgeneau, S. J. Guggenheim, and G. Shirane, *Phys. Rev. B* **1**, 2211 (1970).
174. R. A. Cowley, G. Shirane, R. J. Birgeneau, and H. J. Guggenhein, *Phys. Rev. B* **15**, 4292 (1977).
175. J. A. Wilson and A. D. Yoffe, *Adv. Phys.* **18**, 193 (1969).
176. M. Steiner, J. Villain, and C. G. Windsor, *Adv. Phys.* **25**, 87 (1976).
177. H. E. Stanley and T. A. Kaplan, *Phys. Rev. Lett.* **17**, 913 (1966).
178. T. M. Rice, *Phys. Rev. A* **140**, 1889 (1965).
179. V. L. Ginzburg and L. D. Landau, *Zh. Eksp. Teor. Fiz.* **26**, 1064 (1950).
180. V. L. Ginzburg and L. P. Pitaevskii, *Zh. Eksp. Theor. Fiz.* **34**, 1240 (1958).
181. K. G. Wilson and J. Kogut, *Phys. Rep.* **12C**, 75 (1974).
182. S.-K. Ma, *Modern Theory of Critical Phenomena*, Benjamin, New York (1976).
183. A. R. Harris, T. C. Lubensky, W. K. Holcomb, and C. Dasgupta, *Phys. Rev. Lett.* **35**, 327 (1975).
184. R. G. Priest and T. C. Lubensky, *Phys. Rev. B* **13**, 4159 (1976).
185. B. Widom, *J. Chem. Phys.* **43**, 3892, 3898 (1965).
186. R. B. Griffiths, *J. Chem. Phys.* **43**, 1958 (1965).
187. L. P. Kadanoff, *Physics* **2**, 263 (1966).
188. K. G. Wilson, *Phys. Rev. Lett.* **28**, 584 (1972).
189. K. G. Wilson and M. E. Fisher, *Phys. Rev. Lett.* **28**, 240 (1972).
190. K. G. Wilson, *Physica.* **73**, 119 (1974).
191. Th. Niemeyer and J. M. J. Van Leeuwen, *Physica* **71**, 17 (1974).
192. Th. Niemeyer and J. M. J. Van Leeuwen, in: *Phase Transitions and Critical Phenomena* (C. Domb and M. S. Green, eds.), Vol. 6, Academic Press, New York (1976).
193. S.-K. Ma, *Rev. Mod. Phys.* **45**, 589 (1973).

194. M. N. Barber, *J. Phys. C* **8**, L203 (1975).
195. L. P. Kadanoff and A. Houghton, *Phys. Rev. B* **11**, 377 (1975).
196. C. Domb, *Adv. Phys.* **9**, 149 (1960).
197. M. E. Fisher, *Am. J. Phys.* **32**, 343 (1964).
198. R. B. Stinchcombe, *J. Phys. C* **12**, 2625 (1979).
199. R. B. Stinchcombe, *J. Phys. C* **13**, L133 (1980).
200. A. A. Migdal, *Sov. Phys. JETP (Engl. Transl.)* **42**, 743 (1976).
201. L. Sneddon and M. N. Barber, *J. Phys. C* **10**, 2653 (1977).
202. A. P. Young and R. B. Stinchcombe, *J. Phys. C* **8**, L535 (1975).
203. P. J. Reynolds, W. Klein, and H. E. Stanley, *J. Phys. C* **10**, L167 (1977).
204. L. G. Marland and R. B. Stinchcombe, *J. Phys. C* **10**, 2223 (1977).
205. A. G. Dunn, J. W. Essam, and D. S. Ritchie, *J. Phys. C* **8**, 4219.
206. A. P. Young and R. B. Stinchcombe, *J. Phys. C* **9**, 4419 (1976).
207. C. Jayaprakash, E. J. Riedel, and M. Wortis, *Phys. Rev. B* **18**, 2244 (1978).
208. J. M. Yeomans and R. B. Stinchcombe, *J. Phys. C* **11**, L525 (1978).
209. R. B. Stinchcombe, *J. Phys. C* **12**, 4533 (1979).
210. J. M. Yeomans and R. B. Stinchcombe, *J. Phys. C* **12**, L169 (1979).
211. S. Kirkpatrick, *Phys. Rev. B* **15**, 1533 (1978).
212. B. Southern and A. P. Young, *J. Phys. C* **10**, 2179 (1977).
213. D. J. Thouless, in: *Ill-Condensed Matter* (R. Balian, R. Maynard, and G. Toulouse, eds.), North-Holland Publ. Co. Amsterdam (1979).
214. S. Kirkpatrick, in: *Ill-Condensed Matter* (R. Balian, R. Maynard, and G. Toulouse, eds.), North-Holland Publ. Co., Amsterdam (1979).
215. See, for example, G. D. Mahan, *Many-Particle Physics*, Plenum Press, New York (1981).
216. Another excellent source is A. L. Fetter and A. D. Walecka, *Quantum Theory of Many-Particle Systems*, McGraw-Hill, New York (1971).
217. P. Nozières and D. Pines, *Nuovo Cimento* **9**, 470 (1958).
218. D. Pines and P. Nozières, *Theory of Quantum Liquids*, Benjamin, New York (1966).
219. P. Hohenberg and W. Kohn, *Phys. Rev. B* **136**, 864 (1964).
220. W. Kohn and L. J. Sham, *Phys. Rev. A* **140**, 1133 (1965).
221. W. Kohn and P. Vashishta, in: *Theory of the Inhomogeneous Electron Gas* (S. Lundqvist and N. H. March, eds.), Plenum Press, London (1983).
222. J. Lindhard, *K. Dan. Vidensk. Selkap, Mat. Fys. Medd.* **28**, 8 (1954).
223. L. Hedin and S. Lundqvist, in: *Solid State Physics*, Vol. 23 (F. Seitz, D. Turnbull, and H. Ehrenreich, eds.), p. 1, Academic Press, New York (1969).
224. J. Hubbard, *Proc. R. Soc. London, Ser. A* **243**, 336 (1957).
225. K. S. Singwi, M. P. Tosi, R. H. Land, and A. Sjolander, *Phys. Rev.* **176**, 589 (1968).
226. M. P. Tosi, *Nuovo Cimento*.
227. K. S. Singwi and M. P. Tosi, *Solid State Physics 36* (H. Ehrenreich, F. Seitz, and D. Turnbull, eds.), Academic Press, New York (1981).
228. J. Hubbard, *Phys. Lett.* **25A**, 709 (1967).
229. L. Hedin and B. I. Lundqvist, *J. Phys. C* **4**, 2064 (1971).
230. See, for example, W. Jones and N. H. March, *Theoretical Solid State Physics*, Vol. 1, Wiley–Interscience, London (1973).
231. L. Hedin, *Phys. Rev. A* **139**, 769 (1965).
232. P. A. Wolff, *Phys. Rev.* **120**, 814 (1964).
233. R. Lobo, K. S. Singwi, and M. P. Tosi, *Phys. Rev.* **136**, 470 (1969).
234. G. Pizzimenti, M. P. Tosi, and A. Villari, *Lett. Nuovo Cimento* **2**, 81 (1971).
235. U. von Barth and L. Hedin, *J. Phys. C* **5**, 1629 (1972).
236. F. Bloch, *Z. Phys.* **57**, 545 (1929).
237. D. Ceperley and B. J. Alder, *Phys. Rev. Lett.* **45**, 568 (1980).
238. E. P. Wigner, *Phys. Rev.* **46**, 1002 (1934); *Trans. Faraday Soc.* **34**, 678 (1938).

239. W. J. Carr, *Phys. Rev.* **122**, 1437 (1961).
240. R. A. Coldwell-Horsfall and A. A. Maradudin, *J. Math. Phys.* **4**, 582 (1963).
241. A. W. Overhauser, *Phys. Rev.* **128**, 1437 (1962); *Phys. Rev.* **167**, 167 (1968).
242. F. Stern and W. E. Howard, *Phys. Rev.* **163**, 816 (1967).
243. F. Stern, *Phys. Rev. B* **5**, 4891 (1972).
244. M. Jonson, *J. Phys. C* **9**, 3055 (1976).
245. F. Stern, *Phys. Rev. Lett.* **30**, 278 (1973).
246. M. Gell-Mann and K. A. Brueckner, *Phys. Rev.* **106**, 364 (1957).
247. A. K. Rajagopal and J. C. Kimball, *Phys. Rev. B* **18**, 2339 (1978).
248. G. Srinivasan and M. Jonson, *J. Phys. C* **8**, L37 (1975).
249. J. Hubbard, *Proc. R. Soc. London, Ser. A* **276**, 238 (1963).
250. J. Hubbard, *Proc. R. Soc. London, Ser. A* **281**, 401 (1964).
251. J. Hubbard, *Proc. R. Soc. London, Ser. A* **285**, 542 (1964).
252. N. F. Mott, *Proc. Cambridge Philos. Soc.* **32**, 281 (1949).
253. N. F. Mott, *Proc. Phys. Soc. A* **62**, 416 (1956).
254. S. Lundqvist and A. Sjölander, *Ark. Fys.* **26**, 17 (1963).
255. A. Many, Y. Goldstein, and N. B. Grover, *Semiconductor Surfaces*, North-Holland Publ. Co., Amsterdam (1965).
256. D. R. Frankl, *Electrical Properties of Semiconductor Surfaces*, Pergamon Press, Oxford (1967).
257. E. H. Rhoderick, *Metal-Insulator Contacts*, Oxford University Press (1978).
258. M. A. Lampert, *Current Injection in Solids*, Academic Press, New York (1970).
259. R. A. Smith, *Semiconductors*, 2nd edn., Cambridge University Press (1978).
260. For examples in the context of surfaces, see R. W. Keyes, *Comments Solid State Phys.* **7**, 53 (1976) and p. 236 of Ref. 263.
261. B. Vinter, *Phys. Rev. B* **17**, 2429 (1978).
262. S. M. Sze, *Physics of Semiconductor Devices*, 2nd edn., Wiley–Interscience, New York (1981).
263. N. D. Lang, in: *Solid State Physics*, Vol. 28 (H. Ehrenreich, F. Seitz and D. Turnbull, eds.), p. 225, Academic Press, New York (1973).
264. L. Ley and M. Cardona (eds.), *Photoemission in Solids*, Vol. II, Springer-Verlag, Berlin (1978).
265. J. A. Appelbaum and D. R. Hamann, *Rev. Mod. Phys.* **48**, 479 (1976).
266. G. Chiarotti, in: *Surface Science*, Vol. I, p. 423, International Atomic Energy Agency, Vienna (1975).
267. W. H. Brattain and J. Bardeen, *Bell Syst. Tech. J.* **32**, 1 (1953).
268. A. Many, *Crit. Rev. Solid State Sci.* **4**, 515 (1974).
269. W. Göpel and U. Lampe, *Phys. Rev. B* **22**, 6447 (1980).
270. D. E. Eastman and W. D. Grobman, *Phys. Rev. Lett.* **28**, 1378 (1972).
271. L. F. Wagner and W. E. Spicer, *Phys. Rev. Lett.* **28**, 1381 (1972).
272. N. F. Mott and R. W. Gurney, *Electronic Processes in Ionic Crystals*, 2nd edn., p. 176, Oxford University Press (1950).
273. S. Kurtin, T. C. McGill, and C. A. Mead, *Phys. Rev. Lett.* **22**, 1433 (1969).
274. J. M. Andrews and J. C. Phillips, *Phys. Rev. Lett.* **35**, 56 (1975).
275. J. Bardeen, *Phys. Rev.* **71**, 717 (1947).
276. V. Heine, *Phys. Rev. A* **138**, 1689 (1965).
277. S. G. Louie, J. R. Chelikowsky, and M. L. Cohen, *Phys. Rev. B* **15**, 2154 (1977).
278. L. J. Brillson, *Surf. Sci. Rep.* **2**, 123 (1982).
279. G. Ottaviani, K. N. Tu, and J. W. Mayer, *Phys. Rev. Lett.* **44**, 284 (1980).
280. W. E. Spicer, I. Lindau, P. Skeath, C. Y. Su, and P. Chye, *Phys. Rev. Lett.* **44**, 420 (1980).
281. J. L. Freeouf, *Solid State Commun.* **33**, 1059 (1980).
282. R. Ludeke and G. Landgren, *J. Vac. Sci. Technol.* **19**, 667 (1981).

283. A. S. Grove, *Physics and Technology of Semiconductor Devices*, John Wiley and Sons, New York (1967).
284. M. W. Cole, *Rev. Mod. Phys.* **46**, 451 (1974).
285. V. B. Shikin and Yu. P. Monarkha, *Fiz. Nizk. Temp.* **1**, 957 (1975) [*Sov. J. Low Temp. Phys.* **1**, 459 (1975)].
286. R. S. Crandall, *Surf. Sci.* **58**, 266 (1976).
287. C. C. Grimes, *Surf. Sci.* **73**, 379 (1978).
288. V. B. Shikin, in: *Physics of Low-Dimensional Systems* (Proceedings of Kyoto Summer Institute, 1979) (Y. Nagaoka and S. Hikami, eds.), p. 177, Publication Office, Progress of Theoretical Physics, Kyoto (1979).
289. F. I. B. Williams, *J. Phys. (Paris) Colloque* **41**, C3–249 (1980).
290. C. C. Grimes and G. Adams, *Phys. Rev. Lett.* **42**, 795 (1979); *Surf. Sci.* **98**, 1 (1980).
291. A. Many, in: *Surface Science*, Vol. I, p. 447, International Atomic Energy Agency, Vienna (1975).
292. M. J. McNutt and C. T. Sah, *Solid-State Electron.* **17**, 377 (1974); *J. Appl. Phys.* **45**, 3916 (1974); *Appl. Phys. Lett.* **26**, 378 (1975); *Solid-State Electron.* **19**, 255 (1976); *IEEE Trans. Electron Devices* **ED-25**, 847 (1978).
293. T. W. Collins, J. N. Churchill, F. E. Holmstrom, and A. Moschwitzer, in: *Advances in Electronics and Electron Physics* (L. Marton, ed.), Vol. 47, p. 267, Academic Press, New York (1978).
294. K. Fuchs, *Proc. Cambridge Philos. Soc.* **34**, 100 (1938).
295. R. F. Greene, in: *Solid State Surface Science* (M. Green, ed.), Vol. I, p. 87, Marcel Dekker, Inc., New York (1969).
296. A. B. Fowler and A. Hartstein, *Surf. Sci.* **98**, 169 (1980).
297. P. A. Muls, G. J. Declerck, and R. J. Van Overstraeten, in: *Advances in Electronics and Electron Physics* (L. Marton, ed.), Vol. 47, p. 197, Academic Press, New York (1978).
298. N. B. Goodman and H. Fritzsche, *Philos. Mag. B* **42**, 149 (1980).
299. Y. Takada, *J. Phys. Soc. Jpn.* **45**, 786 (1978).
300. W. Hanke and M. J. Kelly, *Phys. Rev. Lett.* **45**, 1203 (1980).
301. J. R. Maltby, C. E. Reed, and C. G. Scott, *Surf. Sci.* **93**, 287 (1980).
302. B. Goldstein and D. J. Szostak, *Surf. Sci.* **99**, 235 (1980).
303. R. Williams, *Phys. Rev. A* **140**, 569 (1965); *J. Vac. Sci. Technol.* **14**, 1106 (1977).
304. R. C. Hughes, *Phys. Rev. Lett.* **30**, 1333 (1973).
305. R. J. Powell and C. N. Berglund, *J. Appl. Phys.* **42**, 4390 (1971).
306. D. J. DiMaria, *J. Appl. Phys.* **47**, 4073 (1976).
307. D. R. Young, in: *Insulating Films on Semiconductors, 1979* (G. G. Roberts and M. J. Morant, eds.), p. 28, Institute of Physics, Bristol (1980).
308. S. I. Raider and R. Flitsch, in: *The Physics of SiO$_2$ and its Interfaces* (S. T. Pantelides, ed.), p. 384, Pergamon Press, New York (1978).
309. A. Ishizaka, S. Iwata, and Y. Kamigaki, *Surf. Sci.* **84**, 355 (1979).
310. T. E. Jackman, Jack R. MacDonald, L. C. Feldman, P. J. Silverman, and I. Stensgaard, *Surf. Sci.* **100**, 35 (1980).
311. O. L. Krivanek and J. H. Mazur, *Appl. Phys. Lett.* **37**, 392 (1980).
312. T. Sugano, *Surf. Sci.* **98**, 145 (1980).
313. T. Sugano, J. J. Chen, and T. Hamano, *Surf. Sci.* **98**, 154 (1980).
314. G. Dorda, in: *Festkörperprobleme/Advances in Solid State Physics* (H. J. Queisser, ed.), Vol. XIII, p. 215, Pergamon-Vieweg, Braunschweig (1973).
315. T. Sugano, K. Hoh, H. Sakaki, T. Iizuka, K. Hirai, K. Kuroiwa, and K. Kakemoto, *J. Fac. Eng., Univ. Tokyo, Ser. B* **XXXII**, 155 (1973).
316. F. Stern, *Crit. Rev. Solid State Sci.* **4**, 499 (1974).
317. G. Landwehr, in: *Festkörperprobleme/Advances in Solid State Physics* (H. J. Queisser, ed.), Vol. XV, p. 49, Vieweg, Braunschweig (1975).

318. J. F. Koch, in: *Festkörperprobleme/Advances in Solid State Physics* (H. J. Queisser, ed.), Vol. XV, p. 79, Vieweg, Braunschweig (1975).
319. M. Pepper, *Contemp. Phys.* **18**, 423 (1977).
320. L. J. Sham, in: *Physics of Low-Dimensional Systems* (Proceedings of Kyoto Summer Institute, 1979) (Y. Nagaoka and S. Hikami, eds.), p. 195, Publication Office, Progress of Theoretical Physics, Kyoto (1979).
321. R. K. Kalia, G. Kawamoto, J. J. Quinn, and S. C. Ying, *Solid State Commun.* **34**, 423 (1980).
322. F. Stern, *J. Comput. Phys.* **6**, 56 (1970).
323. T. Ando, *Phys. Rev. B* **13**, 3468 (1976).
324. B. Vinter, *Phys. Rev. B* **13**, 4447 (1976); *Phys. Rev. B* **15**, 3947 (1977).
325. A. B. Fowler, in: *Handbook on Semiconductors* (T. S. Moss, ed.), Vol. I, *Band Theory and Transport Properties* (W. Paul, ed.), p. 599, North-Holland, Amsterdam (1982).
326. F. F. Fang and A. B. Fowler, *Phys. Rev.* **169**, 619 (1968).
327. H. Ezawa, S. Kawaji, and K. Nakamura, *Jpn. J. Appl. Phys.* **13**, 126 (1974); **14**, 921 (E) (1975).
328. D. K. Ferry, *Surf. Sci.* **57**, 218 (1976).
329. K. M. Cham and R. G. Wheeler, *Phys. Rev. Lett.* **44**, 1472 (1980).
330. F. Stern, *Phys. Rev. Lett.* **44**, 1469 (1980).
331. Y. Kawaguchi, T. Suzuki, and S. Kawaji, *Solid State Commun.* **36**, 257 (1980).
332. A. Hartstein, A. B. Fowler, and M. Albert, *Surf. Sci.* **98**, 181 (1980).
333. S. Mori and T. Ando, *Phys. Rev. B* **19**, 6433 (1979).
334. J. R. Brews, *J. Appl. Phys.* **46**, 2193 (1975).
335. C. C. Shiue and C. T. Sah, *Surf. Sci.* **58**, 153 (1976).
336. A. Yagi and M. Nakai, *Surf. Sci.* **98**, 174 (1980).
337. R. E. Prange and T. W. Nee, *Phys. Rev.* **168**, 779 (1968).
338. Y. Matsumoto and Y. Uemura, *Jpn. J. Appl. Phys.*, Suppl. **2**, Pt. 2, 367 (1974).
339. T. Ando, *J. Phys. Soc. Jpn.* **43**, 1616 (1977).
340. T. Ando, *J. Phys. Soc. Jpn.* **38**, 989 (1975).
341. S. Das Sarma, *Solid State Commun.* **36**, 357 (1980).
342. A. B. Fowler, F. F. Fang, W. E. Howard, and P. J. Stiles, *Phys. Rev. Lett.* **16**, 901 (1966).
343. F. F. Fang and P. J. Stiles, *Phys. Rev.* **174**, 823 (1968).
344. J. F. Janak, *Phys. Rev.* **178**, 1416 (1969).
345. T. Ando and Y. Uemura, *J. Phys. Soc. Jpn.* **37**, 1044 (1974).
346. J. L. Smith and P. J. Stiles, *Phys. Rev. Lett.* **29**, 102 (1972).
347. T. K. Lee, C. S. Ting, and J. J. Quinn, *Solid State Commun.* **16**, 1309 (1975).
348. B. Vinter, *Phys. Rev. Lett.* **35**, 1044 (1975).
349. F. J. Ohkawa, *Surf. Sci.* **58**, 326 (1976).
350. T. K. Lee, C. S. Ting, and J. J. Quinn, *Phys. Rev. Lett.* **35**, 1048 (1975); *Surf. Sci.* **58**, 246 (1976).
351. G. Dorda, *J. Appl. Phys.* **42**, 2053 (1971).
352. D. C. Tsui and G. Kaminsky, *Surf. Sci.* **58**, 187 (1976).
353. I. Eisele, *Surf. Sci.* **73**, 315 (1978).
354. P. Stallhofer, J. P. Kotthaus, and G. Abstreiter, *Solid State Commun.* **32**, 655 (1979).
355. Y. Takada and T. Ando, *J. Phys. Soc. Jpn.* **44**, 905 (1978).
356. B. Vinter, *Solid State Commun.* **32**, 651 (1979).
357. K. v. Klitzing, G. Dorda, and M. Pepper, *Phys. Rev. Lett.* **45**, 494 (1980).
358. P. Kneschaurek and J. F. Koch, *Phys. Rev. B* **16**, 1590 (1977).
359. B. D. McCombe and T. Cole, *Surf. Sci.* **98**, 469 (1980).
360. T. Ando, *Surf. Sci.* **73**, 1 (1978).
361. D. A. Dahl and L. J. Sham, *Phys. Rev. B* **16**, 651 (1977).
362. A. Eguiluz and A. A. Maradudin, *Surf. Sci.* **73**, 437 (1978); *Ann. Phys. (N. Y.)* **113**, 29 (1978).

363. J. P. Kotthaus, *Surf. Sci.* **73**, 472 (1978).
364. R. J. Wagner, T. A. Kennedy, B. D. McCombe, and D. C. Tsui, *Phys. Rev. B* **22**, 945 (1980).
365. R. H. Ritchie, *Phys. Rev.* **106**, 874 (1957).
366. C. C. Grimes and G. Adams, *Surf. Sci.* **58**, 292 (1976); *Phys. Rev. Lett.* **36**, 145 (1976).
367. S. J. Allen, Jr., D. C. Tsui, and R. A. Logan, *Phys. Rev. Lett.* **38**, 980 (1977).
368. T. N. Theis, *Surf. Sci.* **98**, 515 (1980).
369. A. Pinczuk, J. M. Worlock, H. L. Störmer, R. Dingle, W. Wiegmann, and A. C. Gossard, *Solid State Commun.* **36**, 43 (1980).
370. G. Abstreiter, Ch. Zeller, and K. Ploog, in: *Gallium Arsenide and Related Compounds, 1980* (H. W. Thim, ed.), p. 741. Institute of Physics, Bristol (1981).
371. E. Burstein, A. Pinczuk, and D. L. Mills, *Surf. Sci.* **98**, 451 (1980).
372. F. Stern, *Phys. Rev. B* **9**, 2762 (1974).
373. N. F. Mott, *Commun. Phys.* **1**, 203 (1976).
374. E. Arnold, *Surf. Sci.* **58**, 60 (1976); *Phys. Rev. B* **17**, 4111 (1978).
375. C. J. Adkins, *J. Phys. C* **11**, 851 (1978); **12**, 3389, 3395 (1979).
376. E. Abrahams, P. W. Anderson, D. C. Licciardello, and T. V. Ramakrishnan, *Phys. Rev. Lett.* **42**, 673 (1979).
377. B. L. Altshuler, A. G. Aronov, and P. A. Lee, *Phys. Rev. Lett.* **44**, 1288 (1980).
378. B. L. Altshuler, D. Khmel'nitzkii, A. I. Larkin, and P. A. Lee, *Phys. Rev. B* **22**, 5142 (1980).
379. P. W. Anderson, E. Abrahams, and T. V. Ramakrishnan, *Phys. Rev. Lett.* **43**, 718 (1979).
380. P. W. Anderson, D. J. Thouless, E. Abrahams, and D. S. Fisher, *Phys. Rev. B* **22**, 3519 (1980).
381. H. Fukuyama, *J. Phys. Soc. Jpn.* **48**, 2169 (1980); **49**, 644, 649 (1980).
382. S. Hikami, A. I. Larkin, and Y. Nagaoka, *Prog. Theor. Phys.* **63**, 707 (1980).
383. A. I. Larkin, *Pis'ma Zh. Eksp. Teor. Fiz.* **31**, 239 (1980) [*JETP Lett.* **31**, 219 (1980)].
384. P. A. Lee, *Phys. Rev. Lett.* **42**, 1492 (1979); *J. Non-Cryst. Solids* **35**, 21 (1980).
385. C. M. Soukoulis and E. N. Economou, *Phys. Rev. Lett.* **45**, 1590 (1980).
386. D. J. Thouless, *Phys. Rev. Lett.* **39**, 1167 (1977); *Solid State Commun.* **34**, 683 (1980).
387. D. Vollhardt and P. Wölfle, *Phys. Rev. B* **22**, 4666 (1980).
388. D. J. Bishop, D. C. Tsui, and R. C. Dynes, *Phys. Rev. Lett.* **44**, 1153 (1980).
389. P. Chaudhari and H. U. Habermeier, *Phys. Rev. Lett.* **44**, 40 (1980).
390. G. J. Dolan and D. D. Osheroff, *Phys. Rev. Lett.* **43**, 721; 1690 (E) (1979).
391. R. C. Dynes, J. P. Garno, and J. M. Rowell, *Phys. Rev. Lett.* **40**, 479 (1978).
392. N. Giordano, *Phys. Rev. B* **22**, 5635 (1980).
393. Y. Kawaguchi and S. Kawaji, *J. Phys. Soc. Jpn.* **48**, 699 (1980).
394. M. J. Uren, R. A. Davies, and M. Pepper, *J. Phys. C* **13**, L985 (1980).
395. M. Pepper, S. Pollitt, C. J. Adkins, and R. A. Stradling, *Crit. Rev. Solid State Sci.* **5**, 375 (1975).
396. D. C. Tsui and S. J. Allen, Jr., *Phys. Rev. Lett.* **34**, 1293 (1975).
397. A. Hartstein and A. B. Fowler, *J. Phys. C* **8**, L249 (1975).
398. N. Mott, M. Pepper, S. Pollitt, R. H. Wallis, and C. J. Adkins, *Proc. R. Soc. London, Ser. A* **345**, 169 (1975).
399. D. C. Licciardello and D. J. Thouless, *J. Phys. C* **8**, 4157 (1975); **11**, 925 (1978).
400. D. J. Thouless and M. E. Elzain, *J. Phys. C* **11**, 3425 (1978).
401. J. Stein and U. Krey, *Z. Phys. B* **34**, 287 (1979); **37**, 13 (1980).
402. M. Pepper, *Proc. R. Soc. London, Ser. A* **353**, 225 (1977).
403. A. Hartstein and A. B. Fowler, *Phys. Rev. Lett.* **34**, 1435 (1975).
404. A. Hartstein, A. B. Fowler, and M. Albert, in: *Physics of Semiconductors, 1978* (B. L. H. Wilson, ed.), p. 1001, Institute of Physics, Bristol (1979).
405. K. J. Hayden and P. N. Butcher, *Philos. Mag. B* **38**, 603 (1979).

406. K. J. Hayden, *Philos. Mag. B* **41**, 619 (1980).
407. B. G. Martin and R. F. Wallis, *Phys. Rev. B* **18**, 5644 (1978).
408. G. M. Kramer and R. F. Wallis, in: *Physics of Semiconductors, 1978* (B. L. H. Wilson, ed.), p. 1243, Institute of Physics, Bristol (1979).
409. N. O. Lipari, *J. Vac. Sci. Technol.* **15**, 1412 (1978).
410. B. Vinter, *Solid State Commun.* **28**, 861 (1978); *Phys. Rev. B* **26**, 6808 (1982).
411. O. Hipolito and V. B. Campos, *Phys. Rev. B* **19**, 3083 (1979).
412. M. Baus and J. P. Hansen, *Phys. Rep.* **59**, 1 (1980).
413. L. I. Schiff, *Quantum Mechanics*, 2nd edn., p. 120, McGraw-Hill, New York (1955).
414. S. Kawaji and Y. Kawaguchi, *J. Phys. Soc. Jpn. Suppl.* **21**, 336 (1966).
415. L. D. Landau and E. M. Lifshitz, *Quantum Mechanics*, 3rd edn., p. 163, Pergamon Press, Oxford (1977).
416. L. V. Keldysh, *Pis'ma Zh. Eksp. Teor. Fiz.* **29**, 716 (1979) [*JETP Lett.* **29**, 658 (1979)].
417. J. Avery, *Creation and Annihilation Operators*, McGraw-Hill, New York (1976).
418. R. P. Feynman, *Statistical Mechanics, Frontiers of Physics*, Benjamin, New York, (197?).
419. L. J. Sham and J. M. Ziman, in: *Solid State Physics* (F. Seitz and D. Turnbull, eds.), Vol. 15, Academic Press, New York (1963).
420. R. D. Parks, *Superconductivity*, Academic Press, New York (1965).
421. G. Eilenberger and A. E. Jacobs, *J. Low Temp. Phys.* **20**, 516 (1975).
422. D. Götzel, D. Rainer, and H. R. Schoter, *Z. Phys. B* **35**, 317 (1979).
423. W. Baltensperger and S. Strassler, *Phys. Kondens. Mater.* **1**, 20 (1963).
424. W. A. Little, *Phys. Rev. A* **134**, 1416 (1964).
425. D. Davis, H. Gutfreund, and W. A. Little, *Phys. Rev. B* **13**, 4766 (1976).
426. D. Allender, J. Bray, and J. Bardeen, *Phys. Rev.* **137**, 1026 (1973).
427. M. L. Cohen and P. W. Anderson in: *Superconductivity in d- and f-Band Metals* (D. H. Douglass, ed.), AIP Conf. Proc. No 4, AIP, New York (1973).
428. J. C. Phillips, *Phys. Rev. Lett.* **29**, 1551 (1972).
429. H. Gutfreund and W. A. Little, in: *Highly Conducting One-Dimensional Solids* (J. Y. Devreese, R. P. Evrard, and V. E. van Doren, eds.), pp. 305–372, Plenum Press, New York (1979).
430. T. M. Rice, *Phys. Rev. A* **140**, 889 (1965).
431. P. C. Hohenberg, *Phys. Rev.* **158**, 383 (1967).
432. H. Fröhlich, *Proc. R. Soc. London, Ser. A* **223**, 296 (1959).
433. L. B. Coleman, J. A. Cohen, A. P. Garito, and A. J. Heeger, *Phys. Rev. B* **7**, 2122 (1973).
434. D. Allender, J. W. Bray, and J. Bardeen, *Phys. Rev. B* **9**, 119 (1974).
435. M. J. Rice, *Solid State Commun.* **16**, 1285 (1975).
436. A. J. Heeger, in: *Highly Conducting One-Dimensional Solids* (J. Y. Devreese, R. P. Evrard, and V. E. van Doren, eds.), pp. 69–146, Plenum Press, New York (1979).
437. J. Bardeen, in: *Highly Conducting One-Dimensional Solids* (J. Y. Devreese, R. P. Evrard, and V. E. van Doren, eds.), pp. 373–404, Plenum Press, New York (1979).
438. P. Brüesch, S. Strässler, and H. R. Zeller, *Phys. Rev. B* **12**, 219 (1975).
439. P. Monceau, N. P. Ong, A. M. Portis, A. Meerschat, and J. Rouxel, *Phys. Rev. Lett.* **37**, 602 (1976).
440. R. M. Fleming, in: *Physics in One-Dimension* (J. Bernasconi and T. Schneider eds.), Springer Series in Solid-State Sciences, Vol. 23, p. 253, Springer-Verlag, Berlin (1981).
441. J. Bernasconi and T. Schneider (eds.), *Physics in One-Dimension*, Springer Series in Solid-State Sciences, Vol. 23, p. 339, Springer-Verlag, Berlin (1981).
442. D. Jérome, A. Mazaud, M. Ribault, and K. Bechgaard, *J. Phys. (Paris) Lett.* **41**, L95 (1980).
443. K. Bechgaard, C. S. Jacobsen, K. Mortensen, H. J. Pedersen, and N. Thorup, *Solid State Commun.* **33**, 1119 (1980).
444. C. S. Jacobsen, D. B. Tanner, and K. Bechgaard (to appear).

IV

Special Topics

Biopolymer Electronic Phenomena

J. Ladik, S. Suhai, and M. Seel

The quantum-mechanical study of the electronic structure of biopolymers (such as nucleic acids and proteins) is a challenging physical problem due to the complexity of these systems (many orbitals in the unit cell, aperiodicity, environmental effects, etc.) and requires the combination of different techniques with rather large-scale computations. One can expect, however, that the results of these calculations will enable one to compute their different physical and chemical properties (charge distribution and reactivity indices of the constituent molecules, density of states, spectra, different transport properties, etc.). The final aim of these investigations is to understand on the basis of these properties the different biological functions (e.g., the roles of point mutations in DNA, the mechanism of duplication of DNA, cell differentiation regulated by DNA–protein interactions, and the mechanism of carcinogenesis caused by chemical carcinogens).

As a first step the *ab initio* self-consistent-field linear-combination-of-atomic-orbitals crystal-orbital (SCF LCAO CO) theory is formulated both for simple translational and combined (for instance, helical) symmetry operations and it is shown how the different semiempirical CO theories can be deduced from it. Next, the incorrect virtual levels (in Hartree–Fock theory) are corrected with the help of the excitation Hamiltonian ($\hat{O}\hat{A}\hat{O}$

J. Ladik, S. Suhai, and M. Seel · Theoretical Chemistry Department and Laboratory of the National Foundation for Cancer Research, University of Erlangen-Nürnberg, D-8520 Erlangen, FRG.

method) and long-range correlation is introduced with the help of the simple electron–polaron model. Applications to different periodic DNA (like polycytosine, polyguanine, polycytidine, etc.) and protein models (like polyglycine, polyalanine, etc.) close this part.

In the next step, we discuss the treatment of excited states of polymers with the aid of the intermediate exciton theory and its application to periodic DNA models. This is followed by a general discussion of the possibilities of the treatment of short-range correlation both in insulator and conductive polymers.

The subsequent part contains a discussion of different methods for the treatment of aperiodicity of polymers (virtual crystal approximation, SCF resolvent method, negative-factor counting method in its simple and SCF form). An application to an aperiodic protein chain is presented.

In the final part, a simple theory of transport properties of polymers with narrow bands based on a k-dependent relaxation-time $\tau(k)$ approximation of Boltzmann's transport equation and its application to periodic DNA models is discussed. A short review of possibilities to interpret the different biological functions of DNA, proteins, etc., on the basis of these quantum-mechanical investigations closes the chapter.

17.1. *Ab Initio* SCF-LCAO Crystal-Orbital Formalism

17.1.1. Simple Translational Symmetry

If we have m orbitals in the unit cell of a three-dimensional polymer or molecular crystal and the number of unit cells in the direction of each crystal axis is $2N + 1$, we can write the delocalized crystal orbitals (COs) in the LCAO approximation in the form

$$\Phi_h^p = \sum_q \sum_{g=1}^m C(\mathbf{p})_{h;q,g} X_g^q \tag{17.1}$$

where $\mathbf{p} = (p_1, p_2, p_3)$, $\mathbf{q} = (q_1, q_2, q_3)$, the integers p_j and q_j $(j = 1, 2, 3)$ run from $-N, \ldots, 0, \ldots, N$, and Σ_q denotes summation over all cells, $\Sigma_{q_1=-N}^N \Sigma_{q_2=-N}^N \Sigma_{q_3=-N}^N$. Further $X_g^q = X_g(\mathbf{r} - \mathbf{R}_q - \mathbf{r}_{gA})$ is the gth AO (which belongs to the atom with position vector $\mathbf{r}_{gA}$ of the cell characterized by the vector $\mathbf{R}_q = q_1\mathbf{a}_1 + q_2\mathbf{a}_2 + q_3\mathbf{a}_3$.

Writing down the expectation value

$$\frac{\langle \Phi_h^p | \hat{F} | \Phi_h^p \rangle}{\langle \Phi_h^p | \Phi_h^p \rangle} = \varepsilon(\mathbf{p})_h \tag{17.2}$$

of the Fock operator $\hat{F}$ and performing a Ritz-variation procedure, we obtain in the standard way for the whole polymer the matrix equation

$$FC(\mathbf{p})_h = \varepsilon(\mathbf{p})_h SC(\mathbf{p})_h \tag{17.3}$$

The hypermatrices F and S have the dimensions $M \times M$ with $M = m \times (2N+1)^3$, and their elements are defined as

$$F_{fg}^{\mathbf{pq}} = \langle X_f^{\mathbf{p}} | \hat{F} | X_g^{\mathbf{q}} \rangle \tag{17.4a}$$

$$S_{fg}^{\mathbf{pq}} = \langle X_f^{\mathbf{p}} | X_g^{\mathbf{q}} \rangle \tag{17.4b}$$

respectively. (The $m \times m$ blocks of F and S give the interactions between the orbitals belonging to the different unit cells.)

Taking into account the translational symmetry and introducing periodic boundary conditions, it is easy to show that F and S are cyclic hypermatrices, i.e., they are cyclic in their blocks (a detailed derivation is given elsewhere[1,2]). It is possible to show[3] that with the aid of the unitary matrix U, which has the $m \times m$ blocks

$$U^{\mathbf{p,q}} = (2N+1)^{-3/2} \exp\left(\frac{i2\pi\mathbf{pq}}{2N+1}\right) \mathbf{1} \tag{17.5}$$

we can block-diagonalize F and S. Defining

$$F' = U^+FU, \qquad S' = U^+SU, \qquad D(\mathbf{p})_h = U^+C(\mathbf{p})_h \tag{17.6}$$

the pth diagonal block of the blockdiagonal matrices F' and S' is given by the expression[1,2]

$$F'(\mathbf{p}) = \sum_{\mathbf{q}} \exp[i2\pi\mathbf{pq}/(2N+1)]F(\mathbf{q}) \tag{17.7a}$$

$$S'(\mathbf{p}) = \sum_{\mathbf{q}} \exp[i2\pi\mathbf{pq}/(2N+1)S(\mathbf{q}) \tag{17.7b}$$

respectively, where $F(\mathbf{q})$ and $S(\mathbf{q})$ denote the submatrices of the original cyclic hypermatrices.

Using the fact that F' and S' are blockdiagonal, one readily obtains

$$F'(\mathbf{p})\mathbf{d}(\mathbf{p})_h = \varepsilon(\mathbf{p})_h S'(\mathbf{p})\mathbf{d}(\mathbf{p})_h \tag{17.8}$$

corresponding to the different blocks $F'(\mathbf{p})$ and $S'(\mathbf{p})$.

If $N \rightarrow \infty$ we can introduce the continuous variables

$$k_j = \frac{2\pi p_j}{a_j (2N + 1)} \qquad (j = 1, 2, 3) \tag{17.9}$$

Since the p_js assume the values $-N, \ldots, N$, the k_j will have values between $-\pi/a_j$ and π/a_j. Defining the vector $\mathbf{k} = k_1\mathbf{b}_1 + k_2\mathbf{b}_2 + k_3\mathbf{b}_3$ where the $\mathbf{b}_j$s are the basis vectors of the reciprocal space (by definition $\mathbf{a}_i\mathbf{b}_j = 2\pi\delta_{ij}$* we can now write instead of (17.8)

$$F(\mathbf{k})\mathbf{d}(\mathbf{k})_h = \varepsilon(\mathbf{k})_h S(\mathbf{k})\mathbf{d}(\mathbf{k})_h \qquad (h = 1, 2, \ldots, m) \tag{17.10}$$

with

$$F(\mathbf{k}) = \sum_q \exp(i\mathbf{k}\mathbf{R}_q)F(\mathbf{q}) \tag{17.11a}$$

$$S(\mathbf{k}) = \sum_q \exp(i\mathbf{k}\mathbf{R}_q)S(\mathbf{q}) \tag{17.11b}$$

To solve equations (17.10) we can eliminate the overlap matrix $S(\mathbf{k})$ with the aid of Löwdin's symmetric orthogonalization procedure in analogy to the molecular case. The only difference is that both $F(\mathbf{k})$ and $S(\mathbf{k})$ are now not real, but Hermitian complex matrices (the details can be found elsewhere[2]). This approach yields the eigenvalue equation

$$\tilde{F}(\mathbf{k})\mathbf{b}(\mathbf{k})_h = \varepsilon(\mathbf{k})_h \mathbf{b}(\mathbf{k})_h \tag{17.12}$$

where

$$\tilde{F}(\mathbf{k}) = S(\mathbf{k})^{-1/2}F(\mathbf{k})S(\mathbf{k})^{-1/2}, \qquad \mathbf{b}(\mathbf{k})_h = S(\mathbf{k})^{1/2}\mathbf{d}(\mathbf{k})_h \tag{17.13}$$

Substituting into (17.12), respectively, (17.4a), the expression

$$\hat{F} = -\tfrac{1}{2}\Delta - \sum_q \sum_\alpha^{M_A} \frac{Z_\alpha}{|\mathbf{r} - \mathbf{R}_\alpha^q|} + \sum_p \sum_{h=1}^{n^*} [2\hat{J}(\mathbf{p}, h) - \hat{K}(\mathbf{p}, h)]$$

$$= \hat{H}^N + \sum_p \sum_{h=1}^{n^*} [2\hat{J}(\mathbf{p}, h) - \hat{K}(\mathbf{p}, h)] \tag{17.14}$$

of the Fock operator (where M_A is the number of atoms in the unit cell, n^* is the number of filled bands,

$$\hat{J}(\mathbf{p}, h; \mathbf{r}_1)\Phi(\mathbf{r}_1) = \left\langle \Phi_h^p(\mathbf{r}_2) \left| \frac{1}{r_{12}} \right| \Phi_h^p(\mathbf{r}_2) \right\rangle \Phi(\mathbf{r}_1) \tag{17.15a}$$

$$\hat{K}(\mathbf{p}, h; \mathbf{r}_1)\Phi(\mathbf{r}_1) = \left\langle \Phi_h^p(\mathbf{r}_2) \left| \frac{1}{r_{12}} \right| \Phi(\mathbf{r}_2) \right\rangle \Phi_h^p(\mathbf{r}_1) \tag{17.15b}$$

---

* By identifying the vectors $\mathbf{k}$ with the crystal momentum we demand that they have to belong to the first Brillouin zone of the crystal.

are the Coulomb and exchange operators, respectively), the LCAO form (17.1) of the crystal orbitals, and introducing the charge-bond-order matrix of the polymer as

$$P = 2 \sum_{\mathbf{p}} \sum_{h=1}^{n^{\bullet}} \mathbf{C}(\mathbf{p})_h \mathbf{C}(\mathbf{p})_h^+ \tag{17.16}$$

one can derive[1,2] for the $[F(\mathbf{q})]_{r,s}$ matrix elements the expression

$$[F(\mathbf{q})]_{r,s} = \langle X_r^{\circ} | \hat{H}^N | X_s^q \rangle + \sum_{\mathbf{q}_1} \sum_{\mathbf{q}_2} \sum_{u,v=1}^{m} p(\mathbf{q}_1 - \mathbf{q}_2)_{u,v}$$
$$\cdot (\langle X_r^{\circ} X_u^{q_1} | X_s^q X_v^{q_2} \rangle - \tfrac{1}{2} \langle X_r^{\circ} X_u^{q_1} | X_v^{q_2} X_s^q \rangle) \tag{17.17}$$

Here $p(\mathbf{q}_1 - \mathbf{q}_2)_{u,v}$ is the u,vth element of the submatrix $p(\mathbf{q}_1 - \mathbf{q}_2) = p(\mathbf{q}_1, \mathbf{q}_2)$ for which, if we take into account the $\mathbf{C}(\mathbf{p})_h = \mathbf{U}\mathbf{D}(\mathbf{p})_h$ transformation, the definition (17.5) of U, and introduce again instead of the vector $\mathbf{p}$ with discrete components the vector $\mathbf{k}$ with continuously varying components [see equation (17.9)], we can write

$$p(\mathbf{q}_1 - \mathbf{q}_2) = \frac{2}{\omega} \int_{\omega} \sum_{h=1}^{n^{\bullet}} \mathbf{d}(\mathbf{k})_h \mathbf{d}(\mathbf{k})_h^+ \exp[i\mathbf{k}(\mathbf{R}_{\mathbf{q}_1} - \mathbf{R}_{\mathbf{q}_2})] d\mathbf{k} \tag{17.18}$$

where ω is the volume of the first Brillouin zone.

Equations (17.10), (17.11), (17.17), and (17.18) then define the *ab initio* SCF-LCAO crystal-orbital (Hartree–Fock) method with a nonlocal exchange for crystals or polymers possessing simple translational symmetry.

It should be mentioned that the outlined crystal-orbital method can be easily generalized to different orbitals with different spins[2] and to the case when in a chain (or crystal) of weakly interacting units the constituent molecules are radicals ("open-shell" SCF-LCAO-CO method[2]). Finally, the developed method can also be formulated relativistically (relativistic Hartree–Fock CO method).[4] The discussion of these methods is, however, outside the scope of this chapter.

17.1.2. The SCF-LCAO-CO Method in the Case of a Combined Symmetry Operation

We can apply the formalism developed in the preceding section also in the case of a combined symmetry operation. To show this let us assume that we have a helix in which we get from one unit to the next by a translation τ and a simultaneous rotation α. We can then introduce the helix operator

$$\hat{S}(\alpha, \tau) = \hat{D}(\alpha) + \tau \tag{17.19}$$

where $\hat{D}(\alpha)$ stands for the operator of the rotation around the main axis of the helix through an angle α.[5] For the sake of simplicity let us assume further that after n repetitions of the helix operation we obtain the "large" translation $\hat{T}$,* where

$$\hat{S}^n(\alpha, \tau) = \hat{T} \tag{17.20}$$

We can again introduce the Born–von Kármán periodic boundary conditions in the form

$$\hat{S}^{2N+1} = \hat{1} \tag{17.21}$$

where N is a large integer and measures the number of unit cells.

If $\hat{F}$ is the Fock operator of the helix we also have

$$[\hat{S}, \hat{F}] = \hat{0} \tag{17.22}$$

and so we can classify the eigenfunctions of $\hat{F}$ according to the one-dimensional representations of the finite Abelian group $G = \{\hat{S}^m ; m = 1, \ldots, 2N+1\}$. The kth representation of this group is thus $\xi_{lk} = \exp[i2\pi lk/(2N+1)]$. This means that the eigenvalue equation

$$\hat{S}^m \psi_k = \exp[i2\pi mk/(2N+1)]\psi_k = \xi_{mk}\psi_k \tag{17.23}$$

has to be fulfilled, where ψ_k may have again an LCAO form [see equation (17.1)], but k is now defined on the combined symmetry operation.

To generate the eigenfunctions ψ_k of $\hat{S}^m$ we can introduce the projection operator $\hat{O}_k$[5] given by

$$\hat{O}_k = (2N+1)^{-1} \sum_{m=1}^{2N+1} \xi_{mk}\hat{S}^{-m} \tag{17.24}$$

which fulfills the relation $\hat{S}^m \hat{O}_k = \xi_{mk}\hat{O}_k$, $\hat{O}_k \hat{O}_l = \delta_{kl}$, and $\Sigma_k \hat{O}_k = 1$. If $X_g^q = X_g(\mathbf{r} - \mathbf{R}_q - \mathbf{r}_{gA})$ stands again for the gth AO of the qth cell, we can generate the generalized LCAO Bloch orbitals of the helix with the aid of the expression

$$\psi_{k,q}(\mathbf{r}) = \hat{O}_k X_g(\mathbf{r} - \mathbf{R}_q - \mathbf{r}_{gA}) = \hat{O}_k X_g(\mathbf{r} - \mathbf{R}_q^{gA})$$

$$= (2N+1)^{-1} \sum_{m=1}^{2N+1} \xi_{mk}\hat{S}^{-m} X_g(\mathbf{r} - \mathbf{R}_q^{gA}), (\mathbf{R}_q^{gA} = \mathbf{R}_q + \mathbf{r}_{gA}) \tag{17.25}$$

---

* It should be noted that the following considerations hold also in the case when (17.20) is not fulfilled, i.e., $\alpha/2\pi$ is not an integer.

The same procedure can be applied also if we have in our reference cell not only a single AO but a linear combination of them (LCAO MO).

To be able to apply (17.25) we must express

$$\hat{S}^{-m} X_g (\mathbf{r} - \mathbf{R}_q^{gA}) = X_g [\hat{S}^m (\mathbf{r}) - \mathbf{R}_q^{gA}] \tag{17.26}$$

where the right-hand side of (17.26) follows from the well-known relation that $\hat{S}^{-m}$ applied to a function is identical with the transformation of the coordinate system under the inverse operation.[6] Taking into account that $\hat{S}^m \mathbf{r} = \hat{D}(m\alpha)\mathbf{r} + m\boldsymbol{\tau}$ we can further write

$$X_g [\hat{S}^m (\mathbf{r}) - \mathbf{R}_q^{gA}] = X_g [\hat{D}(m\alpha)\mathbf{r} + m\boldsymbol{\tau} - \mathbf{R}_q^{gA}] \tag{17.27}$$

Using the identity[5]

$$\hat{D}(m\alpha)\mathbf{r} - \hat{D}(m\alpha)\hat{S}^{-m}(\mathbf{R}_q^{gA}) = \hat{D}(m\alpha)\mathbf{r} - \hat{D}(m\alpha)[\hat{D}(-m\alpha)\mathbf{R}_q^{gA} - m\boldsymbol{\tau}]$$
$$= \hat{D}(m\alpha)\mathbf{r} - \mathbf{R}_q^{gA} + m\boldsymbol{\tau}$$

we can write down our final result

$$\hat{S}^{-m} X_g (\mathbf{r} - \mathbf{R}_q^{gA}) = X_g \{\hat{D}(m\alpha)[\mathbf{r} - \hat{S}^{-m}(\mathbf{R}_q^{gA})]\} \tag{17.28}$$

This means that by applying the helix operator $\hat{S}^{-m}$ to an AO we have to (1) perform m times the helix operation on the position of the nucleus and (2) we have to rotate the argument of the AO through on angle $m\alpha$ around the axis of the helix.

17.2. Semiempirical SCF-LCAO Crystal-Orbital Methods

17.2.1. The Semiempirical SCF-Electron (Pariser–Parr–Pople) Crystal-Orbital Method

Starting from equations (17.10), (17.11), (17.17), and (17.18), and using the same approximations and parametrizations as the semiempirical SCF-LCAO π-electron theory [the Pariser–Parr–Pople (PPP) method[7]] employs in the case of molecules, one can formulate a PPP-CO theory.[8]

To derive the PPP-CO method we take into account explicitly, as in the case of molecules, only the π-electrons of the system. Putting $S(k) = 1$, equation (17.10) can be replaced by the matrix eigenvalue equation

$$H(k)\mathbf{d}(k)_h = \varepsilon(k)_h \mathbf{d}(k)_h \tag{17.29}$$

If differential overlap is neglected, as in the PPP method[7] in the case of molecules, and allowing for only first-neighbor interactions, then instead of equation (17.17) we have

$$[H(0)]_{r,r} = \langle X_r^0 | \hat{H}^{eff} | X_r^0 \rangle + \sum_{u=1}^{m} \sum_{q_1=-1}^{+1} P(0)_{u,u} \langle X_r^0 X_u^{q_1} | X_r^0 X_u^{q_1} \rangle$$
$$- \tfrac{1}{2} P(0)_{r,r} \langle X_r^0 X_r^0 | X_r^0 X_r^0 \rangle \tag{17.30}$$
$$[H(q)]_{r,s} = \langle X_r^0 | \hat{H}^{eff} | X_s^q \rangle - \tfrac{1}{2} P(q)_{r,s} \langle X_r^0 X_s^q | X_r^0 X_s^q \rangle$$
$$(q = -1, 0, 1; \text{ if } q = 0, \; r \neq s)$$

for the elements of the matrices $H(q)$, where X_r^0 now denotes the rth π-orbital in the reference cell and

$$\hat{H}^{eff} = \hat{H}^N + \sum_{q_1=-1}^{+1} \sum_{\mu=1}^{m_\pi} V_{el\,\mu}^{eff\,q_1} \tag{17.31}$$

Here $V_{el\,\mu}^{eff\,q_1}$ is the potential of the inner shell and σ electrons of the μth atom with a π-orbital in the cell characterized by q_1; $\hat{H}^N$ is the one-electron part of the Fock operator, but now q_1 runs only from -1 to $+1$; m_π is the number of π-orbitals in the unit cell. We now introduce the usual approximations and notations of the PPP-MO method generalized for our case:

$$\langle X_r^0 X_s^{q_1} | X_r^0 X_s^{q_1} \rangle = \gamma(q_1)_{r;s} \tag{17.32a}$$

$$\langle X_r^0 X_r^0 | X_r^0 X_r^0 \rangle = I_r - E_r \tag{17.32b}$$

$$\langle X_r^0 | \hat{H}^{eff} | X_r^0 \rangle = -I_r - \sum_{q_1=-1}^{+1} \sum_{s=1}^{m} Z_s \gamma(q_1)_{r,s} \qquad (s \neq r \text{ if } q_1 = 0) \tag{17.32c}$$

$$\langle X_r^0 | \hat{H}^{eff} | X_s^q \rangle = \beta(q)_{r,s} \tag{17.32d}$$

Hence we can write

$$[H(0)]_{r,r} = -I_r + \tfrac{1}{2} P(0)_{r,r} (I_r - E_r) + \sum_{\substack{s=1 \\ s \neq r}}^{m} [P(0)_{s,s} - Z_s] \gamma(0)_{r,s}$$

$$+ \sum_{s=1}^{m_\pi} [P(0)_{s,s} - Z_s][\gamma(1)_{r,s} + \gamma(-1)_{r,s}] \tag{17.33a}$$

$$[H(q)]_{r,s} = \beta(q)_{r,s} - \tfrac{1}{2} P(q)_{r,s} \gamma(q)_{r,s} \qquad \text{if } q = 0, \; r \neq s) \tag{17.33b}$$

where I_r and E_r are the ionization potential and electron affinity, respectively, of the rth atom in its appropriate valence state and the generalized charge-bond orders have been defined previously by equation (17.18). The

Coulomb integrals $\gamma(q)_{r,s}$ can be calculated with the aid of the expression

$$\gamma(q)_{r,s} = \frac{1}{q_{r,s} + R(q)_{r,s}}, \qquad \frac{1}{q_{r,s}} = \tfrac{1}{2}(I_r + I_s - E_r - E_s) \qquad (17.34)$$

given by Mataga and Nishimoto,[9] where $R(q)_{r,s}$ is the distance between atom r in the reference cell and atom s in the qth cell. For the core integrals $\beta(0)_{r,s}$ between atoms within the same molecule the usual PPP values can be used, while the intercell $\beta(1)_{r,s}$ core integrals can be taken proportional to the corresponding overlap integrals. In the first-neighbor-interaction approximation we finally have to solve the eigenvalue problem of the Hermitian complex matrix

$$\boldsymbol{H}(k) = \boldsymbol{H}(0) + \boldsymbol{H}(1)e^{ika} + \boldsymbol{H}(-1)e^{-ika} \qquad (17.35)$$

at about 7–9 different points in the first Brillouin zone. We form the matrices $\boldsymbol{P}(q)$ from the eigenvectors by numerical integration [see equation (17.18)] and repeat this procedure until self-consistency is attained for all the elements $P_{rs}(q)$, $q = 0, \pm 1$.

17.2.2. The Semiempirical SCF All-Valence Electron (CNDO/2) Crystal-Orbital Method

In the case of the CNDO/CO (complete neglect of differential overlap/2) scheme for the valence electrons of one-dimensional periodic systems, we can write equation (17.29) with expression (17.35) for $\boldsymbol{H}(k)$ if we again apply the first-neighbor-interaction approximation. In this case the expressions for the elements of the matrices $\boldsymbol{H}(q)$ (which we can again derive from the corresponding *ab-initio* equations by introducing the approximations of the CNDO/2 method[10]) will be[11]

$$[\boldsymbol{H}(0)]_{r,r} = U_r + [P(0)_{A,A} - \tfrac{1}{2}P(0)_{r,r}]\gamma(0)_{A,A} + \sum_B [P(0)_{B,B} - Z_B]\gamma(0)_{A,B}$$

$$+ \sum_B [P(0)_{B,B} - Z_B][\gamma(1)_{A,B} + \gamma(-1)_{A,B}] \qquad (B \neq A, r \in A) \qquad (17.36a)$$

$$[\boldsymbol{H}(q)]_{r,s} = \beta^0_{A,B}S(q)_{r,s} - \tfrac{1}{2}P(q)_{r,s}\gamma(q)_{A,B} \qquad (\text{if } q = 0, r \neq s; r \in A, s \in B) \qquad (17.36b)$$

where

$$P(0)_{A,A} = \sum_{r \in A} P(0)_{r,r} \qquad (17.37)$$

See equation (17.18) for the expressions of $P(q)_{r,s}$. With a generalization of the integral expression occurring in the CNDO method,

$$\gamma(q)_{A,B} = \int |X_A^0(1)|^2 \frac{1}{r_{12}} |X_B^q(2)|^2 dV_1 dV_2 \qquad (q = -1, 0, 1) \quad (17.38)$$

where X_A^0 is an appropriate valence s-orbital centered on atom A in the reference cell. Further, the parameter $\beta_{A,B}^0 = \frac{1}{2}(\beta_A^0 + \beta_B^0)$, where the constants β_A^0 and β_B^0 are given numerically for the elements of the first two rows of the periodic table,[12] Z_B is the core charge of atom B, and the core integrals $U_r = \langle X_r^0 | H_{A,0}^{eff} | X_r^0 \rangle$ can be calculated with the aid of the valence-state ionization potentials and electron affinities of the atom A ($r \in A$) and of its positive and negative ions. Finally, $H_{A,0}^{eff} = -\frac{1}{2}\Delta + V_{A,0}^{eff}$ where $V_{A,0}^{eff}$ is the effective potential of the atomic core (nucleus and inner-shell electrons) A in the reference cell.

It should be pointed out that in a similar way all other semiempirical molecular-orbital methods (such as simple-Hückel, extended-Hückel-MO, INDO, MINDO, and MNDO-MO methods) can be easily generalized to the case of periodic boundary conditions, thus obtaining the corresponding crystal-orbital methods.[13]

17.3. Correction of the Virtual Levels and for Long-Range Correlation

17.3.1. Application of the Excitation Hamiltonian ($\hat{O}\hat{A}\hat{O}$) Method to Polymers

If the polymer contains weakly interacting subunits, like the cytosine molecules in the stacked chain of polycytosine, we can approximate the excited states of the polymer on the basis of those of its constituents. The simplest way to calculate these monomeric excited levels is given by the $\hat{O}\hat{A}\hat{O}$ method.[14] Before discussing this method we should like briefly to point out why it is so difficult to calculate excitation energies in polymers. In the closed-shell Hartree–Fock (HF) theory the singlet excitation energy from a filled level with energy ε_i to a virtual level with energy ε_a is given by the well-known expression

$$^1\Delta E_{i \to a} = \varepsilon_a - \varepsilon_i - J_{ia} + 2K_{ia} \qquad (17.39)$$

where

$$J_{ia} = \left\langle \phi_i(1)\phi_a(2) \left| \frac{1}{r_{12}} \right| \phi_i(1)\phi_a(2) \right\rangle$$

$$K_{ia} = \left\langle \phi_i(1)\phi_a(2) \left| \frac{1}{r_{12}} \right| \phi_a(1)\phi_i(2) \right\rangle$$

Expression (17.39) is, of course, only an approximation to the correct excitation energy, because (1) it does not take into account the change in the distribution of the other electrons (this so-called relaxation energy can be treated if one makes a separate open-shell calculation for the excited state and takes the difference between the total energies of the excited state and of the ground state; the Δ SCF method), and (2) it does not take into account the change of correlation energies in the excited and ground state, respectively. To take into account both effects one has to make a separate calculation together with correlation (and possibly, if known, also taking account of the different geometry of the excited state) for the excited and ground states, respectively, and then one must compute the difference between the total energies of both states.

For polymers and solids we have the additional, easily demonstrated[15] problem that the integrals J_{ia} and K_{ia} vanish if the number of electrons n goes to infinity. This means that we are left with the singlet-excitation energy

$$^1\Delta E_{i \to a}(\mathbf{k}) = \varepsilon_a(\mathbf{k}) - \varepsilon_i(\mathbf{k}) \tag{17.40}$$

of the solid or polymer which, of course, is too large.

To overcome this difficulty one can calculate more correct virtual-energy levels for the constituent molecules of a polymer than are provided by the Hartree–Fock method if, instead of using the incorrect n-particle V^n potential of the HF method, one employs the correct $n-1$ particle V^{n-1} potential with the aid of the modified Fock operator[14]

$$\hat{F}' = \hat{F} + \hat{O}\hat{A}\hat{O} \tag{17.41}$$

In this expression the projection operation $\hat{O}$, defined by

$$\hat{O} = \hat{1} - \hat{\rho} = \hat{1} - \sum_{i=1}^{n^*} \left| \phi_i^{HF}(1) \right\rangle\!\left\langle \phi_i^{HF}(1) \right| = \sum_{a=n^*+1}^{n} \left| \phi_a^{HF}(1) \right\rangle\!\left\langle \phi_a^{HF}(1) \right|$$

$$= \sum_{a=n^*+1}^{n} \left| \psi_a(1) \right\rangle\!\left\langle \psi_a(1) \right| \tag{17.42}$$

projects into the subspace of the virtual orbitals. The HF orbitals $\left| \phi_i^{HF}(1) \right\rangle$ are defined by the equation

$$\hat{F} \left| \phi_i^{HF}(1) \right\rangle = \varepsilon_i^{HF} \left| \phi_i^{HF}(1) \right\rangle \tag{17.43}$$

while the orbitals $\left| \psi_a(1) \right\rangle$ are the eigenfunctions of the modified operator $\hat{F}'$. In the case of a singlet–singlet $i \to a$ excitation we can choose our $\hat{A}$

operator as

$$\hat{A}_i = -\left\langle \phi_i^{\mathrm{HF}}(2) \left| \frac{1}{r_{12}} (1 - 2\hat{P}_{12}) \right| \phi_i^{\mathrm{HF}}(2) \right\rangle \tag{17.44}$$

Expanding the orbitals $| \psi_a (1) \rangle$ in terms of the HF orbitals

$$| \psi_a (1) \rangle = \sum_{i=1}^{m} C_{a,i} \, | \phi_i^{\mathrm{HF}}(1) \rangle$$

we obtain the matrix equation

$$\mathbf{F}^{(i)}\mathbf{C}_a = \varepsilon_a' \mathbf{C}_a \tag{17.45}$$

where the elements of the matrix $\mathbf{F}^{(i)}$ are defined as

$$(\mathbf{F}^{(i)})_{k,l} = \langle \phi_k^{\mathrm{HF}} | \hat{F} + \hat{O}\hat{A}^{(i)}\hat{O} | \phi_l^{\mathrm{HF}} \rangle \tag{17.46}$$

From equation (17.46) it is easy to see that, for the filled orbitals, $\hat{F}^{(i)}$ has the same eigenvalues ε_i as the Fock matrix of the unmodified Fock operator $\hat{F}$; the virtual levels, however, will be changed. The corrected excitation energy (the difference between the total energies in the excited and in the ground state) can be obtained in this way directly as the difference between the corrected one-electron energies[14]:

$$^1\Delta\tilde{E}_{i \to a} = E_{i \to a} - E_G^{\mathrm{U}} = \varepsilon_a' - \varepsilon_i \tag{17.47}$$

Since we can assume to a rather good approximation that in a stacked polymer like polyC an excitation occurs locally on a single cytosine molecule, after performing an $\hat{O}\hat{A}\hat{O}$ calculation for the single molecule one can shift the centers of the empty bands to the corresponding corrected positions of the virtual levels. Applying the $\hat{O}\hat{A}\hat{O}$ method for excitation from the HOMO level, one thus obtains a corrected excitonic gap between the valence band and the approximated exciton band (assuming it has the same width as the conduction band).

17.3.2. Correlation Corrections to the Hartree–Fock Bands on the Basis of the Electron–Polaron Model

As a further step toward more realistic band structures one can introduce corrections to the Hartree–Fock valence and conduction bands to allow for long-range correlation effects using the electron–polaron model.[16–18] According to this method one obtains the following k-

dependent energy shifts for these bands:

$$\Delta E_{\text{cond}}(\mathbf{k}) = \overset{\text{1st Brill. zone}}{\underset{\mathbf{K}}{\sum}} \frac{|V_{\mathbf{K}}^{n^{\bullet}\to n^{\bullet}+1}|^2}{\varepsilon_{\text{cond}}^{\text{HF}}(\mathbf{k}) - {}^1\Delta E_{n^{\bullet}\to n^{\bullet}+1} - \varepsilon_{\text{cond}}^{\text{HF}}(\mathbf{k}-\mathbf{K})} \qquad (17.48a)$$

$$\Delta E_{\text{val}}(\mathbf{k}) = \overset{\text{1st Brill. zone}}{\underset{\mathbf{K}}{\sum}} \frac{|V_{\mathbf{K}}^{n^{\bullet}\to n^{\bullet}+1}|^2}{\varepsilon_{\text{val}}^{\text{HF}}(\mathbf{k}) + {}^1\Delta E_{n^{\bullet}\to n^{\bullet}+1} - \varepsilon_{\text{val}}^{\text{HF}}(\mathbf{k}-\mathbf{K})} \qquad (17.48b)$$

where ${}^1\Delta E_{n^{\bullet}\to n^{\bullet}+1}$ is the energy of a singlet excitation from the valence band to the first exciton band [which can be approximated with the aid of the $\hat{O}\hat{A}\hat{O}$ procedure; see equation (17.40)], and $\varepsilon_{\text{cond}}^{\text{HF}}(\mathbf{k})$ and $\varepsilon_{\text{val}}^{\text{HF}}(\mathbf{k})$ are the corresponding Hartree–Fock eigenvalues. Furthermore, $V_{\mathbf{K}}^{n^{\bullet}\to n^{\bullet}+1}$ is given by

$$V_{\mathbf{K}}^{n^{\bullet}\to n^{\bullet}+1} = \frac{ie}{|\mathbf{K}|}\left[\frac{2\pi\,{}^1\Delta E_{n^{\bullet}\to n^{\bullet}+1}(1-1/\varepsilon_{\infty})}{V_c}\right]^{1/2}$$

$$\times \int |\phi_{(\mathbf{r})}^{\text{Wannier}}|^2 e^{i\mathbf{k}\cdot\mathbf{r}}d\mathbf{r} \approx \frac{ie}{|\mathbf{K}|}\left[\frac{2\pi\,{}^1\Delta E_{n^{\bullet}\to n^{\bullet}+1}(1-1/\varepsilon_{\infty})}{V_c}\right]^{1/2} \qquad (17.49)$$

[since the integral is, to a good approximation, equal to 1, the matrix element $V_{\mathbf{K}}^{n^{\bullet}\to n^{\bullet}+1}$ is simplified to the expression given on the right-hand side of equation (17.49)]. Finally, ε_{∞} is the high-frequency dielectric constant of the system.

In the actual calculations one can substitute the summations over the vector $\mathbf{K}$ by an integration over $\mathbf{K}$. In that case one obtains a factor of $V_c/8\pi^3$ before the integral and in this way V_c (the crystal volume) is eliminated from equations (17.48).

17.4. Applications to Periodic DNA and Protein Models

17.4.1. Periodic DNA Models

We have applied the *ab-initio* SCF-LCAO-CO method to the four nucleotide base stacks (polycytosine, polythymine, polyadenine, and polyguanine; see Figure 17.1), to the sugar-phosphate chain of DNA (polySP), and to polycytidine, which contains one cytosine-sugar phosphate unit in the elementary cell (polyCSP[19]; see Figure 17.2).

For the calculations an STO-3G[20] basis set and an *ab-initio* CO program has been used that takes into account also the necessary rotation of the basis functions[5] if one gets from a unit to the next one not via a simple translation, but if a combined symmetry (for instance, helix) operation has to be applied. The four nucleotide base stacks and the

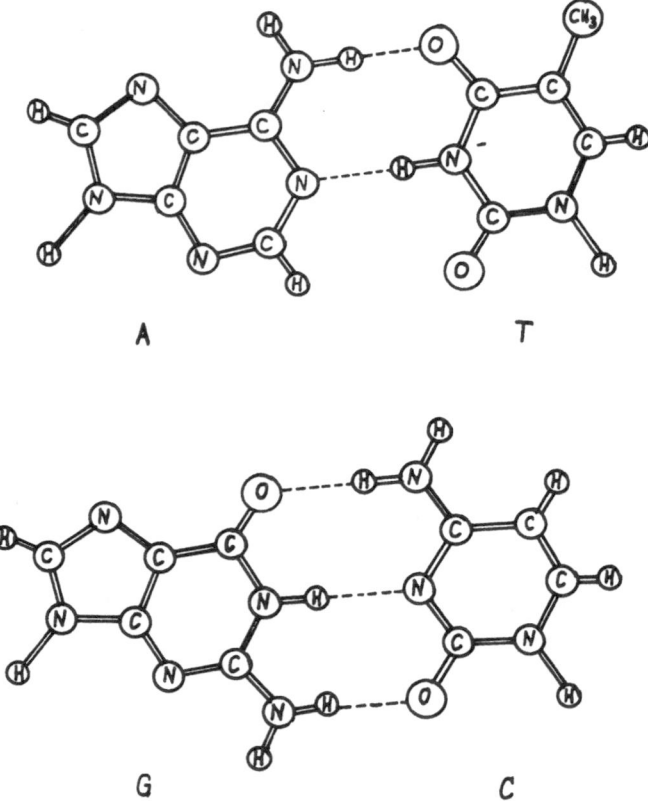

Figure 17.1. Nucleotide base pairs. A = adenine, T = thymine, G = guanine, and C = cytosine.

sugar-phosphate (polySP) chain as well as the polycytidine (polyCSP) chain were taken in the same geometrical arrangement as in DNA-B.[21] In the case of polySP and polyCSP, instead of a K^+ ion, a proton has been attached to the PO_4^- group, thus keeping these chains neutral. In the calculations, which in the case of the polycytidine chain (with 19 nonhydrogen atoms, 11 hydrogens, and 110 contracted Gaussians in the unit cell) took approximately 80 hours on a Cyber 172 computer, second-neighbor interactions have been included with a correct (electrostatically balanced) cutoff.[22] This means that the two-electron repulsion integrals $\langle X_a X_b | 1/r_{12} | X_c X_d \rangle$ have been neglected so as to balance neglect of the $-\langle X_a | 1/r_b | X_c \rangle$-type nuclear-attraction integrals. Therefore, a part of those two-electron integrals had to be retained that contain centers three units apart from each other.

It should be noted that all the valence and conduction bands of the four nucleotide base stacks originate from the π HOMO and LEMO levels of the constituent molecules. Since the SP and CSP units are not planar, such a classification in the case of the polySP and polyCSP chains is not possible.

With the help of Mulliken's population analysis the amount of transferred charge from the sugar-phosphate chain to the cytosine chain has been computed using the results of the polyCSP superchain calculation.

The results, presented in Tables 1 and 2, show that the valence and conduction bands of the stacked bases and of the polySP chain are several tenths of an eV wide (0.16–0.86 eV) indicating that there is a possibility for Bloch-type conduction in these systems if free-charge carriers are generated in them. However, the gap in all cases is more than 10 eV. Though one knows that a Hartree–Fock calculation gives too large a gap for conduction (not to be confused with the considerably smaller first singlet-excitation energy, which can be obtained with the aid of an intermediate exciton

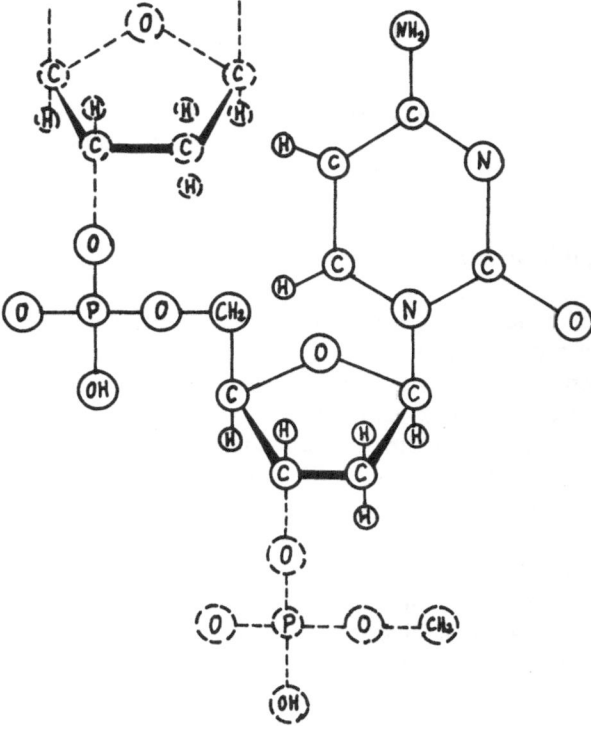

Figure 17.2. Chemical formula of cytidine (CSP).

Table 1. Limits and Widths of the Valence and Conduction Bands of the Four Nucleotide Base Stacks (in eV)[a]

Chain	Valence band				Conduction band			
	ε^{MO}	ε^{CO}_{min}	ε^{CO}_{max}	$\delta\varepsilon$	ε^{MO}	ε^{CO}_{min}	ε^{CO}_{max}	$\delta\varepsilon$
PolyC	−5.61	−5.51	−4.65	0.86	6.00	6.07	6.91	0.84
PolyT	−6.66	−6.48	−5.88	0.60	5.97	6.02	6.33	0.31
PolyA	−6.10	−6.04	−5.57	0.47	6.46	6.56	6.86	0.29
PolyG	−5.08	−5.16	−4.34	0.82	6.45	6.41	7.15	0.74

[a] For comparison, the locations of the corresponding MOs are also given.

Table 2. Band Limits and Widths of the Valence and Conduction Bands of PolyC and PolySP and of the Two Highest Filled and Lowest Unfilled Bands of the Polycytidine Chain (in eV)

	PolyC			PolySP			PolyCSP		
	ε^{CO}_{min}	ε^{CO}_{max}	$\delta\varepsilon$	ε^{CO}_{min}	ε^{CO}_{max}	$\delta\varepsilon$	ε^{CO}_{min}	ε^{CO}_{max}	$\delta\varepsilon$
Conduction band	6.07	6.91	0.84	7.43	8.00	0.57	7.40	7.96	0.56
							6.55	7.38	0.83
Valence band	−5.51	−4.65	0.86				−5.19	−4.36	0.83
				−6.44	−6.28	0.16	−6.79	−6.71	0.08

theory[23]), this certainly rules out the possibility of intrinsic semiconduction in DNA.

On the other hand, our calculation of the polyCSP superchain has resulted in a charge transfer (CT) of 0.19 e per molecule pair from the sugar-phosphate unit to cytosine. Due to the method, one obtains in a restricted Hartree–Fock superchain calculation only completely filled bands. Nevertheless, if there is considerable CT from one chain to the other the calculated CT also indicates the possibility of the creation of free-charge carriers in the system. Though the valence band of polySP according to this calculation is rather narrow (0.16 eV), one should bear in mind that the presence of the positive ions may increase the bandwidths to a significant degree (in some cases by a factor of approximately 2 or 3), as indicated by previous π-electron-band-structure calculations.[24] Therefore, if one were to repeat the polySP and polyCSP band-structure calculations (now in progress) assuming the presence of K^+ ions around the PO_4^- groups with no proton chemically bound to them, one would expect a rather large broadening of the valence band of polySP.

In connection with the periodic DNA-model calculations, it should be mentioned that in the 1960s and at the beginning of the 1970s a large number of semiempirical SCF π-electron (PPP) crystal-orbital calculations were performed for the nucleotide base stacks, for the poly(A–T) and poly(G–C) base pairs, and for more complicated periodic DNA models like poly($^{A-T}_{G-C}$). (A review can be found elsewhere.[25]) In the case of the nucleotide base stacks and of poly(A–T) and poly(G–C), the bandwidths were about half of those found in the *ab-initio* case (for the base stacks), while if a different base was always superimposed over another one [as in poly($^{A-T}_{G-C}$)], the bandwidths were one order of magnitude smaller. Of course, this fact can be easily interpreted on the basis of perturbation theory. (Further details are given elsewhere.[25])

Let us now consider the correction to the virtual levels. Table 3 gives the uncorrected and, with the help of the $\hat{O}\hat{A}\hat{O}$ procedure, the corrected (assuming a singlet–singlet transition starting from the highest filled level) lowest unfilled levels of the four nucleotide bases. Comparison of the corrected $\varepsilon_{LEMO} - \varepsilon_{HOMO}$ energy differences with the experimental values indicates that though they are still too large (especially in the case of C), the $\hat{O}\hat{A}\hat{O}$ procedure has corrected the major part of the error, if we compare the band gap with the experimental excitation energy (as must be done in the case of a polymer).

In Section 17.3.2 we described briefly the electron–polaron model for the treatment of long-range correlation. This method was first applied to the band structure of polyC.[28] The value of ε_∞ was estimated on the basis of the expression

$$\frac{\varepsilon_\infty - 1}{\varepsilon_\infty + 2} = \tfrac{4}{3}\alpha_\perp \frac{1}{v_m} \qquad (17.50)$$

where $\alpha_\perp$ is the polarizability of the cytosine molecule in the direction perpendicular to the molecular plane and v_m is the volume of the cytosine molecule. For $\alpha_\perp$ we have used the value 1.33 Å^3, previously calculated by

Table 3. Minimal Basis[26] HOMO and LEMO Levels and $\hat{O}\hat{A}\hat{O}$-Method-Corrected LEMO Levels (in eV) of the Four Nucleotide Bases

Level	Cytosine (C)	Guanine (G)	Thymine (T)	Adenine (A)
LEMO	1.931	3.612	3.203	2.332
LEMO (corrected)	−2.816	−3.495	−4.488	−3.582
HOMO	−9.768	−8.209	−9.602	−9.288
Corrected gap	6.952	4.714	5.154	5.706
Experimental HOMO–LEMO excitation energy[27]	4.3	4.2	4.7	4.9

Seprödi et al.,[29] and for v_m the value $3.36 \cdot \pi \cdot 1.75^2 = 32.33$ Å³, where 3.36 Å is the stacking distance in polyC and 1.75 Å is the estimated molecular radius in the plane of the molecule. With these values, equation (17.50) yields ε_∞ equal to about 1.5. The results are presented in Table 4.

On comparing these correlated band structures with the Hartree–Fock values, we see that the gap and widths of the valence and conduction bands decrease by about 10%. (A similar effect is also found in the case of the other three nucleotide base stacks.) It should be mentioned that this effect of long-range correlation[30] is much larger in the case of polyacetylene.[31]

17.4.2. Periodic Protein Models

Table 5 contains the ab-initio valence and conduction bands of a polyglycine polypeptide (main) chain and a polyglycine hydrogen-bonded perpendicular chain[32] (see Figure 17.3). The Gaussian-lobe minimal basis set discussed elsewhere[26] has been applied again in the calculation.

Table 4. Conduction and Valence Bands of PolyC Corrected with the Help of the Electron–Polaron Model with $\varepsilon_\infty \sim 1.5$ (in eV)

	ε_{min}^{CO}	ε_{max}^{CO}	$\Delta\varepsilon^{CO}$
Conduction band	1.116	2.275	1.158
	(1.535)[a]	(2.775)	(1.240)
Valence band	−9.190	−8.674	0.516
	(−9.665)	(−9.113)	(0.552)

[a] Uncorrected Hartree–Fock values using the basis set[26] are given in parentheses.

Table 5. Valence and Conduction Bands of Polyglycine Calculated by the ab initio SCF-LCAO-CO Method (in eV).[a]

	Polyglycine main chain			Polyglycine H-bonded perpendicular chain		
	ε_{min}^{CO}	ε_{max}^{CO}	$\delta\varepsilon$	ε_{min}^{CO}	ε_{max}^{CO}	$\delta\varepsilon$
Conduction band	3.817	5.195	1.378	3.755	3.896	0.141
	(3.102)	(4.357)	(1.255)			
Valence band	−11.252	−9.154	2.098	−11.382	−11.091	0.291
	(−10.373)	(−8.465)	(1.908)			

[a] The bands belonging to the main chain are corrected for long-range correlation effects with $\varepsilon_\infty = 3.5$ (numbers in parentheses).[33]

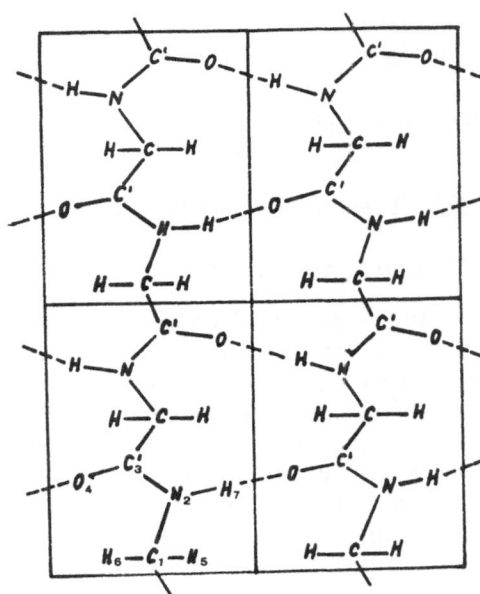

Figure 17.3. Two-dimensional polypeptide network in the parallel-chain β-pleated sheet conformation.

It can be seen from Table 5 that in the case of the main chain the bands are rather broad (the width of the valence band is about 1.2 eV and that of the conduction band is approximately 1.4 eV), while the bandwidths of the hydrogen-bonded chain are one order of magnitude smaller. These results agree qualitatively with those of the previous semiempirical SCF all-valence electron calculations.[34] The long-range correlation decreases the bandwidths by about 10%, as in the case of polyC in the previous section. The Hartree–Fock gap is about −12.4 eV for the main chain (which decreases to approximately 11.6 eV due to long-range correlation), while for the hydrogen-bonded chain its value (due to the smaller bandwidths) is even larger (about 14.8 eV).

From Figure 17.3 and Table 5 it is obvious that for a more realistic description of a polypeptide one has to take into account both interactions (along the main chain and across the hydrogen bonds) simultaneously. Semiempirical SCF all-valence electron calculations on the polyformamide $(NH_2 . HC{=}O)_n$ system have shown[35] that the influence of the second dimension can cause considerable changes in the band structure of the main chain.

Finally, it should be mentioned that a recent *ab-initio* calculation of the polyalanine chain[36] has resulted in a band structure very similar to that of polyglycine. Despite this a calculation of the periodic poly(glyala) mixed chain resulted in a gap of 0.1–0.2 eV in the valence and conduction bands, showing that already a small degree of aperiodicity (in alanine, a H atom of glycine is substituted by a CH_3 group) can cause considerable changes in the band structure of a polymer.[36]

17.5. Treatment of Correlation in Polymers

17.5.1. Intermediate Exciton Theory for Excited States of Polymers

In Section 17.3.1 we discussed briefly the simple excitation Hamiltonian ($\hat{O}\hat{A}\hat{O}$) method for the correction of the virtual levels. If one wants to go further and take into account at least the correlation between the electron promoted to the conduction band and the remaining hole in the valence band, one must use exciton theory,[37] which gives at the same time also the width and dispersion of the exciton band.

The calculation of such states in polymers is complicated by the fact that neither of the two usually applied models of exciton theory can be used for these systems. Both models assume a limiting case: the Frenkel picture is valid only for nearly localized excitations (within the same elementary cell), while the Wannier model applies to strongly delocalized ones. By inspection of the electronic indices of most polymers it is clear, however, that excitations between neighboring elementary cells should play an important role but it is also evident that simple effective mass theory (continuum model) would not work for them. This conceptual difficulty was removed by Takeuti,[23] who devised the so-called "intermediate-exciton" scheme based on the mathematical procedure proposed by Slater and Koster[38] to treat localized impurity levels in semiconductors. We shall shortly describe here his scheme for the case when the ground state is a completely filled valence band represented by a Slater determinant containing doubly filled Wannier functions and there is only one empty conduction band (generalization of his expressions to the case of more valence and conduction bands is obvious and would only complicate the notation).

The many-particle wave function of Takeuti's method is constructed in the form

$$^{M}\Psi^{K} = \sum_{j} {}^{M}U^{K}(\mathbf{R}_{j}) \cdot {}^{M}\Phi_{v,c}^{K}(\mathbf{R}_{j}) \tag{17.51}$$

where the function $^{M}\Phi_{v,c}^{K}(\mathbf{R}_{j})$ itself is a symmetry-adapted linear combination of singly excited configurations with appropriate multiplicity M:

$$^{M}\Phi_{v,c}^{K}(\mathbf{R}_{j}) = N^{-1/2} \sum_{h} \exp(i\mathbf{K} \cdot \mathbf{R}_{h}) \cdot {}^{M}\Lambda_{v,c}^{h,h+j} \tag{17.52}$$

The Slater determinant $^M\Lambda_{v,c}^{h,h+j}$ can be obtained if we substitute the Wannier function $\Phi_v^h(\mathbf{r})$ in the ground-state determinant by $\Phi_c^{h+j}(\mathbf{r})$, where $\Phi_v^h(\mathbf{r})$ is a valence-band Wannier function centered on the cell characterized by $\mathbf{R}_h$ and $\Phi_c^{h+j}(\mathbf{r})$ is the conduction-band Wannier function belonging to the cell $\mathbf{R}_{h+j}$; superscript M denotes the spin multiplicity of the excited state. Each function $^M\Phi_{v,c}^K(\mathbf{R}_j)$ thus represents in the exciton state $^M\Psi^K$ an excitation wave corresponding to an electron–hole separation by a lattice vector $\mathbf{R}_j$ and moving with wave vector $\mathbf{K}$. This "exciton representation," proposed by Wannier,[39] combines the expected explicit dependence of the matrix elements on the electron–hole separation in direct space with the fact that the total momentum of the electron–hole pair has to remain a good quantum number ($\mathbf{K}$-dependence). By treating the electron–hole interaction as a perturbation and using expansion (17.51) for the perturbed-wave functions, the Schrödinger equation of this simple configuration-interaction problem

$$\hat{H}\,^M\Psi^K = {}^ME^K\,^M\Psi^K \tag{17.53}$$

can be rewritten (applying the Slater–Koster idea) in the form[38]

$$^MU^K(\mathbf{R}_j) = \sum_h \sum_l G^K(\mathbf{R}_j, \mathbf{R}_l, {}^ME^K) \cdot {}^MV^K(\mathbf{R}_l, \mathbf{R}_h) \cdot {}^MU^K(\mathbf{R}_h) \tag{17.54}$$

The Green's function of the electron–hole pair is defined here by

$$G^K(\mathbf{R}_j, \mathbf{R}_l, {}^ME^K) = N^{-1} \sum_{\mathbf{k}}^{\mathrm{BZ}} \frac{\exp[i\mathbf{K}\cdot(\mathbf{R}_j - \mathbf{R}_l)]}{{}^ME^K - [\varepsilon_c(\mathbf{k}) - \varepsilon_v(\mathbf{k} - \mathbf{K})]} \tag{17.55}$$

where $\varepsilon_c(\mathbf{k})$ and $\varepsilon_v(\mathbf{k})$ are the energy dispersions in the conduction and valence bands, respectively. Finally, the matrix elements of the electron–hole interaction are given by

$$^MV^K(\mathbf{R}_l, \mathbf{R}_h) = \sum_m [\exp(-i\mathbf{K}\cdot\mathbf{R}_m)]$$
$$\times (-\langle \phi_v^0 \phi_c^{m+1} | \phi_v^m \phi_c^h \rangle + 2\delta_M \langle \phi_v^0 \phi_c^{m+1} | \phi_c^h \phi_v^m \rangle) \tag{17.56}$$

In equation (17.56) all the basis functions are, of course, Wannier functions (for an advantageous choice of the phase factor of the Bloch functions that influence the properties (localization) of the Wannier functions see Blount's review[40]) with appropriate band index (the constant δ_M in the exchange part is equal to unity in the case of singlet excitons and to zero for triplet ones). The most time-consuming step in exciton calculations (as well as in CI ones) is the transformation of the two-electron interaction integrals (which have been evaluated during the band-structure calculations in the atomic basis) to the Wannier basis. The matrix elements in equation (17.56) must then be calculated for each value of K, and the zero values of the determinant corresponding to the system of homogeneous linear equations (17.54) as a function of $^M E^K$ provide the solutions to equation (17.53). More details of this method, including an extension to doubly excited configurations, will be presented elsewhere.[41]

By constructing from approximate Hartree–Fock Bloch orbitals the Wannier functions of the valence and conduction bands of alternating trans-polyacetylene (a —CH=CH— unit forms the elementary cell), which is in itself an insulator, the method described above has been applied[41] to compute the first excitonic band. Table 6 presents the excitation energies thereby obtained as a function of the number N of cells separating the excited electron and the hole, and of the crystal momentum K of the exciton.

It can be seen that the excitation energy decreases rapidly with N and converges only at $N = 8$. The exciton band has a width of 3.7 eV. The smallest excitation energy from the ground state to the exciton band at $K = 0$ with $N = 8$ is still by a factor of 1.5 larger than the experimental value of 2.24 eV (this discrepancy will certainly be smaller if a better basis set is employed), which has been determined by extrapolating to an infinite number of units of a carotinoid chain.[42] It is very small, however, in comparison to the gap of 9.74 eV between the upper limit of the valence band of polyacetylene (at $k = \pi$) and the lower limit of its conduction band (at $k = \pi$ again[31]; for further details see Suhai[41]). Further applications to periodic DNA and protein models are in progress.

Table 6. Excitation Energies of Alternating Trans-Polyacetylene as a Function of K and N Calculated with the Aid of Intermediate Exciton Theory (in eV)

$K = 0$	$\pi/2$	π	N
7.35			0
5.71	7.34	9.52	1
4.96			2
3.94			4
3.76			6
3.74			7
3.73	5.51	7.38	8

17.5.2. Discussion of the Correlation in the Ground State of Polymers

17.5.2.1. Insulators

In the case of a polymer with a completely filled valence band (and correspondingly empty conduction band), if the gap is considerable then the ground-state correlation can be subdivided into long-range and short-range parts. For the treatment of the long-range correlation we have discussed in Section 17.3.2 the electron–polaron model, and we have shown some applications for periodic DNA and protein models in Section 17.4.

The short-range correlation in the ground state of a polymer can be treated by Fourier transforming the delocalized Hartree–Fock Bloch orbitals into localized Wannier functions[39]

$$\phi_n^j(\mathbf{r}) = N^{-1/2} \sum_{\mathbf{k}}^{BZ} \psi_n^k(\mathbf{r}) \exp(-i\mathbf{k}.\mathbf{R}_j) \qquad (17.57)$$

where the Wannier function $\phi_n^j(\mathbf{r}) = \phi_n(\mathbf{r} - \mathbf{R}_j)$ is centered around the cell at $\mathbf{R}_j$. Though this localization process does not eliminate the "tails" of the Wannier functions,[43] it certainly localizes the overwhelming part of the crystal orbitals around a site (molecule). If these Wannier functions are used as a basis instead of the MOs of the free molecules, one can apply the usual quantum-chemical methods, such as configuration interaction (CI), the coupled electron-pair approximation (CEPA[44]), or the coupled-cluster expansion method of Čižek and Paldus,[45] to obtain the correlation energy per unit cell due to the short-range correlation. In this procedure, one must use not only excitations within that cell, to which the Wannier function is localized, but also charge-transfer-type excitations to the neighboring cells.

One should point out that if we disregard the possibility of internal charge transfer (CT) in DNA, or CT between a DNA and a polypeptide chain, the ground-state correlation of periodic DNA and protein models can be treated in this way. This work is presently in progress.

17.5.2.2. Metallic Polymers

The problem of ground-state correlation becomes much more difficult if (1) the subunits are strongly coupled (as in a $(CH)_x$ chain] where, to achieve the desirable accuracy, charge-transfer-type excitations to more distant neighbors cannot be neglected, and (2) if the valence band is partially filled [as in $(SN)_x$], because in this latter case the localized Wannier functions could be formed only if one were to take into account in the Fourier transformation also the unfilled part of the band (i.e., one would have to mix excited Bloch functions to the single ground-state HF Slater determinant). Furthermore, in this case the ground-state correlation energy cannot be subdivided into long- and short-range contributions. In such cases in solid-state physics, electron-gas methods are customarily used.[46] Instead of starting from HF Bloch orbitals, the density functional $E[\rho]$ formalism is applied to describe all contributions to the electronic energy. Since, however, the Hohenberg–Kohn theorem[47] is only an existence theorem and the explicit form of the exact $E[\rho]$ functional is unknown, it seems to be difficult to find a systematic way to improve the results obtained with this formalism. This problem becomes especially difficult in the case of metallic organic polymers containing hetero (noncarbon) atoms (like the TCNQ-TTF system, or biopolymers between which CT occurred), because one has to deal with polymers in which the density is small and the density gradients (due to bonds between different atoms) are large.

On the other hand, starting with the (numerically harder to use) Hartree–Fock method, one can systematically improve the results if a comparatively simple and accurate method can be developed to treat the ground-state correlation. Such a method could be provided for polymers also by the further development of an approximate CI technique[48] that used not excitation from single levels to single levels, but excitations from a region of a given band to other regions of the same band or to regions of other bands. To achieve this, one must subdivide each band into specific regions by analyzing the density-of-state curves of the bands in question and taking into account the dependence of the partial charge-density distribution of these particular bands on $\mathbf{k}$.[48] In this connection the problem of size consistency is important, and the question to how many units the orbitals representing the excited regions should be extended requires further investigations, which are in progress.

17.6. Methods for Treatment of Aperiodic Polymers

17.6.1. The Virtual Crystal, Coherent Potential Approximation (CPA), and SCF Resolvent Method

It is well known that most polymers, especially biopolymers, are aperiodic. Besides possible structural (geometric) disorder and local impurities, they are also aperiodic in their sequences. The sequence of the four different nucleotide bases (A, T, G, C) determines the genetic information carried by DNA, and proteins contain 20 different amino acids in an aperiodic sequence. Therefore, one needs methods that can treat structural and compositional (chemical) disorder simultaneously.

The virtual crystal approximation treats the potential of the system as the average of the potentials of the constituents. At the Hartree–Fock level, this corresponds to the averaging of the Fock matrices of the reference cells according to the composition of the polymer, while matrix blocks describing the interactions between different cells are averaged according to the first-, second-, etc., neighbor frequencies:

$$\tilde{F}(0) = \sum_r w_r F_r(0) \tag{17.58}$$

$$\tilde{F}(\mathbf{q}) = \sum_{r,s} w_{r,s}(\mathbf{q}) F_{r,s}(\mathbf{q}) \tag{17.59}$$

where w_r gives the weight (probability of occurrence) of the rth component in the polymer, and $w_{r,s}(\mathbf{q})$ denotes the probability that if we have in the reference cell the rth type of unit, we find in the cell characterized by $\mathbf{R_q}$ the sth type ($\mathbf{q}$th-neighbor frequency). Solving the equations

$$\tilde{F}(\mathbf{k})\tilde{\mathbf{d}}_i(\mathbf{k}) = \tilde{\varepsilon}_i(\mathbf{k})\tilde{S}(\mathbf{k})\tilde{\mathbf{d}}_i(\mathbf{k}) \tag{17.60}$$

with

$$\tilde{F}(\mathbf{k}) = \sum_q [\exp(i\mathbf{k}\cdot\mathbf{R_q})]\tilde{F}(\mathbf{q}), \qquad \tilde{S}(\mathbf{k}) = \sum_q [\exp(i\mathbf{k}\cdot\mathbf{R_q})]\tilde{S}(\mathbf{q}) \tag{17.61}$$

one can define an SCF procedure[49] in the same way as in the periodic case [see equations (17.10), (17.11), and (17.18)].

This method, briefly outlined above, has been applied in the CNDO/2 level to the calculation of the band structures of different poly(ABC)-type model polymers.[50] Comparison of the results thereby obtained with direct periodic calculations indicates that this method (in agreement with previous experience gained by simpler tight-binding or OPW-type calculations) works tolerably well only when the units A, B, and C are very similar (possibly isoelectronic), which is certainly not the case with the biopolymers.

As the next step, one can try to apply the coherent potential approximation (CPA) method[51] to the calculation of the density of states (and band structure) of disordered polymers. This method, which assumes an effective medium for the disordered system such that the average fluctuation from it should give zero ($\langle T \rangle = 0$, where T is the scattering matrix element), is by its very nature more adequate for the treatment of a 3D disordered system than for an aperiodic 1D-polymer. Further, if one wants to apply the simpler one-site CPA (no second scattering of an electron on a given site occurs before it has been scattered on another site;

this is a plausible assumption in the case of an atomic or ionic lattice) one encounters conceptual difficulties in the case of a polymer or molecular crystal that has larger molecules (like the nucleotide base or the TCNQ-TTF stacks) in the unit cell. Therefore, we have applied the single-site CPA (in the 2-component, 1-band approximation) only to a 1D mixed $(SN)_x$—(SN) chain [in $(SN)_x$ there is 4–8% hydrogen impurity[52] where there are only 2, respectively 3, atoms in the unit cell[53]]. Since the density-of-state curves of the two chains have completely different shapes, no constant self-energy $\Sigma(E)$ approximation[51] has been applied, but one has to work with a k- and energy-dependent $\Sigma(k, E)$. By solving the problem in an iterative way,[53] the SCF solution to the problem has already provided, at 3% H, small gaps and spikes in the density-of-state curve of the mixed chain.[53]

These calculations for DNA and proteins (with larger molecules in the unit cell) can be extended by using the cluster CPA formalism.[54] Since in this theory all the quantities occurring in the CPA formalism are matrices,[54] this method could be easily coupled with the recently formulated different-orbitals-for-different-spins (DODS) CPA method in which all the relevant quantities (self-energy Σ, potential-energy difference Δ, Green's function of the effective medium G_e, etc.) are also matrices.[55] The formulation of the DODS-cluster CPA method is in progress.

A further difficulty in applying the CPA method to biopolymers is that they contain not two but four (DNA), respectively 20 (proteins), different components and due to the positions of the (sometimes overlapping) bands one cannot usually apply the one-band approximation. The generalization of the CPA method to these problems is also in progress at Erlangen.

In many cases a periodic polymer contains a local perturbation due to a cluster of impurities, a conformational change, as a result of binding of additional molecules to the polymers (like carcinogen binding to DNA and proteins; see below). To treat this problem, one can formulate the Koster–Slater resolvent method[38] also in the *ab-initio* SCF-LCAO (Hartree–Fock) level including overlap.[49] One starts from the solution of the periodic problem and, applying in an SCF way the Green's-matrix formalism,[56] one can obtain the energy levels and wave functions of the cluster of impurities, or the chemisorbed or chemically bound admolecule imbedded in the periodic polymers. The generalization of this formalism to the *ab-initio* calculation of surface-energy bands is also straightforward.[57]

An essential shortcoming of this method is that it cannot describe correctly the change in the energy bands of the host-polymer crystal due to the presence of the local perturbation (the so-called bulk distortion). To be able to obtain a mutually consistent solution both for the impurity and for the host polymer, one can apply a Green's-function formalism developed first in the simple tight-binding approximation by Callaway[58] both for the so-called evanescent states (falling into gaps between the bands) and for resonant states (states lying inside the bands of the bulk crystal). As a next step this method has been formulated in an SCF way, but using a simple pseudopotential for the perturbation (a vacancy in a Si lattice) developed by Baraff and Schlüter.[59] Berholc et al.[60] then extended it further for the Hartree–Fock–Slater approximation (local exchange). Finally, Kaspar[61] in Erlangen has given a full *ab-initio* SCF-LCAO (Hartree–Fock with nonlocal exchange) formulation of Callaway's method,[58] which can be applied again also to surface bands.[61] The coding of this most sophisticated form of the resolvent method is in progress.

17.6.2. Negative-Factor Counting Techniques

The basic idea of this method was proposed by Dean[62] to interpret vibrational spectra in disordered solids. It can be easily applied also to a simple linear chain (one orbital per site) consisting of N units. This chain is described in the framework of a simple tight-binding scheme with first-neighbor interactions (in the absence of periodic boundary conditions) by the secular equation

$$|H(\lambda)| = \begin{vmatrix} (a_1 - \lambda) & b_2 & 0 & 0 & \cdots & 0 \\ b_2 & (a_2 - \lambda) & b_3 & 0 & & \\ 0 & b_3 & (a_3 - \lambda) & b_4 & & \\ \vdots & & & (a_{N-1} - \lambda) & b_N & \\ 0 & & \cdots & & b_N & (a_N - \lambda) \end{vmatrix} = 0 \tag{17.62}$$

Here a_i $(i = 1, 2, \ldots, N)$ and b_i $(i = 2, 3, \ldots, N)$ are the diagonal and off-diagonal matrix elements, respectively, of an effective one-electron Hamiltonian and λ is its eigenvalue. The above secular determinant can be factorized as

$$|H(\lambda)| = \prod_{i=1}^{N} (\lambda_i - \lambda) \tag{17.63}$$

Assuming that some other convenient factorization is found in the form

$$|H(\lambda)| = \prod_{i=1}^{N} \varepsilon_i(\lambda) \tag{17.64}$$

Dean's negative eigenvalue theorem[62] states that the number of eigenvalues less than a particular λ value is equal to the number of negative factors $\varepsilon_i(\lambda)$. If we transform (17.62) with the aid of successive Gaussian eliminations into an upper-triangular form, the $\varepsilon_i(\lambda)$s are given by the simple recurrence relation

$$\varepsilon_1(\lambda) = a_1 - \lambda$$

$$\varepsilon_i(\lambda) = a_i - \lambda - b_i^2/\varepsilon_{i-1}(\lambda) \qquad (i = 2, 3, \ldots, N) \quad (17.65)$$

The eigenvalue distribution of the polymer can thus be calculated simply by counting the numbers of the negative factors $\varepsilon_i(\lambda)$. By giving λ different values throughout the range of the energy spectrum of interest, and then taking the difference between the number of negative factors belonging to consecutive values of λ, the distribution of the eigenvalues of H (density of states) can be determined to any desired accuracy.

This method was first applied to a mixed glycine (A)-alanine (B) chain assuming different sequences and 1000 units[63] (at 50% alanine concentration the results for 10,000 units show little difference in the major characteristics of the spectrum of the chain with 1000 units). The a_i and b_i parameters were chosen to match the locations and widths of the valence bands of the corresponding periodic polyglycine and polyalanine chains[36]:

$$a_A = -9.67 \text{ eV} \quad b_{AA} = -0.12 \quad a_B = -9.40 \quad b_{BB} = -0.12$$

$$b_{AB} = \tfrac{1}{2}(b_{AA} + b_{BB})$$

The random two-component chains were generated employing a random-number generator but keeping the composition prespecified. The sequence of numbers $\{\varepsilon_i(\lambda)\}$ is then computed for $n = 200$ values of λ in the energy region between -9.920 eV and -8.925 eV (step width 0.005 eV) and the eigenvalue spectra are plotted by noting the numbers of negative ε_is. The computer time for the calculation of a spectrum of a chain of length 1000 units was about 6 s on a Cyber 172, including the random-chain generator (about 5 s without the Monte-Carlo routine).

Figure 17.4 gives the computed eigenvalue spectra of (a) a periodic glycine chain (—AAAAAA—), (b) a periodic alanine chain (—BBBBBB—), and (c) a periodic glycine-alanine chain (-ABABAB-), each chain consisting of 1000 units. The histograms in Figures 17.4a and 17.4b match the density-of-state curves of the valence bands of polyglycine and polyalanine of Stanley,[36] where the SCF-CO method for infinite periodic chains[1] was used. The bandwidth in both chains is about 0.5 eV and the shift in the position is 0.27 eV, i.e., about half the bandwith. This shift is enough to produce a splitting in the ordered —ABABAB— chain. Figure 17.4c shows that these are two narrow bands of about one-third of the original width that are separated by a gap of about 0.25 eV.

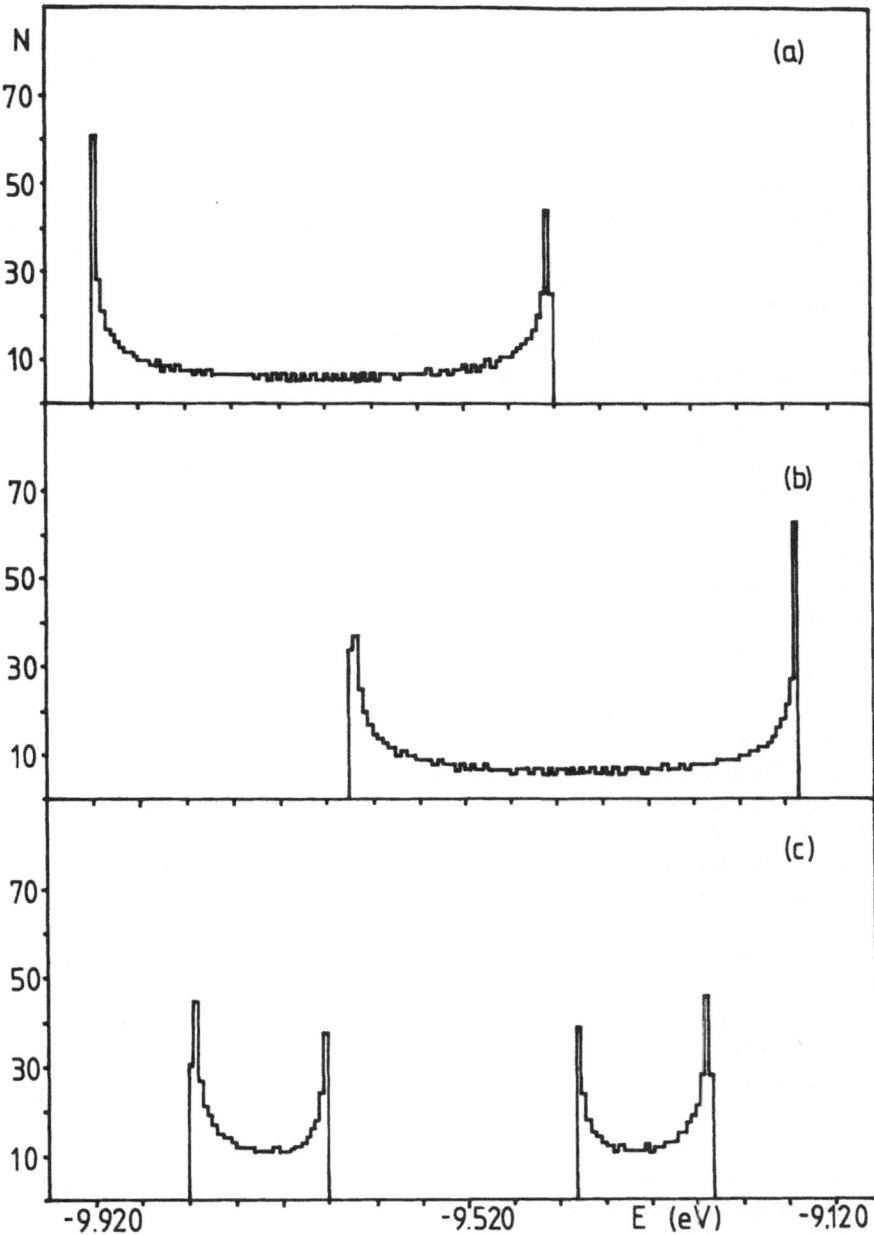

Figure 17.4. Eigenvalue spectra for a periodic chain (a) of 1000 glycine units —AAAAAA—, (b) of 1000 alanine units —BBBBBB—, and (c) for the ordered glycine–alanine chain —ABABAB— of the same length. The histogram interval is 0.005 eV and N is the number of eigenvalues in one interval.

Figures 17.5 and 17.6 show the eigenvalue spectra for randomly generated chains, again of length 1000 units. The percentage of alanine residues varies from 1% in Figure 17.5a to 50% in Figure 17.6b, as indicated on the diagrams. The histograms exhibit a complex structure, a complicated system of peaks, while valleys at the upper-energy end of the spectrum start to develop, the whole energy range spanned by periodic (A) and (B) becomes covered with increasing concentration, and the eigenvalue spectrum for the case of 50% alanine residues (Figure 17.6b) has no resemblance whatever to the ordered —ABABAB— chain shown in Figure 17.4c.

The explanation for the existence of well-defined peaks at the upper-energy end of the spectrum, some of which are denoted by letters A to D, is the same as in the case of vibration spectra[62] and is as follows. Consider, for example, peak A in Figure 17.5a. This peak is composed of eigenvalues of states that are highly localized at single alanine residues surrounded by glycine units. At each point in the chain where the local structure —AAABAAA— occurs, an eigenvalue is contributed to the peak A. This statement can be made without calculating the eigenvectors on the basis of the spectra depicted in Figure 17.7. Figure 17.7a shows the spectrum of a glycine chain of 1000 monomers that contains a single alanine unit at position 500. One eigenvalue appears in the histogram interval between -9.305 eV and -9.300 eV (interval, respectively peak, A). In Figure 17.7b, the spectrum of a glycine chain that contains nine alanine residues at positions $100, 200, \ldots, 900$ is given. In this case, nine eigenvalues lie in the histogram interval A. If the glycine chain contains 99 alanine residues at positions $10, 20, \ldots, 990$, then 99 eigenvalues appear in the interval A. We therefore see that the eigenstate of an alanine residue surrounded locally by nine glycine units on each side is very little affected by the sequence of the chain elsewhere; the eigenstates lie still in the energy interval of 0.005 eV as in the case of a single alanine residue in an otherwise monomolecular chain of glycine units. A similar situation holds for the other clusters containing alanine units surrounded locally by glycine units. The other peaks labeled in the spectra in Figures 17.5 and 17.6 are identified by similar calculations[63] and are associated with the following types of local sequence: peak B corresponds to the local sequence —AAABBAAA—, C to the local sequence —AAABBBAAA—, and D to the local sequence AAABABAAA—. This simple picture of associations between peaks of the spectrum and local chain sequence accounts fully for the changes that occur at the upper-energy end of the spectrum as the concentration of alanine residues is increased. Thus at low concentration of alanine units (Figure 17.5) the spectrum consists mainly of the well-known spectrum of the glycine valence band with a little structure in the alanine region dominated by the peak A due to isolated alanine

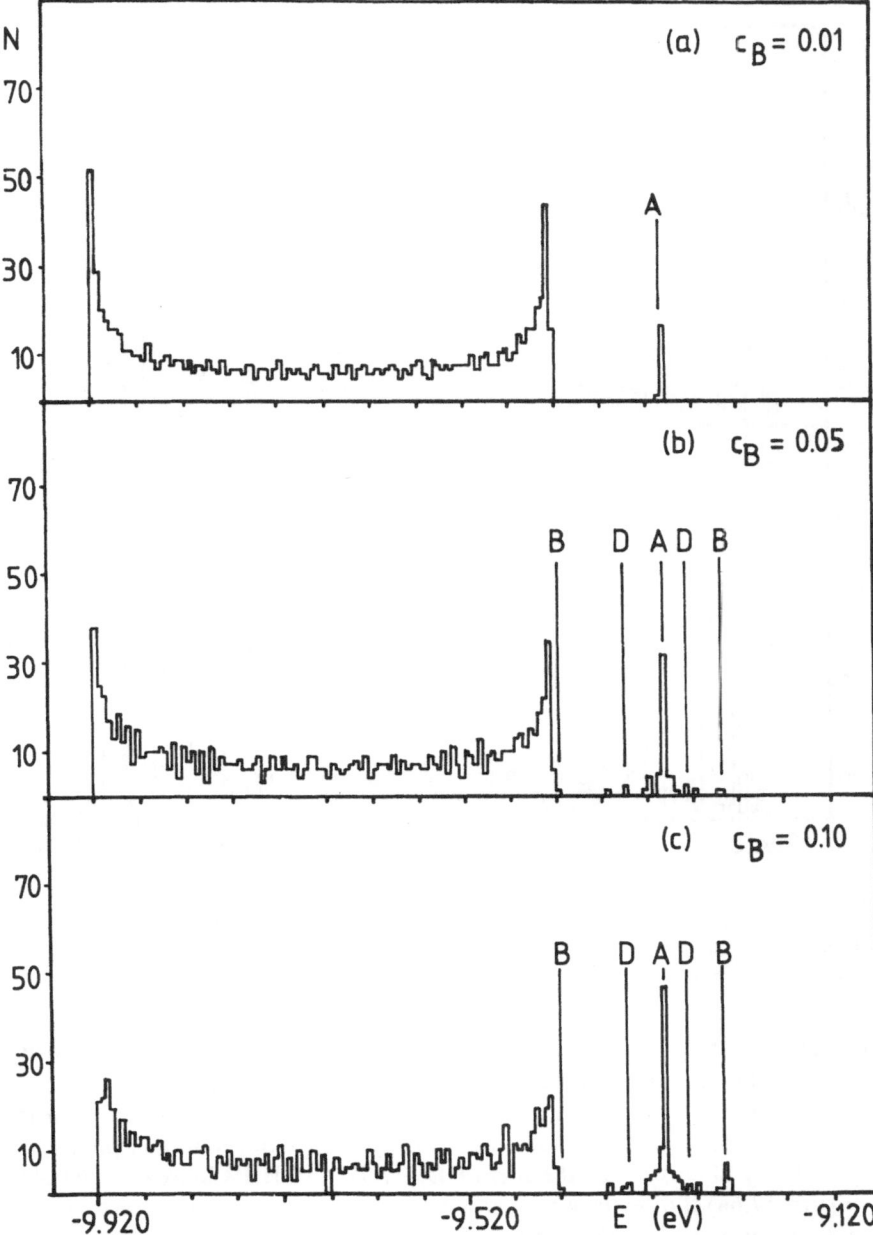

Figure 17.5. Eigenvalue spectra for two-component (glycine–alanine) disordered chains of length 1000 units. Parameter c_B refers to the fraction of alanine residues in the chain. The spectral lines A to D are associated with particular local-chain sequences and are explained in the text. N is the number of eigenvalues in a histogram interval of 0.005 eV.

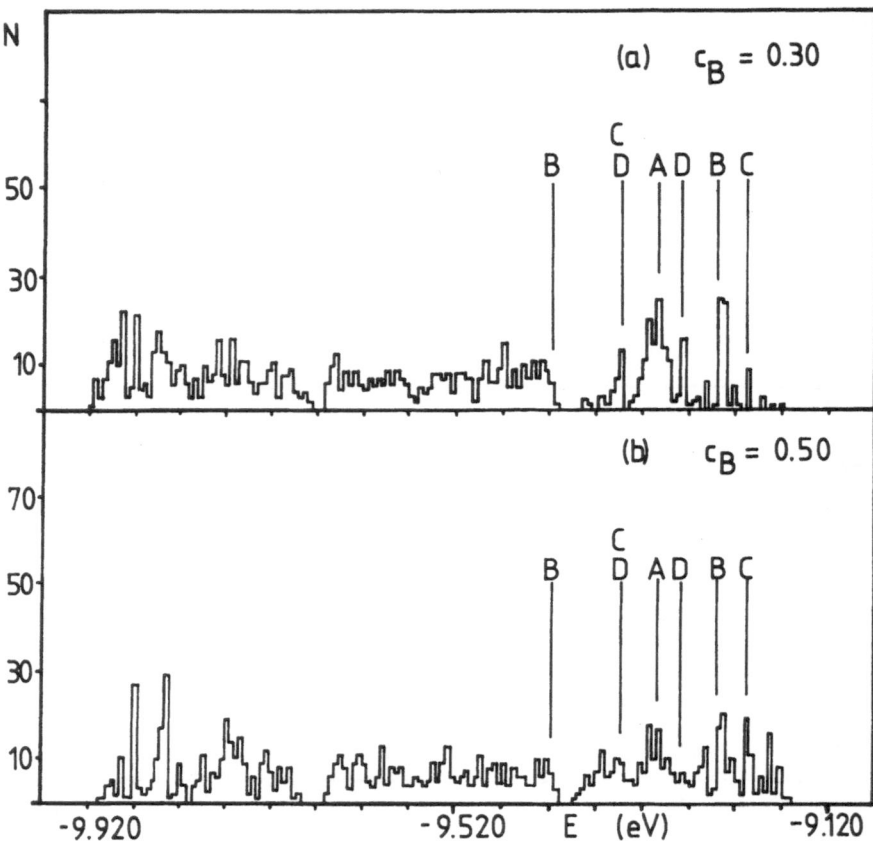

Figure 17.6. Eigenvalue spectra for two-component (glycine–alanine) disordered chains of length 1000 units. Parameter c_B refers to the fraction of alanine residues in the chain. The spectral lines A to D are explained in the text. N is the number of eigenvalues in a histogram interval of 0.005 eV.

residues. At 5% alanine concentration, the peak A increases in intensity. Also, peaks such as B and D due to clusters of BB and BAB appear. At 10% alanine concentration, peak A reaches its maximum intensity in the computed spectra and then declines at the expense of secondary peaks; the probability of clusters of the form BB, BAB, and BBB increases. At 30% (Figure 17.6a) the tertiary peaks due to three-alanine clusters are already quite pronounced, and at 50% (Figure 17.6b) they are as high as the primary and secondary peaks. At this concentration the spectrum becomes very complicated indeed, although it is clear that the identity of the individual peaks still holds.

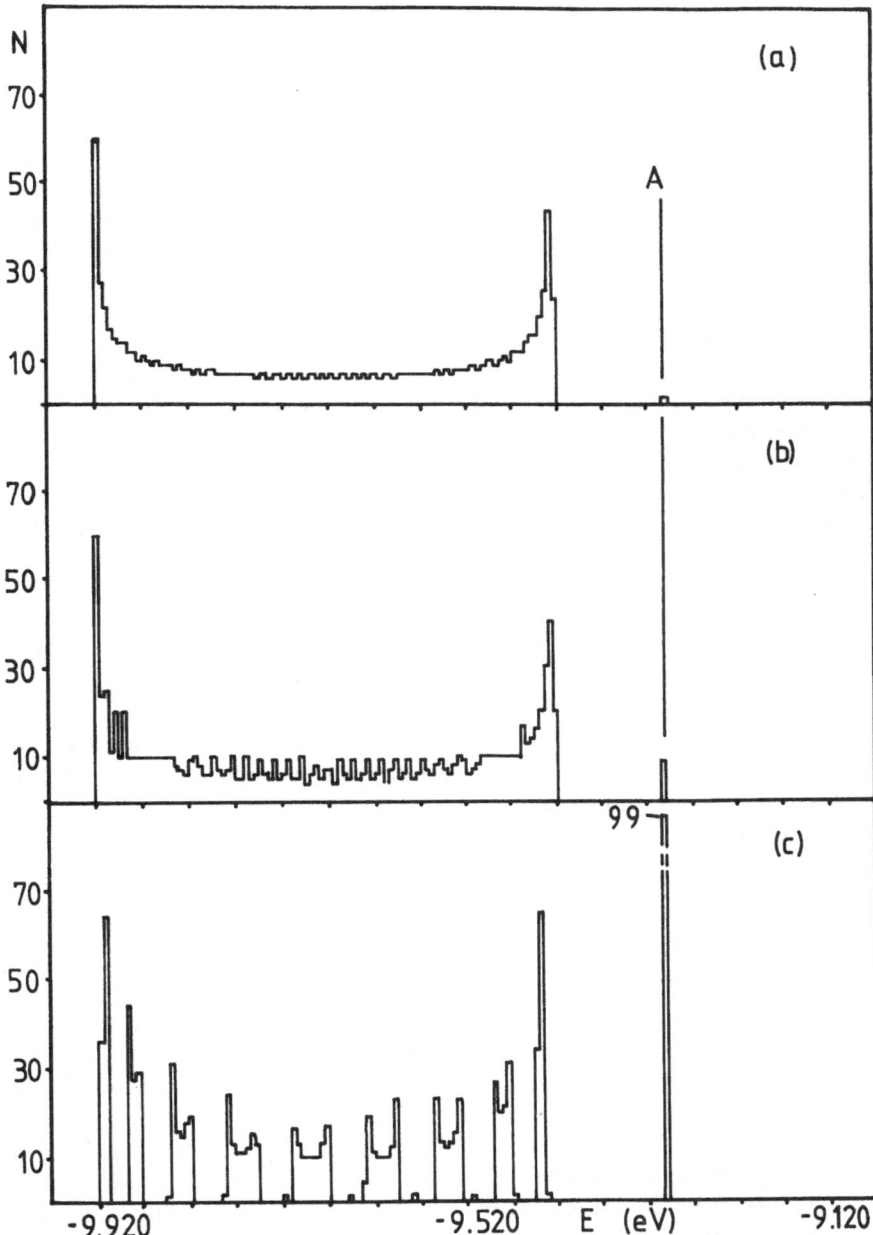

Figure 17.7. Eigenvalue spectra for glycine chain of 1000 units that contains (a) a single alanine residue at position 500, (b) nine alanine residues at positions $100, 200, \ldots, 900$, and (c) 99 alanine units at positions $10, 20, \ldots, 990$. The spectral line A is associated with the local chain sequence —AAABAAA—.

It is tempting to use this direct numerical approach for an *ab-initio* self-consistent treatment of aperiodic polymer chains. One has to replace the a_i and b_i parameters by the appropriate Fock matrices, which describe the interactions within the elementary cell and the first-neighbor interactions. These Fock matrices can be determined from appropriate SCF cluster calculations, as was done in the calculation of a polyacetylene chain with structural disorder.[64] Dean's negative-eigenvalue theorem holds also for tridiagonal block matrices; the recurrence relations are replaced by matrix relations.[62] Numerical methods of computing the distribution and exact location of eigenvalues are discussed by Dean.[62] The computation of eigenvectors with the help of the inverse iteration method[65] has also been outlined by Dean.[62] This calculation is not as straightforward as the determination of eigenvalues due to possible numerical instabilities.[62,65] Eigenvectors of chains of 512 atoms have been computed[62] by the method of inverse iteration.[65] This means that charge-bond-order matrices can be calculated and the *ab-initio* self-consistent treatment is possible, at least in principle. If the chain length is long and therefore very many eigenvectors had to be stored at each iteration step, one can perform a partially SCF calculation for the whole chain by changing only the eigenvectors belonging to the physically most interesting valence band at the iterations. The question of how long polymer chains can be treated in this way can be answered if numerical calculations are conducted and more experience in the numerical procedure, in their stability, and in computation times is obtained. Another possibility would be to calculate the eigenvalues and eigenvectors of chains of shorter length and average over the possible structural configurations that appear, for example, in a protein chain. This method reproduces almost exactly the correct spectrum for a very long chain in the region where the highly localized states are situated, but is unreliable in the region where the spectrum has a band character.[62]

17.7. Transport Properties of Biopolymers

Having derived electronic-band structures for biopolymers, one can use in the zeroth-order approximation either simple-harmonic-oscillator wave functions as phonon wave functions, or extract phonon wave functions from the experimental vibrational spectra, as has been done by Prohofsky and van Zandt for poly(A–T) and poly(G–C).[66] The electron and phonon wave functions can be used to directly calculate (as conducted previously using simple tight-binding band structures for different periodic DNA models and Hermite polynomials for the phonon wave functions[67]) the electron–phonon matrix elements. On substituting these into the approximate expression of time-dependent perturbation theory, one derives an expression for the probability per unit time of the scattering of an electron from state k' to k via absorption or emission of a phonon with momentum q (Suhai[67] has provided the details and mathematical formulation).

When the electronic bands of a system at the equilibrium nuclear positions are narrow (the bandwith is comparable with the $h\nu_q$ phonon energy), the electron–phonon scattering must be regarded as inelastic. In this case, a general relaxation time does not exist,[68] but one can still define approximate k-dependent relaxation times $\tau(k')$ with the aid of the transition probabilities. From the $\tau(k')$ values one can compute an approximate average mean-free path at temperature T by averaging over the states of a band using Boltzmann statistics:

$$\bar{\Lambda}^{T}_{\text{appr}} = \frac{\int v(k')\tau_{\text{appr}}(k')\exp[\varepsilon(k')/k_0 T]dk'}{\int \exp[\varepsilon(k')/k_0 T]dk'} \tag{17.66}$$

where $v(k') = (1/\hbar)\partial\varepsilon(k')/\partial k'$ and k_0 is the Boltzmann constant.

On applying the variational method of transport theory one can obtain an expression for the specific electric conductivity σ in terms of the transition probabilities for phonon absorption and emission, and of a variational trial function (by changing the appropriate expressions for metals to nondegenerate semiconductors[69]). Instead of performing a complete variational calculation, one can use the relaxation-time approximation for the solution of the Boltzmann transport equation also with the k-dependent approximate relaxation times $\tau(k)$. After calculating σ in this way, one can obtain the mobility values μ from the well-known expression

$$\mu = \sigma/ne, \qquad n = \sum_k f^0[\varepsilon(k)] \tag{17.67}$$

where f^0 is the Boltzmann distribution function.

Table 7 presents the characteristic quantities of charge-carrier transport in different periodic DNA models obtained with the aid of the method sketched above.[67] ($\delta\varepsilon$ stands for the widths of the conduction and valence bands, respectively.)

Although the mechanism of conduction in DNA is not yet entirely clear, there are many indications from the experimental side that the possibility of electron (and hole) transport cannot be excluded (stationary current is reached within a short time after application of the electric field,[70] photoconductivity is observed,[71] there are nonlinear effects characteristic of space-charge-limited conduction,[71] and so on). To investi-

Table 7. Transport Properties of Different Periodic DNA Models

Periodic DNA model		$\delta\varepsilon$ (eV)	$\tilde{\Lambda}_{appr}$ 300 °K (Å)	μ 180 °K	μ 240 °K (cm² V⁻¹ s⁻¹)	μ 300 °K
polyT	electron	0.072——12		189	57	30
	hole	0.274	42	297	132	82
polyA	electron	0.244	32	227	93	65
	hole	0.318	54	352	132	94
polyC	electron	0.142	15	183	77	54
	hole	0.310	41	270	116	87
poly(A–T)	electron	0.268	25	142	65	40
	hole	0.254	32	150	65	43
poly(G–C)	electron	0.142	18	146	71	42
	hole	0.268	27	254	105	77
poly(T,C)	hole	0.056	10	82	30	16
poly(T,A)	hole	0.030	6	78	18	12
poly(G,A)	hole	0.040	6.8	64	18	9.5
poly(A–T, G–C)	electron	0.046	8.5	45	16	7
	hole	0.053	8	63	18	10
poly(A–T, T–A)	hole	0.037	5.2	82	18	13
poly(G–C, C–G)	electron	0.027	5	24	8	4
	hole	0.020	3.5	28	9	4

gate the consistency of a band-type description of this assumed electron transport in DNA, we have to analyze our calculated results and compare them with experimental data.

The consistency of the delocalized description now requires that equivalently $\Lambda > a$ must be fulfilled. From the second column of Table 7, we see that the periodic DNA models with bandwidths between 0.1 and 0.3 eV [the nucleotide base stacks and the poly(A–T), poly(G–C) chains] have Λ values at 300 °K of 20–50 Å; hence the band model appears adequate for them. In the case of the other group with mean-free paths of the order of one or two lattice distances, the use of the delocalized description may be very questionable. So for them and for DNA models with even narrower bands, the mobility calculations must be performed in the localized representation too.

The last three columns of Table 7 give the mobility values calculated with the variational method at 180, 240, and 300 °K, respectively. There are two significant differences between the results obtained for the above-mentioned two groups. DNA models with relatively broader bands have mobilities between 30 and 100 cm² V⁻¹ s⁻¹, while in the case of the very narrow bands these values are smaller by one order of magnitude. All values follow the T^{-n} law, but for the former group $n = 1.5$–2, while for the latter $n = 2$–2.5. Mobilities much larger than 1 cm^2 V^{-1} s^{-1} indicate delocalized motion,[72] while in the case of values comparable to 1 cm^2 V^{-1} s^{-1} the band description may be problematic.

Unfortunately, no direct experimental data are available for the drift mobility in DNA. (It should be noted here that only high-frequency AC measurements can give information about the real charge-carrier dynamics because in DC measurements the current also flows through the junctions between the macromolecules and between the electrodes and the macromolecules, and this has a very strong disturbing effect.) There seems, however, to be some possibility of estimating the order of magnitude of the mobility from the AC conductivity measurements of O'Konski et al.,[70] who observed saturation of the conductivity at about 10^8 s^{-1}. If the conductivity is assumed to be electronic, this saturation must be the consequence of the fact that at such high frequencies the charge carriers have no time to accumulate at the ends of the macromolecular chains. Assuming the average number of base pairs in one chain to be 10^6 and using the relation $l = Ft$ (where l is the length of the individual chains, F is the field strength, and t is the period of the external field), mobility values of about 10^2 cm^2 V^{-1} s^{-1} are obtained.

This estimate is subject to uncertainty, because l is not known very well. (In DNA of higher organisms the number of base pairs is 10^8–10^9, but in the course of the preparation of the samples the macromolecules might have broken.) Therefore this value may be accepted only for orientation. To be able to compare more seriously the experimental and theoretical results, both the calculations and the experiments must be refined. On the theory side one should use ab-initio band structures, more realistic phonon wave functions, and it will be necessary to take into account also the effect of the intramolecular vibrations, uncorrelated hopping-type transitions, and the effect of various impurities (such as ions and surrounding water, which can very strongly influence the band structures[24] and therefore the transport process). The experiments (first in vitro) must be performed on strictly periodic DNA models (like the Ap–Tp or Gp–Cp double helixes), trying to determine the upper limits of the different impurity concentrations (especially ions). It will be necessary to orient such samples and study their transport properties over a wide frequency range (from $\nu = 0$ up to $\nu = 10^9$ s^{-1}) both in dark and under illumination (photoconductivity) as a function of T and the pressure. Only after such a refinement of both theory and experiment could one expect that comparison of theoretical and experimental numbers would have quantitative meaning.

In the case of proteins, since their structure is still more complicated and diverse one can expect such a development only later. However, both experimentalists and theoreticians still have to direct their efforts toward this final goal.

17.8. From Electronic Structures to Biological Functions

One can hope that, in the future, when all the necessary methods (including methods for the calculation of different physical properties of biopolymers, and methods not referred to here for the treatment of their different interactions[73] have been satisfactorily refined and corresponding corresponding experimental work has proceeded further, then one will be in a position to attack seriously different biological problems. For instance, it should be possible to understand the probability of point mutations in DNA (which is, besides the electronic structure of the nucleotide base pairs and their stacking interactions, also strongly dependent on the structure of water and on the presence of ions around DNA), the detailed mechanism of DNA duplication, the explanation of the catalytic activity with high selectivity of enzymes (proteins), etc., much better than today.

As a more detailed example, we shall discuss here some ideas about the mechanism of action of chemical carcinogens after their ultimate (metabolized) form has been bound to different parts of the DNA molecule.

17.8.1. Hypotheses for Tumor Development on the Molecular Level

There are several proposed theories of tumor development at the molecular level. According to the mutagenic model, the cancerous information can develop by chance through the accumulation of randomly distributed mutations in the DNA molecule of a cell. This cancerous information will then be transcribed to messenger RNA and translated to proteins. The difficulty with this hypothesis is that, unless the cancerous information in DNA is very unspecific, it has a rather low probability that it will be developed by chance through random mutations.

A large amount of experimental evidence points to the so-called tumor-virus theory. In this connection, it is well established that if the so-called oncogenic (tumor) viruses are taken out from a cancerous cell and injected into the same kind of normal cells, they very quickly cause cancer.[74] On the other hand, it seems rather improbable that most of these tumor viruses are of external origin. (A recent statistical analysis of the different causes of tumor occurrence has shown that probably less than 5% of tumor occurrences are due to external viruses.[75])

According to our present knowledge, the most realistic hypothesis for tumor development on the molecular level is the so-called "reading-error theory" first proposed by Bush.[76] According to this theory, the cancerous information is already contained in the DNA molecules of the normal cells, but it cannot be expressed because this dangerous part (or parts) of DNA is suppressed by suitable suppressor proteins. If a carcinogen binds to the suppressor protein, according to Bush this protein will be released and so the cancerous part of DNA will become free to transcribe its information to RNA, which will then be translated to proteins. Experimental findings have shown, however, that practically all chemical carcinogens bind also directly to DNA.[77] It was established, for instance, that the ultimate metabolite of 3,4-benzopyrene is bound to the amino group of guanine (see Figure 17.8).

The same compound and other carcinogenic hydrocarbons, besides binding to guanine, bind to adenine and to cytosine but not to thymine, showing that the ultimates of all of them most probably bind to the amino groups of these nucleotide bases. It was further established that the metabolites of carcinogenic hydrocarbons can bind also to the phosphate groups of DNA. It was further discovered that a completely different class of chemical carcinogens, the so-called alkylating agents, alkylate the N_7 atoms of the purine-type nucleotide bases. So we have to modify Bush's original hypothesis and assume that carcinogens bound directly to DNA can also cause the release of suppressor proteins.

Even if we accept the reading-error theory, there remain several open questions. First of all, it is unclear how specific is this so-called cancerous information in DNA. We know that in differential cells of higher organisms, the major part of the genetic information is suppressed by proteins (nucleohistones) while in an undifferentiated embryonic cell at the early

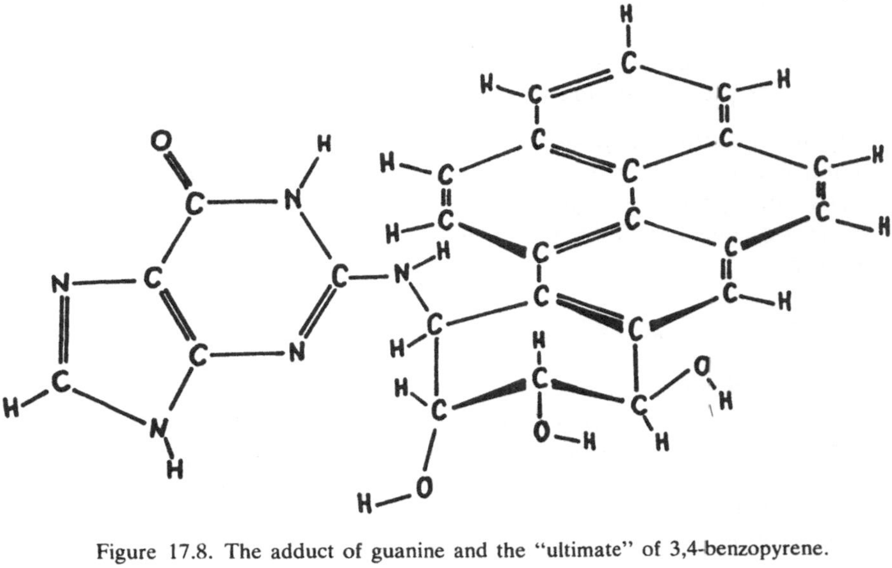

Figure 17.8. The adduct of guanine and the "ultimate" of 3,4-benzopyrene.

stage of the development, very probably the overwhelming majority of the genes are free. So the question arises whether there exists a specific cancerous information, or whether many of the genes, suppressed in a differentiated cell, can act as cancer genes. (The latter possibility is in close relation to Szent-Györgyi's point of view according to which cancerous cells and undifferentiated cells are very similar to each other.[78]). If we accept for a moment the existence of specific cancerous genes, located at certain parts of the DNA macromolecule, the question arises whether it is really necessary to assume that chemical carcinogens bind to these parts of DNA, or whether the effect of their binding to any part of DNA can be transmitted to those regions where the cancerous genes are situated.

The aforementioned analysis of the relative probabilities of different causes of cancer[75] has also shown that, most probably in over 90% of the cases, the cause of tumor development is of chemical origin. Therefore, one has to concentrate primarily on understanding the mechanism of action of chemical carcinogens. As mentioned above, carcinogens can bind directly to DNA. Therefore, if one accepts the most realistically looking "reading-error theory" of tumor development, one should try to answer the question as to how the binding of different chemical carcinogens to different types of DNA constitutents can cause the release of the suppressor proteins from those parts of DNA that contain the cancerous information. In approaching the solution of this problem, one has to consider basically two possibilities:

1. The carcinogens bound to certain parts of DNA can cause only the release of such suppressor proteins that are bound to the same DNA sections (local effects).
2. Carcinogens bound to DNA can also cause the release of suppressor proteins bound to other parts of DNA (nonlocal effects).

The understanding of both possible effects (especially the nonlocal effect) requires a fair knowledge of the electronic and vibrational structure of DNA and proteins, including their transport properties. Finally, the interaction between DNA and proteins in a nucleoprotein has to be studied in detail.

17.8.2. Possible Local Effects of Carcinogens

On suitably chosen model systems, like the ultimate of 3,4-benzopyrene or N_7-alkylated purine rings, first the change in the charge distribution of the nucleotide bases caused by the carcinogen binding can be investigated. Special emphasis should be put on the possibility of charge transfer between the carcinogen and DNA constituents (not necessarily that DNA constituent to which the carcinogen is chemically bound but more probably the neighboring ones; there is also a possibility for charge transfer between the carcinogen and, according to the primary structure, a far-lying constituent of DNA if the tertiary structure allows a bending back of this DNA part to the position of the carcinogen). For these investigations, *ab-initio* SCF-LCAO-MO supermolecule calculations could be performed. Finally, the major part of the correlation in the ground state of these molecules could be calculated (for instance, with the aid of suitable pair-correlation or cluster-expansion methods) to assess its effect on the above-mentioned phenomena.

One should carefully study how the aforementioned changes in the charge distribution can affect the Watson–Crick base pairing during DNA duplication. In other words, one should look into the possibility whether the charge-distribution changes due to the binding of the carcinogens could cause mispairing and, in this way, point mutations.

One should study, especially in the case of bulky carcinogens, the probable conformational changes that their binding can cause in DNA. The usual quantum-chemical methods could be used for these investigations. Besides carcinogens bound to nucleotide bases, the possibility of their binding to the phosphate groups of DNA should also be included in these studies. Such conformational studies can again throw light on the problem of the possible change of the genetic information in DNA due to conformational changes induced by the binding of carcinogens. Further, by performing a normal-mode analysis one could also investigate the change in vibrations of those DNA constituents to which the carcinogens are bound. One can think also of local changes in the tertiary structure of DNA (and/or proteins) at the spot where the carcinogen is bound. Finally, it should be mentioned that some chemical carcinogens have (as has been experimentally established) bond breaking or forming effects on DNA.

17.8.3. Possible Nonlocal Effects of Carcinogens Bound to DNA

One must also keep in mind the possibility already mentioned that a chemical carcinogen bound to a certain point of DNA can exert a long-range effect and in this way interfere with the DNA–protein interaction in another part of DNA. To investigate this possibility, one has to perform not only the *ab-initio* Hartree–Fock energy-band-structure calculations of periodic DNA models, but also investigations on correlation and aperiodicity effects with the help of the methods described above.

A more realistic description of the electronic structure of DNA can be obtained by making allowance also for the effects of bound water and ions. In addition, one should not forget that due to the double layer formed by the PO_4^- and K^+ ions, the electrons of DNA (with the exception in its symmetry axis) feel a strong inhomogeneous electric field, which very probably considerably influences their distribution and with it the band structure, as previous semiempirical band-structure calculations have shown.[79]

One has to perform the above-described investigations also on the protein macromolecules. Here, as compared to DNA, we have two additional complications. First, since the number of different amino acids in proteins is now not four (the number of different nucleotide bases) but 20, the aperiodicity problem becomes still more important than in the case of DNA. To solve this rather intricate problem, the negative-factor counting techniques referred to above seem to afford the most promising approach (at least until the problem is 1D). A further complication arises from the fact that, as already mentioned, a protein (either in the α-helix or in the β-pleated sheet) forms a two-dimensional periodic system due to the interactions along its chemically bound polypeptide chain and due to the hydrogen-bonded interactions between different polypeptide chains (or between different segments of the same chain). Previous semiempirical band-structure calculations have shown that to obtain a realistic band structure one must take into account both interactions simultaneously. One should point out in this respect that the *ab-initio* crystal-orbital method is valid also for two- and three-dimensional periodic systems, so the treatment of the two-dimensional protein problem would not cause bigger difficulties (the corresponding programs are under test).

In the investigations of the electronic structure of biopolymers, one should also look at the possibility of the existence of collective electronic states (such as Mott insulator states, Peierls instabilities, charge and spin density waves, plasmon-type states, excitonic insulators, and possible new types of collective states) using many-body techniques.[80] The effects of the nonlinear terms (both electronic and vibrational) in the many-body equations need to be included also in these investigations on biopolymers. It is quite possible that collective states and the above-mentioned nonlinear terms play an important role in energy and charge transfer in biopolymers, as was pointed out by Davydov[81] in the case of delocalized nonlinear vibrations (solitons). Therefore, they could be very important in long-range effects caused by the binding of a carcinogen.

These calculations must be supplemented by the determination of the phonon wave functions and, with their help, the calculation of the transport properties of DNA and protein (see above).

17.8.4. DNA–Protein Interaction and Its Possible Change Due to the Binding of a Carcinogen

Research on DNA–protein interaction in nucleoproteins plays a key role in understanding the blocking and deblocking of the genetic information. For such investigations, one requires primarily everything known about the geometrical structure of nucleoproteins (nucleohistone + DNA complexes) based on X-ray diffraction experiments.

In a theoretical treatment of the interaction between DNA and a protein, one should first investigate the possibility of charge transfer between these two polymers, as this determines whether the valence bands of these systems become partially filled (in the case of a charge transfer) or remain completely filled. In other words, one can show readily from perturbation theory that the polarization and dispersion interaction terms are larger between such polymers that have partially filled bands than between those with completely filled ones.[82] For these studies, as a first step one could use DNA base–amino acid or phophate–amino acid supermolecule calculations. As a further step one could perform DNA–protein superchain calculations to better establish the amount of transferred charge.

When the relative geometry and the amount of transferred charge between the DNA double helix and a nucleohistone chain are known, one can perform direct calculations for the interaction energy between them. For this, either perturbation theoretical schemes[83] or the newly developed mutually consistent field (MCF) method[73] could be used in a suitably adapted form. One should mention that in the DNA–protein interaction the surrounding water molecules, ions, and the charged units of both DNA and proteins (PO_4^- groups in DNA and positively charged arginine molecules in the nucleohistone) very probably play a crucial role and therefore must be included in the investigations.

After performing all the aforementioned interaction calculations between DNA and protein chains, one can look into the changes caused in these interactions by the binding of a chemical carcinogen to DNA. In other words, if the carcinogen can cause a charge transfer, this would change the position of the Fermi level of the macromolecular chains and hence change also the interaction energy between them. Further possible local conformational changes due to the carcinogen would certainly cause first of all local changes in the interaction between DNA and that protein directly bound to the region where the carcinogen attack has occurred. Further, one can also consider nonlocal effects of a local conformational change (primarily by a delocalized perturbation of the vibrations of the chain[81] or by causing changes in the tertiary structure). A further possibility for nonlocal interference with the DNA–protein interaction is offered by the probable formation of collective states, which arise from the solutions of the many-body equations (see above). Last, but not least, one can study the perturbation of the periodicity of the sugar–phosphate chain, which can influence its conduction properties and with it the interaction energy between this chain and a polypeptide chain. In the case of 7-N methyl guanine (which is carcinogenic), recent calculations have shown[81] that the methyl substitution hardly influences the charge distribution of guanine in the region of its H-bonds with cytosine. Since space-filling model investigations have shown that no steric hindrance is due to the additional methyl group, the most probable explanation seems to be the mentioned perturbation of the periodicity.

It is clear that all these proposed investigations require substantial effort. On the other hand, in our opinion this is the only way to gain a deeper insight at the submolecular level into the different possible effects and mechanisms caused by chemical carcinogens bound to DNA. One hopes to clear up whether a carcinogen can interfere with the DNA–protein interaction only locally, or whether nonlocal effects must also be taken into account. Since chemically, rather different carcinogens bound to chemically different parts of DNA may have the same effect by deblocking the previously blocked cancerous information in DNA (if the reading-error theory of tumor development is correct), one supects that the second case is more probable. If one really could determine on the basis of the proposed quantum-mechanical investigations how different carcinogens bound to DNA could cause the deblocking of the previously masked cancerous genetic information, one would be taking a very important step toward understanding the detailed mechanism of tumor development. This would also provide better chances of prevention, with the help of combined chemical and physical methods.

Acknowledgments

One of the authors (J. L.) would like to express his gratitude to the International Center for Theoretical Physics (Trieste) for inviting him to give a series of lectures in their Spring College on the Physics of Polymers, Liquid Crystals, and Low-Dimensional Solids. We are further very much indebted to Professors J. Čížek, T. C. Collins, G. Del Re, and F. Martino, and to Drs. P. Otto and R. Day for their continuous cooperation and for the many illuminating discussions. The financial support of the "Fond der Chemischen Industrie" is gratefully acknowledged.

Topological Defects and Disordered Systems

J. Vannimenus

The first and most obvious evolution in the study of defects in ordered systems is toward a much greater diversity, but it is counterbalanced by a powerful effort toward unification. The diversity stems from the experimental study of new materials, such as the liquid heliums, liquid crystals, or two-dimensional systems. These materials contain defects which play an important physical role, and are different from the well-known defects of metals and semiconductors. On the other hand, recent theoretical work based on topological concepts has restored unity by providing deeper understanding and a broader framework that includes many *a priori* unrelated cases (though not all the old classics!). The topological theory also leads to the prediction of new phenomena and makes contact with other fields of physics where similar concepts are used (e.g., solitons or monopoles may be viewed as defects). Here is a beautiful example of how progress occurs in physics!

It is tempting to push one's advantage and consider disordered systems just as the limit of ordered systems when the number of imperfections becomes large, with maybe some complications due to the interactions between defects. The defects are then taken into account by computing their contribution to some physical property (e.g., the conductivity) in a self-consistent way. Finally, the disordered system is replaced by an equivalent (or "effective") medium with the same conductivity, plus

J. Vannimenus · Groupe de Physique des Solides, Ecole Normale Supérieure, 24 rue Lhomond, 75231 Paris 05, France.

some measure of the fluctuations. This conventional approach is often successful and in many cases disorder just shifts and broadens the experimental curves.

Often, but not always! First, if the original defects have somewhat unusual properties, it is not trivial at all to describe a medium containing many of them, and it is worse if several kinds of defects coexist. Open problems subsist in that direction. What is more, much work now focuses on situations where strong disorder gives rise to original behavior, not reducible to that of ordered systems. The (rewarding) surprise is that some simple, reproducible effects have been put into evidence and they occur in a wide range of materials. A typical example is given by dilute systems (say, a mixture of conducting and insulation materials), where a well-defined "percolation threshold" exists (the material ceases to be conducting).

Any such universal behavior faces theory with a real challenge. New concepts are needed, a framework must be built and be broad enough to explain the facts ... and make predictions. A case of present interest is the class of materials known as spin glasses. Their definition is rather loose and it is too early to say which approach will best explain their common properties, and even whether a single approach will do. It is clear, though, that the concept of frustration introduced for that particular problem is of universal value, and is worth studying by itself. In particular, it is directly related to the notions of gauge symmetry and gauge fields that are now central in particle physics.

Each one of the topics touched upon could be the subject matter for a complete chapter (actually, most have been), so this chapter will stay at an introductory level, from a technical point of view. We begin with a short review of the well-known classical theory of dislocations in crystalline materials. This is to stress that the modern theory based on homotopy groups is a natural — and necessary — extension of the old theory, and that its basic concepts are generalizations of more conventional ideas. More complete material may be found in several recent reviews or lecture notes, and the relevant references are quoted with each section.

18.1. Classical Theory of Dislocations

18.1.1. Two Everyday Examples

Defects are ubiquitous in nature; they exist as soon as order itself is recognized. Just look at your fingertips: each of them bears a special pattern of lines, which may be characterized by smooth regions and a few singular points. Some of these points are accidental, due to a small scar for instance, but others are more fundamental. What is meant by fundamental is that you cannot eliminate them by distorting or reshaping a bit of skin. This displaces them, but they remain quite recognizable! They are a stable feature, and that is what makes fingerprints so useful for detective stories.

Another familiar example is provided by a phone cord. After some time it often develops a specific pattern: in some regions it spirals with one direction of rotation, in others the rotation is opposite. The intermediate, nearly straight regions may be viewed as defects. To eliminate them it is necessary to unfold the chord until another defect is encountered, or even to go all the way to the earphone.

In both cases, the stability of the defects is due to deep topological reasons: even if the deformation itself is localized in a small and well-defined area, its suppression involves a global operation. This notion of topological defect is a very useful one, which may be generalized to a considerable extent.

18.1.2. Translation Dislocation Lines

Dislocations are certainly the most studied among topological defects. They were already analyzed by Volterra in 1907 within the frame of elasticity theory, and their importance in explaining the deformation properties of metals was suggested as early as 1934, nearly 20 years before their direct observation.

To obtain a picture of the simplest dislocation in an isotropic, crystalline solid, imagine inserting an extra half-plane of atoms in the periodic array. Figure 18.1a shows a cross-section of the resulting atomic configuration; the very distorted region near the end of the extra half-plane is called the core of the dislocation (marked by an inverted T). The distortion gradually decreases with distance from the core and far away the residual strain is uniformly distributed, so there is no way of knowing

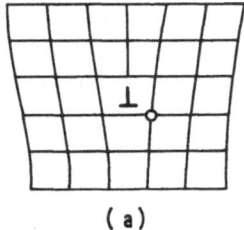
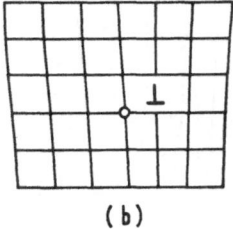

(a)   (b)

Figure 18.1. Schematic representation for the cross-section of a dislocation line in a cubic lattice. The core is marked by the inverted T. The displacement of an atom (∘) in the core region does not remove the dislocation, it just amounts to its moving in the opposite direction (from a to b).

where the extra atoms have been added. One cannot even tell whether there is a dislocation at all by looking at a small piece of crystal; the strain might as well be due to external forces.

Now let us try to remove the dislocation by rearranging atoms in the core region. Adding or removing atoms at best amounts to a displacement of the defect. If atoms are moved as in Figure 18.1b, the dislocation just moves in the opposite direction. Repeating the process, it will finally be driven out at the surface of the crystal, but this requires a huge number of moves and is very unlikely to happen spontaneously. Once created, the dislocation is therefore metastable and the density of dislocations in a crystal is not an equilibrium property but depends on the way the sample has been prepared.

On the other hand, when a stress is applied on the crystal, preexisting dislocations respond by moving in a definite direction. Each time a dislocation glides through, a part of the crystal is displaced by one atomic distance with respect to the rest. This is the basic mechanism for plastic deformation in many materials and costs much less energy than to displace one plane uniformly with respect to another.

18.1.3. Burgers Circuit and Burgers Vector

The existence of a dislocation cannot be recognized by inspecting the crystal locally at a distance from its core, but maybe it leaves some global mark, and there exists a relation similar to Gauss theorem for the charge in a closed region of space? This is indeed the case, as shown by a simple construction due to Burgers.

Consider two pieces of the same crystal, piece (a) contains a dislocation, while piece (b) is perfectly periodic (Figure 18.2). Far enough from the core, the atomic array of (a) is nearly perfect, and a one-to-one correspon-

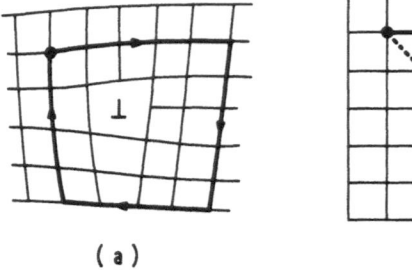

(a)   (b)

Figure 18.2. Closed Burgers circuit in the dislocated crystal (a) and its image in the perfect crystal (b). The image path fails to close by an amount independent of the circuit chosen. This misfit (the Burgers vector) is a topological invariant of the dislocation line.

dence between the (a) and (b) lattices may be established without ambiguity, at least locally. A path in (a) is thus mapped on to a unique path in (b), as long as it stays clear of the dislocation core. If this path is a closed circuit, however, the image path in (b) fails to close when the circuit surrounds the dislocation line. The misfit vector $\mathbf{B}$ joining the ends of the image path is called the Burgers vector and it possesses several important properties:

1. Since it joins two points of the perfect lattice, it may be written as

$$\mathbf{B} = n_1\mathbf{a}_1 + n_2\mathbf{a}_2 + n_3\mathbf{a}_3 \qquad (18.1)$$

   where n_1, n_2, and n_3 are integers and $\mathbf{a}_1$, $\mathbf{a}_2$, and $\mathbf{a}_3$ are primitive translation vectors of the lattice.
2. It is easy to see that the misfit is unaffected by a deformation of the initial loop. The sign of the vector depends on the orientation of the loop (Figure 18.2) and one has to introduce a conventional choice, much as in electromagnetism. Once this is done, the Burgers vector is independent of the circuit considered and is a *topological invariant* of the dislocation line.
3. If the Burgers circuit encloses several dislocations, the total Burgers vector is the sum of the individual Burgers vectors and each component is independently additive.

The existence of such an invariant has deep consequences:

1. A dislocation line cannot have an endpoint inside the crystal. If it had one, the loop could be distorted and shrunk down to an elementary square without ever crossing the core region, but then the Burgers vector must vanish and all three integers n_i must have been zero at the start. So the dislocation lines may only end at the surface of the crystal or be closed lines.
2. Any dislocation line has its "antidefect," namely the dislocation line with opposite Burgers vector (all three components with opposite sign).
3. A dislocation line with Burgers vector $\mathbf{B}_1$ may branch into lines with vectors $\mathbf{B}_2$ and $\mathbf{B}_3$. The same reasoning shows that there exists a conservation law

$$\mathbf{B}_1 = \mathbf{B}_2 + \mathbf{B}_3 \qquad (18.2)$$

   which is very similar to Kirchhoff's law for current conservation in electrical circuits.

What is remarkable is that these results do not depend on any detailed calculation, only some basic symmetry properties are involved. Clearly, many important properties of dislocations depend on the precise atomic configuration near the core that is very difficult to compute accurately, and entire books have been devoted to the various aspects of the problem. Nevertheless, the basic ideas are quite simple and powerful.

18.1.4. Disclinations

The dislocations considered so far were characterized as defects in the translation symmetry of the lattice. Topological defects also exist in the rotation symmetry of the lattice and are called rotation dislocations or disclinations, referred to in another context in Chapter 9.

To get a picture of such a defect, cut a disk out of a sheet of cross-ruled paper, remove a sector, and glue the two sides of the cut together to get a paper cone. In general the square lattices on both sides do not fit together, unless the removed sector is exactly one quadrant for instance Figure 18.3. In that case a defect has been created with all the desired properties — localized distortion, topological stability, and so on.

What is the analogue for this new structure of the Burgers vector? Here rotation is the important thing to keep track of. Again, choose a loop around the core of the defect (the Nabarro circuit), but now choose also a direction relative to the local orientation of the lattice and mark it by an arrow. Then proceed along the loop, keeping the arrow pointing in a fixed direction with respect to the *local* lattice (this is called "parallel transport" of a vector). On completion of the circuit the arrow makes a 90° angle with its original direction (Figure 18.3). This rotation angle is independent of the loop and is a topological invariant of the disclination. It plays the same role as the Burgers vector, but its nature and symmetry properties are different.

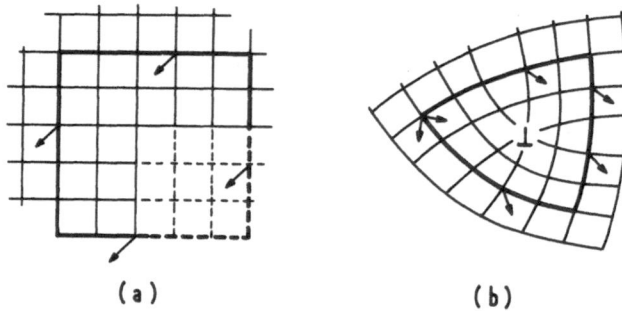

(a)   (b)

Figure 18.3. A disclination is created by removing a quadrant from a piece of perfect crystal and glueing the two sides of the cut together (a). An arrow is used to mark a fixed direction relative to the local orientation of the lattice. When one goes around a complete loop, the arrow comes back to its initial position in the perfect crystal (a). In the crystal containing a disclination, it makes a 90° angle with its initial direction (b).

As shown by the example, a disclination corresponds to a large deformation of a crystal. The minimum possible rotation angle in usual structures is 60°, and such large distortions would give rise to enormous stresses, so they do not occur in ordinary three-dimensional crystals. They are observed in two-dimensional structures and occur readily in liquid crystals; we will return to them within the general theory.

In a two-dimensional lattice, a pair of nearly opposite disclinations are (topologically) equivalent to a single dislocation (Figure 18.4). Such pairs are tightly bound at low temperatures, but their dissociation may be important in the melting process.[1,2] This is due to the unusual properties of two-dimensional "solids," which show no long-range translational order but only long-range orientational order (see Chapter 13). If the transition to an isotropic liquid is unique, it cannot be continuous and must be first-order, as in three-dimensional melting.

The alternative possibility is that *two* continuous transitions exist, as pointed out recently by Halperin and Nelson[1] and by Young.[2] The corresponding phase diagram is sketched in Figure 18.5. At the first

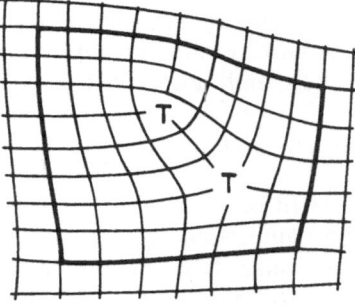

Figure 18.4. Two nearby disclinations of opposite sign ($+\pi/2$ and $-\pi/2$) on a square lattice are equivalent at large distance to a single dislocation.

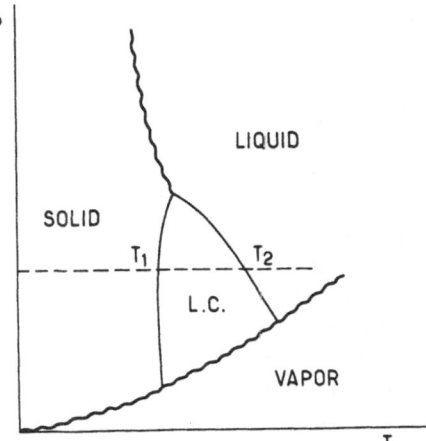

Figure 18.5. Schematic phase diagram for a two-dimensional system. The transition from solid (with long-range orientational order) to liquid may be first order (wavy line) or may involve two continuous transitions, with an intermediate liquid-crystal phase.

transition T_1 dislocation pairs start to unbind and the long-range orientational order is lost, but the orientational correlations decay very slowly above T_1 (with a power law rather than an exponential law). The second transition T_2 corresponds to the dissociation of disclination pairs, and above T_2 all correlations decay exponentially. The hypothetical intermediate phase is of liquid-crystal type and has been called the "hexatic phase" for the case of triangular lattices.

Numerical simulations have been performed on Lennard-Jones systems, but they seem to favor the first-order transition mechanism.[3] Other types of defects may play a role, e.g., for the electron lattice at the surface of liquid helium[4] (see also Chapter 15). In any case, the role of topological defects in phase transitions is an active research area.

18.2. Point Defects

This section is devoted to the study of two systems showing topologically stable point defects. In one case ("magnetic" systems) the conventional theory may be extended to describe these defects. In the other case (nematics) a direct extension leads to paradoxes, which reveal the need for a new and more general approach.

18.2.1. Defects in "Magnetic" Systems

In real magnets the magnetization points preferentially along certain crystallographic directions. This anisotropy leads to the existence of a particular type of topological defects that are not linear but two-dimensional, the Bloch *walls*. On one side of the wall the magnetization points along an easy axis, on the other side it points in the opposite direction. The way it rotates in between depends on the specific material, but the existence of the walls is essential for hysteresis properties. The walls may in turn have line defects, whose study is very active in connection with the technology of magnetic bubbles.

For our purposes, it is useful to consider idealized "magnetic" systems which are perfectly isotropic. They may be represented by a vector field $V(r)$, the direction and magnitude of which vary slowly, on a scale much larger than the lattice size. In the perfectly ordered state, the magnetization V_0 is uniform and points in a fixed direction. The system is no longer isotropic and only retains rotational symmetry around this direction — one says it shows spontaneously broken symmetry and V_0 is the order parameter (see Chapter 13 for other examples). In general, it costs much more free energy to change the magnitude of V than its direction, so $|V|$ tends to remain fixed everywhere, even when imperfections are present. The regions where this modulus vanishes are in particular as small as possible, and what prevents them from disappearing completely is the existence of topological constraints.

What are these topological defects? First, are there stable lines, analogous to the dislocation lines of crystals, on which the magnetization vanishes? A few drawings are necessary to be convinced that a singular line may be removed by continuous rotation of the magnetization in the core region, so it is *not* topologically stable. This phenomenon is called "escape in the third dimension." Indeed, the line defect is stable if the vectors are constrained to lie in a plane — this shows the importance of the dimensionality of the order parameter.

On the other hand, there exist stable configurations where the magnetization vanishes at one point only. Some configurations giving rise to such point defects are shown in Figure 18.6, where only the magnetization along these axes has been represented for clarity. A topological

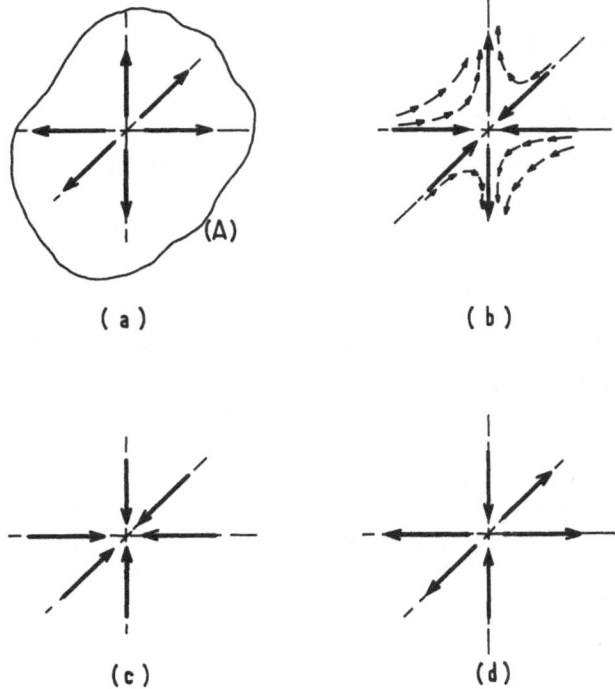

(a)   (b)

(c)   (d)

Figure 18.6. Point defects in isotropic magnetic systems. The large arrows indicate the magnetization direction along 3 orthogonal axes. Configurations (a) and (b) have index ($+1$) and are topologically equivalent, (c) and (d) have index (-1) and are equivalent but distinct from (a) and (b).

invariant of these defects is given by the "index" of the vector field, and its definition involves consideration of a closed surface (A) surrounding the defect, instead of the loop used for a line defect. For the reasons given above, the modulus of **V** may be considered as fixed on this surface, so **V** may be specified by a point on the unit sphere S_2 in three dimensions. A point of S_2 then corresponds to each point of A and this defines a natural mapping of A onto S_2 and, roughly speaking, the index is the number of times S_2 is algebraically covered by this mapping when A is completely described.

The precise definitions and the proof that the index is an integer, independent of the surface chosen and with simple additive properties, belong to homotopy theory, but drawings are sufficient to realize that cases (a) and (b) in Figure 18.6 may be transformed continuously into each other (they both have index $+1$), as well as cases (c) and (d), which have index -1. Configuration (a), on the contrary, cannot be transformed into (c) without creating another singularity. More generally, the index is sufficient to characterize the topology of the defects and predict their interactions. Here, the use of homotopy and its language is convenient; it justifies one's intuition and simple reasoning but does not lead to really new predictions.

18.2.2. Nematic Liquid Crystals

The physics of liquid crystals has been treated in detail in Part II and only a few basic properties will be recalled here. The nematics are organic materials made of elongated molecules, in which the centers of gravity of the molecules have no long-range order but their axes are preferentially aligned in one direction. There is no preferred orientation along that direction, so the order parameter is not a vector as in magnetic systems but a "director" — a vector without arrow — whose magnitude is related to the proportion of aligned molecules.

This (apparently) minor change in the nature of the order parameter has deep consequences on the topological defects and leads to a breakdown of the conventional theory!

The configurations of Figure 18.6 may be turned into director configurations by erasing the vector arrows and the four stable point defects so obtained are indistinguishable. More generally, the absolute value of the index I_V for a vector point defect is a topological invariant I_D for the corresponding director defect:

$$I_D = |I_V| \tag{18.3}$$

What are the composition rules for the director defects? By analogy with the vector case, it would seem that combining configurations (a) and

(b) in Figure 18.6 yields a defect of index 2, while combining (b) and (c) yields an index 0, i.e., no singularity. But this is paradoxical, since (a) and (c) are just the same defect for directors!

One may try different schemes in order to escape this paradox, but they all fail in some way or another. In fact it is simply not possible to predict the interaction of two point defects D_1 and D_2 from the knowledge of their individual configurations. The index of the resulting defect D_3 may have two values:

$$I(D_3) = I(D_1) + I(D_2) \quad \text{or} \quad |I(D_1) - I(D_2)| \qquad (18.4)$$

depending on how D_1 and D_2 are brought together.

The shortcomings of both intuition and the conventional approach are obvious in this case — and the need for more powerful tools becomes compelling.

18.3. Homotopy Theory of Defects

18.3.1. The Problems

Once it is realized that the convential theory of defects cannot be extended in a simple fashion, one may try to devise specific approaches for each case or even resort to detailed energy calculations. But the diversity of defects is so great that this rapidly becomes very complicated, as is obvious when looking back at the literature on liquid crystals in the early 1970s.[5]

The goal of a general theory of defects is then to restore some order among them and to classify "elementary" defects according to basic symmetry and dimensionality properties.

The preceding examples have suggested precise questions such a theory should answer:

1. What is the dimensionality of stable defects: points, lines, walls? Can they be fully characterized by some numbers?
2. Can different types of defects coexist?
3. Does every defect possess an antidefect?
4. What are the composition rules for defects? When and why do the simple additive rules fail?

To be more systematic, the problems may be grouped in three successive steps:

1 defect: classification of elementary objects (dimensionality, topological invariants).

2 defects: study of the interactions (coalescence, crossing).

N defects: description of a system containing many defects.

We shall see that the modern theory gives elegant and general answers to the first two classes of problems. In particular, it has led to the discovery of striking effects, such as how the existence of stable nonsingular distortions affects the possibility of topological obstructions to the crossing of line defects. The many-defect problem has been considered more recently and is connected with gauge-field theories, thus providing a new point of contact between condensed-matter physics and a very active area of particle physics.

18.3.2. Mathematical Tools

In the conventional theory, a central question is "How can this defect be created?" This approach leads one to consider Volterra processes or their generalizations as basic concepts. In the preceding section the relevant question turned out to be: "How can this defect be removed? What are the possible obstructions?" The branch of mathematics that deals with obstruction problems is homotopy theory. The connection between homotopy and defects in ordered systems was discovered by Toulouse and Kleman[6] and independently by Volovik and Mineyev.[7]

Some definitions are necessary before studying applications of their theory.

18.3.2.1. Manifold of Internal States

The fundamental concept is the "manifold of internal states," namely the ensemble M of points in the order-parameter space that correspond to all possible states of the system with the same energy.

For the "magnetic" systems considered previously the order parameter is a three-dimensional vector and the manifold of internal states is the ordinary sphere S_2:

$$M = S_2 \quad \text{(isotropic magnet)} \tag{18.5}$$

More generally, if the order parameter is a vector with n components then the manifold M is the sphere S_{n-1}, e.g., a circle if $n = 2$.

What is the manifold of internal states for nematics? The order parameter is a director, so two opposite points on the sphere represent the same internal state (Figure 18.7). Topologically, the sphere whose diametrically opposed points are identified is equivalent to the projective plane P_2 and

$$M = P_2 \quad \text{(nematic liquid crystals)} \tag{18.6}$$

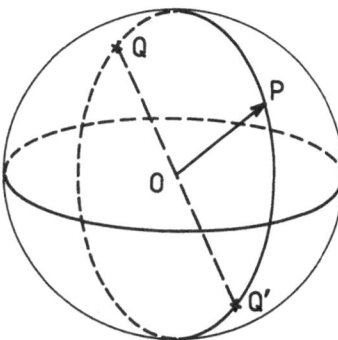

Figure 18.7. The state of a perfect isotropic magnet may be represented by a vector OP and the manifold of internal states is the sphere S_2. For nematic liquid crystals, two opposite vectors OQ and OQ' represent the same state and the manifold has the topology of the projective plane P_2.

A direct definition of the manifold M in terms of the symmetry elements of the system is also useful. Let G be the symmetry group of the disordered (high-temperature) phase and H the symmetry group of the ordered (low-temperature) phase. Now H is a closed subgroup of G, and M may be identified with the "coset space" (i.e., the set of left cosets gH):

$$M = G/H \qquad (18.7)$$

As an example, consider the isotropic magnet. The disordered phase has the full symmetry of the rotations in 3 dimensions, so $G = SO(3)$, but the ordered phase has only rotation symmetry around the direction of magnetization, so $H = SO(2)$, and we recover

$$M = SO(3)/SO(2) = S_2 \qquad (18.8)$$

18.3.2.2. The Fundamental Group

Consider a closed loop L in the ordered medium that avoids all singular points. The order parameter at every point of L is represented by a point in the manifold of internal states, so there is a natural mapping of L into M. This is a direct generalization of the idea introduced in Section 18.1.4 to study disclinations (see Figure 18.3).

Topologically, the loop L is equivalent to the circle S_1 and by choosing a bijection between L and S_1 one may define a mapping f of S_1 into M:

$$S_1 \xrightarrow{\ f\ } M \qquad (18.9)$$

Now, consider another map g such that the order parameter is fixed at a given base point R_0: $f(R_0) = g(R_0)$. The maps f and g are called *homotopic* if one can deform f into g continuously. All such maps may be grouped into homotopy classes $[f]$ and two maps are homotopic if, and only if, they belong to the same class.

There exists a natural composition law for loops with the same base point: the product loop is obtained by going around the first loop, starting from the base point, then around the second one. This directly translates into a composition law for the maps and for the homotopy classes. The fundamental mathematical result is that the set of homotopy classes of M, with this composition law, has a group structure. It is called the fundamental group of M, or the first homotopy group, and is denoted by $\pi_1(M, R_0)$. As a formal group it does not depend on R_0, so it is usually denoted by $\pi_1(M)$ for short.

18.3.3. Connection with Defects

If a map f is homotopic to the constant (or trivial) map f_0, it can be continued smoothly on any surface bounded by L. If it is not, the continuation cannot be made without introducing a singularity. This shows the connection between the first homotopy group and the existence of line singularities.

Stable line singularities exist if $\pi_1(M)$ is not reduced to the homotopy class of the constant map, denoted 0 for simplicity:

$$\pi_1(M) \neq 0 \Leftrightarrow \text{There exist stable line defects} \qquad (18.10)$$

Suppose that M is a circle S_1, a case encountered in spin systems with two components only, or in superfluid ^4He. The result from homotopy theory is that

$$\pi_1(S_1) = Z \qquad (18.11)$$

where Z is the additive group of signed integers.

To interpret this result, note that one can associate to every map f the number of turns N effected around M (in order-parameter space) when one turn is made around L (in real space); see Figure 18.8. This number (counted algebraically) is an integer that cannot be changed by continuous deformation, so it is the same for two homotopic maps. The converse is also true and in particular a map is homotopic to the constant map if N is zero. It is clear that maps may be combined and that the number of turns is algebraically additive — this is jut what equation (18.11) expresses.

The corresponding line defects are well known; they are usually called

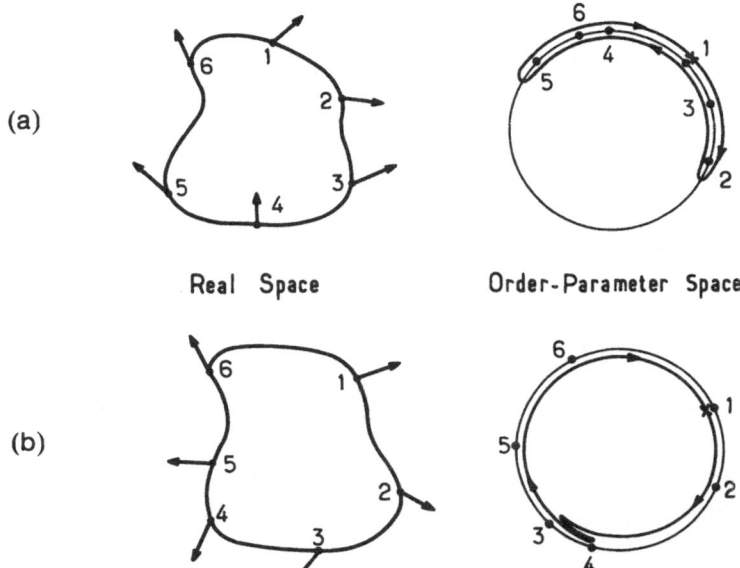

Real Space   Order-Parameter Space

Figure 18.8. In case (a), the path described by the representative point in the order-parameter space may be deformed continuously into a single point. In case (b) this is not possible and the loop in real space must surround a defect.

vortices and have recently been observed directly in liquid helium,[8] long after their theoretical prediction.

18.3.4. General Homotopy Groups

The extension of these notions to other types of defects is quite direct. In a d-dimensional space, a defect of dimensionality d' is to be surrounded by a sphere S_r, with r such that

$$d' + r + 1 = d \tag{18.12}$$

Two maps of S_r into M belong to the same homotopy class if they can be continuously deformed into each other. A natural composition law may again be defined, and the set of homotopy classes then acquires a group structure. It is called the rth homotopy group and is denoted $\pi_r(M)$.

Stable defects of dimensionality d' exist if the group π_r is not reduced to the class of the constant map. Specifically, in three-dimensional space, one has

$$\pi_0(M) = 0 \Leftrightarrow \text{no stable walls}$$

$$\pi_1(M) = 0 \Leftrightarrow \text{no stable lines} \qquad (18.13)$$

$$\pi_2(M) = 0 \Leftrightarrow \text{no stable points}$$

More precise mathematical statements may be found in the general references for the present chapter, especially in the work by V. Poenaru.

18.4. Applications of the Topological Theory

The previous analysis answers some of the questions formulated at the start:

1. The homotopy groups provide a classification of defects based only upon general properties of the system, as embodied in the manifold of internal states.
2. Their group structure contains the existence of antidefects and the composition rules for defects of the same type. In particular, a simple additive rule, as for dislocations, cannot hold if the relevant homotopy group is not isomorphic to Z.

Let us see how this works for specific physical systems.

18.4.1. Vector Order Parameter

When the order parameter is a vector with n components, the manifold M is the sphere S_{n-1}. Topology tells us that

$$\pi_r(S_m) = 0 \qquad (r < m)$$

$$\pi_m(S_m) = Z \qquad (18.14)$$

so there are stable defects if

$$r = n - 1$$

that is,

$$\boxed{d' = d - n} \qquad (18.15)$$

[using relation (18.12); remember that $d' = 0$ for points].

In three-dimensional space:

1. For $n = 1$, the stable defects are surfaces ($d' = 2$). This corresponds

to gas–liquid interfaces, for instance, or to Bloch walls in magnetic systems with one preferred direction.

2. For $n = 2$, stable defects are lines. This corresponds to vortices in superfluid $^4\mathrm{He}$ or in superconductors.

3. For $n = 3$, we recover the existence of stable points. Also, since a closed loop on a sphere S_2 may always be reduced to a point, this implies the absence of stable lines, as already noted.

The important result (18.15) is conveniently displayed in the same (n, d)-diagram as used in the theory of phase transitions (Figure 18.9).

18.4.2. Nematics

The manifold of internal states is the projective plane P_2 and its homotopy groups are

$$\pi_0(P_2) = 0$$
$$\pi_1(P_2) = Z_2 \qquad (18.16)$$
$$\pi_2(P_2) = Z$$

where Z_2 is the two-element group of the integers module 2.

So there are no stable walls, but stable point defects with an integer charge exist, as for the vector case. The lines are remarkable: there is only one kind of singular line which is its own antidefect!

This result is best interpreted by considering the manifold of internal states as the sphere S_2 with opposite points identified (Figure 18.10a). A path like PQP' is then closed but cannot be shrunk continuously to a point.

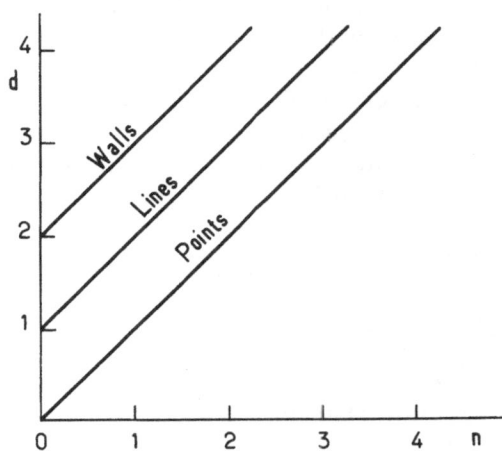

Figure 18.9. (n, d)-diagram of stable topological defects for systems with vector-type order parameter (d = dimension of space, n = number of order-parameter components).

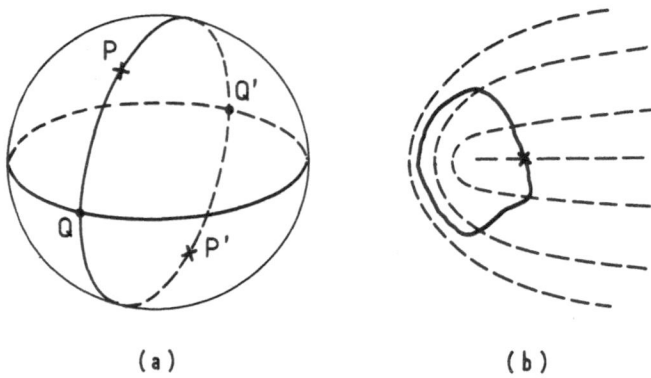

(a)   (b)

Figure 18.10. The stable disclination line for nematics. (a) In order-parameter space, PQP' is a closed loop since P and P' are equivalent, but it cannot be contracted to a point. (b) The corresponding director configuration in real space.

The corresponding singularity is the disclination line shown in Figure 18.10b. But if we combine two paths like PQP' and $P'Q'P$, the resulting closed loop *can* be contracted to a point, so the combination of two disclinations yields a nonsingular configuration.

One might have suspected that the coalescence of two disclinations would give a singular line with a 2π rotation of the director. Such lines are the counterparts of line singularities in the vector case, and they are unstable through "escape in the third dimension." This shows clearly the meaning of the topological result.

The topological analysis also reveals what mathematicians call the "nontrivial action of π_1 on π_2," i.e., the fact that the group element associated with a point defect is not always defined absolutely, and may depend upon the presence of line defects.[7] Here, for nematics, the action of π_1 on π_2 is nontrivial; it is a change of sign: only the absolute value of the integer associated with a point defect is well-defined. As a consequence, a disclination line can "catalyze" the removal of point defects!

This action of lines on point defects explains the apparent paradox pointed out in Section 18.2. It is gratifying that the framework of homotopy groups allows a complete study of this effect and provides general theorems for its existence (see, e.g., the lectures by Poenaru).

18.4.3. New Problems

As with any new theory, the topological theory of defects raises new questions. Here two such problems are mentioned that are particularly fascinating.

18.4.3.1. The Crossing of Line Defects

A general theorem on homotopy groups states that

$$\pi_n(v) \text{ is commutative} \qquad (n \geq 2) \qquad (18.17)$$

but the fundamental group $\pi_1(v)$ may be noncommutative.

Poenaru and Toulouse[9] have pointed out that the noncommutativity of π_1 has deep physical consequences, namely that line defects cannot cross as usual!

If one tries to pull a defect line across another, a singular kink is created and can be removed afterward if π_1 is commutative. If it is not, there remains a filament between the lines (Figure 18.11) whose energy grows linearly with the separation. This is a kind of "confinement" mechanism, which drastically modifies the mobility of the defects and should affect the macroscopic properties of the medium.

Candidates to observe this effect are materials where the order parameter is an ellipsoid with three different axes, e.g., the "biaxial" nematics, made of ellipsoidal molecules rather than the rod-shaped molecules of usual (uniaxial) nematics. In this case

$$\pi_0(v) = \pi_2(v) = 0$$
$$\text{(biaxial nematics)} \qquad (18.18)$$
$$\pi_1(v) = Q$$

where Q is the noncommutative 8-element group of quaternions. The experimental test of these startling predictions should become possible in

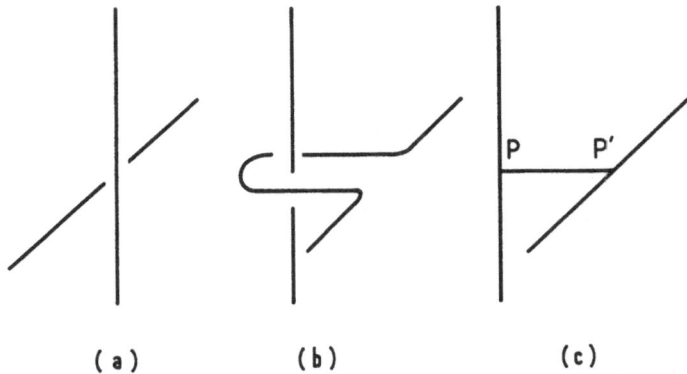

(a)   (b)   (c)

Figure 18.11. Crossing of two line defects: if the first homotopy group $\pi_1(M)$ is noncommutative, the filament PP' cannot be removed and the lines remain entangled.

the near future, thanks to the very recent discovery[10] of two classes of compounds for which the order parameter is of the biaxial-nematic type.

18.4.3.2. Textures

The word "texture" is used here to denote a nonsingular configuration that is topologically stable. Such configurations exist whenever the third homotopy group $\pi_3(v)$ is not trivial (for $d = 3$). The simplest example is

$$\pi_3(S_2) = Z \tag{18.19}$$

Consider a vector field $\mathbf{V}(\mathbf{r})$, which is nonzero everywhere and is equal to a constant outside a fixed sphere. Such a field defines a map of S_3 into S_2 [the unit vector pointing along $\mathbf{V}(\mathbf{r})$]. If this map cannot be deformed into the constant map without creating a singularity, then the vector field is a texture classified by an integer, the Hopf index. Figure 18.12 shows an example of a nontrivial Hopf texture.

One may wonder how such "objects" can be observed: they are both localized and nonsingular, which does not help! In fact, they appear to exist in cholesteric liquid crystals but the interpretation of the observations remained rather mysterious until the application of topological ideas.[11] Their existence and role in magnetic systems is not established, and there is some debate at the present time.[12-14]

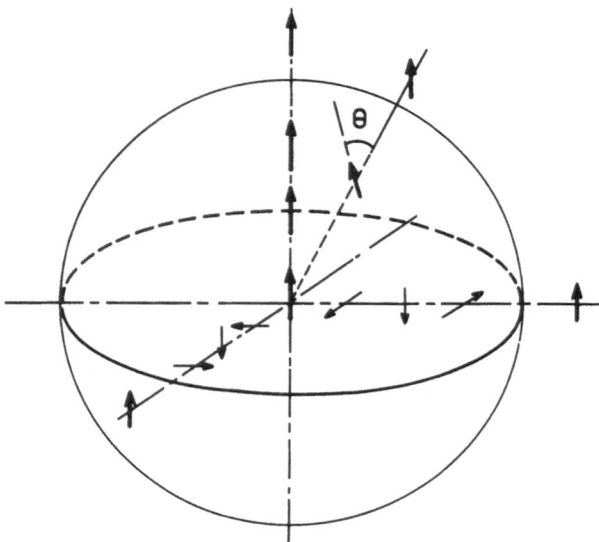

Figure 18.12. A "Hopf texture" for a vector field. Along any radius the vector rotates around the radius through an angle $\theta = 2\pi r/R$. At the center the vector is equal to its (constant) value outside the sphere.

18.4.4. The Many-Defect Problem

The final step in the program outlined above is the description of a system containing a macroscopic number of defects. It is natural to expect new effects that might consist of the existence of additional long-lived modes, with respect to the pure system.[15]

Two origins are usually recognized for the existence of such hydrodynamic modes: conservation laws (for sound waves, thermal diffusion), and continuous broken symmetries (for Goldstone modes, such as spin waves in ferromagnets). Topological stability of the defects provides another mechanism that prevents the rapid relaxation of a local distortion, and it may be relevant in some cases to a hydrodynamic theory.

To see how this works, a convenient model system is provided by superfluid ^{4}He.[16,17] Here the order parameter is a complex number

$$\psi = \rho e^{i\phi}$$

which corresponds to the case $n = 2$ in the vector classification, and the defects are vortex lines. In the absence of vortices, the phase ϕ is defined uniquely everywhere and the superfluid velocity v_s is given by

$$v_s = \nabla \phi \implies \text{curl } \mathbf{v}_s = 0 \tag{18.20}$$

In the presence of many vortices, the phase becomes multivalued and its circulation along a closed contour no longer vanishes; it is related to the number of enclosed vortices. The detailed microscopic behavior is very complicated, and when an average is taken over a region containing many vortices the phase loses physical meaning. The relevant object becomes the superfluid velocity $\mathbf{v}_s$, i.e., there are now three independent hydrodynamic variables instead of one. One may define a (vector) density of vortices ρ, where

$$\boldsymbol{\rho} = \text{curl } \mathbf{v}_s \tag{18.21}$$

This vorticity obeys the conservation law

$$\int \rho \, dS = \oint \mathbf{v}_s \cdot d\mathbf{l} = N \tag{18.22}$$

where N is the (algebraic) number of vortices enclosed by the contour. The current of vortices is given by

$$\mathbf{j} = \partial \mathbf{v}_s / \partial t \tag{18.23}$$

and the following relations hold:

$$\operatorname{div} \rho = 0$$

$$\operatorname{curl} j + \partial \rho / \partial t = 0 \tag{18.24}$$

In this form the analogy with the Maxwell equations is clear, with ρ playing the role of "magnetic" field, j of "electric" field, and v_s of vector potential. The first equation expresses the fact that vortices cannot terminate inside the superfluid, the second one corresponds to the conservation law for vortices. The superfluid velocity v_s is a gauge field that contains the relevant macroscopic information about the vortex network.

For a complete description, these equations have to be supplemented by a constitutive equation for the vortex current that depends on the detailed properties of the system. This equation may in general be linearized for small **j**, leading to the prediction of new slow modes for the vorticity.

The topological approach thus provides a compact derivation for the usual hydrodynamic equations of superfluid ^{4}He. Dzyaloshinskii and Volovik[15] have shown that elasticity theory may also be derived in the same spirit. When disclinations are included in the theory, it becomes isomorphic to the well-known Yang–Mills gauge-field theory — a most elegant synthesis!

The theory outlined for superfluid helium has two attractive features: the conservation law for the topological invariants of vortices is automatically satisfied, and when the vorticity vanishes the pure case is recovered in a natural fashion. The extension of this approach to other systems encounters some difficulties.[17] For instance, if the homotopy group associated with the defects is Z_2, two identical defects can annihilate together and there are no macroscopic conservation laws for them. A hydrodynamic description might still be useful, but it loses its topological foundation.

A clear and detailed discussion of these problems is given in the lectures by Dzyaloshinskii quoted in the general references. Here, theory is clearly ahead of experiment but the mathematical structure that has been brought into play (differential forms and exterior calculus) is so powerful and elegant that we have faith in its future.

18.4.5. Limitations of the Method

The beauty and successes of homotopy methods should not lead one to believe that they provide the solution to all problems in the theory of defects.

A first limitation comes from the absence of quantitative energy considerations. Some defects that are topologically stable may cost too much energy to exist (this is the case for line disclinations in most solids), or may collapse because there is no energy barrier associated to the topological barrier (this happens for textures in isotropic three-dimensional magnets). Also, the stability of defects with high topological numbers with respect to dissociation depends on the detailed energetics of the system.

Other limitations appear when studying systems with broken translation symmetry, such as crystals or smectic liquid crystals. An obvious case is the vacancy in a crystal: this point defect is stable, yet it does not enter the topological classification because it leaves no "signature" at a distance. A more serious problem is that the description of such systems with a *local* order parameter is not really adequate. A map in the order-parameter space may then be unphysical: the corresponding configuration cannot be realized in the real material. This has two consequences: first, some homotopy classes may not be realized at all; second, two homotopic configurations may correspond to physically distinct defects because an intermediate distortion is not realizable.

Attempts at incorporating metric considerations in the theory have been made, but they lose the initial simplicity and involve a detailed analysis of particular cases (see, e.g., the review article by Mermin). New mathematical tools may be needed here.[18]

18.5. Disordered Systems: Dilution and Competition

The realm of disordered systems is quite large, from amorphous silicon to metallic glasses, from window glass to polymer gels ..., and there is no general approach to disordered materials, comparable to the band theory of crystalline solids. Yet, for some important and general problems, the essential effects of disorder may be reduced to a well-defined formulation — a nice Hamiltonian — and the methods of modern statistical mechanics may then be applied. The focus here will be on a few such problems, among the most studied recently.

18.5.1. Dilution-Type Disorder

The paradigm of these "good" cases has become the kind of disorder encountered in alloys of a magnetic substance with a nonmagnetic one, in mixtures of a metal and an insulator, in porous systems: there the "active" parts are diluted within the "inert" ones and one is interested in a property like the net magnetization, or the resistance of the material. As emphasized in the general introduction, a deep change in point of view has occurred in the last ten years. The early work tried to account for the disorder by introducing an average susceptibility or conductivity in the "best" possible way; interest has turned now to the specific features of the disordered systems, which cannot be studied by some kind of perturbation theory and need new concepts.

The problems must then be idealized and simplified to bare-bone models that can be studied exhaustively. The most popular model for dilution disorder is the pure *percolation* model, introduced by Broadbent and Hammersley in 1957. Consider a cubic lattice where atoms are placed at random, the occupation probability of a site being p. Two atoms are connected when they are nearest neighbors. When p is small enough the atoms form isolated clusters, but for p near 1, most atoms belong to a macroscopic cluster. There exists a critical value p_c, the percolation threshold, where the macroscopic cluster first appears. At the percolation threshold there occurs a *geometric* transition, very analogous to usual thermal phase transitions, where the fraction $P(p)$ of sites belonging to the infinite cluster plays the role of order parameter. The threshold p_c depends on details of the problem (the structure of the lattice, the connectivity rules), as does a critical temperature, but the qualitative behavior of $P(p)$ near the threshold is universal: it may be described by a number, which is the counterpart of a critical exponent and depends only on the space dimensionality. The connection with thermal transitions has in fact been made quantitative and this exponent is known with good accuracy.

Percolation ideas and the wealth of results obtained on this simple model prove very useful in more realistic situations. For instance, a ferromagnet containing a fraction $(1 - p)$ of nonmagnetic impurities may be described by the Hamiltonian

$$\mathscr{H} = - \sum_{(i,j)} J_{ij} S_i S_j \tag{18.25}$$

Here the spins S_i are located on the periodic lattice and the exchange interactions J_{ij} between nearest neighbors are of the form

$$J_{ij} = J q_i q_j \tag{18.26}$$

where $q_i = 1$ if site i is occupied by a magnetic atom, and $q_i = 0$ if it is occupied by a nonmagnetic atom. If there are no correlations between the positions of nonmagnetic atoms then, at zero temperature, the calculation of the total magnetization reduces exactly to the percolation problem and long-range magnetic order can exist only for p above the threshold p_c. At nonzero temperatures, there exists a critical line in the plane (T, p) joining the transition point $(T_c, 1)$ of the pure system to the percolation transition

$(0, p_c)$. This line has been investigated in detail, both by numerical experiments and by renormalization group methods (see the lectures by Kirkpatrick and the review by Korenblit and Shender).

The model defined by interactions of the form (18.26) is called the *site* problem, because the randomness bears on the site occupancy. Another simple choice is often considered, where the bonds are the random variables. This *bond* problem models materials with interactions due to indirect exchange and of very different strengths; it is defined by

$$J_{ij} = \begin{cases} J & \text{with probability } p \\ 0 & \text{with probability } 1 - p \end{cases} \qquad (18.27)$$

Section 18.6 is devoted to a more detailed presentation of percolation theory and to some related questions.

18.5.2. Competition-Induced Disorder

In human interactions, both hate and indifference are opposed to love, and generations of poets have devoted much talent to describe their interplay. The corresponding effort in physics is surprisingly recent, and the effects of weakened interactions have received much more attention than the effects of contradictory ones. It is striking that some very general results on Ising systems are proved only for interactions that may be random but must be noncompeting, such as the Yang–Lee theorem on the zeroes of the partition function,[19] or the existence of a boundary free energy.[20]

The experimental discovery in 1975 of sharp and unexpected features in the susceptibility of a class of disordered magnetic alloys, the spin glasses (e.g., CuMn), really marked the starting point of a new era in the study of disorder. In these materials the interactions between the magnetic atoms are rapidly oscillating with the distance and have fairly long range, so their distribution is quite large and covers positive as well as negative values, in contrast with the situation in dilute ferromagnets. Here again the real system is very complex, the position disorder leading to correlations between the interactions, and one must find adequate models to understand what is going on.

A natural attitude is to follow closely the framework set for dilution disorder, with due allowance for the possibility of negative interactions. In this spirit, the magnetic atoms are placed on a cubic lattice and interact via the Hamiltonian

$$\mathcal{H} = - \sum_{(i,j)} J_{ij} S_i S_j$$

and the analogue of site percolation corresponds to the choice

$$J_{ij} = J\varepsilon_i\varepsilon_j \tag{18.28}$$

where

$$\varepsilon_i = \begin{cases} +1 & \text{with probability } x \\ -1 & \text{with probability } 1-x \end{cases}$$

This is called the Mattis model[21] The site disorder may be eliminated altogether by the simple change of variables $\sigma_i = \varepsilon_i S_i$, which reduces the problem to the pure ferromagnetic case, at least for thermal properties in zero external field. The predicted properties are not in agreement with experiment and a better model is needed. Toulouse[22] pointed out that the essential new feature in spin glasses comes from the impossibility to satisfy all the interactions simultaneously, and that this competition effect is eliminated in (18.28). So it is more relevant to study the bond problem defined by

$$J_{ij} = \begin{cases} J & \text{with probability } x \\ -J & \text{with probability } 1-x \end{cases} \tag{18.29}$$

This is now known as the *frustration* model, and numerical simulations have shown that it contains much of the physics of real spin glasses. Of course, in materials like CuMn the interactions have a broad distribution, and both dilution and competition effects are present. Other models with a continuous distribution of J_{ij} may be nearer to the experiments on these systems, but it is useful to be able to disentangle the different effects. In addition, there exist other classes of spin glasses that are better described by a distribution of the form (18.29).

What do we expect? The essential difference with percolation comes from the existence of a new phase, as shown schematically in Figure 18.13. The nature of this spin-glass phase is not completely clear at present, and its very existence in the three-dimensional case is still debated, though it is certainly present in high enough dimensionalities. Clearly, a novel type of thermodynamic phase needs new concepts and the disorder cannot be considered as a perturbation in that case!

One thing we can do, by analogy with the pure percolation problem, is to study the system along the $T = 0$ axis and look at what happens near the frustration threshold x_c, where ferromagnetism disappears. The problem is unfortunately not reducible to a purely geometrical one, and x_c will *a priori* depend on the type of spins considered. Even for the simplest case of Ising spins, the problem proves much harder than percolation and our knowl-

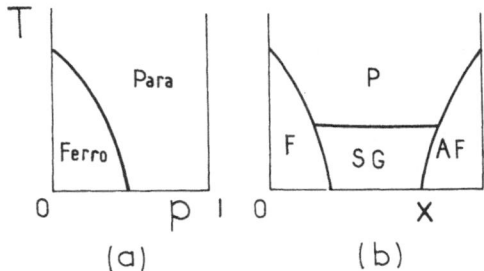

Figure 18.13. Qualitative phase diagrams: (a) for the dilute spin system as a function of the concentration p of missing bonds, (b) for the frustrated system as a function of the concentration x of negative bonds.

edge on the frustration threshold is not very firm, in spite of much recent work. To quote the game inventor Piet Hein, "Problems worthy of attack prove their worth by hitting back"! This should give us motivation for further efforts, and some of the proposed approaches will be discussed in Section 18.7.

The frontal attack on the frustration model is a steep path, because the interactions involve both competition and disorder. The study of periodic frustrated systems provides an intermediate step and attracts growing attention, in particular the "fully frustrated" lattices that generalize the triangular lattice with antiferromagnetic nearest-neighbor bonds[23-27] (e.g., the fcc lattice of CuAu alloys). They share some important qualitative properties with spin glasses, including a large ground-state degeneracy that opposes ordering and the failure of naive mean-field theory.[25]

Strong competition effects also seem to play an important part in the complicated phase diagram of materials like cerium antimonide (CeSb), which are somewhat intermediate between ordered and disordered systems. A popular model used to explain their unusual properties is the anisotropic next-nearest-neighbor Ising (ANNNI) model, where the nearest-neighbor ferromagnetic interactions are in competition along one direction with strong NNN antiferromagnetic interactions. The predicted phase diagram is surprisingly rich, with an infinity of commensurate and incommensurate phases, and it may be a signal that more surprises are to come.[28-31]

18.6. Percolation and Related Problems

Several reviews on percolation theory written by leading scientists in the field have appeared recently, and taken together they provide a good survey of this extremely active area (see the general references). We shall not try to compete with them, but percolation ideas are more and more useful in other contexts, from polymer gelation[32,33] to microemulsions[34,35] (and even water[36]!), so a simple presentation is quite appropriate here. Moreover, the results obtained for percolation provide a guide and a reference as to what should be achieved for a less well-understood problem like frustration.

18.6.1. The "Classical" Theory

There exist a number of exact results on percolation thresholds in two-dimensional systems, but even there the complete solution cannot be obtained in a closed form comparable to Onsager's solution of the two-dimensional Ising model. A useful rigorous solution can be obtained in the special case of the "Bethe lattice," i.e., an infinite tree-like system (Figure 18.14). This system is in a sense infinite-dimensional and for thermal transitions it may be used to derive the usual mean-field equations. For percolation it provides a simple "classical" theory,[37] which is the counterpart of mean-field theory and may be used as a starting point for more refined calculations.

Let us consider the bond problem, with a number $(K + 1)$ of nearest neighbors for each site. Call $P(p)$ the probability that a given site belongs to the infinite cluster and $Q(p)$ the probability that, when one given branch is cut off, this site is connected to only a finite number of other sites. The site does *not* belong to the infinite cluster if for each branch either the first bond is absent (probability $1 - p$), or it is present but the neighboring site is not connected to an infinite cluster through the following bonds (probability pQ). The different branches are independent (this is the crucial

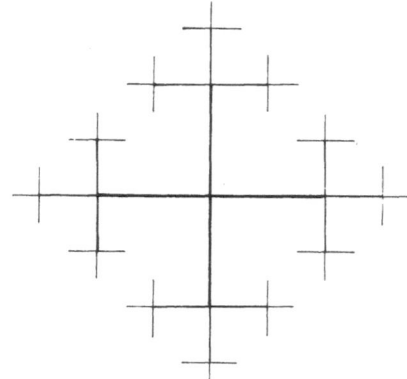

Figure 18.14. Portion of a Bethe lattice with coordination number equal to 4 (branching ratio $K = 3$). An infinite regular loopless structure of this type cannot be embedded in a lattice of finite dimensionality.

property of the tree), so

$$P(p) = 1 - (1 - p + pQ)^{K+1} \qquad (18.30)$$

A similar reasoning, on K branches this time, yields

$$Q(p) = (1 - p + pQ)^K \qquad (18.31)$$

These equations have only the solution $Q = 1$, $P = 0$ for $p < p_c = 1/K$. A bifurcation occurs at p_c and a new solution appears, which is the stable one for $p > p_c$ [this may be seen by interpreting (18.31) as the fixed point of a recurrence relation for successive generations on the tree]. Just above p_c this solution reads

$$P(p) = 2 \left(\frac{K+1}{K-1} \right) \frac{p - p_c}{p_c} + \cdots \qquad (18.32)$$

The linear dependence on $(p - p_c)$ is a universal feature, which does not depend on the branching ratio K and is also valid for the site problem.[37] In the modern language the behavior of $P(p)$ is described by an exponent β:

$$P(p) \sim A (p - p_c)^\beta \qquad (18.33)$$

and the classical theory predicts

$$\beta = 1$$

The mean number of sites in the *finite* clusters $S(p)$ diverges when p approaches p_c:

$$S(p) \sim |p_c - p|^{-1} \qquad (18.34)$$

in the same way as the mean number of generations in a cluster $N(p)$. To compare with usual lattice systems, the successive bonds must be thought of as lying in perpendicular directions, so the characteristic linear dimension $L(p)$ varies as

$$L(p) \sim N^{1/2} \sim |p_c - p|^{-1/2} \qquad (18.35)$$

The exponents γ and ν defined through

$$S(P) \sim |p_c - p|^{-\gamma}, \qquad L(P) \sim |p_c - p|^{-\nu} \qquad (18.36)$$

are here $\gamma = 1$, $\nu = \frac{1}{2}$.

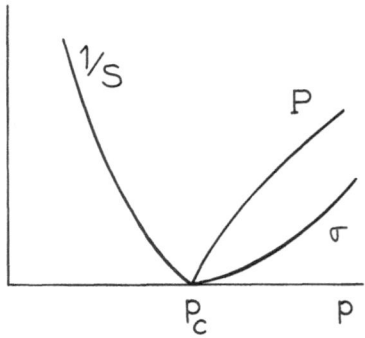

Figure 18.15. Qualitative behavior of different properties near the percolation threshold p_c on the Bethe lattice: mean cluster size (S), fraction of sites belonging to the infinite cluster (P), and conductivity (σ).

The calculation of the conductivity is more subtle and has been carried out only recently.[38] The proper formulation consists in considering the tree in a weak uniform field (*not* under a given potential difference). The conductivity is found to vanish rapidly near p_c (Figure 18.15):

$$\sigma \sim (p - p_c)^3 \qquad\qquad (18.37)$$

and the exponent $t = 3$. The fact that $\sigma \ll P$ is a general result, due to the large number of dead ends in the infinite cluster that do not contribute to the conductivity.

18.6.2. Limitations of the Classical Theory

The predictions of the theory outlined above are not satisfactory. For instance, the threshold for the bond problem on a square lattice is exactly $p_c = \frac{1}{2}$, while the Bethe-lattice solution gives $p_c = \frac{1}{3}$ for a system with four nearest neighbors.

This may be improved, but there is a worse shortcoming (at least for theorists!): the behavior near threshold comes out qualitatively wrong, i.e., the critical exponents do not agree with the values found in experiments or numerical simulations, which depend on one quantity, the dimensionality d of the lattice, as shown in Table 1. (For a critical evaluation of the numbers quoted, see the different review articles.)

An analogous situation is encountered in the mean-field theory of magnetic transitions. There exists in fact a precise correspondence between certain quantities in both problems:

$$P(p) \leftrightarrow \text{magnetization } M(T)$$

$$S(p) \leftrightarrow \text{susceptibility } \chi(T)$$

$$L(p) \leftrightarrow \text{correlation length } \xi(T)$$

Table 1

d	β	γ	ν	t
2	0.14	2.4	1.333	1.3[b]
3	0.4	1.7	0.88[a]	1.9[c]
Bethe lattice	1	1	1/2	3

[a] D. W. Heermann and D. Stauffer, Z. Phys. B **44**, 339 (1981).
[b] B. Derrida and J. Vannimenus, J. Phys. A **15**, L557 (1982).
[c] B. Derrida, D. Stauffer, H. J. Herrmann, and J. Vannimenus, J. Phys. Lett. **44**, L701 (1983).

and the notation for exponents has been chosen in order to agree with the standard one in critical phenomena.

The only difference between percolation and ferromagnetism on the Bethe lattice is that $\beta = \frac{1}{2}$ for the latter case, instead of 1 as found above. This difference has a deep origin and is not just due to the choice of the order parameter. In particular, the critical exponents are found to be correctly given by mean-field theory above a certain dimensionality, called the upper critical dimensionality d_+, given by

$$d_+ = (\gamma + 2\beta)/\nu \qquad (18.38)$$

For spin systems $d_+ = 4$, while this formula yields $d_+ = 6$ for percolation,[39] a result verified by other approaches and by numerical experiments. The large differences between the classical values of the exponents and their values in the physical world are thus not too surprising.

18.6.3. Approaches That Work

A different approach is clearly needed and a rich variety of methods have been proposed. Experience shows it is not too difficult to obtain good results with rather simple-minded approaches and systematic improvements are often possible without prohibitive calculations. Let us say a few words about several fruitful methods:

1. Series expansions (e.g., of the mean cluster size as a function of the bond probability) provide very accurate results when analyzed carefully, but they are brute-force tools and need a large effort without giving much physical insight.
2. The field-theoretic formulation uses the identification of percolation with a certain problem in standard phase transitions (in technical tems, the $s \to 1$ limit of the s-state Potts model). A

justification for the universality of exponents may then be given by renormalization-group theory, and these exponents may be computed in the vicinity of the upper critical dimensionality.[40] One obtains, with $\varepsilon = 6 - d$,

$$\beta = 1 - \varepsilon/7 + \cdots, \qquad \gamma = 1 + \varepsilon/7 + \cdots, \qquad \nu = \tfrac{1}{2} + 5\varepsilon/84 + \cdots$$

but such expansions are not very reliable when $d = 3$.

3. Numerical (Monte-Carlo) simulations contain a lot of microscopic information and help to understand "what is going on," even outside the critical regime. The analysis of the size dependence of the results is now made with the help of renormalization-group theory and has become a powerful tool in the hands of the experts.

4. Real-space renormalization-group methods are presently very popular because they offer several nice features: one can build physical intuition into the theory from the start; approximate results are easily obtained for many problems; the methods are well-adapted to the cases of most interest, $d = 2$ and 3. This class of theories is itself subdivided into several branches: decimation (see Chapter 13), Migdal recursion relations (see the les Houches lectures by Kirkpatrick), cluster renormalization,[41] phenomenological renormalization. This last method seems promising and may be worth a more detailed presentation.

18.6.4. Phenomenological Renormalization

The basic idea consists in studying the size dependence of a physical property in a finite system. A true transition cannot exist, naturally, but the singularity should progressively show up when the size increases. This sounds similar to the analysis of Monte-Carlo simulations, but here we consider systems infinite in one dimension, i.e., strips or tubes, on which the calculations can be carried out exactly. The price to pay is that the tractable transverse dimensions remain relatively small, but it is compensated by the elimination of statistical uncertainties and the method is useful if the convergence with size is rapid enough.

Consider a strip comprising n rows and N columns, on which bonds are present with probability p. The probability P_N that the Nth column will be connected to the first one by a continuous path is of the form

$$P_N \sim \exp(-N/\xi_n) \tag{18.39}$$

for large N, where ξ_n is the characteristic connectivity length of the strip.

When the width n tends to infinity one expects this length to behave very differently below and above the percolation threshold:

$$\xi_n \begin{cases} \to L(p) & p < p_c \\ \sim Cn & p = p_c \\ \sim \exp[nf(p)] & p > p_c \end{cases} \qquad (18.40)$$

This size dependence allows the determination of p_c, but a more sophisticated analysis is possible using the phenomenological renormalization-group equations proposed by Nightingale[42] for spin systems. These are based on a scaling assumption, namely that near p_c only one length scale is relevant — here the connectivity length. A correspondence may then be established between strips of different widths n and m, provided the bond probability is such that the width remains fixed in units of the connectivity length. This gives a renormalization equation for p:

$$\frac{\xi_n(p')}{n} = \frac{\xi_m(p)}{m} \qquad (18.41)$$

Near threshold ξ diverges as $|p_c - p|^{-\nu}$, so p_c is a fixed point of this equation and the convergence rate toward p_c is directly related to ν:

$$\xi_n(p_c) = \frac{n}{m} \xi_m(p_c) \qquad (18.42)$$

and

$$\log\left(\frac{dp'}{dp}\bigg|_{p_c}\right) = \frac{1}{\nu} \log\left(\frac{n}{m}\right) \qquad (18.43)$$

These equations are exact in the limit of very large n and m. For the accessible values of n and m, they yield a series of successive approximations.

Last but not least, a technique is needed to compute ξ_n. This may be achieved by adapting the transfer-matrix method to the present problem.[43] Let us consider the simplest possible examples.

(a) For a chain, the probability that the Nth column is connected to the first one is $P_N = p^N$, so

$$1/\xi_1 = \log(1/p) \qquad (18.44)$$

(b) For the $n = 2$ strip, there are three possible connecting configurations for a column: both sites may be connected to the first column (probability q_2), only site 1 or only site 2 are connected. By symmetry the last two cases have equal probability $q_1/2$. The probabilities for column

$(N + 1)$ are easily obtained from those at column N and, assuming periodic boundary conditions, we obtain

$$q_1' = p(1-p)^2 q_1 + 2p(1-p)^3 q_2$$
$$q_2' = p^2(2-p)q_1 + p^2(5-6p+2p^2)q_2 \qquad (18.45)$$

These equations define a transfer matrix and for large N both probabilities decay as λ^N, where λ is the largest eigenvalue of the matrix and verifies

$$\lambda^2 - \lambda p(1+3p-5p^2+2p^3) + p^3(1-p)^2 = 0 \qquad (18.46)$$

The connectivity length is given by

$$1/\xi_2 = \log(1/\lambda) \qquad (18.47)$$

Introducing now these results into equations (18.41) and (18.42) gives

$$p_c = 0.5026, \qquad \nu = 1.24$$

The agreement with the accepted values for the square lattice is encouraging and the method may be carried out on much larger strips, the transfer-matrix eigenvalue being obtained on a computer with as much accuracy as needed.

For the bond-percolation problem in two dimensions, one may use the known exact value of $p_c = \frac{1}{2}$ to short-cut the calculations. The results then provide the very accurate value[44]

$$\nu = 1.333 \pm 0.001$$

This strongly suggests that the exact value is $\nu = \frac{4}{3}$, as proposed by den Nijs.[45] Will another Onsager prove it?

The transfer-matrix method may be generalized to other systems. Successful applications have already been made to the Potts model,[44] to the self-avoiding random walk[46] and many other statistical problems. In fact, this method is now considered as the most reliable one for two-dimensional problems, and even for some three-dimensional problems.[72,73]

18.6.5. Related Problems

The basic questions of percolation theory are by now well understood, so the tendency is to consider "second-generation" problems where the rules of the game are significantly altered. A strong motivation comes from experiments: in many cases, the simple random-bond and site models are too crude for a quantitative understanding.

An important example is found in the gelation of polymers. The classical Flory theory of gelation exactly corresponds to the theory of percolation on a Bethe lattice (and antedates it by 20 years!). Conversely, the modern theory of gelation relies heavily on the adaptation of percolation concepts.[32,33] The random-bond model is not really adequate, however, because the solvent molecules play an important role that necessitates the introduction of site disorder. Also the various interactions (monomer–monomer, monomer–solvent, and solvent–solvent) are very different, and this leads to strong correlations in the system.

Detailed models that include all these effects are currently investigated,[47] a clear discussion of the different possible situations being given by de Gennes in his book (see the general references). Rather than this vast domain, I will consider two simpler but suggestive "variations on a percolation theme," directed percolation and percoloration.

18.6.5.1. Directed Percolation

Suppose the available bonds are not symmetric, so that the path between two sites is allowed in only one direction (e.g., they are diodes rather than resistances): how will the system properties be affected? Broadbent and Hammersley already mentioned this problem but it received little subsequent attention, in spite of its relevance in many practical situations.

Recent work has involved series methods[48] and Monte-Carlo simulations.[49] For instance, one obtains on the directed square lattice (orientated along a diagonal)

$$P_c \simeq 0.645, \qquad \nu \simeq 1.73$$

(see Kinzel[74] for a discussion of the available results and a detailed review of the problem). Both the threshold and the critical exponents are different from the unoriented case, so directed percolation belongs to a different universality class.

This result is strong, since most perturbations should modify only p_c and leave the exponents unaltered, according to renormalization-group theory. It may be intuitively understood by considering once more the Bethe lattice. Bond orientation would seem to play no role here, since there are no loops. However, if we visualize a portion of the tree as embedded in a cubic lattice of high dimensionality, it is clear that the connectivity length $\xi_\parallel$ along the preferred direction diverges as the mean number of generations in a cluster. This gives

$$\xi_\parallel \sim |p - p_c|^{-1}$$

while

$$\xi_\perp \sim |p - p_c|^{-1/2}$$

in the remaining directions.

The existence of two length scales with different critical exponents modifies the argument that yields the upper critical dimensionality d_+. It is now expected to verify

$$\nu_\parallel + (d_+ - 1)\nu_\perp = \gamma + 2\beta \tag{18.48}$$

with the mean-field exponents, which gives $d_+ = 5$ in agreement with field-theoretic calculations.[50]

18.6.5.2. Percoloration [51]

Another generalization consists in "coloring" the sites, that is, to allow several possible states instead of the simple binary choice — empty or occupied. Random coloring yields "polychromatic percolation,"[52] which is relevant in a number of situations.

Here we consider another problem: take a site-diluted lattice and color the occupied sites in such a way that neighboring sites have different colors. The central question is now: does this rule imply the existence of long-range order in the color pattern when the number of colors is minimal? (In technical terms, this is the antiferromagnetic Potts model on a dilute lattice at zero temperature.)

Color order cannot exist below the percolation threshold since disconnected clusters can be colored independently, but there may exist a critical concentration p_{cc}, different from p_c, for the onset of color long-range order. This may be proved for the triangular lattice with three colors and numerical simulations give the estimate[51]

$$p_{cc} = 0.72 \pm 0.03$$

while $p_c = \frac{1}{2}$ for site percolation on this lattice.

The really new feature of percoloration is in the nature of order propagation. In normal percolation, being a member of the infinite cluster is a local property, transmitted from neighbor to neighbor. Color order, on the contrary, may be due to *global* constraints in addition to local ones, as shown in Figure 18.16: the simultaneous existence of three distant bonds is

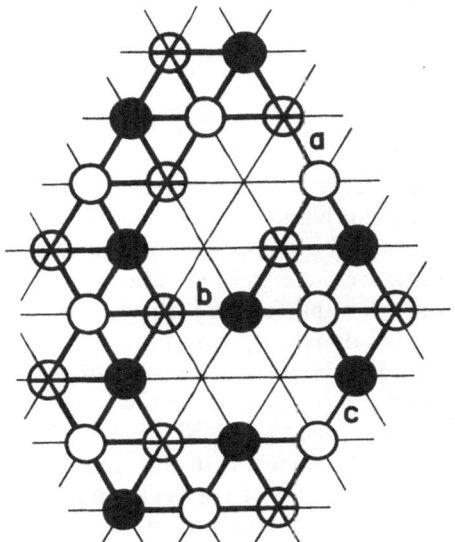

Figure 18.16. A color-periodic cluster on the site-diluted lattice. The accupied sites may be white, black, or gray. The propagation of color ordering is nonlocal here, since it is due to the simultaneous presence of three distant bonds (a, b, c). The figure is reproduced by courtesy of M. Devoret[51] and the Institute of Physics.

crucial to decide how the cluster is colored. In particular, this makes it very difficult to write a coloring algorithm. The problem has been proposed as a model for the ground states of a class of materials such as solid hydrogen, which display a quadrupolar spin-glass phase. Indeed, this effect of global constraints is reminiscent of what happens in frustrated systems and percoloration provides an interesting link between percolation and frustration.

18.7. Frustration

18.7.1. Local Gauge Invariance

Consider again the Hamiltonian of the frustration model defined in Section 18.5:

$$\mathscr{H} = -\sum_{(i,j)} J_{ij} S_i S_j \tag{18.49}$$

where the J_{ij} take the values ± 1 randomly on a cubic lattice and only Ising spins $S_i = \pm 1$ will be studied for simplicity. Starting from a given configuration of spins and bonds, the value of the Hamiltonian is not altered if one spin is reversed and the sign of the bonds with its neighbors is changed simultaneously. Such operations:

$$S_i \rightarrow -S_i$$

$$J_{ij} \rightarrow -J_{ij} \qquad \text{(all } j \text{ adjacent to } i\text{)} \tag{18.50}$$

are local transformations and for the Mattis model defined in Section 18.5.2 they can eliminate the disorder completely. More generally, two systems that can be mapped onto one another by such transformations have the same free energy in the absence of an external magnetic field. The kind of disorder that can be eliminated in this way is therefore irrelevant for thermodynamic properties.

The important features of transformation (18.50) are that it is both local and mixed (i.e., involves spins and bonds) by opposition to the global spin reversal, $S_i \rightarrow -S_i$ for all i. This is analogous to the gauge symmetry of electrodynamics, with the spin playing the role of the wave function and the bonds that of the vector potential.[22] The connection with gauge-field theories on a lattice is in fact close and will be discussed in more detail below.

The remaining relevant disorder is due to the competition between positive and negative bonds. The simplest example consists in an elementary square (called a plaquette in the present context) with one negative bond: there is no way to satisfy all four bonds simultaneously. This "frustration" effect is measured for every plaquette by the frustration function

$$\Phi = J_{12}J_{23}J_{34}J_{41} \tag{18.51}$$

If $\Phi = +1$ the plaquette is not frustrated and all bonds may be satisfied; if $\Phi = -1$ it is frustrated and has four different ground states with one unsatisfied bond. The need to consider a function defined on a loop is natural in a gauge theory — it is the discrete counterpart of the curl operator in usual electrodynamics.

The thermodynamic properties only depend on the quantities Φ_p, and not on all the J_{ij} themselves. This remark may be exploited to give a useful geometric picture for the ground states of Ising systems.[22,53] Mark all frustrated plaquettes by a cross at their center, and draw for all unsatisfied bonds a line joining the centers of the adjacent plaquettes. In two dimensions this gives a set of line defects, with the network of frustrated squares acting as point sources; the total length of these defects is proportional to the number of unsatisfied bonds, and must be minimal in a ground state. In three dimensions the defects are planar and bounded by the linear frustration network; the ground-state determination consists in minimizing their total area and is a discrete analogue of the Plateau problem for membranes.

In general there will be many solutions to the minimization problem that correspond to a high ground-state degeneracy. The simplest degenerate situation, with two unsatisfied bonds and two frustrated plaquettes, is shown in Figure 18.17.

18.7.2. Is There a Spin-Glass Transition?

The fundamental question about the frustration model is the existence of a spin-glass phase. Monte-Carlo simulations[53] clearly show that the model does contain much of the physics of real spin glasses and a freezing temperature is observed for the magnetic susceptibility and for time-dependent properties, both in two and three dimensions. Still, such calculations cannot conclude decisively that a sharp transition exists, because of inherent limitations on the size of the samples and the length of the runs. Here a considerable difference with dilution is manifest: the ground states of the system are not known, and their number increases so rapidly with the size that we cannot hope to find them in a systematic fashion for a reasonably large system (say 80×80).

One may compare spin glasses to democracy: nobody in the system is fully satisfied, but the path to better overall states is unclear and some experts have a "reformist" point of view (they think successive local changes are enough), while others are "radicals" (they believe a global move is needed). For a discussion of computational problems, see Binder's review article.[54]

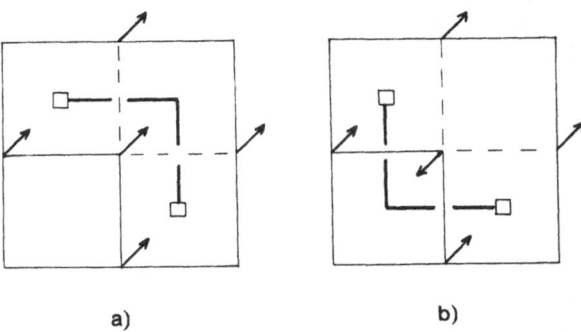

a)   b)

Figure 18.17. String picture of Ising ground states. The negative bonds are denoted by dotted lines, the frustrated plaquettes by little squares. The strings (solid lines) join the centers of the frustrated plaquettes through the unsatisfied bonds. States (a) and (b) have the same string length, i.e, they are degenerate in energy.

The methods that proved successful for percolation may then be tried. A simple idea is to compute the ground-state energy, entropy, magnetization ... as a series in x, the concentration of negative bonds.[55] Unfortunately it rapidly involves a lot of work and only the fourth order has been reached, which is very short compared to similar work for percolation. The frustration thresholds so obtained are only rough estimates. An interesting point is that the series is well behaved when the concentration of frustrated plaquettes is used as the expansion variable, rather than x itself.

Real-space renormalization methods of different types have also been applied to the problem and they seem to favor a transition at finite temperature in three dimensions, but there has been a lot of discussion and the conclusions are not firm.[56]

Field-theoretic formulations lead to expansions around dimension 6 for the critical exponents, as for percolation. However, they are now seriously questioned[57] because their starting point — the mean-field theory of spin glasses — has serious problems of its own, as will be discussed in Section 18.8.

The problem hits back! It seems reasonable in these conditions to be pragmatic and try to understand what is going on, literally by looking at the system. The geometric picture is very useful for that purpose; it allows in particular an analysis of the correlations among the ground states, either by direct inspection[58] or by an algorithm adapted from linear-programming techniques.[59] An example of such an analysis for a random sample is shown in Figure 18.18. An important finding in this type of work is that ferromagnetism disappears when the energy of a topological defect vanishes: at the frustration threshold the local magnetization remains strong but there exist "fracture lines" and distant parts of the system may take opposite orientation at no energy cost.

The most powerful attack on the problem to date consists in an *exact* numerical calculation of the free energy for random samples with sizes up to 18×18 in two dimensions and $4 \times 4 \times 10$ in three dimensions.[60] The samples are rather small but a careful analysis of the size effects shows that the (gauge-invariant) correlation function defined by

$$\Gamma(R) = \sum_i [\langle S(i)S(i+R)\rangle^2] \qquad (18.52)$$

decays exponentially with R (the average is taken over thermal fluctuations at temperature T). This is observed even for T much below the freezing temperature found in Monte-Carlo simulations and implies the absence of long-range spin-glass order, at least of the type initially proposed by Edwards and Anderson.[61]

For the two-dimensional frustration model, at $T = 0$, there seems to

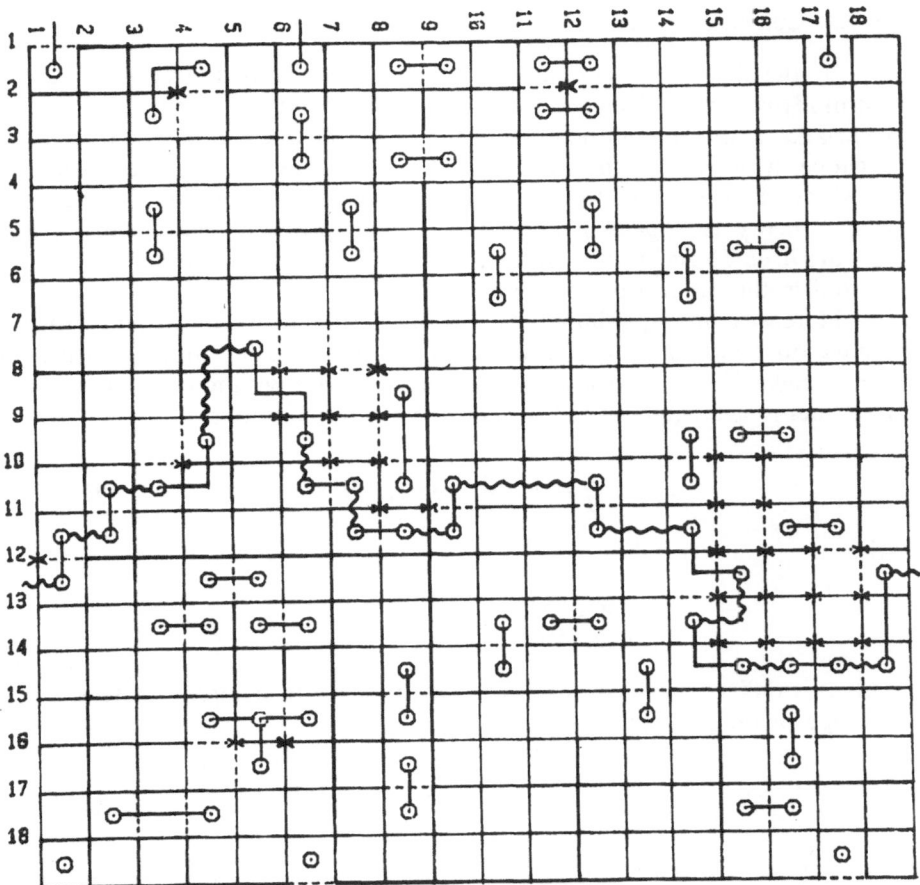

Figure 18.18. Ground state of an 18×18 sample with concentration $x = 0.09$ of negative bonds (periodic boundary conditions). The crosses ($\times$) indicate the 32 spins that may flip in at least one of the other ground states. The system is quite rigid and its magnetization is large, but there exists a fracture line (denoted by wiggly strings) along which one half of the sample may be reversed at no energy cost (picture obtained by L. de Seze).

exist quasi-long-range order (with power-law decay of the correlations) above a threshold given by

$$x_c = 0.12 \pm 0.02$$

The question now is to understand why the Monte-Carlo simulations show strong effects at a temperature where the static correlations are short-range and perfectly regular. The present situation is rather unsatisfactory and some new idea is needed!

18.7.3. Connection with Lattice Gauge-Field Theories

In the disordered systems of interest here, only the spins S_i are thermodynamic variables and the bonds J_{ij} are fixed ("quenched"), so the invariance of the Hamiltonian under the local transformation (18.50) is not a gauge symmetry of the model in the sense now familiar to particle physicists. Still, the idea that spin glasses may be described by a standard gauge theory is appealing and several such proposals have been put forward (see Hertz[62] and the lectures by Toulouse and Dzyaloshinskii quoted in the general references).

Here we consider a simple connection between frustration and a gauge theory on a lattice, of the type recently introduced in studies of quark confinement.[63] In the frustration problem, what we wish to compute is the free energy averaged over all possible bond distributions:

$$\frac{F}{T} = - \sum_{J_{ij} = \pm 1} \log Z(\{J_{ij}\}) \tag{18.53}$$

with

$$Z = \sum_{S_i = \pm 1} \exp\left(\frac{1}{T} \sum_{i,j} J_{ij} S_i S_j\right)$$

This is a difficult problem, and it is much easier to compute instead the average of the partition function:

$$\text{Log } \bar{Z} = \log\left[\sum_{J_{ij} = \pm 1} Z(\{J_{ij}\}) \right] \tag{18.54}$$

This is known as the "annealed" disorder case because it corresponds to letting the bond distribution be at thermal equilibrium with the spins, rather than fixed. The free energy per spin F_a is then simply given by

$$F_a = - T \log \bar{Z} = - T \log 2 - T \frac{z}{2} \log(\cosh 1/T)$$

where z is the number of nearest neighbors.

Unfortunately, it is not very helpful because the problem of physical interest is really the quenched one. To see what happens in the annealed case, let us compute the thermal average of the frustration function

$$\langle \Phi \rangle_a = \left\langle \sum_P J_{ij} J_{jk} J_{kl} J_{li} \right\rangle$$

$$= (\tanh 1/T)^4 \tag{18.55}$$

while in the quenched system ($J_{ij} = \pm 1$ are equiprobable)

$$\langle \Phi \rangle_q = 0 \qquad (18.56)$$

In the annealed system the frustration function is temperature-dependent and tends to 1 at low T, so very few plaquettes are frustrated. The reason is that unfrustrated configurations are energetically favored and dominate (18.54) in spite of their low probability, while in the quenched case the logarithm kills them.

It is then a natural idea to impose a constraint on the annealed model to correct for this effect. The canonical procedure in such cases is to introduce a Lagrange multiplier for the constrained quantity (here, the frustration) and choose it so as to verify the constraint on the average (one thus introduces a chemical potential to satisfy a constraint on the number of particles). In this case one obtains the "gauge-annealing" model,[64] defined by

$$\frac{F_G}{T} = -\log \left[\sum_{S_i = \pm 1} \sum_{J_{ij} = \pm 1} \exp \left(\beta_p \sum_p JJJJ + \frac{1}{T} \sum J_{ij} S_i S_j \right) \right] \qquad (18.57)$$

with the Lagrange multiplier β_p fixed by

$$\frac{1}{T} \frac{\partial F_G}{\partial \beta_p} = \langle \Phi \rangle_G = 0 \qquad (18.58)$$

Equation (18.57) is exactly the Z_2 gauge theory on a cubic lattice[63]; in the usual language β_p is an inverse plaquette temperature and $1/T$ is an inverse link temperature, denoted β_L. Equation (18.58) represents a trajectory in the (β_p, β_L)-plane that is a first approximation to the frustrated system, and a spin-glass transition would correspond to this trajectory crossing a phase transition line. The remarkable point is that β_p must be *negative* to satisfy this equation. This is easily understood since the natural tendency to eliminate frustration must be countered.

The phase diagram for the gauge theory at negative plaquette couplings has been investigated very recently by Monte-Carlo simulations.[65] The trajectory defined by equation (18.58) does not cross phase boundaries in dimensions 3 and 4, so the model contains no spin-glass transition. For $d = 4$, however, the trajectory is very close to a boundary, at low T, as shown in Figure 18.19. This lends support to the idea that $d = 4$ is a special dimension and that a transition exists only for higher dimensions, as other evidence also suggests.

The arguments given above are simple and it is certainly possible to build better approximations involving, for example, more complicated

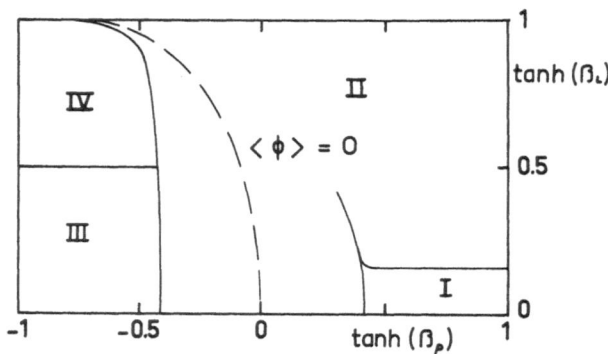

Fgigure 18.19. Phase diagram of the Z_2 matter + gauge theory ($d = 4$). The frustrated system is approximately represented by the dashed trajectory (no distinction is made between first- and second-order transition lines).

terms that take into account the correlations between frustrated pla- quettes. I think the important point to be made is that the existence of constraints on *circuits* leads naturally to the introduction of gauge fields. This idea might well prove fruitful in other areas of statistical mechanics.

18.8. Mean-Field Theory of Spin Glasses

Much effort has been devoted for several years to the elaboration of a mean-field theory of spin glasses. The problem has proved surprisingly complex, with many "exact solutions" that failed for subtle reasons and "laws" that were violated — all the ingredients of a good scientific thriller. A completely satisfactory solution is still lacking, but the overall picture seems reasonably well understood, at least for Ising spins.

Of course, a mean-field theory is not absolutely necessary and anyway it is not expected to describe real materials precisely, as pointed out for the case of percolation. Still, such theories are valuable: they provide a general qualitative understanding and they are the starting point for more refined calculations, such as renormalization-group methods. Indeed, the early renormalization work on spin glasses is on dubious ground,[57] because the underlying mean-field theory is incorrect. "Il ne faut pas mettre la charrue avant les boeufs!"

18.8.1. The Sherrington–Kirkpatrick Model

The first step consists in giving a proper definition of mean-field theory for spin glasses. For ferromagnetic systems, several methods may be used that lead to identical results. Most do not work here: for instance, the Bethe lattice approach, so useful for percolation, is not adequate because the absence of loops on the tree suppresses the fundamental effect of frustration.

An alternative approach consists in letting the interaction range go to infinity.[66] In an external magnetic field H the Hamiltonian is then

$$\mathscr{H} = -\sum_{i,j} J_{ij} S_i S_j - H \sum_i S_i \tag{18.59}$$

where the first summation bears over all $N(N-1)/2$ pairs of spins. For a meaningful thermodynamic limit to exist, the interactions must decrease with system size: the mean value must scale as J_0/N and the variance as $1/\sqrt{N}$. The detailed form of the bond distribution $P(J_{ij})$ is not important and one may choose a Gaussian form or two delta functions, whichever is most convenient.

The calculation of the free energy may then be performed using the now famous "replica trick."[61] One uses the identity

$$\log Z = \lim_{n \to 0} \frac{Z^n - 1}{n} \tag{18.60}$$

and computes the average of Z^n over the distributions $P(J_{ij})$, for integer n. This is much simpler than for $\log Z$, but the limit $n \to 0$ proves tricky. (Mathematically, the difficulty arises because the moments $\overline{Z^n}$ increase too rapidly with n for uniqueness theorems to apply.) Indeed, the results initially obtained are manifestly incorrect, since the entropy becomes negative at low temperatures, an impossible situation for Ising systems (though not for classical vector spins). Monte-Carlo simulations reveal other discrepancies between the "naive" theory and the true results.

Also, linear-response theory predicts that the magnetic susceptibility in zero field is directly related to the order parameter q defined by Edwards and Anderson[61] through

$$\chi = \frac{1-q}{T} \tag{18.61}$$

In fact, it appears that this relation cannot hold if the zero temperature entropy is nonnegative, so linear response must be violated!

18.8.2. Replica-Symmetry Breaking

To resolve this paradox several ingenious schemes have been proposed, involving "replica-symmetry breaking." Roughly speaking, the n

systems appearing in Z^n are no longer considered as equivalent. The technical details are rather difficult and the physics behind this new type of symmetry breaking remained obscure for a long time and involves such unusual concepts as the ultrametric topology of phase space,[75] but the most sophisticated scheme, due to Parisi,[67] is quite successful and agrees very well with the Monte-Carlo data.

The essential predictions of the theory are the following:

1. A transition occurs at $T = 1$ (for $J_0 = 0$, $H = 0$), but contrary to usual transitions the low-temperature free energy per spin is *higher* than the continuation of the high-temperature branch (Figure 18.20). This shows the profound originality of this transition, and prevents the topological theory of defects from applying in spin glasses (the core energy of the defects would be negative!).

2. Below T_c the order parameter is a function $q(x)$, which may be obtained variationally, and the susceptibility is a constant. Instead of relation (18.61) one has

$$\chi = \frac{1 - \int_0^1 q(x)\,dx}{T} = 1 \qquad (18.62)$$

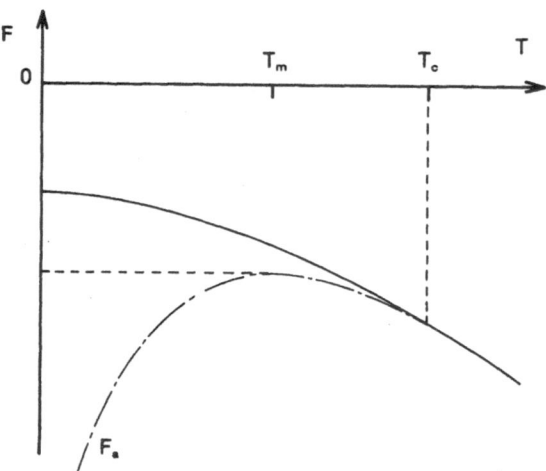

Figure 18.20. Free energy vs temperature for the $S-K$ model of spin glasses. Below T_c, $F(T)$ lies *above* the analytic continuation F_a of the high-temperature branch (dash–dot line). In the random-energy model, the free energy is equal to F_a above T_m and constant below (dotted line).

3. In the presence of an external field there exists a critical line $H_c(T)$ for replica-symmetry breaking that separates the spin-glass phase from the high-temperature paramagnetic phase (see Figure 18.22 below).
4. The spin-glass phase is stable for $J_0 < 1$; for $J_0 > 1$ ferromagnetic order sets in, but at low temperatures there exists a mixed ferromagnetic–spin-glass phase where replica symmetry is broken (see Figure 18.21 below).

18.8.3. The Random-Energy Model

A virtue of a mean-field theory is usually its simplicity, both mathematical and physical. Here both seem to be lost if we want an exact solution, and it is useful to look for other models that keep the essential features of the Sherrington–Kirkpatrick (S–K) system and remain simple.

Such a model has been proposed by Derrida,[68] as a limit of a family of disordered systems that contains the S–K case. It may be defined through three basic properties:

1. The system has 2^N energy levels E_i.
2. The probability $P(E)$ that a given configuration of spins has energy E is Gaussian, of variance $1/\sqrt{N}$.
3. The energy levels E_i are independent random variables.

The last property is quite strong and is not verified by the S–K model, since configurations with many indentical spins are correlated in energy, hence the name "random-energy model."

The problem so defined is exactly solvable and a phase transition occurs at a temperature $T_c^{-1} = 2(\text{Log } 2)^{1/2}$. The existence of a transition is *a priori* surprising. If one just writes down the partition function

$$Z = \sum_i \exp(-E_i/T) \tag{18.63}$$

it is diffcult to see how a singularity comes out of this sum of random variables. In fact this transition is very different from what we are used to and its nature is probabilistic: below a certain energy almost all samples have no levels at all. A few samples do have levels, so that the average density of states $\langle n(E) \rangle$ is nonzero and shows no singularity, but they are negligible for the free-energy average. Essentially the system is "frozen" below T_c; its specific heat and entropy vanish identically.

The calculations can be carried out with an external field, showing the existence of a critical line, and also with a nonzero mean value for the interactions J_0. The corresponding phase diagram contains a phase that is

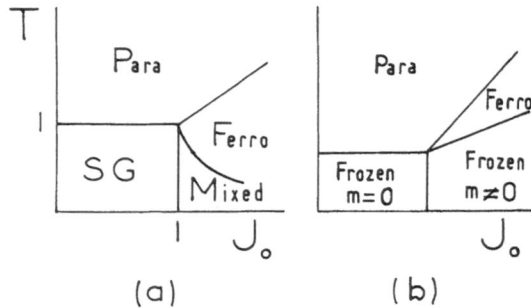

Figure 18.21. Phase diagrams of (a) the S–K model and (b) the random-energy model, as a function of the mean value J_0 of the interactions. The most striking feature is the existence in both cases of a mixed phase, with both ferromagnetic and spin-glass character, for Ising spins.

both ferromagnetic and frozen, and is compared in Figure 18.21 with the phase diagram of the S–K model. The similarity is striking, and the random-energy model certainly contains some clues on the nature of the spin-glass state.

18.8.4. The Projection Hypothesis for the S–K Model

Returning now to the S–K system one may wonder whether some kind of freezing process also takes place there. A simple hypothesis has been put forward in this spirit,[69] namely that in the spin-glass phase the temperature and field dependence of the free energy decouple:

$$F(T, H) = F_i(T) + F_2(H) \qquad (18.64)$$

or equivalently

$$M(T, H) = M(H), \qquad S(T, H) = S(T) \qquad (18.65)$$

where M and S are then just given by projecting their values from the critical line (see Figure 18.22). All calculations may be carried out unambiguously, since they do not involve replica-symmetry breaking.

The results are in very good agreement with the predictions of Parisi's theory and with the Monte-Carlo simulations, in particular the ground-state energy, the constant susceptibility below T_c, the behavior for $J_0 \neq 0$.[70] The hypothesis lacks a rigorous basis and must be regarded as a useful guess, providing the simple picture one expects from a mean-field theory.

As an example, consider the behavior of the magnetization as a function of H for different values of T. The projection hypothesis leads to

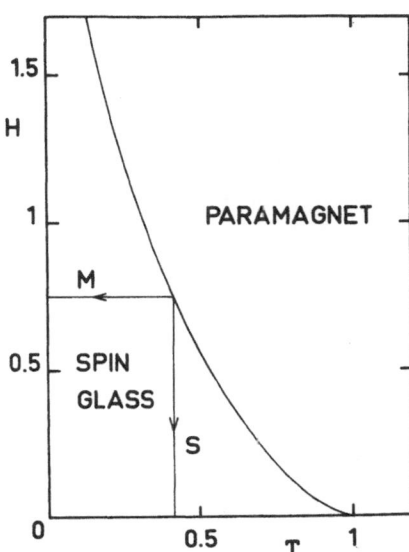

Figure 18.22. Critical line $H_c(T)$ for the S-K model and projection hypothesis for the entropy S and magnetization M.

the prediction of a singularity at T_c with

$$M/H \sim 1 - AH^{4/3} + \cdots \qquad T < T_c$$
$$\sim 1 - H\sqrt{2} + \cdots \qquad T = T_c$$
$$\sim 1/T - O(H^2) \qquad T > T_c \qquad (18.66)$$

The corresponding curves are plotted in Figure 18.23. A qualitatively similar singularity has indeed been observed in recent experiments on Cu–Mn alloys.[71] The interpretation should be made with care, because the Mn spins have vector rather than Ising character, but it is extremely encouraging. We think many features of the mean-field theory will have their counterpart in real materials, once properly interpreted, and this is a justification for working hard on it.

18.9. Conclusion

The area of disordered systems is active and diverse and we have tried to present a few landmarks, with their history and present state. The pitfall is to give the impression of a hurried guided tour, in the "If it's Tuesday, this must be Belgium" style. We hope the problems evoked seem interesting enough for a more complete study, and the general references quoted below provide useful guides.

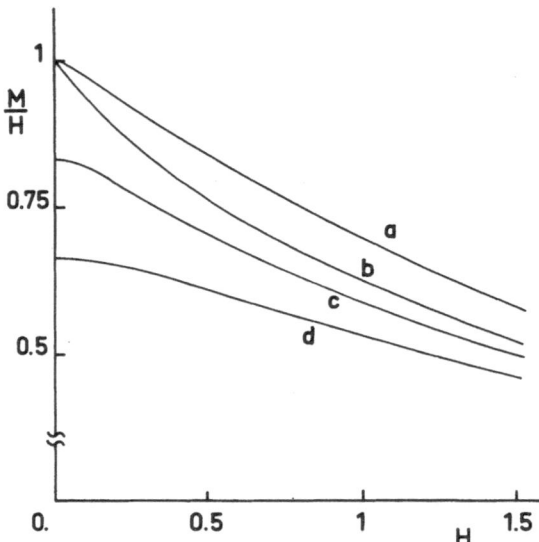

Figure 18.23. Mean-field theory results for the magnetization (using the projection hypothesis): (a) $T = 0$, (b) $T = T_c = 1$, (c) $T = 1.2$, (d) $T = 1.5$.

In some cases we have deliberately emphasized research work at an early stage to give some feeling for the variety of possible approaches, at the risk of being belied by future progress. It is quite possible in fact that big surprises are awaiting us, even that basic statistical-mechanics ideas will need a deep reappraisal for disordered systems — the definition of entropy for glasses and spin glasses is presently being scrutinized by various people.

Also, the most fruitful ideas might well emerge in "peripheric" areas, with the realization that certain approaches developed in one case — such as spin glasses — apply in a very different domain, just as percolation concepts are useful for gelation. In particular, gauge concepts have appeared at different places in these sections and it is reasonable to expect them to invade progressively other domains in condensed-matter physics.

I am grateful to the Trieste Center and the College directors for their warm hospitality, and to G. Toulouse for his continuous encouragement in the preparation and redaction of this chapter.

References to Part IV

Chapter 17

1. G. Del Re, J. Ladik, and C. Biczó, *Phys. Rev.* **155**, 967 (1967); J.-M. André, L. Gouverneur, and G. Leroy, *Int. J. Quantum Chem.* **1**, 427, 451 (1967); R. N. Euwema, D. L. Wilhite, and G. T. Surrat, *Phys. Rev. B* **7**, 818 (1973).
2. J. Ladik, in: *Electronic Structure of Polymers and Molecular Crystals* (J.-M. André and J. Ladik, eds.), p. 23, Plenum Press, New York–London (1975).
3. G. Biczó, unpublished result.
4. J. Ladik and J. Cizek, *Int. J. Quantum Chem.* (submitted).
5. I. I. Ukrainski, *Theor. Chim. Acta* **30**, 139 (1975); C. Merkel, in: *Electronic Properties of Molecular Crystals* (in German), Thesis, Technical University of Munich (1977).
6. See, for instance: M. Hamermesh, *Group Theory and its Application to Physical Problems* p. 80, Addison-Wesley Publ., Reading, Mass. (1964).
7. R. Pariser and R. G. Parr, *J. Chem. Phys.* **21**, 466, 707 (1953); J. A. Pople, *Trans. Faraday Soc.* **49**, 1375 (1953).
8. J. Ladik, *Acta Phys. Acad. Sci. Hung.* **18**, 185 (1965); J. Ladik, in: *Electronic Structure of Polymers and Molecular Crystals* (J.-M. André and J. Ladik, eds.), p. 663, Plenum Press, New York–London (1975).
9. N. Mataga and K. Nishimoto, *Z. Phys. Chem.* **13**, 140 (1957).
10. J. A. Pople, D. P. Santry, and G. A. Segal, *J. Chem. Phys.* **43**, 129 (1963); J. A. Pople and G. A. Segal, *J. Chem. Phys.* **45**, 136 (1965).
11. H. Fujita and A. Imamura, *J. Chem. Phys.* **53**, 4555 (1970); J. Ladik and G. Biczó, *Acta Chim. Acad. Sci. Hung.* **67**, 297 (1971); K. Morokuma, *J. Chem. Phys.* **54**, 1962 (1971).
12. J. A. Pople and D. C. Beveridge, *Approximate Molecular Orbital Theory*, McGraw-Hill, New York (1970).
13. For further details see J. Ladik, in: *Electronic Structure of Polymers and Molecular Crystals* (J.-M. André and J. Ladik, eds.), p. 663, Plenum Press, New York–London (1975).
14. T. C. Collins, A. B. Kunz, and P. W. Deutsch, *Phys. Rev. A* **10**, 1034 (1974).
15. J. Avery, J. Packer, J. Ladik, and G. Biczó, *J. Mol. Spectrosc.* **29**, 194 (1969).
16. Y. Toyozawa, *Prog. Theor. Phys. Kyoto* **12**, 421 (1954).
17. A. B. Kunz, *Phys. Rev. B* **6**, 606 (1972).
18. J. T. Devreese, A. B. Kunz, and T. C. Collins, *Solid State Commun.* **11**, 673 (1972).
19. J. Ladik and S. Suhai, *Int. J. Quantum Chem.* **QBS7**, 181 (1980).
20. W. Hehre, R. F. Stewart, and J. A. Pople, *J. Chem. Phys.* **51**, 2657 (1969).
21. S. Arnott, S. D. Dover, and A. J. Wonacott, *Acta Crystallogr., Sect. B* **25**, 2192 (1969).
22. J. Ladik and S. Suhai, in: *Theoretical Chemistry* (C. Thompson, ed.), p. 49, The Royal Society of Chemistry, London (1981).
23. Y. Takeuti, *Prog. Theor. Phys. Kyoto* **18**, 421 (1957).
24. B. F. Rozsnyai and J. Ladik, *J. Chem. Phys.* **52**, 5711 (1970); **53**, 4325 (1970).

25. J. Ladik, in: *Advances in Quantum Chemistry* (P. O. Löwdin, ed.), Vol. 7, p. 397, Academic Press, New York–London (1970).
26. B. Mely and A. Pullman, *Theor. Chim. Acta* **13**, 278 (1969); S. Huzinaqa, *J. Chem. Phys.* **42**, 1293 (1965).
27. L. B. Clark, G. G. Peschel, and I. Tinoco, Jr., *J. Chem. Phys.* **69**, 3615 (1965).
28. J. Ladik, S. Suhai, P. Otto, and T. C. Collins, *Int. J. Quantum Chem.* **QBS4**, 55 (1977).
29. L. Seprödi, G. Biczó, and J. Ladik, *Int. J. Quantum Chem.* **3**, 62 (1969).
30. See, for instance: T. C. Collins, in: *Electronic Structure of Polymers and Molecular Crystals* (J.-M. André and J. Ladik, eds.), p. 405, Plenum Press, New York–London (1975).
31. S. Suhai, *Phys. Rev. B* **27**, 3506 (1983).
32. S. Suhai, T. C. Collins, and J. Ladik, *Biopolymers* **18**, 899 (1978).
33. D. A. Breen and P. J. Flory, *J. Am. Chem. Soc.* **87**, 279 (1965).
34. S. Suhai, *Theor. Chim. Acta* **34**, 157 (1974).
35. S. Suhai and J. Ladik, *Theor. Chim. Acta* **28**, 27 (1972).
36. S. Suhai, J. Kaspar, and J. Ladik, *Int. J. Quantum Chem.* **17**, 995 (1980).
37. See, for instance: H. Ehrenreich and R. S. Knox, in: *The Theory of Excitons, Solid State Physics* (F. Seitz and D. Turnbull, eds.), Suppl. 5, p. 1, Academic Press, New York–London (1963).
38. J. C. Slater, MIT Technical Report, No. 5 (1953); G. F. Koster and J. C. Slater, *Phys. Rev.* **95**, 1167 (1954); G. F. Koster, *Phys. Rev.* **95**, 1436 (1954).
39. G. H. Wannier, *Phys. Rev.* **52**, 191 (1937).
40. For a review see: E. I. Blount, *Solid State Phys.* **13**, 305 (1963).
41. S. Suhai, *Phys. Rev. B* (in print).
42. B. Hudson and B. Kohler, *Am. Rev. Phys. Chem.* **25**, 437 (1974).
43. S. Suhai, *J. Chem. Phys.* **73**, 3843 (1980).
44. R. Ahlrichs and W. Kutzelnigg, *J. Chem. Phys.* **48**, 1819 (1968); W. Meyer, *J. Chem. Phys* **58**, 1017 (1973).
45. J. Čižek, *J. Chem. Phys.* **45**, 4256 (1966); J. Čižek and J. Paldus, *Int. J. Quantum Chem.* **5**, 359 (1975); J. Paldus and J. Čižek, *Adv. Quantum Chem.* **9**, 105 (1975).
46. For a review see: N. H. March, in: *Quantum Theory of Polymers* (J.-M. André, J. Delhalle, and J. Ladik, eds.), p. 48, D. Reidel Publ. Co., Dordrecht (1978).
47. P. Hohenberg and W. Kohn, *Phys. Rev.* **136**, 3864 (1964).
48. J. Ladik, in: *Recent Advances in the Quantum Theory of Polymers* (J.-M. André, J.-L. Brédas, J. Delhalle, J. Ladik, G. Leroy, and C. Moser, eds.), p. 155, Springer-Verlag, Berlin–Heidelberg–New York (1979).
49. J. Ladik and M. Seel, *Phys. Rev. B* **13**, 5338 (1976).
50. M. Seel and J. Ladik, *Chem. Phys.* **45**, 349 (1980).
51. S. Kirkpatrick, B. Velicky, and H. Ehrenreich, *Phys. Rev. B* **1**, 3250 (1977).
52. B. Györffy and S. Faulkner, personal communication (1976).
53. M. Seel, T. C. Collins, F. Martino, D. K. Rai, and J. Ladik, *Phys. Rev. B* **18**, 6460 (1978).
54. R. J. Elliott, J. A. Krumhansl, and P. L. Leath, *Rev. Mod. Phys.* **46**, 465 (1974); F. Martino in: *Quantum Theory of Polymers* (J.-M. André, J. Delhalle, and J. Ladik, eds.), p. 169, D. Reidel Publ. Co., Dordrecht (1978).
55. F. Martino and J. Ladik, *Phys. Rev. B* **22**, 1092 (1980).
56. G. Del Re and J. Ladik, *Chem. Phys.* **49**, 321 (1980).
57. J. Ladik, *Phys. Rev. B* **17**, 1663 (1978).
58. J. Callaway, *J. Math. Phys.* **5**, 783 (1964).
59. G. A. Baraff and M. Schlüter, *Phys. Rev. Lett.* **41**, 892 (1978); *Phys. Rev. B* **19**, 4969 (1979).
60. J. Berholc, N. D. Lipari, and S. T. Pantelides, *Phys. Rev. Lett.* **41**, 895 (1978).
61. J. Kaspar (unpublished results).

62. P. Dean, *Proc. R. Soc. London, Ser. A* **254**, 507 (1960); *Proc. R. Soc. London, Ser. A* **260**, 263 (1961); *Rev. Mod. Phys.* **44**, 127 (1972).

63. M. Seel, *Chem. Phys.* **43**, 103 (1979); see also M. Seel, in: *Recent Advances in the Quantum Theory of Polymers* (J.-M. André, J.-L. Brédas, J. Delhalle, J. Ladik, G. Leroy, and C. Moser, eds.), p. 271, Springer-Verlag, Berlin–Heidelberg–New York (1979).

64. R. Day and F. Martino, *Chem. Phys. Lett.* **84**, 86 (1981).

65. J. H. Wilkinson, *The Algebraic Eigenvalue Problem*, Clarendon Press, Oxford (1965).

66. E. W. Prohofsky and L. L. van Zandt, personal communication (1980).

67. S. Suhai, *J. Chem. Phys.* **57**, 5599 (1972); S. Suhai, in: *Quantum Theory of Polymers* (J.-M. André, J. Delhalle, and J. Ladik, eds.), p. 335, D. Reidel Publ. Co., Dordrecht (1978).

68. See, for instance: F. J. Blatt, *Physics of Electronic Conduction in Solids*, p. 121, McGraw-Hill, New York (1978).

69. F. J. Blatt. *Physics of Electronic Conduction in Solids*, pp. 135, 186, McGraw-Hill, New York (1978).

70. C. T. O'Konski, P. Moser, and M. Shirai, *Biopolymers Symp.* **1**, 479 (1964).

71. C. Y. Liang and E. G. Scalco, *Nature* **198**, 86 (1963); R. S. Snart, *Trans. Faraday Soc.* **59**, 854 (1963).

72. S. H. Glarum, *J. Phys. Chem. Solids* **24**, 1577 (1963).

73. P. Otto and J. Ladik, *Chem. Phys.* **8**, 192 (1975); *Chem. Phys.* **19**, 205 (1977); P. Otto, *Chem. Phys.* **33**, 407 (1978); J. Ladik, *Int. J. Quantum Chem.* **QBS3**, 51 (1976).

74. I. Berenblum, *Carcinogenesis as a Biological Problem*, p. 211, North-Holland Publ. Co., Amsterdam–Oxford (1973).

75. P. O. Löwdin, *Int. J. Quantum Chem.* **QBS4**, 185 (1977).

76. H. Bush, *Biochemistry of the Cancer Cell*, p. 282, Academic Press, New York–London (1962).

77. R. Daudel, in: *Mutagenesis and Carcinogenesis* (P. Daudel, R. Daudel, Y. Moulé, and F. Zajadela, eds.), C.N.R.S., Paris (1977).

78. A. Szent-Györgyi, *Int. J. Quantum Chem.* **QBS3**, 45 (1976); A. Szent-Györgyi, *Bioenergetics* **4**, 535 (1973); A. Szent-Györgyi, *Electronic Bioology and Cancer*, Marcel Dekker Inc., New York–Basel (1976).

79. A. Karpfen and J. Ladik (unpublished results).

80. T. C. Collins, private communication.

81. A. S. Davydov, *Studia Biophys. (Berlin)* **62** (1977).

82. K. Laki and J. Ladik, *Int. J. Quantum Chem.* **QBS3**, 51 (1976).

83. J. N. Murrel, M. Randic, and D. R. Williams, *Proc. R. Soc. London, Ser. A* **284**, 566 (1965).

84. P. Otto and J. Ladik (unpublished results).

Chapter 18

1. D. R. Nelson and B. I. Halperin, *Phys. Rev. B* **19**, 2457 (1979).

2. A. P. Young, *Phys. Rev. B* **19**, 1855 (1979).

3. F. F. Abraham, *Phys. Rev. Lett.* **44**, 463 (1980); S. Toxvaerd, *Phys. Rev. Lett.* **44**, 1002 (1980).

4. D. S. Fisher, B. I. Halperin, and R. Morf, *Phys. Rev. B* **20**, 4692 (1979).

5. F. R. N. Nabarro, *J. Phys. (Paris)* **33**, 1089 (1972).

6. G. Toulouse and M. Kleman, *J. Phys. Lett.* **37**, L149 (1976).

7. G. E. Volovik and V. P. Mineyev, *Sov. Phys. JETP* **45**, 1186 (1977).

8. E. J. Yarmchuk, M. J. V. Gordon, and R. E. Packard, *Phys. Rev. Lett.* **43**, 214 (1979).

9. V. Poenaru and G. Toulouse, *J. Phys. (Paris)* **38**, 887 (1977).

10. L. J. Yu and A. Saupe, *Phys. Rev. Lett.* **45**, 1000 (1980).

11. Y. Bouligand, B. Derrida, V. Poenaru, Y. Pomeau, and G. Toulouse, *J. Phys. (Paris)* **39**, 863 (1978).
12. M. Luscher, *Nucl. Phys. B* **135**, 1 (1978).
13. H. de Vega, *Phys. Rev. D* **18**, 2945 (1978).
14. C. Kawabata and A. R. Bishop, *Solid State Commun.* **33**, 453 (1980).
15. I. E. Dzyaloshinskii and G. E. Volovik, *J. Phys. (Paris)* **39**, 693 (1978); *Ann. Phys. (N.Y.)* **125**, 67 (1980).
16. G. E. Volovik and V. S. Dotsenko, Jr., *Sov. Phys. JETP* **51**, 65 (1980).
17. B. Julia and G. Toulouse, *J. Phys. Lett.* **40**, L395 (1979).
18. V. Poenaru, Orsay preprint (1980).
19. R. B. Griffiths, Rigorous results and theorems, in: *Phase Transitions and Critical Phenomena* (C. Domb and M. S. Green, eds.), Vol. 1, Academic Press, New York (1972).
20. M. E. Fisher and G. Caginalp, *Commun. Math. Phys.* **56**, 11 (1977).
21. D. C. Mattis, *Phys. Lett.* **56A**, 421 (1976).
22. G. Toulouse, *Commun. Phys.* **2**, 115 (1977).
23. S. Alexander and P. Pincus, *J. Phys. A* **13**, 263 (1980).
24. M. K. Phani, J. L. Lebowitz, M. H. Kalos, and C. C. Tsai, *Phys. Rev. Lett.* **42**, 577 (1979).
25. B. Derrida, Y. Pomeau, G. Toulouse, and J. Vannimenus, *J. Phys. (Paris)* **40**, 67 (1979); **41**, 213 (1980).
26. K. Binder, *Phys. Rev. Lett.* **45**, 811 (1980).
27. M. Gabay, *J. Phys. (Paris), Lett.* **41**, L427 (1980).
28. P. Bak and J. von Boehm, *Phys. Rev. B* **21**, 5297 (1980).
29. M. E. Fisher and W. Selke, *Phys. Rev. Lett.* **44**, 1502 (1980).
30. J. Villain and P. Bak, *J. Phys. (Paris)* **42**, 657 (1981).
31. M. E. Fisher and D. A. Huse, in: *Melting, Localization and Chaos* (R. Kalia and P. Vashishta, eds.), Elsevier, Amsterdam (1982).
32. P. G. de Gennes, *J. Phys. (Paris), Lett.* **36**, 1049 (1975).
33. D. Stauffer, *J. Chem. Soc., Faraday Trans. 2* **72**, 1354 (1975).
34. M. Lagues, *J. Phys. (Paris), Lett.* **40**, L331 (1979).
35. A. M. Cazabat, D. Chatenay, D. Langevin and A. Pouchelon, *J. Phys. (Paris), Lett.* **41**, L441 (1980).
36. H. E. Stanley, *J. Phys. A* **12**, L329 (1979).
37. M. E. Fisher and J. W. Essam, *J. Math. Phys.* **2**, 609 (1961).
38. P. G. de Gennes, *J. Phys. (Paris), Lett.* **38**, L355 (1977).
39. G. Toulouse, *Nuovo Cimento B* **23**, 234 (1974).
40. D. J. Amit, *J. Phys. A* **9**, 1441 (1976).
41. P. J. Reynolds, H. E. Stanley, and W. Klein, *Phys. Rev. B* **21**, 1223 (1980).
42. M. P. Nightingale, *Physica* **83A**, 561 (1976); *Proc. Konink. Ned. Akad. Wetenschap B* **82**, 235 (1979).
43. B. Derrida and J. Vannimenus, *J. Phys. (Paris), Lett.* **41**, L473 (1980).
44. H. W. J. Blote, M. P. Nightingale, and B. Derrida, *J. Phys. A* **14**, L45 (1981).
45. M. P. M. den Nijs, *J. Phys. A* **12**, 1857 (1979).
46. B. Derrida, *J. Phys. A* **14**, L5 (1981).
47. A. Coniglio, H. E. Stanley, and W. Klein, *Phys. Rev. Lett.* **42**, 518 (1979).
48. J. Blease, *J. Phys. C* **10**, 917; **10**, 3461 (1977).
49. J. Kertesz and T. Viczek, *J. Phys. C* **13**, L343 (1980).
50. S. P. Obukhov, *Physica* **101A**, 145 (1980).
51. M. Devoret, *J. Phys. C* **13**, 2257 (1980).
52. R. Zallen, *Phys. Rev. B* **16**, 1426 (1977).
53. S. Kirkpatrick, *Phys. Rev. B* **16**, 4630 (1977).
54. K. Binder, *J. Phys. (Paris), Colloq. C6*, **39**, C6–1527 (1978).

55. G. Grinstein, C. Jayaprakash, and M. Wortis, *Phys. Rev. B* **19**, 260 (1979).
56. W. Kinzel and K. H. Fischer, *J. Phys. C* **11**, 2115 (1978).
57. J. Rudnick, *Phys. Rev. B* **22**, 3356 (1980).
58. J. Vannimenus, J. M. Maillard, and L. de Seze, *J. Phys. C* **12**, 4523 (1979).
59. I. Bièche, R. Maynard, R. Rammal, and J. P. Uhry, *J. Phys. A* **13**, 2553 (1980).
60. I. Morgenstern and K. Binder, *Phys. Rev. B* **22**, 288 (1980); *Z. Phys. B* **39**, 227 (1980).
61. S. F. Edwards and P. W. Anderson, *J. Phys. F* **5**, 965 (1975).
62. J. A. Hertz, *Phys. Rev. B* **18**, 4875 (1978).
63. R. Balian, J. M. Drouffe, and C. Itzykson, *Phys. Rev. D* **11**, 2098 (1975).
64. G. Toulouse and J. Vannimenus, *Phys. Rep.* **67**, 47 (1980).
65. G. Banhot and M. Creutz, *Phys. Rev. B* **22**, 3370 (1980).
66. S. Kirkpatrick and D. Sherrington, *Phys. Rev. B* **17**, 4384 (1978).
67. G. Parisi, *J. Phys. A* **13**, 1101 (1980); **13**, 1887 (1980).
68. B. Derrida, *Phys. Rev. Lett.* **45**, 79 (1980).
69. G. Parisi and G. Toulouse, *J. Phys. (Paris), Lett.* **41**, L361 (1980).
70. J. Vannimenus, G. Toulouse, and G. Parisi, *J. Phys. (Paris)* **42**, 565 (1981).
71. P. Monod and H. Bouchiat, in: *Disordered Systems and Localization* (C. Castellani, C. di Castro, and L. Peliti, eds.), Springer-Verlag, Berlin (1981).
72. M. P. Nightingale, *J. Appl. Phys.* **53** (II) 7927 (1982).
73. J. Vannimenus and J. P. Nadal, Random systems, *Phys. Rep.* **103**, 47 (1984).
74. W. Kinzel, in: *Percolation Structures and Processes* (G. Deutscher, R. Zallen, and J. Adler, eds.), Adam Hilger, Bristol (1982).
75. N. Mezard, G. Parisi, N. Sourlas, G. Toulouse, and M. Virasoro, *Phys. Rev. Lett.* **52**, 1156 (1984).

General Bibliography to Chapter 18

R. Balian, R. Maynard, and G. Toulouse (eds.), *Ill-Condensed Matter, Les Houches 1978*, North-Holland Publ. Co., Amsterdam (1979).

K. Binder, Statistical mechanics of Ising spin glasses, in: *Fundamental Problems in Statistical Mechanics* (E. G. D. Cohen, ed.), North-Holland Publ. Co., Amsterdam (1980).

C. Castellani, C. di Castro, and L. Peliti, *Disordered Systems and Localization*, Lectures Notes in Physics **149**, Springer-Verlag, Berlin (1981).

P. G. de Gennes, *The Physics of Liquid Crystals*, Oxford University Press (1974).

P. G. de Gennes, *Scaling Concepts in Polymer Physics*, Cornell University Press (1979).

I. E. Dzyaloshinskii, Gauge theories and densities of topological singularities, in: *Proc. Les Houches 1980 Summer School*.

I. E. Dzyaloshinskii, Macroscopic theory of spin glasses, in: *Modern Trends in the Theory of Condensed Matter* (A. Pekalski and J. Przystawa, eds.), Lecture Notes in Physics **15**, Springer-Verlag, Berlin (1980).

J. W. Essam, Percolation theory, *Rep. Prog. Phys.* **43**, 883 (1980).

B. I. Halperin, Theory of melting and liquid-crystal phases in two dimensions, *Proc. Colloque Pierre-Curie, Paris* (1980).

W. F. Harris, Disclinations, *Sci. Am.*, p. 130 (Dec. 1977).

S. Kirkpatrick, Models of disordered materials, in: *Ill-Condensed Matter, Les Houches 1978* (R. Balian, R. Maynard, and G. Toulouse, eds.), North-Holland Publ. Co., Amsterdam (1979).

I. Ya. Korenblit and E. F. Shender, Ferromagnetism of disordered systems, *Sov. Phys. Usp.* **21**, 332 (1978).

A. P. Malozemoff and J. C. Slonczewski, *Physics of Magnetic Domain Walls in Bubble Materials*, Academic Press, New York (1979).

N. D. Mermin, The topological theory of defects in ordered media, *Rev. Mod. Phys.* **51**, 591 (1979).

L. Michel, Symmetry defects and broken symmetry configurations: Hidden symmetry, *Rev. Mod. Phys.* **52**, 617 (1980).

V. P. Mineyev, Topologically stable inhomogeneous states in ordered media, *Soviet Scientific Reviews A, Physics Reviews,* (I. M. Khalatnikov, ed.), Vol. 2, Harwood Academic Publishers, New York (1980).

F. R. N. Nabarro, *Theory of Crystal Dislocations,* Oxford University Press (1964).

D. R. Nelson, in: *Proc. 1980 Summer School on Fundamental Problems in Statistical Mechanics, Enschede* (June 1980).

V. Poenaru, Elementary algebraic topology related to the theory of defects and textures, in: *Ill-Condensed Matter, Les Houches 1978* (R. Balian, R. Maynard, and G. Toulouse, eds.), North-Holland Publ. Co., Amsterdam (1979).

D. Stauffer, Scaling theory of percolation, *Phys. Rep.* **54**, 1 (1979).

D. J. Thouless, Percolation and localization, in: *Ill-Condensed Matter, Les Houches 1978* (R. Balian, R. Maynard, and G. Toulouse, eds.), North-Holland Publ. Co.. Amsterdam (1979).

G. Toulouse, A lecture on the topological theory of defects in ordered media, in: *Modern Trends in the Theory of Condensed Matter* (A. Pekalski and J. Przystawa, eds.), Lecture Notes in Physics **15**, Springer-Verlag, Berlin (1980).

G. Toulouse, Gauge concepts in condensed matter physics, in: *Proc. of Cargese Institute on Recent Developments in Gauge Theories (Sept. 1979)* (H. Lehmann, ed.), Plenum Press, New York (1980).

J. L. van Hemmen and I. Morgenstern (eds.), *Heidelberg Colloquium on Spin Glasses,* Lecture Notes in Physics **192**, Springer-Verlag, Berlin (1983).

Index